Recent Progress in Botany

Recent Progress in Botany

Editor: Edward Ackles

STATES
ACADEMIC PRESS
www.statesacademicpress.com

States Academic Press,
109 South 5th Street,
Brooklyn, NY 11249, USA

Visit us on the World Wide Web at:
www.statesacademicpress.com

ISBN: 978-1-63989-465-9 (Hardback)

Cataloging-in-Publication Data

Recent progress in botany / edited by Edward Ackles.
p. cm.
Includes bibliographical references and index.
ISBN 978-1-63989-465-9
1. Botany. 2. Plants. I. Ackles, Edward.
QK45.2 .R43 2022
580--dc23

Table of Contents

Preface

Every book is initially just a concept; it takes months of research and hard work to give it the final shape in which the readers receive it. In its early stages, this book also went through rigorous reviewing. The notable contributions made by experts from across the globe were first molded into patterned chapters and then arranged in a sensibly sequential manner to bring out the best results.

Plants provide the necessary oxygen, food and medicine along with various other products. They are important for sustaining life on earth. The branch of biology which studies the life of plants is known as botany. This field studies physical as well as chemical properties associated with the life of plants. Plant biochemistry, plant anatomy, plant ecology, plant taxonomy, plant morphology, plant physiology, etc. are some of the primary branches of botany. It has applications in the diverse areas of agronomy, forestry, pomology, horticulture, plant breeding, arboriculture, etc. The various areas of research in this field include plant reproduction, evolutionary relationships, biochemistry and primary metabolism, and plant structure. Botany has undergone rapid development over the past few decades. This book brings forth some of the most innovative concepts and elucidates the unexplored aspects of this area of study. It aims to serve as a resource guide for students and experts alike and contribute to the growth of the discipline.

It has been my immense pleasure to be a part of this project and to contribute my years of learning in such a meaningful form. I would like to take this opportunity to thank all the people who have been associated with the completion of this book at any step.

Editor

1

Identification and trichothecene genotypes of *Fusarium graminearum* species complex from wheat in Taiwan

Chih-Li Wang* and Yi-Hong Cheng

Abstract

Background: Fusarium head blight (FHB) of wheat caused by *Fusarium graminearum* species complex (FGSC) is a devastating disease worldwide. The pathogens not only reduce the yield of wheat, but also impact the quality of wheat by contamination with trichothecene mycotoxins. A systematic investigation on the pathogens of FHB in Taiwan is lacking. Here, molecular and morphological approaches were used to identify species of the Taiwanese FGSC isolates and determine their trichothecene genotypes.

Results: In this study, a total of 195 isolates of FGSC from diseased wheat were collected from 8 areas of northern and central Taiwan. All isolates were subjected to seedling inoculation for verification of pathogenicity. The pathogenic isolates were genetically characterized by sequence characterized amplified region (SCAR), PCR- restriction fragment length polymorphism (RFLP), phylogenetic analysis and fixed nucleotides to clarify their phylogenetic species, and by PCR assays of *TRI* genes to determine trichothecene genotypes. They were identified as *F. asiaticum*, *F. graminearum* sensu stricto, *F. meridionale* and an unknown species. Isolates of *F. asiaticum* were the major causal agents (98%) in this investigated population and were comprised of SCAR type 5 (75%), SCAR type 4 (21%) and SCAR type 3 (2%). Their trichothecene genotypes were either 15-acetyl-deoxynivalenol (15-ADON) (83%) or nivalenol (NIV) genotype (17%). These genetic characterizations indicated that *F. asiaticum* (15-ADON SCAR type 5) accounts for 60% of this Taiwanese population. Virulence assay on wheat heads indicated virulence of *F. asiaticum* isolates in subpopulations divided by SCAR types or trichothecene genotypes were comparable, suggesting other factors influence the unequal subpopulation sizes.

Conclusions: This is the first study that FGSC isolates in Taiwan were systematically collected and characterized. In addition to *F. graminearum* sensu stricto and *F. meridionale*, *F. asiaticum* with 15-ADON genotype was identified as the predominate species in Taiwan. In contrast to Chinese and Japanese populations that *F. asiaticum* isolates were typically of 3-ADON or NIV genotype, the predominate 15-ADON genotype in Taiwanese population was unique among *F. asiaticum* populations and represented the southernmost 15-ADON genotype population in East Asia.

Keywords: Fusarium head blight, *Fusarium asiaticum*, *Fusarium graminearum*, *Fusarium meridionale*, Deoxynivalenol, Nivalenol

Background

Wheat was an important crop that grew in many counties of Taiwan in 1960s. Due to the impact of world trade, wheat production and plantation areas have shrunk dramatically over the years. There were only limited areas in central Taiwan growing wheat for malt production in the past 40 years. In recent years, the acreages of wheat were gradually increased from central Taiwan to southern and northern Taiwan because of the advocacy of eating locally grown crops and a governmental policy of raising food self-sufficiency rates. In an informal survey, Fusarium head blight (FHB) was found to be prevalent

*Correspondence: clwang@nchu.edu.tw
Department of Plant Pathology, National Chung Hsing University, Taichung 40227, Taiwan

in many counties after the anthesis stage of wheat (Kuo et al. 2014).

FHB could be caused by a number of *Fusarium* species. Some species played a more important role than others, depending on geographic regions. Among them, *F. graminearum* sensu lato was reported as an economically devastating pathogen (Windels 2000). The morphologically characterized *F. graminearum* sensu lato was revealed as a species complex since 2000 (O'Donnell et al. 2000). To date, at least 15 species were contained within *F. graminearum* species complex (FGSC), isolated from various hosts and substrates that have been established based on the genealogical concordance phylogenetic species recognition (O'Donnell et al. 2004, 2008; Sarver et al. 2011; Starkey et al. 2007; Yli-Mattila et al. 2009). Except for *F. graminearum* sensu stricto (s. str.) that distributed worldwide, most species of FGSC were mainly restricted to few geographic regions (O'Donnell et al. 2004). *Fusarium austroamericanum*, for example, was distributed in South America. *Fusarium asiaticum* was reported in Brazil, USA, and mainly in Asian countries. In spite of numerous causal agents of FHB to wheat, particular species usually were more prevalent than others. For instance, among 623 isolates of FHB pathogens from 7 states of United States, 523 isolates belonged to *F. graminearum* s. str. (Zeller et al. 2004). *Fusarium graminearum* s. str. represented 76% of a genotyped Louisiana population (154 isolates) (Gale et al. 2011). These studies illustrated that *F. graminearum* s. str. was the predominate species in the United States. *Fusarium asiaticum* (77%) and *F. graminearum* s. str. (23%) were identified from 299 isolates collected from 12 provinces in China that were mainly located in the Yangtze River valleys (Zhang et al. 2007). Among 530 isolates collected from 6 provinces (Jiangsu, Anhui, Henan, Hebei, Shandong and Hubei) in China, 65% were *F. asiaticum* and 34% were *F. graminearum* s. str. (Shen et al. 2012). In a broader sampling from 15 provinces in China, *F. graminearum* s. str. (36%) and *F. asiaticum* (58%) were the major FHB pathogens of 469 isolates though the other 6 *Fusarium* species were also found (Zhang et al. 2012). From 35 prefectures in Japan, 246 *F. asiaticum* and 50 *F. graminearum* s. str. were identified in 298 collected isolates (Suga et al. 2008). In South Korea, of 356 isolates from rice grown in 5 provinces, 333 were *F. asiaticum* and 23 were *F. graminearum* s. str. (Lee et al. 2009). Of 568 isolates from maize at eight locations in South Korea, *F. graminearum* s. str. (75%) was most common species followed by *F. asiaticum* (12%) (Lee et al. 2012). Above information also indicated that *F. asiaticum* and *F. graminearum* were the major FHB pathogens on wheat in China and Japan, and the major FGSC species in South Korea.

In addition to yield loss, diseased grains of FHB usually accumulated trichothecenes that were associated with toxicoses in humans and farm animals (Desjardins 2006). Species of FGSC were trichothecene B-producing fungi that could be classified into three strain-specific chemotypes based on their trichothecene production: the 3-ADON chemotype isolates produced deoxynivalenol (DON) and 3-acetyl-deoxynivalenol (3-ADON); the 15-ADON chemotype isolates produced DON and 15-acetyl-deoxynivalenol (15-ADON); and the NIV chemotype isolates produced nivalenol (NIV) and 4-acetylnivalenol (Ward et al. 2002; Miller et al. 1991). Individual trichothecene may differ in toxicological effects on plants and animals. DON inhibited root growth and coleoptiles elongation of wheat, but NIV had no effect at similar concentration (Eudes et al. 2000; Shimada and Otani 1990). DON was shown to be a virulent factor of FHB to wheat whereas NIV was a virulent factor of ear rot to maize (Maier et al. 2006; Jansen et al. 2005; Proctor et al. 1995). Compared to DON, some studies observed higher cytotoxicity for NIV (Minervini et al. 2004). These differences in toxic effects made chemotypes of isolates important in geographic distribution and population composition of fungal pathogens.

The conventional method for chemotyping the trichothecenes of *Fusarium* species was the HPLC or GC/MS analysis of culture extracts from individual isolates. PCR methods for genotyping the trichothecene chemotypes of *Fusarium* species were developed according to allelic variants of trichothecene biosynthesis genes in isolates with strain-specific chemotypes. Sequences of *TRI7*, *TRI12* and *TRI13* genes, for instances, were distinct in DON- and NIV-producing isolates (Lee et al. 2001; Chandler et al. 2003; Ward et al. 2002; Quarta et al. 2006). Primers designed on *TRI3* and *TRI12* were used to assess 3-ADON- and 15-ADON-producing isolates (Jennings et al. 2004; Quarta et al. 2006; Ward et al. 2002). These studies also demonstrated that results of genotyping were highly correlated to these of chemical chemotyping. The trichothecene genotyping by PCR assays allowed rapid screening of a large number of isolates, and identification of the increasing occurrence of high toxigenic population such as 3-ADON- and NIV-producing isolates (Pasquali et al. 2010; Puri and Zhong 2010; Zhang et al. 2012).

In Taiwan, FHB of wheat caused by *F. graminearum* sensu lato was first reported in 1919 (Sawada 1919). Although other species of potential FHB pathogens such as *F. poae*, *F. avenaceum* and *F. culmorum* were also reported, none of them were isolated from wheat (Huang and Sun 1997). These earlier studies were based on the morphological characterization which was not able to resolve species in a species complex. Along with the significant occurrence of FHB, there was an increasing need to clarify the composition of the causal agents, contributing to the formulation of control strategies and for evaluation of mycotoxin threat. Trichothecenes, especially the DON (also known as vomitoxin), produced in the infected grains will become a concern for consumers. Unfortunately, there was no systematic survey carried out regarding these concerns in Taiwan. This study was aimed to identify the species of FHB pathogens including species of FGSC, and to determine their trichothecene genotypes. Results will also improve our understanding of the distribution and trichothecene genotypes of FGSC species in wheat production areas in East Asia.

Methods

Strain isolation and morphological identification

Wheat samples were collected from 10 areas of 6 cities and counties from northern Taiwan to southern Taiwan. Potential Fusarium head blight pathogens of wheat were isolated from seeds and wheat heads showing head blight symptoms. In areas of southern Taiwan where Fusarium head blight was not observed, around 100 wheat heads were randomly collected from each field for fungal isolation. For isolating fungi from diseased or symptomless wheat heads, 10 every other kernels were picked from bottom to top of a head, and incubated on 2% (w/v) water agar (WA) at room temperature for 3–5 days. After fungi grew out, one isolate was picked from each head for further study. Therefore, all isolates were obtained from different plants. A modified blotter method (De Tempe 1953) was used for fungal isolation from seeds. Briefly, seeds were treated with 1% sodium chloride for 1 min, rinsed with sterilized water for 3 times, and then incubated on a piece of wet autoclaved filter paper in 9-cm Petri dishes at 24 °C for 7 days. Ten seeds were plated in each plate, and 400 seeds were randomly sampled from each bag of seeds. Fungal colonies secreting carmine red pigment in WA or filter paper and producing characterized spindle-sickle macroconidia were selected for further purification. Single macroconidium of *Fusarium* was isolated for analyses. For morphological analysis, *Fusarium* isolates were grown on potato dextrose agar (PDA; Difco, NJ, USA) to observe culture characteristics and carnation leaf agar (CLA) to examine the morphology of macroconidia (Leslie and Summerell 2006). Thirty macroconidia with 4–5 septa from each isolate were measured to determine the conidial size. Perithecia were induced on carrot medium (5 g of Bacto agar, 400 g freshly sliced carrots, per liter of distilled water) (Cavinder et al. 2012). Ascospores released from asci were measured and characterized.

Molecular identification on FGSC

For genomic DNA isolation, fungal isolates were grown in yeast extract peptone dextrose broth (0.3 g of yeast extract, 1.0 g of peptone and 1.0 g of dextrose in 100 ml of distilled water) at 25 °C for 3 days. Fungal mycelia were harvested with miracloth (pore size: 22–25 μm) (Calbiochem, CA, USA) and DNAs were extracted following the standard procedures (Sambrook and Russell 2001). Primers Fg16F (5′-CTCCG-GATATGTTGCGTCAA-3′) and Fg16R (5′-GGTAGGTATCCGACATGGCAA-3′) were used as a pair of specific PCR primers to determine if a *Fusarium* isolate belongs to a FGSC isolate, and sizes of this PCR amplicons were used to assign the sequence characterized amplified region (SCAR) types to each isolate (Carter et al. 2000; Nicholson et al. 1998; Carter et al. 2002). The PCR program was set according to Nicolson et al. (1998). Except where indicated otherwise all PCR reactions in this study were carried out in a total volume of 25 μl containing 19.7 μl of MilliQ water, 2.5 μl of 10× buffer, 0.5 μl of dNTP (200 μM each of dATP, dTTP, dGTP and dCTP), 0.5 μl of each primer (10 μM), 1 μl of DNA template (100 ng/μl), and 0.3 μl of DNA polymerase (Pro Taq plus; Protech Technology Enterprise, Taipei, Taiwan). PCR amplification was conducted in a thermal cycler (Applied Biosystems 2720; Life Technologies, CA, USA). Electrophoresis was performed in 1.5% (w/v) TAE agarose gel, and ethidium bromide was used for gel staining. PCR-restriction fragment length polymorphism (PCR–RFLP) of partial histone gene was applied to identify *F. graminearum* s. str. and *F. asiaticum*. The partial histone H3 gene fragment with size of 223 bp was amplified using primers H3dStyI (5′-AGCATCACCYGAACATCGCATCATCCCATG-3′) and H3R1 (5′-TTGGACTGG-ATRGTAACACGC-3′), and purified with Plus DNA Clean/Extraction Kit (GeneMark, Taichung, Taiwan). The purified PCR products were digested with *Eco*RV and *Sty*I, respectively (Suga et al. 2008). The digestion results were visualized by electrophoresis in 2.5% (w/v) TAE (Tris-acetate-EDTA) agarose gel.

Trichothecene genotyping

Primers locating at *TRI13* gene were used to differentiate the genotypes of DON or NIV. Primers Tri13F (5′-CATC ATGAGACTTGTKCRAGTTTGGG-3′) and Tri-13DONR (5′-GCTAGATCGATTGTTGCATTGAG-3′) would amplify a PCR band from DON genotype isolates, but not from NIV genotype isolates. In contrast, primers Tri13NIVF (5′-CCAAATCCGAAAACCGCAG-3′) and Tri13R (5′-TTGAAAGCTCCAATGTCGTG-3′) would amplify a band from NIV genotype isolates, but not from DON genotype isolates (Chandler et al. 2003). Primers Tri303F (5′-GATGGCCGCAAGTGGA-3′) and

Tri303R (5′-GCCGGACTGCCCTATTG-3′) were used to detect 3-ADON genotype whereas primers Tri315F (5′-CTCGCTGAAGTTGGACGTAA-3′) and Tri315R (5′-GTCTATGCTCTCAACGGACAAC-3′) were used to determine 15-ADON genotype (Jennings et al. 2004).

Pathogenicity assays

To induce the production of conidia, agar blocks containing mycelia of each isolate were cultured in mung bean broth (40 g of mung beans was steeped in 1 l of just boiled distilled water for 10 min. After beans were removed by cheesecloth, the filtrate was autoclaved for use.) at 25 °C and 100 rpm for 3 days (Wu et al. 2005). Conidia were filtrated through miracloth and counted by hemocytometer. The suspension was centrifuged and adjusted to 10^5 conidia/ml with sterilized water and used for pathogenicity tests. The local wheat cultivar, Taichung Sel. 2, was used as host plants. A seedling coleoptile inoculation method (Wu et al. 2005) was modified to quickly determine the pathogenicity of all isolates. Briefly, seedling coleoptiles were wounded by a needle, and then immersing in a conidia suspension (10^5 conidia/ml) for 30 min. Five seedlings were inoculated with each isolate. Inoculated seedlings were placed on wet paper towel in Petri dishes and put in moister chambers at 24 °C with 12 h light cycle. To evaluate the virulence of each isolate, 5 μl conidia suspension (10^5 conidia/ml) was inoculated at the interior of the two central spikelets of each head at the anthesis stage (Malbrán et al. 2012). The inoculated wheat heads were covered with plastic bags for 48 h to maintain moisture. Three replicates were conducted for each isolate. Virulence was evaluated by the number of symptomatic spikelets at the 12 days post inoculation (dpi).

Phylogenetic analysis

The partial sequences of *RED* gene and *TEF-1α* gene of 19 selected representative isolates were used for resolving their phylogenetic relationships with reference strains of all known species of FGSC. A PCR product of *RED* sequence in a range from 900 to 1000 bp was amplified with primers RED1d (5′- TCTCAGAAAG ACGCATATATG-3′) and RED2 (5′- CGTAACTGCGTC ATTCGGC-3′). *TEF-1α* sequence in a range from 600 to 700 bp was amplified with primers TEF1 (5′- ATGGGT AAGGAGGACAAGAC-3′) and TEF2 (5′- GGAAGT ACCAGTGATCATGTT-3′) (O'Donnell et al. 2000). PCR products were purified for sequencing with the forward and reverse primers. Phylogenetic analysis was performed by MEGA 6 (Tamura et al. 2013). ClustalW software was applied to align the two gene sequences from the 19 isolates in this study and other reference strains belonging to species of FGSC. The concatenate sequences of *RED* and *TEF-1α* genes were analyzed by the maximum likelihood method with 1000 bootstrap replications.

Results

Species diagnosis and SCAR types of FGSC isolates from wheat

Based on secretion of carmine red pigment and sickle-shaped macroconidia of *Fusarium*, a total of 203 potential causal agents of Fusarium head blight were isolated. Almost 100% of fungi that produce the feature pigment and macroconidia could be isolated from wheat heads with typical symptom of FHB from most areas such as Daya, Pingzhen, Guanying and Xinwu where the occurrence of FHB was more frequent than other areas. In areas of southern Taiwan (Chiayi county and Tainan city), the FHB was not observed. Wheat heads with suspicious symptoms were collected for isolation. In this case, the isolation rate of carmine red pigment colony was rare. A seedling coleoptile inoculation method was adapted to quickly assay the pathogenicity of each isolate. Pathogenic isolates would cause necrotic lesions on leaf sheath or leaf blade while the water control or non-pathogenic *Fusarium* isolates to wheat such as the *Fusarium oxysporum* f. sp. *lycopersici* did not produce any lesion on wheat seedlings (Fig. 1a). Among them, 195 isolates (186 isolates from diseased plants and 9 isolates from seeds) collected from 8 areas of northern and central Taiwan were determined as wheat pathogens whereas 8 isolates obtained from 600 wheat heads with suspicious symptoms produced in southern Taiwan (Chiayi county and Tainan city) were not FHB pathogens. Of these 195 isolates, 193 isolates but 2 isolates (Dacheng 1-2 and Dacheng 2-1) showed the positive band in the PCR results using primers Fg16F and Fg16R, indicating that most of them belonged to FGSC isolates. Different sizes of the PCR bands were isolated and cloned for sequencing to determine the exact number of nucleotide base pairs (Fig. 1b). The 193 isolates were assigned into different SCAR types based on their band sizes. The SCAR type 5 (497 bp) isolates (75%) were most prevalent in population. The SCAR type 4 (557 bp) isolates represented 21% of the population. The SCAR type 3 (527 bp) isolates was only 1.5% of this studied population. Both SCAR type 1 and SCAR type 2 were found only in one isolate. Isolates from Daya area (Taichung city, central Taiwan) were most diverse in SCAR types. Isolates of SCAR type 5 and SCAR type 4 consistently presented in 3- 3.5–1 ratio in most geographic areas (Table 1). A PCR–RFLP method that was based on fixed nucleotide character states specific to *F. graminearum* s. str. and *F. asiaticum* has been reliable to diagnose 296 isolates of the two species in Japan (Suga et al. 2008), 530 isolates of the two species in China (Shen et al. 2012), and 41

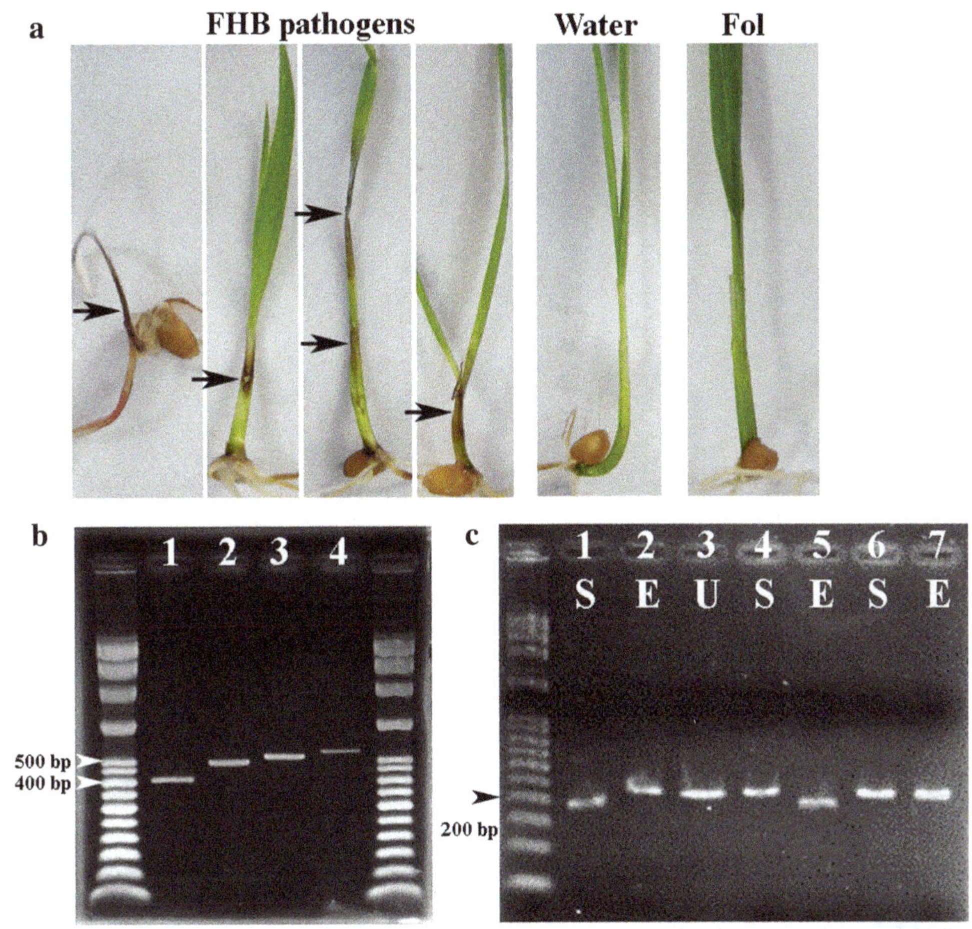

Fig. 1 Representative results of seedling pathogenicity assays, SCAR types, and PCR–RFLP diagnosis. **a** The pathogenicity assay with seedling coleoptile inoculation method was used to test if all isolates are pathogenic to wheat. Necrotic leaf sheath lesions (*arrows*) indicated that the tested isolates were FHB pathogens. No symptom displayed after inoculation with sterilized water or *Fusarium oxysporum* f.sp. *lycopersici* (Fol, not a FHB pathogen), respectively. **b** The variation of SCAR types in studied population. The *first lane* is the DNA ladder; *Lane 1* is SCAR type 1 (410 bp); *Lane 2* is SCAR type 2 or type 5 (497 bp); *Lane 3* is SCAR type 3 (527 bp); *Lane 4* is SCAR type 4 (557 bp). The *last lane* is another DNA ladder. **c** A PCR–RFLP method for diagnosis of *F. asiaticum* and *F. graminearum* s. str. through a partial histone H3 fragment. A *F. asiaticum* isolate will be digested by *Sty*I but not *Eco*RV whereas a *F. graminearum* s. str. isolate will be digested by *Eco*RV but not *Sty*I. Other species will not be digested by the two enzymes. The *first lane* is the DNA ladder; *Lanes 1* to *3* are a *F. asiaticum* isolate (Pingzhen1-49); *Lanes 4* and *5* are a *F. graminearum* s. str. isolate (Daya211-13); *Lanes 6 and 7* are an unknown species isolate (Daya272-3). Different restriction enzyme treatments are indicated: for "S" this is *Sty*I, for "E" this is *Eco*RV, and for "U" this is the uncut control

isolates of *F. asiaticum* in US (Gale et al. 2011). Since *F. graminearum* s. str. and *F. asiaticum* were the most common species of FGSC in Asian countries (e.g. China, South Korea and Japan) (review see van der Lee et al. 2015; Wang et al. 2011), we used the method to diagnose if any of these Taiwanese FGSC isolates belong to the two species. In an earlier report, the partial histone H3 gene fragment of *F. asiaticum* would be cut into 191 and 32 bp by *Sty*I whereas the partial histone H3 gene fragment of *F. graminearum* s. str. would be cut into 195 and 28 bp by *Eco*RV (Suga et al. 2008). The band shifted to a reduced size, after the restriction enzyme treatments, indicating a positive digestion (Fig. 1c). DNAs of a total of 192 out of the 195 isolates were digested into smaller fragments by *Sty*I, suggesting that *F. asiaticum* (99%) was the predominate species of FGSC in Taiwan. Only one isolate (Daya211-13) was diagnosed as *F. graminearum* s. str. Two isolates (Fanyuan1-11 and Daya272-3) that failed to be digested by neither *Sty*I nor *Eco*RV were selected for further identification by phylogenetic analysis.

Phylogenetic analysis

To further confirm the diagnostic results of PCR–RFLP and determine the 2 unidentified isolates, a total of 19 representative isolates including 16 isolates of *F. asiaticum*, 1 isolate of *F. graminearum* s. str., and the 2

Table 1 The geographic distribution of identified species and SCAR types of all isolates

Collection area	No. of isolates	*F. asiaticum*			*F. graminearum* s. str.	*F. meridionale*	Unknown species
		Type[a] 3	Type 4	Type 5	Type 1	Type 5	Type 2
Northern Taiwan							
Taoyuan city							
Guanying	30		7	23			
Xinwu	30		8	22			
Pinzhen	30		3	27			
Central Taiwan							
Miaoli county							
Yuanli	3		1	2			
Taichung city							
Daya	70	3	15	50	1		1
Waipu	4			4			
Changhua county							
Fanyuan	23		5	17		1	
Dacheng	5[b]		2	1			
Total	195[b]	3	41	146	1	1	1

[a] The SCAR types were defined by Carter et al. (2002). Type 1 is 410 bp; Type 2 is 479 bp; Type 3 is 527 bp; Type 4 is 557 bp; Type 5 is 479 bp. Type 2 and Type 5 are different in nucleotide sequences

[b] The SCAR patterns were not amplified from two *F. asiaticum* isolates (Dacheng 1-2 and Dacheng 2-1)

unidentified isolates were further investigated for their phylogenetic relationships with reference strains and type strains of all known FGSC species. The 16 isolates of *F. asiaticum* were randomly selected from each SCAR type and trichothecene genotype. For phylogenetic analysis, *RED* gene and *TEF-1α* gene of the 19 representative isolates were sequenced and deposited at GenBank (http://www.ncbi.nlm.nih.gov/genbank/) with accession numbers (Table 2). A total of 1296 informative nucleotides were used for analysis including 775 nucleotides of *RED* gene and 521 nucleotides of *TEF-1α* gene. The 16 isolates of *F. asiaticum* were well clustered with 3 reference *F. asiaticum* strains with a high bootstrap value (99%). The *F. graminearum* s. str. isolate was also placed in the *F. graminearum* s. str. clade with a bootstrap value of 91%. These results not only confirmed the PCR–RFLP method was reliable, but also illustrated this phylogenetic tree was able to resolve the phylogenetic relationship at species level to some extent. One of the unidentified isolates (Fanyuan1-11) was clearly clustered with 2 reference strains and the type strain of *F. meridionale* with a bootstrap value of 92%. The other unidentified isolate (Daya272-3) was closest to the *F. cortaderiae* clade, but had a low bootstrap value of 52% and this did not support the relationship (Fig. 2).

All sequenced isolates of identified *F. asiaticum* were further compared with the species type strain NRRL 13818 on the defined fixed nucleotides of *RED* sequence, and the results showed that they were identical at all fixed nucleotides. The *RED* and *TEF-1α* sequences of identified *F. graminearum* s. str. and *F. meridionale* were aligned with reference strains *F. graminearum* s. str. NRRL 5883 and *F. meridionale* NRRL 28436, respectively. All fixed nucleotides were also conserved in the investigated strains in the 2 gene alignments. In addition, morphology of all sequenced isolates was characterized and was indistinguishable from each other. All isolates produced carmine red pigment on PDA and CLA, and formed macroconidia and no microconidia. Macroconidia from CLA were gradually curved and 52–63 × 5–6 μm with 5–6 septa. All isolates were homothallic and produced various numbers of dark blue perithecia containing a number of asci on carrot medium. Each ascus bore 8 ascospores. Ascospores were 21–24 × 3–5 μm with 1–3 septa, hyaline in color and fusiform with rounded ends (Additional file 1: Figure S1). These morphological characteristics conformed to the defined characteristics of FGSC.

Trichothecene genotypes

The allelic nucleotide polymorphism of *TRI13* gene was used to distinguish DON genotype or NIV genotype. A PCR band of about 227 bp indicating DON genotype was amplified from 162 isolates (83%) by primers Tri13F and Tri13DONR. On the other hand, a PCR band of about 312 bp indicating NIV genotype was amplified from 33 isolates (17%) by primers Tri13NIVF and Tri13R

Table 2 The list of representative isolates used for phylogenetic analysis

Species of FGSC and isolate	SCAR type[a]	Trichothecene genotype	Accession number	
			Reductase	TEF-1α
F. asiaticum				
Daya350-5	Type 5	NIV	KT372157	KT380119
Daya211-2	Type 5	NIV	KT372158	KT380120
Fanyuan1-13	Type 5	NIV	KT372159	KT380121
R-p-1	Type 5	15-ADON	KT372149	KT380111
Daya350-3	Type 5	15-ADON	KT372162	KT380124
Daya211-6	Type 5	15-ADON	KT372164	KT380126
Xinwu1-1	Type 5	15-ADON	KT372163	KT380125
Daya350-11	Type 3	15-ADON	KT372147	KT380109
Daya350-12	Type 3	15-ADON	KT372152	KT380114
Daya272-14	Type 3	15-ADON	KT372153	KT380115
Daya350-2	Type 4	NIV	KT372154	KT380116
Daya272-6	Type 4	NIV	KT372155	KT380117
Pingzhen1-13	Type 4	NIV	KT372156	KT380118
Daya350-13	Type 4	15-ADON	KT372160	KT380122
k-r-1	Type 4	15-ADON	KT372150	KT380112
Guanying 1-5	Type 4	15-ADON	KT372161	KT380123
F. graminearum s. str.				
Daya211-13	Type 1	15-ADON	KT334553	KT380127
F. meridionale				
Fanyuan1-11	Type 5	NIV	KT372148	KT380110
Unknown species				
Daya272-3	Type 2	NIV	KT372151	KT380113

[a] The SCAR types were defined by Carter et al. (2002). Type 1 is 410 bp; Type 2 is 479 bp; Type 3 is 527 bp; Type 4 is 557 bp; Type 5 is 479 bp. Type 2 and Type 5 are different in nucleotide sequences

(Fig. 3a). The DON genotype isolates were further differentiated as 3-ADON or 15-ADON isolates by primers designed from the different alleles of *TRI3* gene. All DON genotype isolates generated a 864 bp band in the PCR reactions with primers Tri315F and Tri315R, but not a 586 bp band with primers Tri303F and Tri303R, indicating that all DON genotype isolates were of 15-ADON genotype (Fig. 3b). In northern Taiwan, all isolates but one belonged to 15-ADON (99%). In central Taiwan, NIV isolates accounted for 32 and 19% of populations in Taichung city and Changhua county, respectively (Fig. 3c). As with SCAR types, trichothecene genotypes were more diverse in central Taiwan than northern Taiwan.

Combined with the above results of isolate identity and SCAR types, one would conclude that the most predominate species and genotype is *F. asiaticum* 15-ADON SCAR type 5 (60%). *Fusarium asiaticum* 15-ADON SCAR type 4 isolates and *F. asiaticum* NIV SCAR type 5 isolates contributed 19 and 15% of the studied population, respectively. The single *F. graminearum* s. str. isolate was of 15-ADON genotype whereas the single *F. meridionale* isolate and the unknown species isolate were of NIV genotype (Table 3).

Virulence comparison among isolates with different SCAR types or trichothecene genotypes

To determine if virulence variation could account for the unequal subpopulation sizes of *F. asiaticum* 15-ADON SCAR type 5 isolates (60%), *F. asiaticum* 15-ADON SCAR type 4 isolates (19%) and *F. asiaticum* NIV SCAR type 5 isolates (15%), 10 isolates from each of the 3 subpopulations were randomly selected to compare their virulence on wheat heads. Conidial suspension of each isolate was inoculated on the 2 central spikelets of heads. Number of the discolored spikelet was counted to evaluate the virulence of each isolate at the 12 dpi. Although isolates of the *F. asiaticum* 15-ADON SCAR type 5 subpopulation caused slightly more symptomatic spikelets in average than isolates in the other two subpopulations, the differences did not statistically distinguish them in virulence to wheat (Fig. 4). This result suggests that the difference of trichothecene genotypes (15-ADON and NIV)

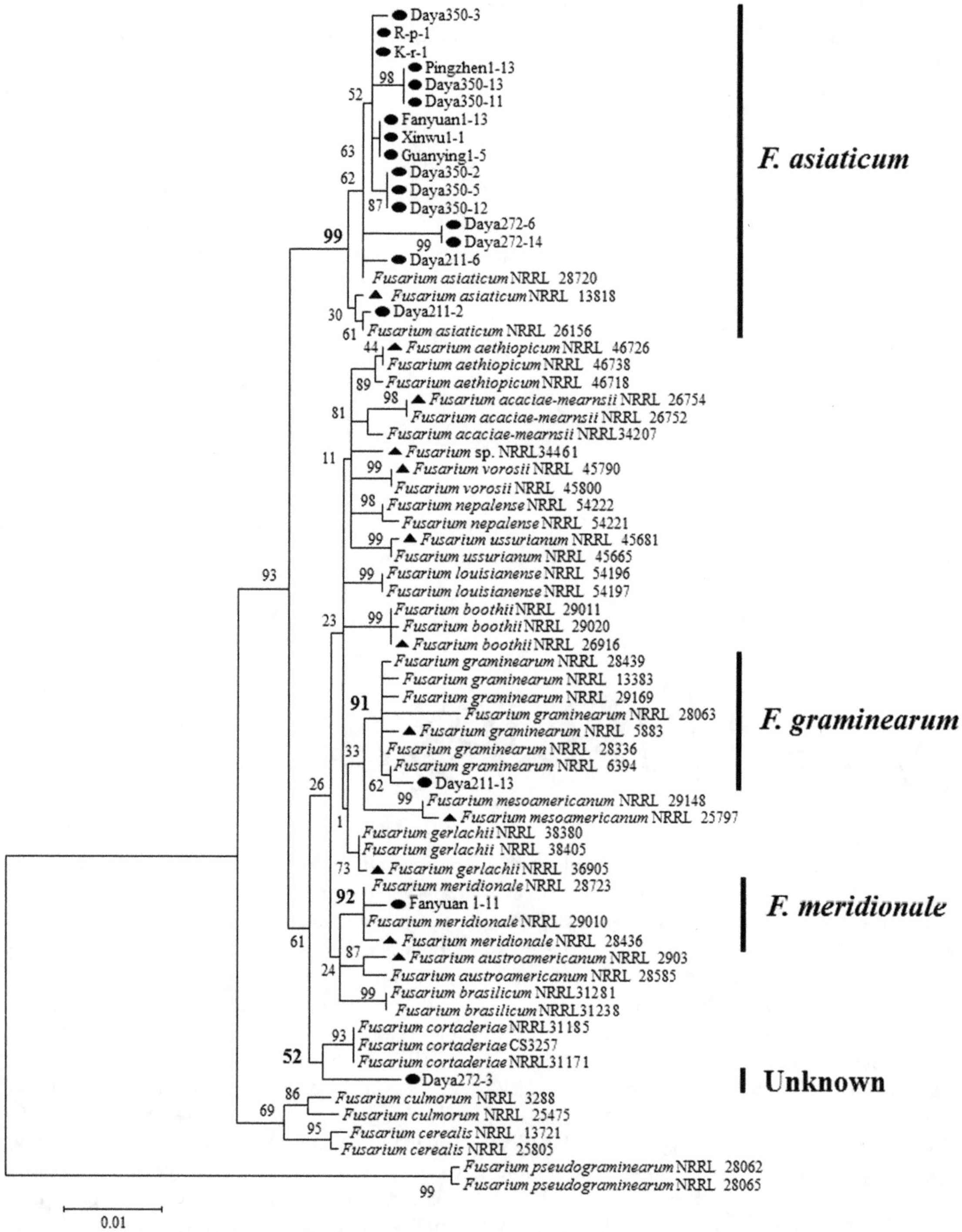

Fig. 2 The phylogenetic analysis. The phylogenetic tree of representative strains of FHB pathogens in Taiwan based on partial sequences of *RED* and *TEF-1α* genes. The maximum likelihood method was used for the analysis with 1000 bootstrap repeats (*black circle* isolates in this study, *filled triangle* type strains of species of FGSC)

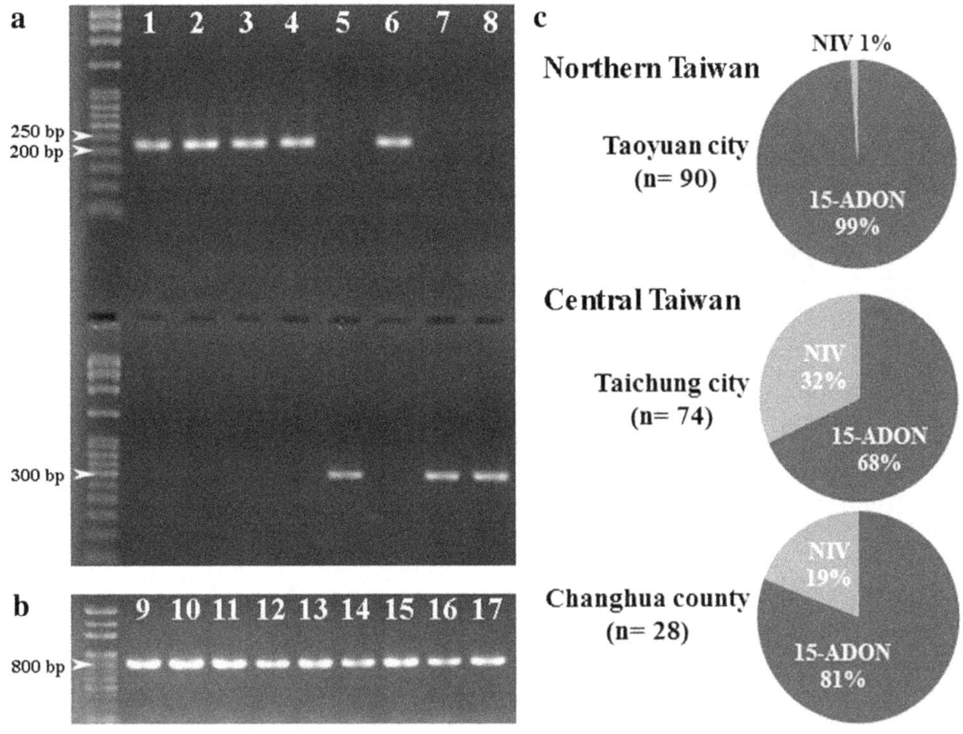

Fig. 3 The identification and distribution of trichothecene genotypes of FGSC isolates. **a** Identification of DON or NIV genotype by the amplification of *TRI13* gene with primers Tri13F and Tri13DONR (*upper panel*), and primers Tri13NIVF and Tri13R (*lower panel*), respectively. The *first lane* is the DNA ladder; *Lane 1* is Daya 350-12; *Lane 2* is Daya 350-13; *Lane 3* is Daya 350-14; *Lane 4* is Daya 350-15; *Lane 5* is Daya 350-16; *Lane 6* is Daya 211-1; *Lane 7* is Daya 211-2; *Lane 8* is Daya 211-3. **b** Identification of 15-ADON genotype by the amplification of *TRI3* gene with primers Tri315F and Tri315R. The *first lane* is the DNA ladder; *Lane 9* is Daya 350-11; *Lane 10* is Daya 350-12; *Lane 11* is Daya 350-13; *Lane 12* is Daya 211-6; *Lane 13* is Xinwu 1-1; *Lane 14* is Pingzhen1-13; *Lane 15* is R-P-1; *Lane 16* is Guanying 1-5; *Lane 17* is Daya 211-13. **c** Distributions of FGSC isolates regarding trichothecene genotypes in northern (Taoyuan city) and central (Taichung city and Changhua county) Taiwan

Table 3 The trichothecene genotypes of all species with reference to SCAR types

Genotype	No. of isolates	*F. asiaticum*			*F. graminearum* s.s.	*F. meridionale*	Unknown
		Type 5[a]	Type 3	Type 4	Type 1	Type 5	Type 2
DON	162[b]						
15-ADON	162[b]	118 (60.5%)	3 (1.5%)	38 (19.4%)	1 (0.5%)		
NIV	33	30 (15.3%)		3 (1.5%)		1 (0.5%)	1 (0.5%)
Total	195[b]	146	3	41	1	1	1

[a] The SCAR types were defined by Carter et al. (2002). Type 1 is 410 bp; Type 2 is 479 bp; Type 3 is 527 bp; Type 4 is 557 bp; Type 5 is 479 bp. Type 2 and Type 5 are different in nucleotide sequences

[b] The SCAR types of two *F. asiaticum* isolates (Dacheng 1-2 and Dacheng 2-1) were not available

or SCAR types (SCAR type 4 and SCAR type 5) did not explain the unequal subpopulation sizes.

Discussion

Fungal isolates collected in this study were from areas of northern and central Taiwan located from North 23°50′ to North 25°02′ which represented one of the wheat production areas at the lowest latitude in the world. FHB pathogens were not detected from samples collected from southern Taiwan near or lower than the Tropic of Cancer. It would be interesting to dissect the reasons of no detection in southern Taiwan, including the presence or absence of pathogens, pathogen survival ability in high temperature, alternative hosts and environmental factors for disease development. In this study, all pathogenic isolates were considered to be members of FGSC

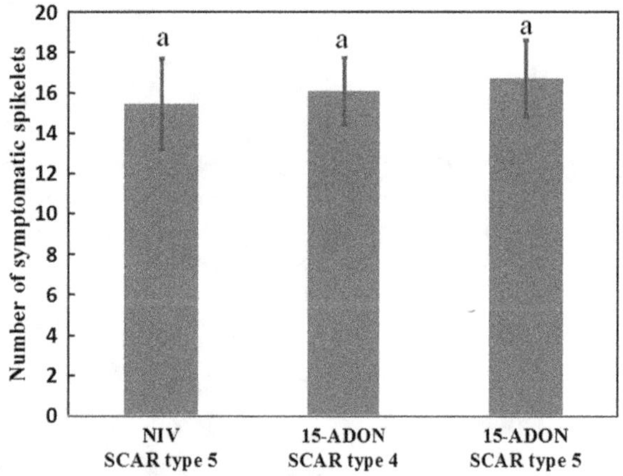

Fig. 4 The virulence comparison of *F. asiaticum* isolates with different genotypes or SCAR types. The pathogenicity assays were performed by point inoculation on central spikelets of wheat heads. Data were analyzed by LSD method with the Software SPSS

by PCR amplification with primers Fg16F and Fg16R and morphological characteristics of asexual and sexual stages. Because species of FGSC were established based on the molecular characteristics and cannot be identified through the morphological characteristics (O'Donnell et al. 2004; Sarver et al. 2011), we used molecular methods to determine the species of each isolate. As mentioned earlier, *F. asiaticum* and *F. graminearum* s. str. have been reported as predominate species of FGSC in China and Japan (Shen et al. 2012; Zhang et al. 2007; Suga et al. 2008; Zhang et al. 2012). A PCR–RFLP method has been reliable to diagnose 867 isolates of *F. graminearum* s. str. and *F. asiaticum* in Japan, China and US (Shen et al. 2012; Gale et al. 2011; Suga et al. 2008). We, therefore, applied the PCR–RFLP diagnosis method to detect *F. asiaticum* and *F. graminearum* s. str. from collected isolates. A phylogenetic analysis based on *RED* and *TEF-1α* sequences complied with the diagnosed results, and further identified a *F. meridionale* isolate in our collection. As shown in studies, *RED* and *TEF-1α* sequences were able to separate most species of FGSC and used at the initial stage to discover a novel species (O'Donnell et al. 2004; Starkey et al. 2007). These identifications of *F. asiaticum*, *F. graminearum* s. str. and *F. meridionale* were also supported by the fixed nucleotides presenting in the 2 gene sequences and by the morphology of the sexual and conidial stages. Nevertheless, there was still one isolate in query. More informative genes such as genes in the mating-type locus are required to resolve the relationship between the unknown isolate and other species of FGSC (O'Donnell et al. 2004).

Fusarium asiaticum was identified as a predominate species with 98% of the population in this study. *Fusarium asiaticum* and *F. graminearum* s. str. were the major species in many Asian countries (van der Lee et al. 2015; Wang et al. 2011). Factors like temperatures and hosts have been ascribed to species distribution. In China, *F. asiaticum* was mainly distributed in the southern provinces, whereas *F. graminearum* s. str. was mostly in the northern provinces (Qu et al. 2008b; Shen et al. 2012). The same trend was observed in an analysis of Japanese isolates (Suga et al. 2008). From a study of FHB of wheat, Qu et al. (2008b) concluded that *F. asiaticum* was mainly obtained from warmer regions where the annual average temperature were above 15 °C, and the majority of *F. graminearum* s. str. was isolated from cooler regions with the annual average temperature at 15 °C or lower. A BIOCLIM analysis suggested that areas where the warmest quarter with the mean temperature above 22 °C and the precipitation over 320 nm favored the occurrence of *F. asiaticum* (Backhouse 2014). In addition, the host preference was implicated by a study of rice population in South Korea in which *F. asiaticum* (FGSC lineage 6) dominated over *F. graminearum* s. str. (FGSC lineage 7) and *F. boothii* (FGSC lineage 3) (Lee et al. 2009). The study showed that *F. asiaticum* produced more perithecia on rice straw than the other two species in an experiment of mixed-species inoculum. Zhang et al. (2012) found a strong association between the predominate crops and the occurrence of *F. asiaticum* and *F. graminearum* s. str. They showed that *F. asiaticum* was more frequently obtained from the middle and low valleys of Yangtze River (southern China) where rice acreages were higher than maize acreages. *Fusarium graminearum* s. str. was prevalent in northern China with higher maize acreages (Zhang et al. 2012). In our study, the average annual temperatures of sampling areas in Taiwan were 22 °C and higher, and the precipitation of the warmest quarter were from 322 to 608 nm. All areas where samples were collected were operated under the rice–wheat rotation system. The result of predominate *F. asiaticum* in Taiwan was consistent with the notion of those previous studies.

Fg16F and Fg16R were designed as specific primers for FGSC isolates (Nicholson et al. 1998). The various amplicons from Fg16F/R PCR were used for the SCAR analysis of FGSC isolates to reveal the genetic variation (Carter et al. 2000, 2002; Qu et al. 2008a, b; Desjardins et al. 2004). Later studies showed that SCAR type 1 and type 5 were fully congruent with *F. graminearum* s. str. and *F. asiaticum*, respectively, and were used to screen these 2 phylogenetic species (Chandler et al. 2003; Qiu et al. 2014; Zhang et al. 2007). SCAR type 2 was congruent with *F. meridionale*, but contained the same PCR band size as SCAR type 5 (Chandler et al. 2003). In this study, isolates of SCAR type 1 and most SCAR type 5 were confirmed as *F. graminearum* s. str. and *F. asiaticum*, respectively, by PCR–RFLP and phylogenetic analysis. Some *F.*

asiaticum isolates were of SCAR type 3 and SCAR type 4, which has been reported elsewhere (Chandler et al. 2003). Notably, the one *F. meridionale* isolate (Fanyuan1-11) was determined to be SCAR type 5, rather than SCAR type 2. The unknown species isolate (Daya272-3) was of SCAR type 2, but was phylogenetically closest to *F. cortaderiae* rather than *F. meridionale*. These results suggested that additional caution was needed to be taken to infer FGSC species by SACR types.

It was surprising to find that *F. asiaticum* isolates were either 15-ADON or NIV genotype in this Taiwanese population. The 3-ADON genotype isolate was not detected in current collection. The 15-ADON genotype isolates were more prevalent than NIV genotype isolates in all geographic areas, especially in northern Taiwan where 15-ADON genotype isolates accounted for 99% of the population. In Japanese and Chinese populations of *F. asiaticum*, all 3 trichothecene genotypes (3-ADON, 15-ADON and NIV genotypes) were detected. Both 3-ADON and NIV genotype isolates usually dominated over 15-ADON genotype isolates (Qiu et al. 2014; Suga et al. 2008). Outside of Asia, NIV genotype isolates of *F. asiaticum* were prevalent in southern Louisiana, United States and southern Brazil (Gale et al. 2011; Gomes et al. 2015). The 15-ADON genotype isolates of *F. asiaticum* were a typical minority and may be absent in culture collections (Karugia et al. 2009; Shen et al. 2012). In China, the 3-ADON and NIV genotype isolates were prevalent in the warmer southern regions whereas the 15-ADON genotype was mainly presented in the cooler northern regions. The geographic difference of trichothecene genotypes were explained by the unevenly distribution of phylogenetic species. *F. asiaticum*, the warmer-region inhabitant, mainly belonged to 3-ADON and NIV genotypes based on 1903 isolates while *F. graminearum* s. str., the cooler-region inhabitant, contained more 15-ADON genotype isolates based on 383 isolates, according to several studies in China and Japan (Karugia et al. 2009; Qiu et al. 2014; Shen et al. 2012; Suga et al. 2008; Zhang et al. 2007). Regarding the geographic locations of 15-ADON genotype isolates which were mostly in the north of East Asia, this Taiwanese *F. asiaticum* 15-ADON isolates represented a unique population locating in the south of East Asia. This result suggested that 3-ADON genotype isolates were possibly not existed in Taiwanese populations or several factors may favor the 15-ADON genotype isolates to establish in Taiwan. It is notable that the wheat cultivar Taichung Sel. 2 derived from the wheat line Au-Maya74"S" was the only wheat cultivar used for wheat production in Taiwan currently. All FGSC isolates were isolated from diseased plants of the cultivar. The pathogenicity assay, however, indicated that the host cultivar was equally susceptible to 15-ADON and NIV genotype isolates. It needed further evaluations if the Taichung Sel. 2 cultivar favors *F. asiaticum* 15-ADON genotype isolates for growth, survival, sporulation, or other biological traits that would facilitate these 15-ADON genotype isolates to outcompete other genotype isolates.

Conclusions

This study was the first report on phylogenetic species and trichothecene genotypes for a systematically collected FGSC population in Taiwan. *Fusarium graminearum* s. str., *F. meridionale*, *F. asiaticum*, and an unknown species were identified, and both *F. meridionale* and *F. asiaticum* were first reported in Taiwan. Of 195 isolates, *F. asiaticum* (n = 192) was the predominate species of FGSC wheat isolates and prevalent in northern and central Taiwan. Most Taiwanese *F. asiaticum* isolates contained 15-ADON trichothecene genotype (n = 162), and the rest of them contained NIV trichothecene genotype (n = 33). The high 15-ADON genotype of *F. asiaticum* population in Taiwan was a sharp contrast to the high 3-ADON genotype of *F. asiaticum* reported in the neighbor countries. The above information also provided valuable information for the global distribution of FGSC in East Asia (van der Lee et al. 2015).

Abbreviations

FHB: Fusarium head blight; FGSC: *Fusarium graminearum* species complex; SCAR: sequence characterized amplified region; RFLP: restriction fragment length polymorphism; DON: deoxynivalenol; 15-ADON: 15-acetyl-deoxynivalenol; 3-ADON: 3-acetyl-deoxynivalenol; NIV: nivalenol; PDA: potato dextrose agar; CLA: carnation leaf agar.

Authors' contributions

CLW carried out writing and revising the manuscript, and provide the concepts of experimental designs. YHC performed experiments. CLW and YHC participated the discussion of the experimental results. Both authors read and approved the final manuscript.

Acknowledgements

The authors thank Dr. Chung-Jan Chang (University of Georgia, USA) and Dr. Wen-Hsiung Ko (National Chung Hsing University, Taiwan) for his critical reading with English editing, Mr. Chien-Chih Kuo (Taichung District Agricultural Research and Extension Station, Taiwan) for sharing 10 isolates, and Mr. Hong-Yue Chen (National Chung Hsing University, Taiwan) for technical assistance. Research grants to CLW were supported by the National Science Council (NSC 101-2313-B-005-049-MY2) and the Ministry of Science and Technology (MOST 103-2313-B-005-028-MY3), Taiwan.

Competing interests

The authors declare that they have no competing interests.

References

Backhouse D (2014) Global distribution of *Fusarium graminearum*, *F. asiaticum* and *F. boothii* from wheat in relation to climate. Eur J Plant Pathol 139:161–173

Carter J, Rezanoor H, Desjardins A, Nicholson P (2000) Variation in *Fusarium graminearum* isolates from Nepal associated with their host of origin. Plant Pathol 49:452–460

Carter J, Rezanoor H, Holden D, Desjardins A, Plattner R, Nicholson P (2002) Variation in pathogenicity associated with the genetic diversity of *Fusarium graminearum*. Eur J Plant Pathol 108:573–583

Cavinder B, Sikhakolli U, Fellows KM, Trail F (2012) Sexual development and ascospore discharge in *Fusarium graminearum*. J Vis Exp 61:e3895. doi:10.3791/3895:e3895

Chandler EA, Simpson DR, Thomsett MA, Nicholson P (2003) Development of PCR assays to *Tri7* and *Tri13* trichothecene biosynthetic genes, and characterisation of chemotypes of *Fusarium graminearum*, *Fusarium culmorum* and *Fusarium cerealis*. Physiol Mol Plant Pathol 62:355–367

De Tempe J (1953) The blotter method of seed health testing. Proc Int Seed Test Assoc 21:133–151

Desjardins AE (2006) *Fusarium* mycotoxins: chemistry, genetics, and biology. American Phytopathological Society. APS Press

Desjardins AE, Jarosz AM, Plattner RD, Alexander NJ, Brown DW, Jurgenson JE (2004) Patterns of trichothecene production, genetic variability, and virulence to wheat of *Fusarium graminearum* from smallholder farms in Nepal. J Agric Food Chem 52:6341–6346

Eudes F, Comeau A, Rioux S, Collin J (2000) Phytotoxicity of eight mycotoxins associated with *Fusarium* in wheat head blight. Can J Plant Pathol 22:286–292

Gale LR, Harrison SA, Ward TJ, O'Donnell K, Milus EA, Gale SW, Kistler HC (2011) Nivalenol-type populations of *Fusarium graminearum* and *F. asiaticum* are prevalent on wheat in southern Louisiana. Phytopathology 101:124–134

Gomes LB, Ward TJ, Badiale-Furlong E, Del Ponte EM (2015) Species composition, toxigenic potential and pathogenicity of *Fusarium graminearum* species complex isolates from southern Brazilian rice. Plant Pathol 64:980–987

Huang JW, Sun SK (1997) The Genus Fusarium from Taiwan. Shih Way Publishers, Taichung

Jansen C, von Wettstein D, Schafer W, Kogel KH, Felk A, Maier FJ (2005) Infection patterns in barley and wheat spikes inoculated with wild-type and trichodiene synthase gene disrupted *Fusarium graminearum*. Proc Natl Acad Sci USA 102:16892–16897

Jennings P, Coates M, Turner J, Chandler E, Nicholson P (2004) Determination of deoxynivalenol and nivalenol chemotypes of *Fusarium culmorum* isolates from England and Wales by PCR assay. Plant Pathol 53:182–190

Karugia GW, Suga H, Gale LR, Nakajima T, Ueda A, Hyakumachi M (2009) Population structure of *Fusarium asiaticum* from two Japanese regions and eastern China. J Gen Plant Pathol 75:110–118

Kuo CC, Lin HS, Liao CT, Huang DC, Cheng YH, Wang CL (2014) The investigation of wheat Fusarium head blight caused by *Fusarium graminearum*:pathogen identification and fungicide screening assay. Plant Pathol Bull 23:192–193

Lee T, Oh DW, Kim HS, Lee J, Kim YH, Yun SH, Lee YW (2001) Identification of deoxynivalenol- and nivalenol-producing chemotypes of *Gibberella zeae* by using PCR. Appl Environ Microbiol 67:2966–2972

Lee J, Chang IY, Kim H, Yun SH, Leslie JF, Lee YW (2009) Genetic diversity and fitness of *Fusarium graminearum* populations from rice in Korea. Appl Environ Microbiol 75:3289–3295

Lee J, Kim H, Jeon JJ, Kim HS, Zeller KA, Carter LL, Leslie JF, Lee YW (2012) Population structure of and mycotoxin production by *Fusarium graminearum* from maize in South Korea. Appl Environ Microbiol 78:2161–2167

Leslie JF, Summerell BA (2006) The *Fusarium* laboratory manual. Blackwell publishing, Oxford

Maier FJ, Miedaner T, Hadeler B, Felk A, Salomon S, Lemmens M, Kassner H, Schaefer W (2006) Involvement of trichothecenes in fusarioses of wheat, barley and maize evaluated by gene disruption of the trichodiene synthase (*Tri5*) gene in three field isolates of different chemotype and virulence. Mol Plant Pathol 7:449–461

Malbrán I, Mourelos C, Girotti J, Aulicino M, Balatti P, Lori G (2012) Aggressiveness variation of *Fusarium graminearum* isolates from Argentina following point inoculation of field grown wheat spikes. Crop Prot 42:234–243

Miller JD, Greenhalgh R, Wang Y, Lu M (1991) Trichothecene chemotypes of three *Fusarium* species. Mycologia 83:121–130

Minervini F, Fornelli F, Flynn KM (2004) Toxicity and apoptosis induced by the mycotoxins nivalenol, deoxynivalenol and fumonisin B1 in a human erythroleukemia cell line. Toxicol In Vitro 18:21–28

Nicholson P, Simpson D, Weston G, Rezanoor H, Lees A, Parry D, Joyce D (1998) Detection and quantification of *Fusarium culmorum* and *Fusarium graminearum* in cereals using PCR assays. Physiol Mol Plant Pathol 53:17–37

O'Donnell K, Kistler HC, Tacke BK, Casper HH (2000) Gene genealogies reveal global phylogeographic structure and reproductive isolation among lineages of *Fusarium graminearum*, the fungus causing wheat scab. Proc Natl Acad Sci USA 97:7905–7910

O'Donnell K, Ward TJ, Geiser DM, Corby Kistler H, Aoki T (2004) Genealogical concordance between the mating type locus and seven other nuclear genes supports formal recognition of nine phylogenetically distinct species within the *Fusarium graminearum* clade. Fungal Genet Biol 41:600–623

O'Donnell K, Ward TJ, Aberra D, Kistler HC, Aoki T, Orwig N, Kimura M, Bjornstad S, Klemsdal SS (2008) Multilocus genotyping and molecular phylogenetics resolve a novel head blight pathogen within the *Fusarium graminearum* species complex from Ethiopia. Fungal Genet Biol 45:1514–1522

Pasquali M, Giraud F, Brochot C, Cocco E, Hoffmann L, Bohn T (2010) Genetic *Fusarium* chemotyping as a useful tool for predicting nivalenol contamination in winter wheat. Int J Food Microbiol 137:246–253

Proctor RH, Hohn TM, McCormick SP (1995) Reduced virulence of *Gibberella zeae* caused by disruption of a trichothecene toxin biosynthetic gene. Mol Plant Microbe Interact 8:593–601

Puri KD, Zhong S (2010) The 3ADON population of *Fusarium graminearum* found in North Dakota is more aggressive and produces a higher level of DON than the prevalent 15ADON population in spring wheat. Phytopathology 100:1007–1014

Qiu J, Xu J, Shi J (2014) Molecular characterization of the *Fusarium graminearum* species complex in Eastern China. Eur J Plant Pathol 139:811–823

Qu B, Li H, Zhang J, Huang T, Carter J, Liao Y, Nicholson P (2008a) Comparison of genetic diversity and pathogenicity of *Fusarium* head blight pathogens from China and Europe by SSCP and seedling assays on wheat. Plant Pathol 57:642–651

Qu B, Li H, Zhang J, Xu Y, Huang T, Wu A, Zhao C, Carter J, Nicholson P, Liao Y (2008b) Geographic distribution and genetic diversity of *Fusarium graminearum* and *F. asiaticum* on wheat spikes throughout China. Plant Pathol 57:15–24

Quarta A, Mita G, Haidukowski M, Logrieco A, Mule G, Visconti A (2006) Multiplex PCR assay for the identification of nivalenol, 3- and 15-acetyl-deoxynivalenol chemotypes in *Fusarium*. FEMS Microbiol Lett 259:7–13

Sambrook J, Russell DW (2001) Molecular Cloning: a laboratory manual. In. Cold Spring Harbor, New York, Cold Spring Harbor Laboratory

Sarver BA, Ward TJ, Gale LR, Broz K, Kistler HC, Aoki T, Nicholson P, Carter J, O'Donnell K (2011) Novel *Fusarium* head blight pathogens from Nepal and Louisiana revealed by multilocus genealogical concordance. Fungal Genet Biol 48:1096–1107

Sawada K (1919) Descriptive Catalogue of the Formosan Fungi Part 1. Formosa agricultural experiment station special bulletin 19

Shen CM, Hu YC, Sun HY, Li W, Guo JH, Chen HG (2012) Geographic distribution of trichothecene chemotypes of the *Fusarium graminearum* species complex in major winter wheat production areas of China. Plant Dis 96:1172–1178

Shimada T, Otani M (1990) Effects of *Fusarium* mycotoxins on the growth of shoots and roots at germination in some japanese wheat cultivars. Cereal Res Commun 18:229–232

Starkey DE, Ward TJ, Aoki T, Gale LR, Kistler HC, Geiser DM, Suga H, Toth B, Varga J, O'Donnell K (2007) Global molecular surveillance reveals novel *Fusarium* head blight species and trichothecene toxin diversity. Fungal Genet Biol 44:1191–1204

Suga H, Karugia GW, Ward T, Gale LR, Tomimura K, Nakajima T, Miyasaka A, Koizumi S, Kageyama K, Hyakumachi M (2008) Molecular characterization of the *Fusarium graminearum* species complex in Japan. Phytopathology 98:159–166

Tamura K, Stecher G, Peterson D, Filipski A, Kumar S (2013) MEGA6: molecular evolutionary genetics analysis version 6.0. Mol Biol Evol 30:2725–2729 van der Lee T, Zhang H, van Diepeningen A, Waalwijk C (2015) Biogeography of *Fusarium graminearum* species complex and chemotypes: a review. Food Addit Contam Part A 32:453–460

Wang JH, Ndoye M, Zhang JB, Li HP, Liao YC (2011) Population structure and genetic diversity of the *Fusarium graminearum* species complex. Toxins (Basel) 3:1020–1037

Ward TJ, Bielawski JP, Kistler HC, Sullivan E, O'Donnell K (2002) Ancestral polymorphism and adaptive evolution in the trichothecene mycotoxin gene cluster of phytopathogenic *Fusarium*. Proc Natl Acad Sci USA 99:9278–9283

Windels CE (2000) Economic and social impacts of *Fusarium* head blight: changing farms and rural communities in the northern great plains. Phytopathology 90:17–21

Wu AB, Li HP, Zhao CS, Liao YC (2005) Comparative pathogenicity of *Fusarium graminearum* isolates from China revealed by wheat coleoptile and floret inoculations. Mycopathologia 160:75–83

Yli-Mattila T, Gagkaeva T, Ward TJ, Aoki T, Kistler HC, O'Donnell K (2009) A novel Asian clade within the *Fusarium graminearum* species complex includes a newly discovered cereal head blight pathogen from the Russian Far East. Mycologia 101:841–852

Zeller KA, Bowden RL, Leslie JF (2004) Population differentiation and recombination in wheat scab populations of *Gibberella zeae* from the United States. Mol Ecol 13:563–571

Zhang JB, Li HP, Dang FJ, Qu B, Xu YB, Zhao CS, Liao YC (2007) Determination of the trichothecene mycotoxin chemotypes and associated geographical distribution and phylogenetic species of the *Fusarium graminearum* clade from China. Mycol Res 111:967–975

Zhang H, Van der Lee T, Waalwijk C, Chen W, Xu J, Xu J, Zhang Y, Feng J (2012) Population analysis of the *Fusarium graminearum* species complex from wheat in China show a shift to more aggressive isolates. Plos ONE 7:e31722

2

Anti-fungal activity, mechanism studies on α-Phellandrene and Nonanal against *Penicillium cyclopium*

Ji-hong Zhang[1*], He-long Sun[1], Shao-yang Chen[1], Li Zeng[1] and Tao-tao Wang[2*]

Abstract

Background: Essential oils from plants have been reported to have wide spread antimicrobial activity against various bacterial and fungal pathogens, and these include α-Phellandrene, Nonanal and other volatile substances. However, biological activities of α-Phellandrene and Nonanal have been reported only in a few publications. Further investigations are necessary to determine the antimicrobial activity of these compounds, especially for individual application, to establish the possible mechanism of action of the most active compound.

Results: The results are shown that α-Phellandrene and Nonanal have a dose-dependent inhibition on the mycelial growth of *Penicillium cyclopium*. The minimum inhibitory concentration (MIC) and minimum fungicidal concentration (MFC) are 1.7 and 1.8 mL/L for α-Phellandrene, 0.3 and 0.4 mL/L for Nonanal, respectively. The volatile compounds altered the morphology of *P. cyclopium* hyphae by causing loss of cytoplasmic material and distortion of the mycelia. The membrane permeability of *P. cyclopium* increased with increasing concentrations of the two volatile compounds, as evidenced by cell constituent release, extracellular conductivity and induced efflux of K^+. Moreover, the two volatile compounds induced a decrease in pH and in the total lipid content of *P. cyclopium*, which suggested that cell membrane integrity had been compromised.

Conclusions: The results demonstrated that α-Phellandrene and Nonanal could significantly inhibit the mycelia growth of *P. cyclopium* by severely disrupting the integrity of the fungal cell membrane, leading to the leakage of cell constituents and potassium ions, and triggering an increase of the total lipid content, extracellular pH and membrane permeability. Our present study suggests that α-Phellandrene and Nonanal might be a biological fungicide for the control of *P. cyclopium* in postharvest tomato fruits.

Keywords: *P. cyclopium*, Antifungal activity, SEM, Mechanism, Membrane permeability

Background

Many plant species, including tomato, synthesize and store numerous volatile terpenoid compounds during normal leaf development (Buttery et al. 1988; Paré and Tumlinson 1999). Tomato is a constitutive emitter of low amounts of mono- and sesquiterpenes under non-stressed conditions, but these emissions become greatly enhanced under stress (Jansen et al. 2009; Maes and Debergh 2003). The volatile blends from *Solanum lycopersicum* leaves detected with SPME GC–MS were mainly terpenoids (i.e., α-Phellandrene), fatty acid derivatives (i.e., Nonanal) and aromatic compounds (Zhang et al. 2008).

Nonanal has been reported to exhibit antimicrobial activity against gram-positive and gram-negative bacteria in the concentration of 100 to more than 800 mg/kg (Muroi et al. 1993). Nonanal is reported to have a MIC of 0.2 μg/mg against *Staphylococcus aureus* (Kubo et al. 1995). α-Phellandrene showed weak inhibitory effects against all tested bacteria at the concentrations of 1 to >4 mg/mL (Demirci et al. 2001; Iscan et al. 2012).

*Correspondence: jihongzh01@xtu.edu.cn; ttwang@mail.hzau.edu.cn
[1] School of Chemical Engineering, Xiangtan University, Xiangtan 411105, People's Republic of China
[2] College of Horticulture Forestry&Sciences, Huazhong Agricultural University, Wuhan 430070, People's Republic of China

α-Phellandrene, β-Phellandrene, ocimene, limonene, myrcene, and α-caryophyllene have shown in vitro activity against *Bacillus* sp., *Candida albicans, Escherichia coli, Pseudomonas aeruginosa*, and *S. aureus* (Perez et al. 1999; Costa et al. 2000). Essential oils are usually mixtures of monoterpene and sesquiterpene, and their oxygenated derivatives. Their composition and proportion depended on species as well as the extraction and separation methods (Fisher and Phillips 2008). Essential oils are aromatic oily liquids, their antimicrobial properties have been empirically recognized for centuries, but scientifically confirmed only recently (Dorman and Deans 2000). α-Phellandrene and Nonanal are present in large quantities in many species such as canola, soybean, *Senecio laetus, Haplophyllum tuberculatum, Minuartia meyeri* and *Apium graveolens* (Inouye et al. 2001; Fernando et al. 2005; Al-Burtamani et al. 2005; Nurettin et al. 2006; Rodriguez-Burbano et al. 2010; Hernández et al. 2013; Pandey et al. 2014). These essential oils had a broad-spectrum antimicrobial activity against various bacterials and pathogenic fungi including *Candida albicans, Alternaria alternate, Bipolaris* sp., *Curvularia lunata, Fusarium oxysporium* (Inouye et al. 2001; Al-Burtamani et al. 2005; Nurettin et al. 2006; Hernández et al. 2013; Sharma et al. 2014).

The lipophilicity of essential oils enable them to preferentially partition from an aqueous phase into membrane structures of the fungi, resulting in membrane expansion, increased membrane fluidity and permeability, disturbance of membrane-embedded proteins, inhibition of respiration, alteration of ion transport processes in fungi and induced leakage of ions and other cellular contents (Burt 2004; Fadli et al. 2012; Khan et al. 2010; Oonmetta-aree et al. 2006).

Biological activities of α-Phellandrene and Nonanal were reported only in a few publications. Further investigations are necessary to determine the anti-microbial activity of these compounds, especially for individual application, to establish the possible mechanism of action of the most active compound to combat resistant pathogenic fungi. This study aims to analyze α- Phellandrene and Nonanal on the mycelial growth of *Penicillium cyclopium*. The effects of different concentrations of α-Phellandrene and Nonanal on surface morphology, cell membrane permeability, and release of cellular material were investigated to elucidate their anti-fungal mechanisms.

Methods

Chemicals

α-Phellandrene (>99%) and Nonanal (>96%) were obtained from Dieckmann company (Shenzhen, China). Cholesterol (95%) and phosphovanillin (98%) were purchased from TCI Shanghai (Shanghai, China). All the chemicals were analytical grade.

Pathogens

Penicillium cyclopium was provided by the Department of Biotechnology and Food Engineering, Xiangtan, China, and the fungal was isolated from infected tomato (*S. lycopersicum*) fruit. The fungal spores concentrations were adjusted to 5×10^5 spores/mL using a haemocytometer before each test. 100 μL fungal suspensions (10^5 spores/mL) were added into the triangle bottle with 40 mL potato dextrose broth (PDB) and incubated in a rotatory shaker at 28 ± 2 °C and 120 rpm for 4 days.

Measurement of mycelial growth

Effects of α-Phellandrene and Nonanal on the mycelial growth of *P. cyclopium* were evaluated in vitro by agar dilution method (Yahyazadeh et al. 2008). PDA (20 mL) was poured to sterilized Petri dishes (90 mm diameter) and measured amounts of α-Phellandrene and Nonanal were added to PDA mediums (plus with 0.05% Tween-80) to give the following concentration of 0, 0.25, 0.50, 0.75, 1.00, 1.25, 1.50, 1.75 and 2.00 mL/L for α-Phellandrene; of 0, 50, 100, 150, 200, 250, 300, 350 and 400 μL/L for Nonanal. 6 mm diameter discs of *P. cyclopium* inocula were cut from the center of an actively growing *P. cyclopium* culture on fresh PDA plates without antibiotics at 28 ± 2 °C for 4 days with a paper punch, and then was placed at the center of each new Petri plate. The culture plates were then incubated at 28 ± 2 °C for 48 h. As controls, PDA dishes were supplemented with the same amount of filtration sterilization alcohol (99.5%) instead of α-Phellandrene and Nonanal. Each treatment was performed in triplicate. The lowest concentration that completely inhibited the growth of the fungus after 24 h of incubation was considered as the minimum inhibitory concentration (MIC). The minimum fungicidal concentration (MFC) was regarded as the lowest concentration which no growth of the pathogen was observed after a 72 h incubation period at 28 ± 2 °C in a fresh PDA plate, indicating more than 99.5% killing of the original inoculum (Talibi et al. 2012).

Different amounts of volatile substances were added to 50 ml PDB liquid medium for the final concentration of α-Phellandrene 0.00, 225.00, 450.00, 900.00, 1350.00, 1575.00 and 1800.00 μL/L, and of Nonanal 0.00, 50.00, 100.00, 200.00, 300.00, 350.00 and 400.00 μL/L. One hundred microlitres *P. cyclopium* conidial spore suspensions (10^5 cfu/mL) were poured into each triangle bottle, which was incubated on 28 ± 2 °C and a rotatory shaker at 120 rpm for 4 days. Fungal growth was estimated gravimetrically by weighting the biomasses after drying at 80 °C to a constant weight. The percentage of mycelial growth inhibition (PGI) was calculated according to the following formula:

$$PGI(\%) = [(d_c - d_t)/d_c] \times 100$$

where d_c (g) is net dry weight of control fungi and d_t (g) is net dry weight of treated fungi (Helal et al. 2006).

Release of cytoplasmic material absorbing at 260 nm

The release of cytoplasmic material absorbing at 260 nm was measured following the method of Yahyazadeh et al. (2008) and Paul et al. (2011), with some modifications. Viable cells of *P. cyclopium* in their exponential logarithmic phase were collected by vacuum filtration, then washed three times with phosphate buffered saline (pH 7.0), and re-suspended with 20 mL of the above buffer solution. The suspensions were then incubated at 28 ± 2 °C in the presence of α-Phellandrene and Nonanal at three different concentrations (0, MIC and MFC) for 0, 30, 60 and 120 min. After incubation, cells were centrifuged at 12,000*g* for 2 min, and the absorbance (260 nm) of the above supernatant (1 mL) was determined with the UV-2450 UV/Vis Spectrophotometer (Shimadzu Corporation).

Measurement of extracellular conductivity and extracellular pH

The measurement of extracellular conductivity of *P. cyclopium* cells was conducted using a DDS-W conductivity meter (Instructions for Shanghai Precision Scientific Instrument Co., Ltd., Shanghai, China) according to the method described previously (Shao et al. 2013). The extracellular pH of *P. cyclopium* cells was determined using a Delta-320 pH-meter (Instructions for the Mettler-Toledo Co., Ltd., Shanghai, China). Initially, 100 μL fungal suspensions (10^5 spores/mL) were added to 50 mL PDB and incubated at 28 ± 2 °C in a thermostatic cultivation shaker for 4 days. The mixtures were collected by vacuum filtration, washed for 2–3 times with sterilized double distilled water, and re-suspended in 20 mL sterilized double distilled water. After the incubation of α-Phellandrene and Nonanal at MIC or MFC for 0, 30, 60 and 120 min, the extracellular conductivity and extracellular pH of *P. cyclopium* were determined. Control flasks without α-Phellandrene or Nonanal were also tested using an equal amount of alcohol instead.

Determination of total lipid content

Total lipid content of *P. cyclopium* cells with α-Phellandrene and Nonanal at three concentrations (0, MIC, MFC) was determined using the phosphovanillin method (Helal et al. 2007). The 3-day-old mycelia from 50 ml PDB was collected by vacuum filtration and dried with a vacuum freeze drier for 6 h. About 0.05 g of dry mycelia were homogenized with liquid nitrogen and extracted with 4.0 mL of methanol–chloroform–water mixture (2:1:0.8, v/v/v) in a clean dry test tube with vigorous shaking for 30 min. The tubes were centrifuged at 4000 rpm for 10 min. The lower phase containing lipids was thoroughly mixed with 0.2 mL saline solution and centrifuged at 4000×*g* for 10 min. Then, an aliquot of 0.2 mL chloroform and lipid mixture was transferred to a novel tube and 0.2 mL concentrated sulfuric acid (H_2SO_4) was added, heated for 10 min in a boiling water bath. Three milliliter phosphovanillin was added, the tube was shaken vigorously, and incubated at room temperature for 10 min. The absorbance at 520 nm was utilized to calculate total lipid contents from the standard calibration curve. Cholesterol served as a standard.

Statistical analysis

All of the experiments were repeated three times. The results were expressed as the mean value ± standard deviation and were compared using an analysis of variance (one-way ANOVA) for multiple comparisons. All dates were processed by SPSS statistical software package release 16.0 (SPSS Inc., Chicago, IL, USA).

Results

Antimicrobial assay of α-Phellandrene and Nonanal

The mycelial growth of *P. cyclopium* was affected by α-Phellandrene or Nonanal in a dose-dependent manner ($P < 0.05$) (Table 1). High concentrations of α-Phellandrene (≥1.35 mL/L) or Nonanal (≥0.35 mL/L) completely inhibited the mycelial growth. The antifungal activity of Nonanal against *P. cyclopium* was more efficient than α-Phellandrene ($P < 0.05$). Low concentrations of α-Phellandrene (0.45 mL/L) or Nonanal (0.10 mL/L) only showed half of the antifungal activity against *P. cyclopium*, with PGI values of 56.90 and 57.50%, respectively. The MIC and MFC values of α-Phellandrene for *P. cyclopium* were 1.7 and 1.8 mL/L, respectively. The MIC and MFC values of Nonanal for *P. cyclopium* were 0.3 and 0.4 mL/L, respectively.

Release of cell constituents

The release of cell constituents when *P. cyclopium* was treated with MIC and MFC of α-Phellandrene and Nonanal for 0, 30, 60 and 120 min are shown in Fig. 1. The optical density values (OD_{260}) produced by α-Phellandrene were higher than those by Nonanal at the same treatment condition. When the *P. cyclopium* cells were treated with MIC or MFC of α-Phellandrene or Nonanal, the release of cell constituents increased. Suspensions of *P. cyclopium* with α-Phellandrene at MIC (1.7 mL/L, v/v) at 120 min reached an optical density value of 0.554, which is higher ($P < 0.05$) than that of the control (0.249), but lower than that of 0.668 at MFC (1.8 mL/L, v/v). Nevertheless, the OD_{260} values of

Table 1 Effect of α-Phellandrene and Nonanal on the mycelial growth of *Penicillium cyclopium*

α-Phellandrene concentration (μL/L)	Growth inhibition (%)[f]	Nonanal concentration (μL/L)	Growth inhibition (%)[f]
00	0.00 ± 0.00[a]	0.00	0.00 ± 0.00[a]
225.00	45.88 ± 0.56[b]	50.00	15.47 ± 0.49[b]
450.00	56.90 ± 0.64[c]	100.00	57.50 ± 1.10[c]
900.00	99.37 ± 0.01[d]	200.00	65.59 ± 0.16[d]
1350.00	100.00 ± 0.00[e]	300.00	99.61 ± 0.04[e]
1575.00	100.00 ± 0.00[e]	350.00	100.00 ± 0.00[e]
1800.00	100.00 ± 0.00[e]	400.00	100.00 ± 0.00[e]

[a–e] Significant differences at $P < 0.05$ level

[f] Values arepresented as mean ± SD

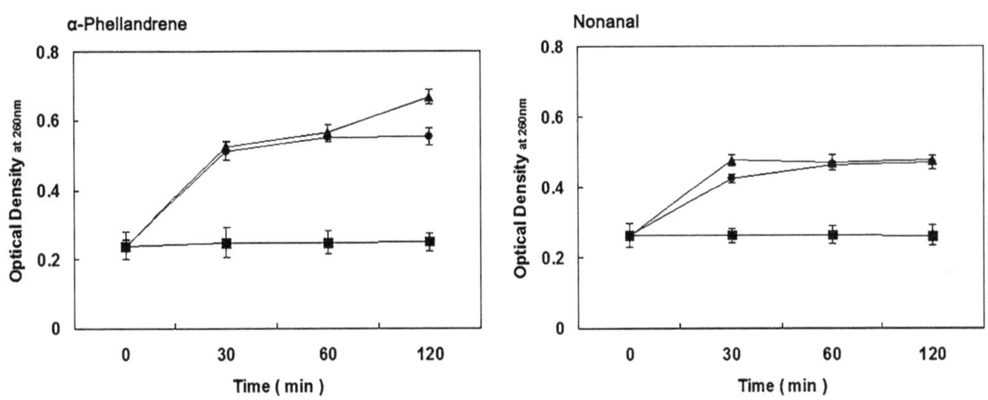

Fig. 1 Leakage of cytoplasmic material (OD at 260 nm) from *Penicillium cyclopium* with α-Phellandrene and Nonanal treatment. (*Closed square*) control; (*closed circle*) at minimum inhibitory concentration (MIC); (*closed triangle*) at minimum fungicidal concentration (MFC). Data presented are the means of pooled data. *Error bars* indicate the SDs of the means (n = 3)

P. cyclopium suspensions with α-Phellandrene treatment were almost identical before 60 min of exposure, and then increased after 60 min incubation. On the other hand, the OD_{260} values of *P. cyclopium* suspensions with Nonanal at MIC (0.3 mL/L, v/v) and MFC (0.4 mL/L, v/v) were higher than those of the control at the same exposure time. After 30 min of exposure, the OD_{260} values at MIC or MFC were almost the same.

Scanning electron microscopy (SEM)

The effect of Nonanal and α-Phellandrene on the morphology of *P. cyclopium* was examined using SEM (Fig. 2). The conidia of *P. cyclopium* grown on PDA plates for 4 days had normal, plump and homogenous morphology (Fig. 2b, d, f, h, j), and the hyphae of control fungus growing on PDA were normal, linearly tubular, regular, and homogeneous (Fig. 2a, c, e, g, i). The growth of *P. cyclopium* on PDA with α-Phellandrene or Nonanal at MIC or MFC treatment for 2 days demonstrated that all fungal mycelia and conidia showed considerable changes in morphology. *P. cyclopium* with Nonanal or α-Phellandrene at MIC concentrations showed slightly depressed conidia, partly distorted and shrunken mycelia (Fig. 2c, d, g, h). In contrast, the conidia of *P. cyclopium* treated with Nonanal and α-Phellandrene at MFC concentrations appeared severely collapsed and depressed possibly because of the lack of cytoplasm (Fig. 2f, j). Moreover, severely shrunken and distorted hyphae were also observed with Nonanal and α-Phellandrene at MFC concentrations (Fig. 2e, i).

Extracellular conductivity

Exposure of *P. cyclopium* cells to different concentrations of α-Phellandrene and Nonanal for 0–120 min caused varying levels of extracellular conductivity (Fig. 3). In general, conductivity increased with exposure time and concentrations of α-Phellandrene and Nonanal. The concentrations for α-Phellandrene at their MIC (1.7 mL/L) and their MFC (1.8 mL/L) significantly affected the extracellular conductivity of *P. cyclopium* cells. Nevertheless, the concentrations for Nonanal at MIC (0.3 mL/L) and at MFC (0.4 mL/L) were only slightly affected with the extension of processing time. At 30 min of exposure, the extracellular conductivity

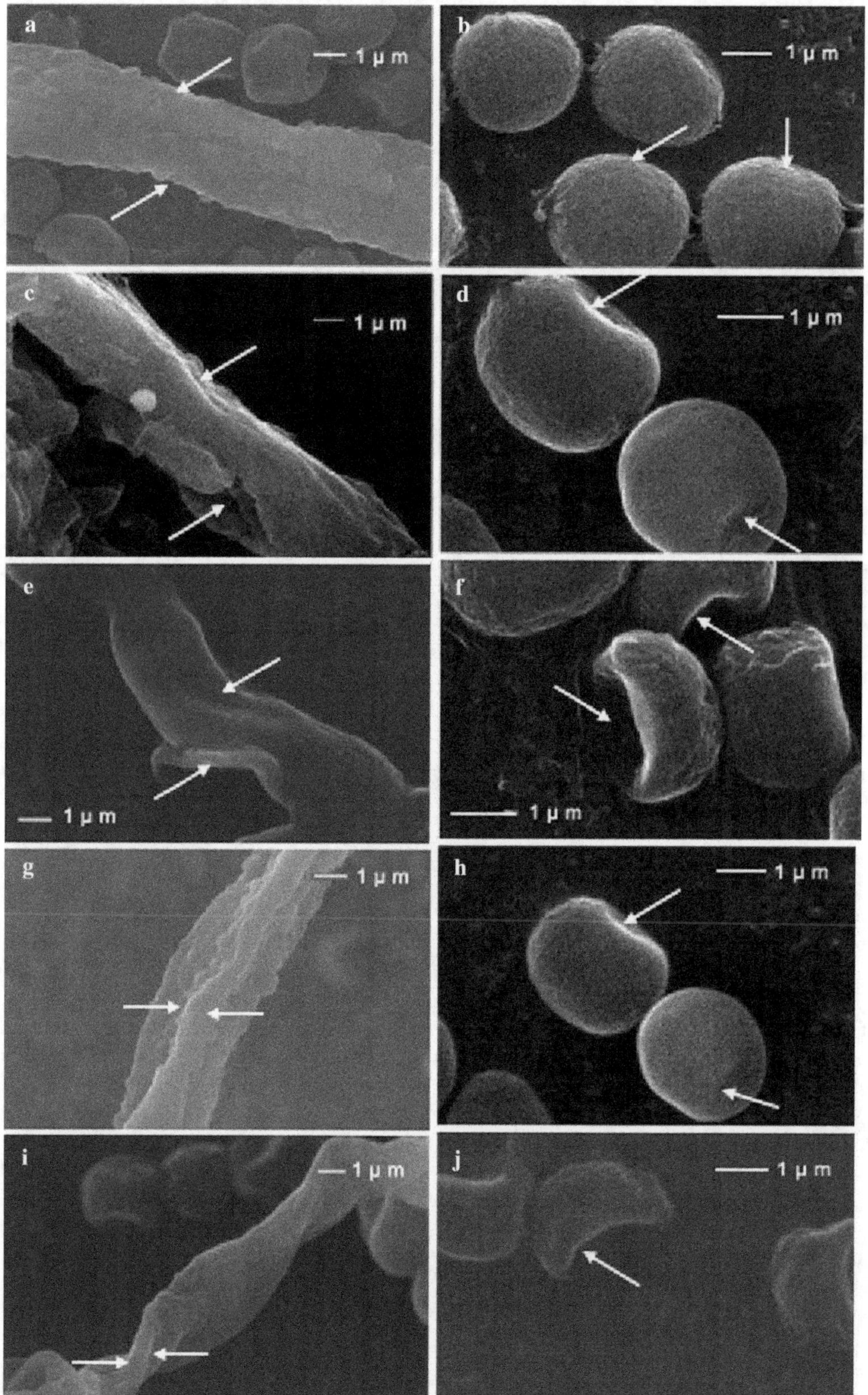

Fig. 2 Effect of Nonanal and α-Phellandrene on mycelium and conidiophore morphological changes of *Penicillium cyclopium*. **a**, **b** *P. cyclopium* without volatile substances (control); **c**, **d** *P. cyclopium* with Nonanal at minimum inhibitory concentration (MIC); **e**, **f** *P. cyclopium* with Nonanal at minimum fungicidal concentration (MFC); **g**, **h** *P. cyclopium* with α-Phellandrene at MIC; **i**, **j** *P. cyclopium* with α-Phellandrene at MFC

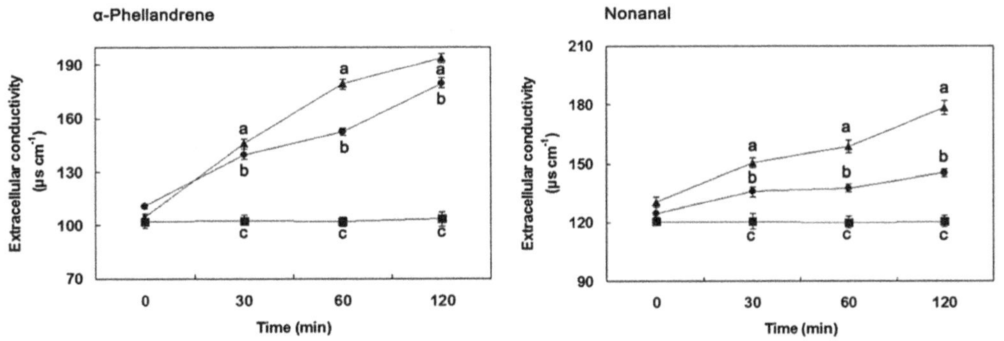

Fig. 3 Effects of α-Phellandrene and Nonanal on the extracellular conductivity of *P. cyclopium* suspensions. (*closed square*) control; (*closed circle*) at minimum inhibitory concentration (MIC); (*closed triangle*) at minimum fungicidal concentration (MFC). Data presented are the means of pooled data. *Error bars* indicate the SDs of the means (n = 3). *Different letters* of *a–c* are significantly different according to Duncan's multiple range test at $P < 0.05$

of *P. cyclopium* suspensions with α-Phellandrene at MIC or MFC remained at almost the same levels, but the values were significantly higher ($P < 0.05$) than the control. By contrast, the conductivity of *P. cyclopium* suspensions with Nonanal at MFC at 30 min of exposure was 150.3 μs/cm, which was significantly higher than that at MIC (135.8 μs/cm) and that in the control (120.5 μs/cm). After 60 min of exposure, the conductivity of *P. cyclopium* suspensions treated with Nonanal remained at a stable increase. On the contrary, the conductivity of *P. cyclopium* suspensions in the presence of α-Phellandrene was markedly increased after 60 min of exposure at MFC. At 120 min, the conductivity of *P. cyclopium* suspensions at MIC reached 179.6 and 193.7 μs/cm for MFC, respectively, which were significantly higher ($P < 0.05$) than those treated with Nonanal for MIC and MFC (145.3 and 178.2 μs/cm, respectively).

Extracellular pH

The extracellular pH of *P. cyclopium* cells exposed to α-Phellandrene and Nonanal is decreased in comparison to the controls (Fig. 4). A gradual decrease in extracellular pH was observed in the control. After 30 min of exposure to α-Phellandrene at MIC and MFC, a sharp reduction in the extracellular pH of the *P. cyclopium* suspensions occurred, whereas the extracellular pH values in the *P. cyclopium* suspensions after incubation with Nonanal gradually decreased. The extracellular pH values in the *P. cyclopium* suspensions after incubation for 120 min with α-Phellandrene at MIC and MFC were 4.72 and 4.33, respectively, which were significantly lower than that of the control (5.3). No significant difference ($P < 0.05$) was found between MIC and MFC after 30 min of exposure, the extracellular pH values in *P. cyclopium* suspensions after incubation with Nonanal at MIC and MFC were 5.06 and 4.81, respectively, and the latter was significantly lower than that of the control (5.25) ($P < 0.05$).

Potassium ion efflux

Potassium ions (K^+) leaked from *P. cyclopium* cells incubated with α-Phellandrene and Nonanal (Fig. 5). MFC and MIC of α-Phellandrene and Nonanal significantly

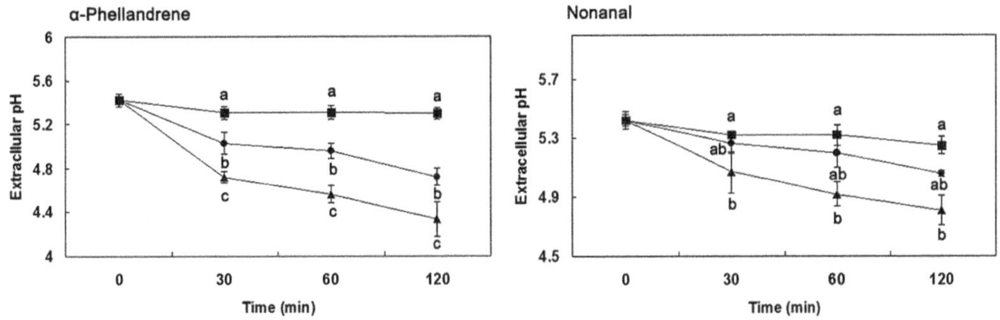

Fig. 4 Effects of α-Phellandrene and Nonanal on the extracellular pH of *P. cyclopium* (*closed square*): control; (*closed circle*): minimum inhibitory concentration (MIC); (*closed triangle*) minimum fungicidal concentration (MFC). Data presented are the means of pooled data. *Error bars* indicate the SDs of the means (n = 3). *Different letters* of *a–c* are significantly different according to Duncan's multiple range test at $P < 0.05$

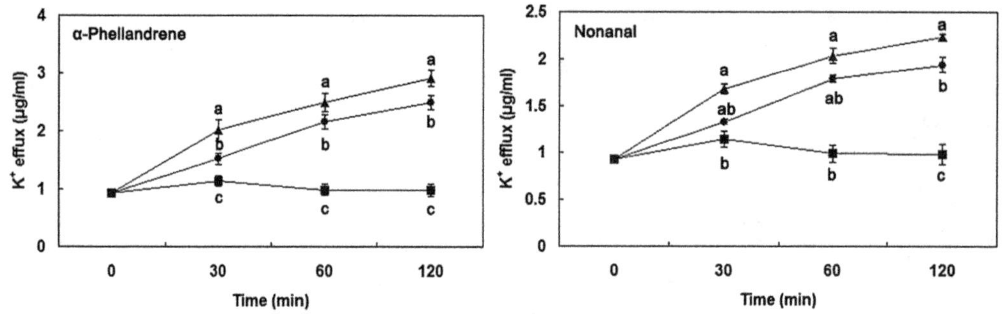

Fig. 5 Effect of α-Phellandrene and Nonanal on the leakage of potassium ions from *P. cyclopium*. (*closed square*) control, (*closed circle*) minimum inhibitory concentration (MIC), (*closed triangle*) minimum fungicidal concentration (MFC). Data presented are the means of pooled data. *Error bars* indicate the SDs of the means (n = 3). *Different letters* of *a–c* are significantly different according to Duncan's multiple range test at $P < 0.05$

induced the release of K^+, and the K^+ concentration after 30 min was 1.520 and 1.330 μg/mL, respectively. When the incubation time increased to 120 min, the release of K^+ concentration continuously increased. By contrast, incubation with MFC of α-Phellandrene and Nonanal resulted in more K^+ release than from the *P. cyclopium* cells with MIC of those. After 120 min of incubation, K^+ released by MFC of α-Phellandrene and Nonanal reached 2.910 and 2.235 μg/mL, respectively.

Total lipid content

α-Phellandrene and Nonanal affected the total lipid contents of *P. cyclopium* cells (Fig. 6). Briefly, α-Phellandrene and Nonanal significantly decreased the total lipid content, especially for α-Phellandrene ($P < 0.05$). The total lipid contents of the *P. cyclopium* cells were 100.3 ± 2.4 and 117.4 ± 2.1 mg/g dry weight for 120 min after incubation with α-Phellandrene at MFC and MIC, respectively. These values are significantly lower ($P < 0.05$) than those of the control (158.1 ± 2.3 mg/g dry weight). By contrast, Nonanal at MFC and MIC remarkably reduced the total lipid content of *P. cyclopium* cells. Moreover, the effect of Nonanal at MFC and MIC was lower than that of α-Phellandrene in *P. cyclopium* cells. The total lipid contents of the *P. cyclopium* cells were 130.5 ± 2.4 and 136.2 ± 3.2 mg/g dry weight, respectively, after incubation with Nonanal at MFC and MIC for 120 min.

Discussion

α-Phellandrene and Nonanal exhibited strong antifungal activity against *P. cyclopium*. The inhibitory effect was positively correlated with the concentration of α-Phellandrene and Nonanal. These results were consistent with those of previous studies describing the antifungal activity of these volatile compounds (Fernando et al. 2005; Rodriguez-Burbano et al. 2010; Pandey et al. 2014; Sharma et al. 2014). At a relatively low concentration (0.1 mL/L), Nonanal reduced the mycelial growth of *P. cyclopium* by half, making it a promising antifungal substance. In addition, the inhibitory effect of Nonanal on *P. cyclopium* was more efficient than that of α-Phellandrene on *P. cyclopium*. The phenomenons were not observed in α-Phellandrene and Nonanal, indicating that the aldehyde compounds are more effective than alcohols and

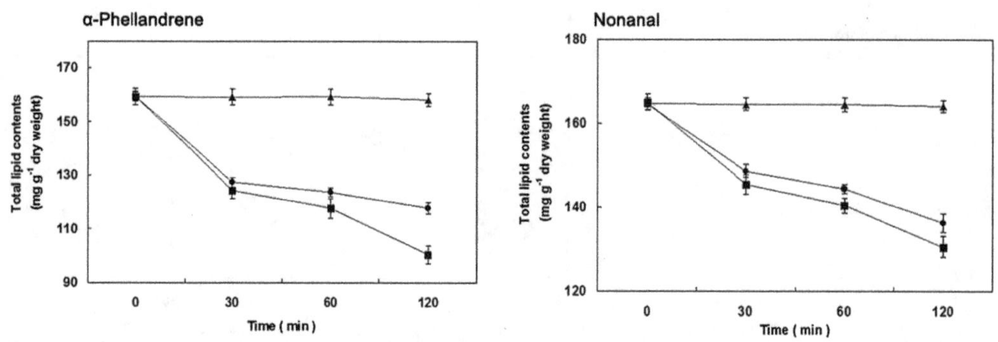

Fig. 6 Total lipid content of *P. cyclopium* cells in the presence of α-Phellandrene and Nonanal. (*closed square*) control, (*closed circle*) at minimum inhibitory concentration (MIC), (*closed triangle*) at minimum fungicidal concentration (MFC). Data presented are the means of pooled data. *Error bars* indicate the SDs of the means (n = 3)

olefine in controlling postharvest pathogens (Droby et al. 2008). Among aldehyde constituents, cinnamaldehyde showed the highest activity, followed by citral, and then perillaldehyde, octanal and Nonanal (Inouye et al. 2001).

The potential mechanisms underlying the anti-microbial activity of aldehydes and terpenes are not fully understood, but a number of possible mechanisms have been proposed. Gram-positive bacteria are known to be more susceptible to essential oils than Gram-negative bacteria (Farag et al. 1989; Smith-Palmer et al. 1998). The weak antibacterial activity against Gram-negative bacteria was ascribed to the presence of an outer membrane (Tassou and Nychas 1995; Mann et al. 2000), which possessed hydrophilic polysaccharide chains as a barrier to hydrophobic essential oils. In the current experiment, in vitro antifungal activity enabled us to hypothesize that the potential antifungal activity of α-Phellandrene and Nonanal against *P. cyclopium* could be closely correlated with the physiology of the hyphae. SEM analysis showed that the volatile compounds could alter the morphology of *P. cyclopium* hyphae, disrupting the membrane integrity (Yahyazadeh et al. 2008; Tyagi and Malik, 2011).

These changes generally occur because of an increase in the permeability of cells, and such changes commonly result in the leakage of small molecular substances, ions, and formation of lesions (Bajpai et al. 2013). The leakage of cytoplasmic membrane was analyzed by determining the release of cell materials including nucleic acid, metabolites and ions which was absorbed at 260 nm in the suspensions (Oonmetta-aree et al. 2006). After the addition of the α-Phellandrene and Nonanal visibly increased with increasing volatile compound concentration. The maximum release of cell constituents was observed in *P. cyclopium* treated with α-Phellandrene at MFC. The methyl ester is able to penetrate to the hydrophobic regions of the membranes and the carboxyl groups pass through the cell membrane, perturbing in the lowering of internal pH and denaturing of proteins inside the cell (Marquis et al. 2003). From the results of the present study combined with the previous studies, we can conclude that the two volatile compounds apparently induced the leakage of intracellular protons. These findings suggest that irreversible damage to the cytoplasmic membranes of *P. cyclopium* occurred, and the ions inside the cells leaked, ultimately leading to apoptosis of the fungus in the presence of volatile compounds.

In addition to cell wall and plasma membrane alteration and disruption, exposure of the hyphae of *P. cyclopium* to α-Phellandrene or Nonanal resulted in K^+ leakage. Our results are in agreement with the reported by Helal et al. (2006). The phenomenon could be explained that the release of ions was only based on their size and/or due to formation of holes or lesions of lipid bilayer of the plasma membrane (Prashar et al. 2003). The fatty acid composition of microbial cell membranes affects their ability to survive in various environments (Ghfir et al. 1994). The decrease in lipid content suggested that membrane stability decreased while the permeability of water-soluble materials increased (Helal et al. 2007). Fumigation of *P. cyclopium* with *C. citratus* essential oil, induced alterations in both the lipid content and the fatty acids methyl esters composition of the cells (Helal et al. 2006). In the present study, the addition of α-Phellandrene and Nonanal significantly decreased the lipid content of *P. cyclopium*. The results have shown that the two volatile compounds had the ability to penetrate lipid structures of the cells and disrupt the cell membrane integrity.

In conclusion, this study showed that α-Phellandrene and Nonanal could significantly inhibit the mycelial growth of *P. cyclopium* cells, disrupt their cell membrane integrity and result in the leakage of cell components. Our present study suggested that α-Phellandrene and Nonanal might be used as fungicides to fight against postharvest fungal diseases.

Conclusions

α-Phellandrene and Nonanal significantly inhibit the mycelia growth of *P. cyclopium*. These changes disrupt the integrity of the fungal cell membrane, leading to the leakage of cell constituent and potassium ions, and triggering an increase of the total lipid content, extracellular pH and membrane permeability. α-Phellandrene and Nonanal might be used as biological fungicides for the control of *P. cyclopium* in postharvest tomato fruits in the future.

Authors' contributions

JZ, HS and TW participated in the discussion and concepts of experimental designs, JZ and HS participated to data analysis and drafted the manuscript; HS, SC and ZL performed experiments. All authors read and approved the final manuscript.

Acknowledgements

This work was supported by National Natural Science Foundation of China (Nos. 31272181 and 31471887), Natural Science Foundation of Hunan Province (No. 2015JJ6108) and Ph.D. Research Fund of Xiangtan University (No. KZ08033). We thank Hu Tian for assistance with technical assistance and providing the bacterial strain.

Competing interests

The authors declare that they have no competing interests.

References

Al-Burtamani SKS, Fatope MO, Marwah RG, Onifade AK, Al-Saidi SH (2005) Chemical composition, antibacterial and antifungal activities of the essential oil of *Haplophyllum tuberculatum* from Oman. J Ethnopharmacol 96:107–112

Bajpai VK, Sharma A, Baek KH (2013) Antibacterial mode of action of *Cudrania tricuspidata* fruit essential oil, affecting membrane permeability and surface characteristics of food-borne pathogens. Food Control 32(2):582–590

Burt S (2004) Essential oils: their antibacterial properties and potential applications in foods-A review. Int J Food Microbiol 94:223–253

Buttery RG, Ling LC, Douglas ML (1988) Tomato leaf volatile aroma components. J Agric Food Chem 35(6):1039–1042

Costa TR, Fernandes OFL, Santos SC, Oliveira CMA, Liao LM, Ferri PH, Paula JR, Ferreira HD, Sales BHN, Silva MRR (2000) Antifungal activity of volatile constituents of *Eugenia dysenterica* leaf oil. J Ethnopharmacol 72:111–117

Demirci F, Kırımer N, Demirci B, Noma Y, Başer KHC (2001) Screening of biotransformation products of carvone enantiomers by headspace-SPME/GC-MS. Z Naturforsch C 56(2):58–64

Dorman HJD, Deans SG (2000) Antibacterial agents from plants: antibacterial activity of plant volatile oils. J Appl Microbiol 88(2):308–316

Droby S, Eick A, Macarisin D, Cohen L, Rafaela G, Stange R, McColum G, Dudai N, Nasser A, Wisniewski M, Shapira R (2008) Role of citrus volatiles in host recognition, germination and growth of *Penicillium digitatum* and *Penicillium italicum*. Postharvest Biol Tech 49(3):386–396

Fadli M, Saad A, Sayadi S, Chevalier J, Mezriouia NE, Pagèsc JM et al (2012) Antibacterial activity of *Thymus maroccanus* and *Thymus broussonetii* essential oils against nosocomial infection-bacteria and their synergistic potential with antibiotics. Phytomedicine 19:464–471

Farag RS, Daw ZY, Hewedi FM, El-Baroty GSA (1989) Antimicrobial activity of some Egyptian spice essential oils. J Food Protect 52(9):665–667

Fernando WGD, Ramarathnam R, Krishnamoorthy AS, Savchuk SC (2005) Identification and use of potential bacterial organic antifungal volatiles in biocontrol. Soil Biol Biochem 37(5):955–964

Fisher K, Phillips C (2008) Potential antimicrobial uses of essential oils in food: is citrus the answer? Trends Food Sci Technol 19(3):156–164

Ghfir B, Fonvieille JL, Koulali Y, Ecalle R, Oargent R (1994) Effect of essential oil of *Hyssopus officinalis* on the lipid composition of *Aspergillus fumigatus*. Mycopathologia 126(3):163–167

Helal GA, Sarhan MM, Abu Shahla ANK, Abou El-Khair EK (2006) Effects of *Cymbopogon citratus* L. essential oil on the growth, lipid content and morphogenesis of *Aspergillus niger* ML2-strain. J Basic Microbiol 46(6):456–469

Helal GA, Sarhan MM, Abu SANK, Abou EEK (2007) Effects of *Cymbopogon citratus* L. essential oil on the growth, morphogenesis and aflatoxin production of *Aspergillus flavus* ML2-strain. J Basic Microb 47(1):5–15

Hernández V, Mora F, Araque M, Montijo SD, Rojas L, Meléndez P, Tommasi ND (2013) Chemical composition and antibacterial activity of *Astronium graveolens* JACQ essential oil. Rev Latinoamer Quím 41(2):89–94

Inouye S, Takizawa T, Yamaguchi H (2001) Antibacterial activity of essential oils and their major constituents against respiratory tract pathogens by gaseous contact. J Antimicrob Chemother 47(5):565–573

Iscan G, Kirimer N, Demirci F, Demirci B, Noma Y, Can Baser KH (2012) Biotransformation of (-)-(R)-α -phellandrene: antimicrobial activity of its major metabolite. Chem Biodivers 9(8):1525–1532

Jansen RMC, Miebach M, Kleist E, van Henten EJ, Wildt J (2009) Release of lipoxygenase products and monoterpenes by tomato plants as an indicator of *Botrytis cinerea* induced stress. Plant Biol 11(6):859–868

Khan A, Ahmad A, Akhtar F, Yousuf S, Xess I, Khan LA et al (2010) *Ocimum sanctum* essential oil and its active principles exert their antifungal activity by disrupting ergosterol biosynthesis and membrane integrity. Res Microbiol 161:816–823

Kubo A, Lunde CS, Kubo I (1995) Antimicrobial activity of the olive oil flavor compounds. J Agric Food Chem 43(6):1629–1633

Maes K, Debergh PC (2003) Volatiles emitted from in vitro grown tomato shoots during abiotic and biotic stress. Plant Cell Tissue Org 75(75):73–78

Mann CM, Cox SD, Markham JL (2000) The outer membrane of *Pseudomonas aeruginosa* NCTC6749 contributes to its tolerance to the essential oil of *Melaleuca alternifolia* (tea tree oil). Lett Appl Microbiol 30(4):294–297

Marquis RE, Clock SA, Mota-Meira M (2003) Fluoride and organic weak acids as modulators of microbial physiology. FEMS Microbiol Rev 26(5):493–510

Muroi H, Kubo A, Kubo I (1993) Antimicrobial activity of cashew apple flavor compounds. J Agric Food Chem 41(7):1106–1109

Nurettin Y, Canan G, Osman Ü, Ahmet Y, Serdar Ü, Kamil C, Salih T (2006) Composition and antimicrobial activities of volatile components of *Minuartia meyeri*. Turk J Chem 30:71–76

Oonmetta-aree J, Suzuki T, Gasaluck P, Eumkeb G (2006) Antimicrobial properties and action of galangal (*Alpinia galangal* Linn.) on *Staphylococcus aureus*. LWT-Food. Sci Technol 39(10):1214–1220

Pandey AK, Mohan M, Singh P, Palni UT, Tripathi NN (2014) Chemical composition, antibacterial and antioxidant activity of essential oil of *Eupatorium adenophorum* Spreng. from Eastern Uttar Pradesh, India. Food. Bioscience 7(9):80–87

Paré PW, Tumlinson JH (1999) Plant volatiles as a defense against insect herbivores. Plant Physiol 121(2):325–331

Paul S, Dubey RC, Maheswari DK, Kang SC (2011) *Trachyspermum ammi* (L.) fruit essential oil influencing on membrane permeability and surface characteristics in inhibiting food-borne pathogens. Food Control 22(5):725–731

Perez C, Agnese AM, Cabrera JL (1999) The essential oil of *Scenecio graveolans* (compositae)-chemical composition and antimicrobial activity tests. J Ethnopharmacol 66:91–96

Prashar A, Hili P, Veness RG, Evans CS (2003) Antimicrobial action of palmarosa oil (*Cymbopogon martini*) on *Saccharomyces cerevisiae*. Phytochem 63(5):569–575

Rodriguez-Burbano D, Quijano-Celis C, Pino J (2010) Composition of the essential oil from leaves of *Astronium graveolens* Jacq grown in Colombia. J Essent Oil Res 22(6):488–489

Shao X, Cheng S, Wang H, Yu D, Mungai C (2013) The possible mechanism of antifungal action of tea tree oil on *Botrytis cinerea*. J Appl Microbiol 114(6):1642–1649

Sharma P, Shah GC, Dhami DS, Chauhan PK, Singh V (2014) Chemical composition, antibacterial and antioxidant activities of *senecio laetus* edgew. From cold desert of western Himalaya. Int J Pharm Res Bio-sci 3(1):188–199

Smith-Palmer A, Stewart J, Fyfe L (1998) Antimicrobial properties of plant essential oils and essences against five important food-borne pathogens. Lett Appl Microbiol 26(2):118–122

Talibi I, Askarne L, Boubaker H, Boudyach EH, Msanda F, Saadi B, Ait Ben Aoumar A (2012) Antifungal activity of Moroccan medicinal plants against citrus sour rot agent *Geotrichum candidum*. Lett Appl Microbiol 55(2):155–161

Tassou CC, Nychas GJE (1995) Antimicrobial activity of the essential oil of Mastic gum (*Pistacia lentiscus* var *chia*) on gram-positive and gram-negative bacteria in broth and model food systems. Int Biodeterior Biodegr 36(3–4):411–420

Tyagi AK, Malik A (2011) Antimicrobial potential and chemical composition of *Mentha piperita* oil in liquid and vapour phase against food spoiling microorganisms. Food Control 22(11):1707–1714

Yahyazadeh M, Omidbaigi R, Zare R, Taheri H (2008) Effect of some essential oils on mycelial growth of *Penicillium digitatum* Sacc. World J Microbiol Biotechnol 24(8):1445–1450

Zhang PY, Chen KS, He PQ, Liu SH, Jiang WF (2008) Effects of crop development on the emission of volatiles in leaves of *Lycopersicon esculentum* and its inhibitory activity to *Botrytis cinerea* and *Fusarium oxysporum*. J Integr Plant Biol 50(1):84–91

3

Small agarics in Taiwan: *Mycena albopilosa* sp. nov. and *Gloiocephala epiphylla*

Yi-Yin Chang[1] and Yu-Ming Ju[2*]

Abstract

Background: Small agarics are poorly documented in Taiwan, with previously reported species either rudimentarily described or lacking a description or diagnosis in most cases. A survey on small agarics in a lowland forest of Taiwan revealed two species previously unrecorded.

Results: One agaric, which is characterized mainly by white hairs overlying the pileus, a conspicuous cup-shaped basal disc surrounding the stipe, and inamyloid basidiospores, fits the genus *Mycena* and appears undescribed. It readily produced abundant basidiomata in culture in three weeks. The other agaric is *Gloiocephala epiphylla*, being characterized by its reduced hymenium and conspicuous pileogloeocystidia. ITS sequences from the two agarics further corroborated the identifications. Their macroscopic and microscopic features and culture morphology are described. A dichotomous key is provided to the species of *Mycena* reported in Taiwan.

Conclusion: The *Mycena* species is newly described as *M. albopilosa* herein. *Gloiocephala epiphylla* is new to Taiwan, being the only species of the genus known on the island.

Keywords: Small agarics, Basidiomata production, *Mycena*, *Gloiocephala*, Taiwan

Background

Small agarics with a pileus less than 1 cm diam pose a challenge for mycobiotic surveys, because they are easily overlooked and, after collected, their delicate, fragile basidiomata need to be measured and recorded in a timely fashion. Most of these small agarics belong to *Mycena* (Pers.) Roussel, *Marasmius* Fr., and *Galerina* Earle, each of which contains hundreds of taxa. Some of the small agarics can be found in genera such as *Gloiocephala* Massee and *Omphalina* Quél. They are saprophytes or in close association with mosses (Davey et al. 2013), while a few of them have been reported as being mycorrhizal (Zhang et al. 2012) or parasitic (Dennis 1957; Baker and Holliday 1957; Sequeira 1958; Bayliss 1911). These overlooked small agarics are actually underestimated. Notable plant pathogenic species include *Mycena citricolor* (Berk. & M.A. Curtis) Sacc., which causes the well-known American leaf spot that decreased annual yields of coffee crops by 20% (Sequeira 1958; Rao and Tewari 1987), and *Marasmius perniciosus* Stahel, which causes witches' broom disease on cacao in South America (Baker and Holliday 1957; Evans 1980). Certain *Marasmius* species contain laccases and other enzymes capable of degrading aromatic compounds, lignin, and β-carotene (Dedeyan et al. 2000; Scheibner et al. 2008). Noticeably, more than 30 species of *Mycena* are known bioluminescent (Desjardin et al. 2007).

There are over 500 *Mycena* species described in the world (Desjardin et al. 2008). In Taiwan, 21 *Mycena* species have been reported, but, in most cases, a species is merely furnished with a rudimentary diagnosis or lacks a description/diagnosis entirely. Shih et al. (2014) described a bioluminescent species *M. kentingensis* Shih et al., which represents the only novel species of the genus documented in Taiwan thus far. Despite the great potential and special features that these small agarics may possess, without extraordinary characters such as bioluminescence and pathogenicity, their taxonomy hardly becomes a research interest for mycologists in Taiwan.

In the present study, *Mycena albopilosa*, which is characterized by a white, minute, hairy pileus and a conspicuous cup-shaped basal disc at the stipe, is described as new. It has characteristics of *Mycena* section *Sacchariferae* Kühner ex

*Correspondence: yumingju@gate.sinica.edu.tw
[2] Institute of Plant and Microbial Biology, Academia Sinica, Taipei, Taiwan
Full list of author information is available at the end of the article

Sing. in general, but its basidiospores are inamyloid. In addition, *Gloiocephala epiphylla*, the type species of the genus, is reported and represents a newly recorded genus in Taiwan. Both *M. albopilosa* and *G. epiphylla* are tiny, growing on substrates, such as fallen leaves or twigs, in humid forests.

Methods

Specimen collecting and culturing

Basidiomata together with the attached substrate were brought back to the laboratory. The substrates were placed in the plastic chambers with high humidity maintained for further basidiomata production. Freshly produced basidiomata were placed onto 2% malt extract agar (MEA) for collecting basidiospores. Germinating basidiospores were transferred to fresh MEA. Cultures were subsequently transferred to potato dextrose agar (PDA), MEA, and oatmeal agar (OA). Growth rate and culture morphology were recorded every 3–5 days.

Both species were also cultured on a formulated soil, which was modified from the one used for *Mycena kentingensis* (Shih et al. 2014), with white rice bran replaced by black rice bran. The formula was as follows: 80% potting soil (Green Orchids Co. #521) thoroughly mixed with 20% black rice bran. The mixture was divided into 10 g per 9-cm Petri dish and autoclaved at 121 °C for 30 min before use. Mycelia were grown in the laboratory at 25 °C for 4 weeks. To maintain the water content, the humidity was kept approximately 50–70%, and 1 ml sterile water was sprayed every 4–7 days on each Petri dish.

Morphological study

Macro- and micro-morphological data of *Mycena albopilosa* were mainly based on cultured fruiting bodies due to exhaustion of original specimens. The morphology of specimens artificially cultivated was similar to those growing in natural habitat. Color terms and notations used for morphological and culture features were compared with Methuen Hand Book of Colour (Konerup and Wanscher 1978). Microscopic observations of *Gloiocephala epiphylla* were rehydrated with distilled water and stained with Melzer's reagent. Hyphal tissue organization was observed from resin fixation and sectioning of basidiomata. Basidiospores were measured with a Leica DM2500 microscope under 100× immersion oil objective and 10× ocular. Basidiospore statistics were calculated by Piximetre 5.8, including the arithmetic mean and range of the spore length and spore width (30 spores in a specimen); Q: the range of spore length to width quotient.

DNA sequencing

Detailed methods were referred to Hsieh et al. (2009). Fungal mycelia were harvested from 1 to 2 weeks culture and transferred into flasks containing malt extract broth, which were incubated in a rotary incubator under 25 °C for 7–10 days. The mycelia were then freeze-dried and stored in a refrigerator at 4 °C. Total DNAs were extracted using a MX-16 automatic nucleic acid extractor (Compacbio Sciences, Burlingame, CA, U.S.A.) with Maxwell® 16 Tissue DNA Purification kit (PROMEGA Corp., Madison, WI, U.S.A.). Polymerase chain reaction amplifications of ribosomal internal transcribed spacers (ITS) were described in detail in Hsieh et al. (2009). The ABI Prism model 3730 × 1 DNA Analyzer® (Applied Biosystems, Foster City, CA, U.S.A.) was used for DNA sequencing.

Results and discussion

***Mycena albopilosa* Y.-Y. Chang & Y.-M. Ju, sp. nov., Figs. 1, 2, 3**

MycoBank number: MB 815859

Etymology: referring to the white hairs on the pileate surface.

Basidiomata gregarious. *Pileus* pure white, bell-shaped, convex to flattened, 1.5–4 mm diam, sometimes sulcate-striate, densely covered with tufts composed of agglutinated pileocystidia and acanthocysts, faded as aged; margin entire or crenate, incurved as aged, with fragile flesh. *Pileipellis* cutis, smooth to verrucose on the hyphal wall. *Pileocystidia* cylindrical, lageniform to lanceolate, sometimes rostrate, 57–180 × 10–34 μm, warty, unevenly distributed. *Acanthocysts* abundant, globose to pyriform, 20–23 μm diam, warty, emerged from pileipellis hyphae. *Lamellae* adnexed, white, alternately arranged with lamellula; *trama* dextrinoid, composed of large globular to subglobular cells. *Basidia* hyaline, subclavate to pyriform, tetrasporic, 22–23 × 9–10 μm, inamyloid, thin-walled; *sterigmata* cone-shaped, 3.1–3.8 μm long, 1.5–1.7 μm at base. *Basidiospores* hyaline, ellipsoid to ovoid, (5–)5.5–7(–7.5) × (3.5–)4–4.7(–5.5) μm, Q = 1.2–1.76, smooth, inamyloid. *Basidioles* present. *Pleurocystidia* absent. *Cheilocystidia* absent. *Stipe* central, white to translucent, 7–15 × 0.4–0.7 mm, filiform, overlain with unevenly distributed caulocystidia, rising from a basal disc originated from remnant of primordia; *trama* turning cream (4A3) when aged or pressed, composed of cylindrical cells tightly and spirally arranged, septate, dextrinoid. *Caulocystidia* cylindrical, obtuse, sometimes clavate, smooth, 24–89 × 5–9 μm, single or 2–6 clustered. The basal disc remaining conspicuous through all growing stages, cup-shaped, 0.8–1.25 × 0.4–0.7 mm, floccose, composed of 4 or more chained cells, smooth, cylindrical to clavate, sometimes subglobose, obtuse to subacute, (19.5–)40–65(–70) × (5.2–)6–9(–14.2) μm, terminal cell sometimes rostrate, branched or warted. *Primordia* oblate to spherical, 0.75–2 mm × 0.25–0.5 mm tall, also covered with chalky white tufts of pileocystidia. *Clamp connections* readily seen in primordia remnant. *Smell* indistinctive.

Fig. 1 *Mycena albopilosa*. **a** Section of a gill. **b** Section of the pileus surface with the *red bar* indicating an acanthocysts. **c** Arrangement of gills. **d** Pileus of a basidioma. **e** Basidiomata produced on commercial soil and black rice bran of 4:1 ratio

Cultures and anamorph

Colonies on formulated soil covering Petri plates at 4 weeks, tomentose, white; basidiomata produced in 4 weeks.

Colonies on PDA 20 mm at 1 week, 32 mm at 2 weeks, 45 mm at 3 weeks, white to cream (4A3), irregular, wrinkled, wooly to tomentose, with powdered patches in places, sometimes radially sulcated and creased, with diffuse margins slightly fimbriate to tentacle-like and submerged; aerial hyphae inconspicuous; advancing zone 1.7–2.3 μm, sometimes 3–5 hyphae aggregated into bundles, aseptate; submerged hyphae branched, with globular enlargement; clamp connections abundant; odor indistinctive. *Basidiomata* produced in 18 days, much as those produced on formulated soil but differing in being extremely crowded, more

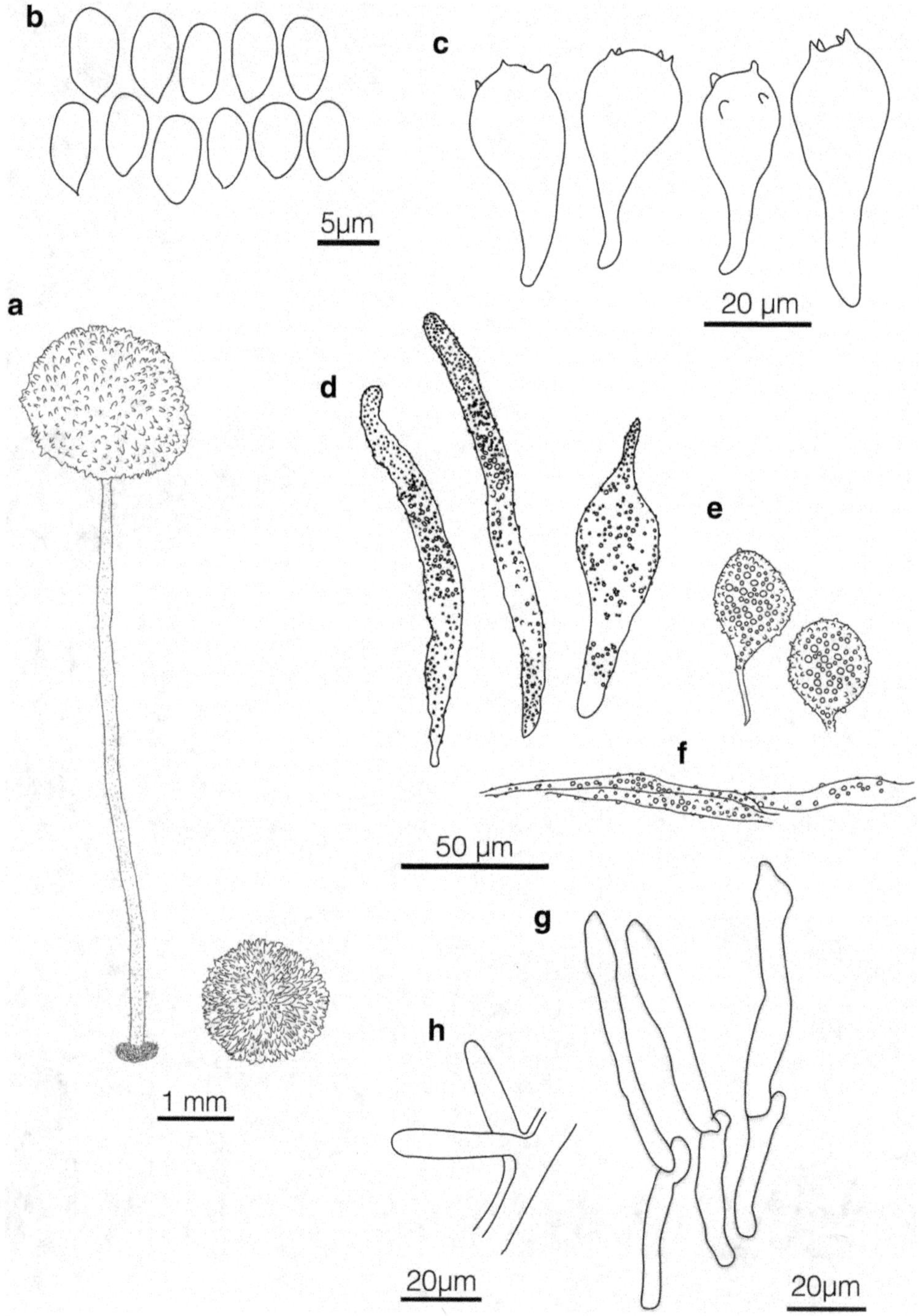

Fig. 2 *Mycena albopilosa*. **a** Basidiomata and primordia overlain with conspicuous white hairs. **b** Basidiospores. **c** Basidia. **d** Pileocystidia. **e** Acanthocysts on pileus. **f** Pileipellis. **g** Cystidia on the basal disc. **h** Caulocystidia

variable in shape and generally smaller 1–1.5 mm diam; partially developed basidiomata sometimes produced.

Colonies on MEA 11 mm at 1 week, 20 mm at 2 weeks, 31 mm at 3 weeks, whitish, plane and waxy, with wavy margins; microscopic features as those on PDA.

Colonies on OA 20 mm at 1 week, 40 mm at 2 weeks, 55 mm at 3 weeks, irregular, partly floccose, with aerial hyphae brush-like, 0.5–1 mm long × 0.1–0.3 mm broad, white, with entire margins; basidiomata sometimes produced in 3 weeks, 3–6 mm diam, with macroscopic and

Fig. 3 Basidiomata production of *Mycena albopilosa*. **a–c** On commercial soil and black rice bran of 4:1 ratio, PDA, and OA, respectively. **d** Partially developed basidiomata on PDA. **e, f** Basidia and basidiospores produced on partially developed basidiomata

microscopic features much as those produced on formulated soil; some primordia lacking further development, ball-shaped, loose and floccose, 0.20–0.65 diam.

Specimens examined

TAIWAN. New Taipei City, Shih-ting district, on dead fern frond on floor of broadleaf forest, 28 Feb 2014, Chang, Y.-Y. (dried culture, HOLOTYPE, HAST 142292), the holotype material composed of basidiomata produced on formulated soil, which are much like those collected from natural substrate, because the material collected from natural substrate was used up during the study.

Commentary

The salient features of *M. albopilosa* are the agglutinated pileocystidia resembling animal's fur, conspicuous cup-like basal disc, lack of cheilocystidia, and inamyloid basidiospores. *Mycena albopilosa* is capable of producing abundant basidiomata on PDA within a short period of time, which may be proven useful for developmental studies of basidiomata. *Mycena albopilosa* can be placed in section *Sacchariferae*, which is characterized by a granular to floccose pileus surface and ascending lamellae (Maas Geesteranus 1992), and in stirps *Adscendens* Desjardin (1995), which possesses smooth caulocystidia but lacks cherocytes.

Twelve taxa have been placed in stirps *Adscendens* thus far. Based on the entirely smooth caulocystidia, *M. albopilosa* most closely resembles *M. adscendens* (Lasch) Maas Geesteranus var. *adscendens*, *M. adscendens* var. *carpophila* (J. E. Lange) Desjardin, and *M. cryptomeriicola* Imazeki & Toki. *Mycena albopilosa* differs from the typical variety and var. *carpophila* of *M. adscendens* in having a fairly floccose pileate surface rather than a somewhat granulose surface and smaller basidiospores. Cheilocystidia are absent in *M. albopilosa* but appear rostrate in the two varieties of *M. adscendens*. Bisporic basidia have never been observed in *M. albopilosa* but are commonly encountered in *M. adscendens* var. *adscendens* (Maas Geesteranus 1992; Desjardin 1995; Aronsen and Læssøe 2016). Furthermore, basidiospores are inamyloid in *M. albopilosa* but amyloid in the two varieties of *M. adscendens* in most cases. *Mycena cryptomeriicola* Imazeki and Toki (1955) was originally described as having inamyloid basidiospores in the section *Sacchariferae*. However, the holotype specimen is in poor condition, and recollected specimens were reported to have amyloid basidiospores (Tanaka and Hongo 2003). *Mycena albopilosa* differs from *M. cryptomeriicola* in mostly lacking ventricose-rostrate cheilocystidia.

Mycena alphitophora (Berk.) Sacc. and its varieties have floccose pilei and thus resemble *M. albopilosa* in gross morphology. However, *Mycena albopilosa* possesses a basal disc at all stages, whereas *M. alphitophora* and its varieties lack a basal disc but instead have a swollen stem base covered with wooly hyphae. In addition, *M. albopilosa* differs *M. alphitophora* and its varieties in lacking cheilocystidia and having smooth rather than spinulose caulocystidia (Table 1).

Mycena albopilosa produces basidiomata in culture easily. In contrast to the basidiomata produced in nature, those formed in culture tend to be more variable in shape, but a great portion of them are normal and fully developed. For those deviating basidiomata, agglutinated hyphal tufts on the pileus and primordia, basal discs, and microscopic characteristics remain the same, but the basidiomata are smaller in diameter (1–1.5 mm) and more fragile. Malformed characters of the basidiomata produced on PDA include partially developed to undeveloped pilei with a wide size range and twisted, smoother stipes. Although primordia are generally normal in microscopic features, they are three to four times smaller than those produced on formulated soil. It should be noticed that a translucent, circular hymenium was found forming directly on the shortened, globular stipe with blackened upper half on OA and PDA. The deformed stipe shows the same tissue construction as the normal stipe and is dextrinoid. The hymenium is inamyloid and composed of only basidia and basidioles. This phenomenon has never been found in other agarics.

An NCBI MEGABLAST query with the ITS sequence of *M. albopilosa* did not show high similarities with the top matches. The top 10 BLAST matches closest to *M. albopilosa* did not bear a species epithet. The next match with a known identity was KF007948, deposited by Baird, R. E. as *M. pura* (query coverage = 100%, identities = 589/691 [85%], gaps = 51/691 [7%]), which did not share a high identity with *M. albopilosa*. *Mycena pura* is a significantly different species from *M. albopilosa* in having a large pileus size of 20–50 mm diam with a pale pink color, having amyloid spores, and lacking pileocystidia and acanthocysts.

Gloiocephala epiphylla **Massee, Grevillea 21: 34. 1892., Fig. 4**

Basidiomata peltate. *Pileus* flat, sunken as aged, rounded, sometimes irregular, 0.3–0.85 mm diam × 0.07–0.16 mm thick, white, turning pale orange (5A4) when aged, sometimes with brown spots at center. *Pileipellis* composed of enlarged globular to subglobular cells, 13–25.5 × 8.6–16.2 μm, scattered with pileogloeocystidia. *Pileogloeocystidia* capitate, 32–66 × 9–11 μm, with capitulum up to 19+ μm, transparent, becoming caramel (6C6-8); exudates hyaline, sticky, turning dark purple (14F8) as aged. *Flesh* consisted of small and irregular cells. *Hymenium* descending, white, powder-like and no gills formed.

Table 1 Comparisons of *M. albopilosa* with similar species

Taxa	Basidiospores	Basidiospores per basidium	Cheilocystidia	Caulocystidia	Basal disc or bulb	Notes
M. albopilosa	Ellipsoid to ovoid, 5.5–7 × 4–4.7 µm, inamyloid	4	Absent	Smooth	Present	See the current study
M. adscendens var. *adscendens*	Ellipsoid, 8.3–10.2 × 5–6 µm, amyloid	2 (or 4)	Spinulose and rostrate	Smooth	Present	See Desjardin (1995)
M. adscendens var. *carpophila*	Ellipsoid, 8.3–10.2 × 4–4.5 µm, amyloid	4	Spinulose and rostrate	Smooth	Present	See Desjardin (1995); fruiting only on pericarps of *Fagus*
M. alphitophora var. *alphitophora*	Ellipsoid, 7.5–10 × 4.5–5.5 µm	4	Slightly spinulose, clamped	Spinulose	Absent	See Desjardin (1995); growing on fern debris
M. alphitophora var. *globispora*	Pyriform, subglobose or globose, 7–9 × 6–8 µm	4	Verrucose on upper half	Spinulose	Absent	See Manimohan & Leelavathy (1989)
M. alphitophora var. *distinda*	Pyriform, subglobose or globose, 6–8 × 3–4 µm	Mainly 2	Verrucose	Spinulose	Absent	See Manimohan and Leelavathy (1989); acanthocysts spinulose and rostrate
M. cryptomeriicola	Ellipsoid, 7–9 × 4–5 µm, inamyloid	4	Ventricose with 1–2 rostrae on top	Smooth	Present	See Desjardin (1995); hyphae unclamped; fruiting on coniferous leaves
M. taiwanensis	Broadly ellipsoid to pyriform, 7–9 × 4.5–6 µm	4	Verrucose	Smooth	Present	See Rexer (1994). Pileus color brownish

Basidia clavate, 19–22 × 6–7 µm, tetrasporic, hyaline, inamyloid. *Basidioles* present, the same shape as basidia. *Sterigmata* horn-shaped, 3.5–4.2 µm. *Cystidioles* clavate to lanceolate, 20–36 µm. *Basidiospores* ellipsoid to fusoid, inequilateral, (7.7) 8.4–10.2 (10.6) × (2.5) 3.2–3.9 (4.1) µm, Q = 2.4–3.1, hyaline. *Stipe* central to slightly eccentric, cylindrical, 2–2.8 × 0.5–0.7 mm, covered with gloeocystidia similar to those on pileipellis, white near the pileus and grading to light yellow (4A3–5) and then brown (6A8–6F8) towards the base.

Cultures and anamorph

Colonies on soil (commercial soil and black rice bran of 4:1 ratio) full at 3 weeks, white, silk-like; basidiomata not produced.

Colonies on PDA 35 mm at 1 week, 70 mm at 2 weeks, full at 3 weeks, irregularly wrinkled, cream (4A3), becoming brown towards center, forming concentric zones of brown shade (7E4-8), occasionally scattered with viscid brown patches and dark brown spots (7F6-8), with submerged, diffuse margins; soluble pigment brown; aerial hyphae conspicuous, acicular, white, mainly distributed near center, bundled, branched. Aerial hyphae variable in diameter, twisted, smooth-walled, septate, sometimes interwoven and forming a thin membrane; submerged hyphae 1.5–2. 5 um diam, twisted, with globose enlargement (mostly at the end of hyphae), rarely septate; advancing zone septate, with short branches, obtuse at tip, 2.2–3.1 µm diam; clamp connections absent; odor strong, resembling burned plastic.

Colonies on MEA 33.5 mm at 1 week, 52.5 mm at 2 weeks, full at 3 weeks, plane, cream (4A3) to pinkish white, forming inconspicuous concentric rings, with irregular, plumose margins; aerial hyphae inconspicuous; microscopic features as those on PDA.

Colonies on OA 35 mm at 1 week, 50 mm at 2 weeks, full at 3 weeks, hyaline to white, with entire margins, covered with conspicuous white, evenly distributed aerial hyphae; microscopic features as those on PDA.

Specimen examined

TAIWAN. New Taipei City, Wulai, Red River Valley, elev. ca. 100 m, scattered on decayed broad leaves, 2 Apr 2014, Chang, Y.-Y. (HAST 142293).

Commentary

Gloiocephala epiphylla, the type species of the genus (Singer 1976), was originally described growing on

Fig. 4 *Gloiocephala epiphylla*. **a** Section of pileus showing distinct layers and pileogloeocystidum. **b, c** Basidiomata in natural habit. **d** Basidioma in detail. **e** Culture on PDA. **f** Culture on MEA. **g** Culture on OA

damp, decaying leaves from Jamaica in Massee (1892), where monosporic basidia and globose basidiospores were recorded. However, specimens collected in Ecuador, Japan and Hawaii have tetrasporic basidia (Desjardin et al. 1992), with which our Taiwan collection agrees. *Gloiocephala epiphylla* is widely distributed, being reported from Jamaica, Puerto Rico, Venezuela, Ecuador, Argentina, Japan, Hawaii (Desjardin et al. 1992), tropical Africa (Antonín 2007), and Taiwan. Features of our Taiwan collection correspond well with those described by Desjardin et al. (1992) and Antonín (2007) except for the pileus size being smaller, i.e., 0.3–0.85 mm diam vs. 1.5–4 mm diam.

Gloiocephala epiphylla is a newly recorded genus and species in Taiwan, being characterized by lamellae reduced or lacking, basidiospores inamyloid, and possession of secretory pileogloeocystidia (Antonín 2007). The secretory pileogloeocystidia on the basidiomata are so conspicuous that they appear spine-like by hand lenses. In Taiwan, the basidiomata were found growing on fallen leaves, twigs or decaying fruits in low-latitude broadleaf forests in spring. Desjardin et al. (1992) recorded that *G. epiphylla* were commonly found along the coastal areas or watercourses. Our collection in Taiwan was made from upper bank forest floor. Nevertheless, the collecting site is humid, and the forest floor is constantly flooded by temporary water flow from the rain fall.

An NCBI MEGABLAST query by using ITS sequence from the Taiwan collection matched well DQ097357 of *G. epiphylla* deposited by Binder et al. (2006) (query coverage = 100%, identities = 770/774 [99%], gaps = 0/774 [0%]).

Key to *Mycena* species reported in Taiwan

A list of *Mycena* species known in Taiwan can be found in *Catalogue of life in Taiwan* (Shao 2009), to which references to the listed species are referred. Five of the species lack a local description and are denoted with an asterisk (*). *Mycena photogena* Kominami is an invalidly published name, which appears in Sawada (1933) and is excluded from this key. The key was adapted mainly from descriptions in Maas Geesteranus (1992), Desjardin (1995) and Aronsen and Læssøe (2016).

1. Pileus viscid; hyphae of pileipellis forming a gelatinous, separable layer.........2
1. Pileus dry or moist, in some cases becoming lubricous; pileipellis not separable as a gelatinous layer.........6
    2. Basidiomata fairly small; pileus rarely exceeding 10 mm; stipe rising from a conspicuous basal disc.........3
    2. Basidiomata medium-sized to large; pileus up to 30 mm; stipe lacking a basal disc.........5
3. Cheilocystidia covered with short, even-length excrescences (warts), broadly clavate to subglobose; caulocystidia widely ellipsoid to subglobular, sometimes rostrate........*M. taiwanensis* Rexer (Rexer 1994) [*Mycena taiwanensis* is not listed in Shao (2009). Although *M. taiwanensis* was described in a Ph.D. dissertation (Rexer 1994), according to *International Code of Nomenclature for algae, fungi, and plants* (*Melbourne Code*) Article 30.8 & Ex. 17, it is considered as a validly published name because the dissertation was effectively published by the German commercial printer Zeeb-Druck]
3. Cheilocystidia smooth, branched, apically narrowed; caulocystidia fusiform.........4
    4. Pileus depressed at center, smooth, luminescent at pileus and stipe; stipe puberulous; pileocystidia present.........*M. chlorophos* (Berk. & Curt.) Sacc. (Maas Geesteranus 1992)
    4. Pileus not depressed at center, with a few spinules, luminescent at mycelium only; stipe glabrous except at the base; pileocystidia absent......... *M. stylobates* (Pers.) P. Kumm. (Maas Geesteranus 1992)
5. Basidiomata reddish orange; lamellar edge not separable as an elastic–tough thread; cheilocystidia clavate, fusiform or forked.........*M. leaiana* (Berk.) Sacc. (Maas Geesteranus 1992, Chou and Wang 2004)
5. Basidiomata brown to pale grey-brown; lamellar edge separable as an elastic–tough thread; cheilocystidia profusely branched.........*M. vulgaris* (Pers.) P. Kumm. (Maas Geesteranus 1992)
    6. Pileus floccose or granular on surface; lamellae ascending; stipe hairy, with base or basal disc; cheilocystidia clavate to fusiform, covered with short, even-length excrescences (warts)......7 (sect. *Sacchariferae*)
    6. Pileus glabrous on surface.........10
7. Stipe base lacking a disc or bulb, rather broadened and covered with white wooly hair; caulocystidia spinulose.........*M. alphitophora* (Berk.) Sacc. (Desjardin 1995) [Liou (1980) recorded this fungus as *M. osmundicola* J. E. Lange, which is treated as a synonym of *M. alphitophora* in Desjardin (1995)]
7. Stipe base with a basal disc or bulb; caulocystidia if present, smooth.........8
8. Pileus extremely floccose; pleurocystidia present; cheilocystidia absent; basidiospores inamyloid.........***M. albopilosa***
8. Pileus granulose; pleurocystidia absent; cheilocystidia distinctively spinulose; basidiospores mostly amyloid.........9
9. Pileus luminescent; caulocystidia absent; basidia tetrasporic.........*M. kentingensis* Y.S. Shih, C.Y. Chen, W.W. Lin and H.W. Kao (Shih et al. 2014)
9. Pileus not luminescent; caulocystidia present; basidia mostly bisporic.........*M. adscendens* Maas Geest. (Desjardin 1995)
    10. Stipe exuding blood red fluid when cut......... *M. haematopus* (Pers.) P. Kumm. (Maas Geesteranus 1992)
    10. Stipe not exuding blood red fluid when cut.......11
11. Lamellar edge more intensely colored than lamellar sides.........12
11. Lamellar edge concolorous with or in a similar color to lamellar sides.........13
    12. Strong nitrous smell; lamellar sides with densely distributed red–brown dots.........*M. capillaripes** (Maas Geesteranus 1992)
    12. Smell indistinct; lamellar sides lacking red–brown dots.........*M. roseomarginata* Hongo (Maas Geesteranus 1992)
13. Basidiospores inamyloid; pileus color bright orange–red.........*M. adonis* (Bull.: Fr.) S. F. Gray (Maas Geesteranus 1992)
13. Basidiospores amyloid; pileus color whitish or brownish.........14

14. Hyphae of the pileipellis, caulocystidia, and cheilocystidia smooth..15
14. Hyphae of the pileipellis, caulocystidia, or cheilocystidia warted, with excrescences or bran ched..16
15. Basidiomata cespitose; odor pungent, yeast-like..........................*M. overholtsii* A.H. Smith and Solheim* (Maas Geesteranus 1992)
[*Mycena overholtsii* is occasionally misspelled as *M. overholtsii.* Both names are listed in Shao (2009)]
15. Basidiomata scattered; odor raphanoid................ *M. pura* (Pers.: Fr.) Kumm.* (Maas Geesteranus 1992)
16. Pileus often less than 2.5 mm; stipe arising from a whorl of radiating dark-brown hyphae........... *M. capillaris* (Schum.: Fr.) Fr. (Maas Geesteranus 1992)
16. Pileus often larger than 5 mm; stipe not arising from a whorl of radiating dark-brown hyp hae..17
17. Lamellar edge homogenous, with cheilocystidia forming a continuous sterile band..............................18
17. Lamellar edge heterogeneous, with cheilocystidia mixed with basidia, forming a discontinuous band...20
18. Basidiomata usually fasciculate; odor strong when cut; cheilocystidia covered with unevenly distributed excrescences..19
18. Basidiomata usually solitary or in small groups; odor indistinct; cheilocystidia mostly covered with evenly distributed excrescences................ *M. filopes* (Bull.) P. Kumm. (Maas Geesteranus 1992, Chou and Wang 2004)
[Chou and Wang (2004) recorded this fungus as *M. amygdalina*, which is listed as a synonym of *M. filopes* in *Species Fungorum* (Kirk 2016)]
19. Pileus margin spotted with vinaceous stains; the stipe not forming a rooting base; pileipellis sparsely diverticulate, not embedded in gelatinous matter............ *M. inclinata* (Fr.) Quèl. (Maas Geesteranus 1992)
19. Pileus margin not spotted, usually upturned when mature; the stipe forming a rooting base; pileipellis covered with densely branched excrescences, embedded in gelatinous matter...........*M. polygramma* (Bull.: Fr.) S. F. Gray (Maas Geesteranus 1992, Chou and Wang 2004)
20. Pleurocystidia numerous; pileipellis simple to highly branched; cheilocystidia fusiform to clavate, unevenly covered with short excrescences..............*M. latifolia* (Peck) A. H. Smith (Maas Geesteranus 1992, Chou and Wang 2004)
20. Pleurocystidia absent; pileipellis the narrower smooth, the wider sparsely warted or diverticulated; cheilocystidia clavate to irregular in shape, covered with a few to many long excrescences that are simple or highly branched...................... *M. maculata* P. Karst.* (Maas Geesteranus 1992) [Shao (2009) adopted the name *M. alcalina* (Fr.) P. Kumm., with which *M. maculata* is considered in synonymy. However, Maas Geesteranus (1992) recognized them as two distinct species and stated that *M. maculata* is often misapplied to *M. alcalina*]

Authors' contributions

Y-YC and Y-MJ collected, recorded, cultured, and photographed the specimens, and prepared the manuscript. Both authors read and approved the final manuscript.

Author details

[1] Department of Life Sciences, National Taiwan University, Taipei, Taiwan.
[2] Institute of Plant and Microbial Biology, Academia Sinica, Taipei, Taiwan.

Acknowledgements

We thank Wann-Neng Jane, Academia Sinica, for resin fixation and sectioning of the mushroom samples, Jia-Ling Wei, National Museum of Natural Science, Taiwan, for helping collect *G. epiphylla*, and Huei-Mei Hsieh, Academia Sinica, for assistance in obtaining ITS sequences. Our gratitude is also extended to an anonymous reviewer for critically reviewing our manuscript, making constructive suggestions, and bringing our attention to the name *Mycena taiwanensis*.

Competing interests

All authors declare that they have no competing interests.

Funding

We greatly appreciate Ministry of Science and Technology of Taiwan for supporting this study with the Grant MOST 104-2621-B-001-008 to Y-MJ.

References

Antonín V (2007) Monograph of *Marasmius, Gloiocephala, Palaeocephala* and *Setulipes* in Tropical Africa. National Botanic Garden, Belgium
Aronsen A, Læssøe T (2016). The genus *Mycena* s.l. Fungi of Northern Europe vol. 5. Danish Mycological Society
Baker RED, Holliday P (1957) Witches' broom disease of cacao (*Marasmius perniciosus* Stahel). Phytopathological Papers, 2
Bayliss JS (1911) Observations on *Marasmius oreades* and *Clitocybe gigantea* as parasitic fungi causing fairy rings. J Eco Biol 6:111–132
Binder M, Hibbett DS, Wang Z, Farnham WF (2006) Evolutionary relationships of *Mycaureola dilseae* (Agaricales), a basidiomycete pathogen of a subtidal rhodophyte. Am J Bot 93:547–556
Chou WN, Wang YZ (2004) Six species of *Tricholomataceae* (Basidiomycotina) new to Taiwan. Fungal Sci 19:89–96
Davey ML, Heimdal R, Ohlson M, Kauserud H (2013) Host-and tissue-specificity of moss-associated *Galerina* and *Mycena* determined from amplicon pyrosequencing data. Fungal Ecol 6:179–186
Dedeyan B, Klonowska A, Tagger S, Tron T, Iacazio G, Gil G, Le Petit J (2000) Biochemical and molecular characterization of a laccase from *Marasmius quercophilus*. Appl Environ Microbiol 66:925–929
Dennis RWG (1957) Some marasminoid fungi allegedly parasitic on leaves and twigs in the tropics. Kew Bull 12:287–292
Desjardin DE (1995) A preliminary accounting of the worldwide members of *Mycena* sect. *Sacchariferae*. Bibliotheca Mycologica 159:1–89
Desjardin DE, Wong GJ, Hemmes DE (1992) Agaricales of the Hawaiian Islands. I. Marasmioid fungi: new species, new distributional records, and poorly known taxa. Can J Bot 70:530–542

Desjardin DE, Capelari M, Stevani C (2007) Bioluminescent *Mycena* species from São Paulo, Brazil. Mycologia 99:317–331
Desjardin DE, Oliveira AG, Stevani CV (2008) Fungi bioluminescence revisited. Photochem Photobiol Sci 7:170–182
Evans HC (1980) Pleomorphism in *Crinipellis perniciosa*, causal agent of witches' broom disease of cocoa. Trans Br Mycol Soc 74:515–523
Hsieh H-M, Ju Y-M, Hsueh P-R, Lin H-Y, Hu F-R, Chen W-L (2009) Fungal keratitis caused by a new filamentous Hyphomycete *Sagenomella keratitidis*. Bot Stud 50:331–335
Imazeki R, Toki S (1955) Contribution to the knowledge of Japanese Agaricales (I). Bull Gov Forest Exp Sta 79:1–14
Kirk PM (2016) Species Fungorum (version Jan 2016). In: Roskov Y, Abucay L, Orrell T, Nicolson D, Kunze T, Flann C, Bailly N, Kirk P, Bourgoin T, DeWalt RE, Decock W, De Wever A, editors. Species 2000 & ITIS Catalogue of Life, 27th June 2016. Digital resource at http://www.catalogueoflife.org/col. Species 2000: Naturalis, Leiden
Konerup A, Wanscher JH (1978) Methuen Handbook of Color, 3rd edn. Methuen & Co., Ltd., London
Liou S-C (1980) A survey of Taiwan fungi. Q J Chin For 13:117–128
Maas Geesteranus RA (1992) Mycenas of the Northern Hemisphere I & II. Tweede sectie. Verhandelingen der Koninklijke Nederlandse akademie van wetenschappen. Afdeling natuurkunde
Manimohan P, Leelavathy KM (1989) Two new varieties of *Mycena alphitophora* from Southern India. Mycol Res 93:118–120
Massee GE (1892) Notes on exotic fungi in the Royal Herbarium, Kew. Grevillea 21:1–6
Rao DV, Tewari JP (1987) Production of oxalic acid by *Mycena citricolor*, causal agent of the American leaf spot of coffee. Phytopathology 77:780–785
Rexer KH (1994) Die Gattung *Mycena*. Tübingen: Zeeb-Druck. Ph. D. Dissertation of Universität Tübingen
Sawada K (1933) Descriptive catalogue of Formosan fungi, vol 6. The Institute Taihoku, Formosa
Scheibner M, Hülsdau B, Zelena K, Nimtz M, De Boer L, Berger RG, Zorn H (2008) Novel peroxidases of *Marasmius scorodonius* degrade β-carotene. Appl Microbiol Biotechnol 77:1241–1250
Sequeira L (1958) The host range of *Mycena citricolor* (Berk & Curt) Sacc. Turrialba (Costa Rica) 8:136–147
Shao KT (2009) Catalogue of life in Taiwan. Web electronic publication at http://taibif.tw/. Accessed 17 Dec 2016
Shih YS, Chen CY, Lin WW, Kao HW (2014) *Mycena kentingensis*, a new species of luminous mushroom in Taiwan, with reference to its culture method. Mycol Prog 13:429–435
Singer R (1976) Marasmieae (Basidiomycetes–Tricholomataceae). Flora Neotropica 17:1–347
Tanaka I, Hongo T (2003) Two new records of *Mycena* sect. *Sacchariferae* from Japan and type study of *Mycena cryptomeriicola* (sect. *Sacchariferae*). Mycoscience 44:421–424
Zhang L, Chen J, Lv Y, Gao C, Guo S (2012) *Mycena* sp., a mycorrhizal fungus of the orchid *Dendrobium officinale*. Mycol Prog 11:395–401

4

The effects of local variation in light availability on pollinator visitation, pollen and resource limitation of female reproduction in *Hosta ventricosa*

Guo-Xing Cao[1,2*], Bi-Xian Wu[1], Xu-Jian Xu[1], Xie Wang[3] and Chun-Ping Yang[1]

Abstract

Background: Light availability may have direct effects on reproduction through resource availability, and indirect effects on female reproduction by influencing plant-pollinator interactions. Floral display size, pollinator visitation per flower, resource and pollen limitation of fruit and seed production were quantified in a forested patch and an adjacent open patch of two populations of the perennial herb *Hosta ventricosa*.

Results: Plants in the open patch produced significantly larger floral displays than those in the forested patch in both populations. Floral display size had a positive effect on pollinator visitation rate per flower in one population, but no effect in the other. Plants in the open patch received approximately 8–11 times more visitation rates and produced significantly more fruit and seeds per flower than those in the forested patch. However, supplemental pollination resulted in significantly more fruit and seed production per flower compared to natural pollination in the forested patch but not in the open patch in one population, and did not enhance fruit and seed production in either the forested or the open patch in the other. In both populations, supplementally pollinated plants in the open patch produced significantly more fruit and seeds per flower than supplementally pollinated plants in the forested patch.

Conclusions: In *H. ventricosa*, local variation in light conditions could affect pollinator activity and influence female reproduction through resource availability; however differences in the degree of pollen limitation between local habitats were found to be population-specific.

Keywords: *Hosta ventricosa*, Local light environment, Pollen limitation, Pollinator visitation rate, Resource limitation

Background

It is commonly observed that not all flowers produce fruit and not ovules in a flower mature seeds in flowering plants (Stephenson 1981; Wiens 1984; Sutherland 1986; Lee 1988). Two main proximate limiting factors, pollen limitation and resource limitation, have been proposed to explain these common phenomena (Haig and Westoby 1988). Pollen limitation, because of insufficient pollinator service, is usually tested by the addition of supplemental pollen to stigmas experimentally; reproduction of plants is considered pollen-limited if fruit or seed set is elevated by hand pollinations relative to natural pollinations (Burd 1994; Larson and Barrett 2000; Ashman et al. 2004; Knight et al. 2005; Wesselingh 2007). Resource limitation can be tested by the comparison of fruit and seed production between plants of different size (Griffin and Barrett 2002) or plants grown at different levels of physical resources (Lee and Bazzaz 1982).

Light availability is potentially an important resource that may constrain reproduction, because light intensity can affect net photosynthetic rate (Kitajima 1994). Experimental studies have shown that overall plant size is higher for plants grown with sun exposure compared to those grown in shade (Cai 2011; Zhao et al. 2012). Light environment

*Correspondence: cgxing@sicau.edu.cn
[1] Department of Forestry, Sichuan Agricultural University, Chengdu 611130, China
Full list of author information is available at the end of the article

can also affect the pattern of resources allocated to reproductive components; plants in low light environments have been shown to divert resources away from reproductive components to parts which can increase light capture, such as stems and leaves (McConnaughay and Coleman 1999). As a result of reductions in plant size and/or relative allocation to reproduction, flower and fruit production and the number of seeds that are matured may decrease with decreasing light availability (Niesenbaum 1993; Cunningham 1997; Kato and Hiura 1999; Cao and Kudo 2008).

Light availability may affect the degree of pollen limitation by plant-pollinator interactions and thus female reproduction indirectly. Because insect thermoregulatory capacity is higher in open than shaded habitats, pollinator visitation rate to flowers may be greater in the former (Herrera 1997; Kilkenny and Galloway 2008). Light conditions can also induce changes in floral display size (number of flowers open at one time on a plant), and thus may indirectly affect visitation rate to flowers. Plants occur in better light environments may produce larger floral displays, and pollinator visitation rate per flower have been shown to increase with increasing floral display size in some studies (Klinkhamer et al. 1989; Grindeland et al. 2005; Kilkenny and Galloway 2008). These differences in visitation rate between habitats, if large enough, could translate into greater pollen limitation of fruit and seed production in shaded habitats unless female reproduction of plants in shaded habitats is resource limited.

Plants of the same species often occur in locally contrasting light environments. When variation in light conditions is sufficient, female reproductive success may be directly affected by light availability, as well as indirectly affected by plant-pollinator interactions. In the present study, variations in pollinator visitation rate, pollen and resource limitation of female reproduction were simultaneously evaluated in the perennial animal-pollinated herb *Hosta ventricosa*, which grows in adjacent forested patches and open patches. Such simultaneous investigation could partition direct effects of light availability on plant reproduction through reproductive resource from indirect effects through changes in pollinator visitation rates. Specifically, the following questions were addressed: (1) Is pollinator visitation rate lower in forested patches than in open patches? (2) And if so, is the degree of pollen limitation of fruit and seed production greater in forested patches than in open patches? (3) Is fruit and seed production lower in forested patches than in open patches in the absence of pollen limitation?

Methods

Study species and populations

Hosta ventricosa Stearn (Hostaceae), a natural tetraploid, can reproduce via seeds and also vegetatively via rhizomes. Seeds are more commonly produced by pseudogamous apomixis, and less frequently through sexual reproduction (Schmid 2010). The species is cultivated widely in China and western countries not only for its ornamental value, but also for its consideration as a medicinal herb and economic value as a food source (Schmid 2010). *H. ventricosa* is a perennial hermaphroditic plant that grows in a variety of habitats, including forest floors and edges, along the edges of roadways, and in open areas (Schmid 2010; Cao et al. 2011). Individual plants produce several ramets and form small clone patches. Each ramet produces a single stem that bears several leaves at the base. In the study area in southern China, leaves emerge in early April and generally mature in mid-May. Bolting of flowering stems begins in early June. Six to greater than 30 flowers may be produced within a single raceme inflorescence. Commonly, one to three flowers open on the same day within an inflorescence. Individual flowers produce, on average, more than 30,000 pollen grains and approximately 40 ovules (Cao et al. 2015). The longevity of individual flowers is 1 day. *H. ventricosa* is adichogamous and seed set of self- and cross-pollinated flowers does not differ (Cao et al. 2015). Therefore, the species is fully self-compatible, although pollinator visitation is necessary for pollination. Bumblebees are the major pollinators of the species, and fruits mature at the end of September.

The experiments in the present study were conducted in one population (30°04′10″N; 102°59′23″E; At 990 m) in the Bifengxia National Nature Reserve, Yaan, Sichuan Province (hereafter, BYS), and one population (29°06′29″N; 107°19′20″E; At 1320 m) at Shanwangping Forest Farm, Nanchuan, Chongqing Municipality (hereafter, SNC), in 2013. Population BYS consisted of two relatively discontinuous patches, one located in the understory of an evergreen broadleaved forest, and the other in the open field. The forested patch was approximately 80 m^2 in area, comprised 107 individuals, and an individual plant produced an average of 2.9 ± 1.0 (SD) flowering racemes. The open patch was approximately 130 m^2 in area, comprised 84 individuals, and an individual plant produced an average of 3.3 ± 1.1 flowering racemes. The two patches were approximately 20 m apart. Relative light intensity was 13.2% in the forest at the flowering season. Flowering commenced in mid-June, and lasted until mid-July. Population SNC was approximately 500 m^2 in area, comprised more than five hundred individuals, and each individual produced an average of 2.5 ± 0.7 flowering racemes. In addition, approximately 60% of the individuals were located in the understory of an evergreen and deciduous broadleaved mixed forest, and the remaining plants were located in the open field. Relative light intensity was 10.8% in the forest at the flowering season. Flowering commenced in early July and lasted until late July. In all experiments conducted in population SNC, plants growing in an approximately 10-m border zone between the forest and open field were eliminated from the study.

Pollinator visitation

Pollinator visitation was observed over four separate sunny days for each population. Observations were conducted in late June and mid-July, 2013, at the peak flowering of populations BYS and SNC, respectively. Each day before observation, 28–32 flowering racemes were randomly tagged at each of the forested and open patch, and the number of open flowers on each tagged flowering raceme was counted. Between 9:00 a.m. and 15:00 p.m. each day, eight 15-min observation periods were carried out at each of the two patches. During each 15-min observation period, three separate observers each observed approximately 10 tagged racemes. Each round of observation was alternated between the two patches within a single day. All pollinator visits to individual open flowers of each tagged raceme were recorded. Visitation rate was calculated as the number of visits per flower per hour.

Pollen and resource limitation

To evaluate whether fruit and seed production were limited by pollen availability, as well as whether pollen limitation differed between the forested patch and the open patch, forty experimental plants were randomly selected at each of the two patches before flowering in each population. Two racemes were randomly tagged from each experimental plant, and one as an experimental raceme and was given supplemental pollen by hand pollination, while the other served as a control raceme and received only natural pollination. For each supplemental raceme on experimental plants, all of the flowers were hand-pollinated daily for each population, with the exception of 2–3 rainy days. Pollen was collected from plants 3 m away from the recipient plant. The hand-pollinated flowers were subsequently exposed to open pollination. In addition, in order to test whether experimental plants could reallocate resources away from control racemes to supplemental racemes, thirty-five control racemes from neighboring control (i.e., unmanipulated) plants at similar flowering stage and similar inflorescence size were randomly tagged at each of the two patches before flowering in each population. All of the infructescences were collected after the fruits had matured, the number of mature fruit or flower scars on each raceme was recorded, and the number of mature seeds in each fruit was assessed.

Statistical analyses

All analyses were carried out using the open-source software R (version 3.32, R Development Core Team 2014). To test the effects of habitat on display size, the lmer function of the 'lme4' package and a Poisson error distribution were used. The model included habitat (forested and open), population (BYS and SNC), and their interaction as fixed factors, and day was included as a random factor. Significance of fixed effects was examined through *F*-tests using the procedure 'anova' in R.

To test for effects of habitat and display size on pollinator visitation rate to flowers, the lmer function of the 'lme4' package and a Poisson error distribution were also used. The model included habitat, display size, and their interaction as fixed factors, and day was included as a random factor. Initially, population was included as a fixed factor, however, because of significant interactions of main effects with population, the two populations were analyzed separately.

Evidence for pollen limitation was assessed by comparing fruit per flower and seeds per flower between the control raceme on experimental plants and the supplemental raceme on experimental plants. Effects of habitat, pollination treatment (natural and supplemental), and their interaction on the two reproductive variables were tested using the lmer function of the 'lme4' package and a Gaussian error distribution. The model included habitat, pollination, and their interaction as fixed factors, and plant was included as a random factor. Initially, population was included as a fixed factor, however, because of significant interactions of main effects with population, the two populations were analyzed separately.

Resource limitation was assessed by testing the effects of habitat on fruit and seed production using only supplemental racemes on experimental plants. The lm function of the 'stats' package and a Gaussian error distribution were used, with habitat as a fixed factor.

To test for the possibility of reallocation of resources to supplementally pollinated racemes within plants, fruit and seed production were compared between the control raceme on experimental plants and the control raceme on control plants using the lm function of the 'stats' package and a Gaussian error distribution. The model included treatment, habitat, population, and all two-way and three-way interactions between these parameters as fixed factors.

Results

Display size

Display size was 2.3 ± 0.6 (SD) and 2.6 ± 0.8 for the forested patch and the open patch, respectively, in population BYS, and was 1.9 ± 0.6 (SD) and 2.1 ± 0.6 for the forested patch and the open patch, respectively, in population SNC. The differences in display size observed between habitats were found to be statistically significant (Table 1).

Pollinator visitation

During the 4-day observation in population BYS, the total number of approaches to tagged racemes was 184 for bumblebees, 39 for honeybees and two for butterflies in the forested patch, and 2370 for bumblebees, 516 for honeybees and six for butterflies in the open patch. The effect of habitat on total visitation rate per flower per hour was significant (Table 2); flowers in the open patch received 11 times more visitation rates than flowers in the forested patch (Fig. 1). Total visitation rate per flower were also found to vary with display size (Table 2); there was a positive correlation between visitation rate per flower (Y) and display size (X) in the forested patch ($Y = 0.35X - 0.25$, $R = 0.46$, $P < 0.001$) and in the open patch ($Y = 0.42X - 5.98$, $R = 0.19$, $P < 0.01$).

During the 4-day observation in population SNC, the total number of approaches to tagged racemes was 162 for bumblebees and 22 for honeybees in the forested patch, and 1423 for bumblebees and 176 for honeybees in the open patch. Visitation rate per flower differed significantly between habitat, but did not vary with display size (Table 2). Flowers in the open patch received approximately eight times more visitation rates than flowers in the forested patch (Fig. 1).

Pollen and resource limitation of female reproduction

In population BYS, pollination treatment had significant effects on fruit and seed production, although a significant interaction term between pollination treatment and habitat was detected (Table 3), indicating that the effect of pollination treatment on fruit and seed production was dependent on habitat. In the open patch, pollination treatment did not affect either fruit per flower ($F_{1,39} = 0.578$, $P = 0.452$; Fig. 2a) or seeds per flower ($F_{1,39} = 0.659$, $P = 0.385$; Fig. 2c), while in the forested patch, pollination treatment significantly affected both fruit per flower ($F_{1,39} = 50.167$, $P < 0.001$) and seeds per flower ($F_{1,39} = 111$, $P < 0.001$); supplemental pollination increased fruit and seed production over natural pollination (Fig. 2a, c). Habitat also had a significant effect on female reproduction (Table 3). Supplementally pollinated flowers in the open patch produced more fruit per flower ($F_{1,78} = 5.994$, $P = 0.017$; Fig. 2a) and seeds per flower ($F_{1,78} = 28.869$, $P < 0.001$; Fig. 2c) than supplementally pollinated flowers in the forested patch.

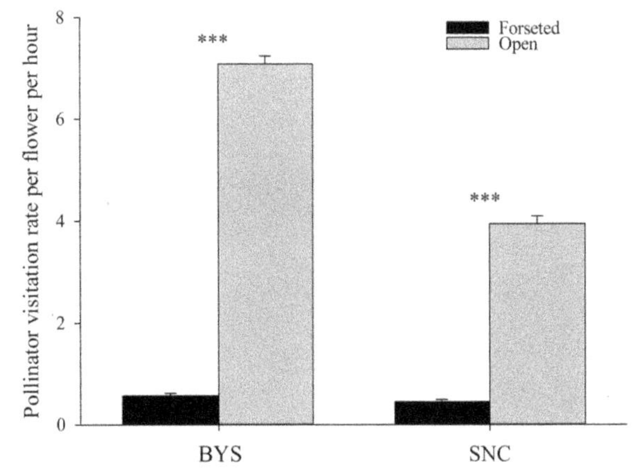

Fig. 1 Total pollinator visitation rate per flower per hour (mean ± SE) in the forested and open patches in two populations of *H. ventricosa*. ***$P < 0.001$

Table 1 Results of the linear mixed model test of the effects of habitat (forested vs. open patch) and population on display size in *H. ventricosa*

Factor	$F_{1,476}$	P
Habitat	27.886	<0.001
Population	52.219	<0.001
Population × habitat	0.000	0.998

Table 2 Results of the linear mixed model test of the effects of habitat (forested vs. open patch) and display size on pollinator visitation rate per flower per hour in the two populations of *H. ventricosa*

Factor	BYS		SNC	
	$F_{1,236}$	P	$F_{1,236}$	P
Habitat	114.000	<0.001	36.089	<0.001
Display size	11.578	<0.001	0.037	0.848
Display size × habitat	0.083	0.773	0.001	0.981

Table 3 Results of the linear mixed model test of the effects of habitat (forested vs. open patch) and pollination treatment (natural vs. supplemental) on fruit and seed production in the two populations of *H. ventricosa*

Factor	BYS		SNC	
	$F_{1,78}$	P	$F_{1,78}$	P
Fruit per flower				
Habitat	21.207	<0.001	23.471	<0.001
Pollination	19.565	<0.001	0.114	0.737
Pollination × habitat	30.335	<0.001	0.001	0.983
Seeds per flower				
Habitat	80.301	<0.001	71.044	<0.001
Pollination	74.637	<0.001	0.640	0.426
Pollination × habitat	108.341	<0.001	0.100	0.753

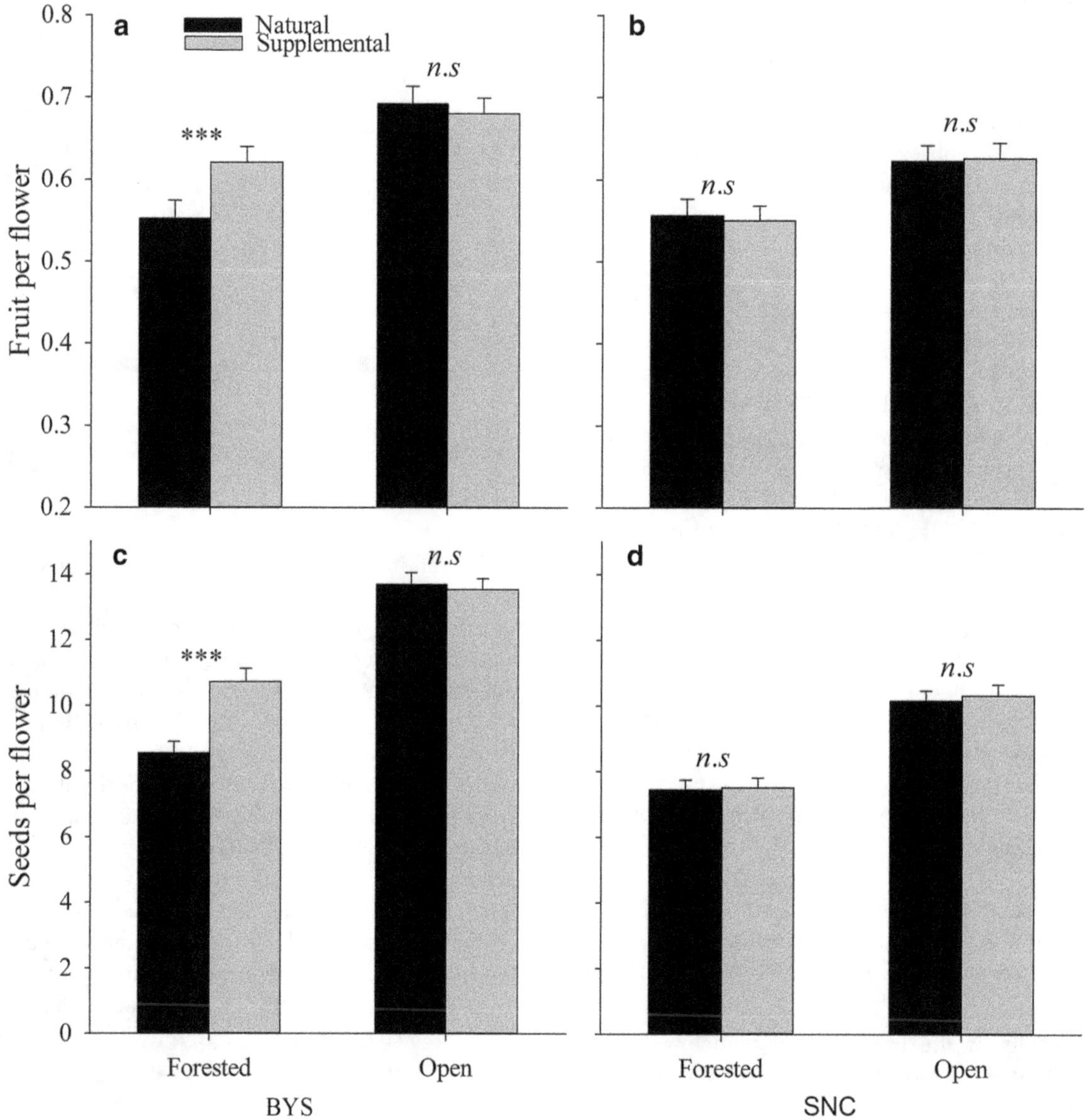

Fig. 2 Fruit (**a**, **b**) and seeds (**c**, **d**) per flower (mean ± SE) for natural and supplemental pollination in the forested and open patches in two populations of *H. ventricosa*. *n.s* = not significant, ****P* < 0.001

In population SNC, neither the main effect of pollination treatment nor the interaction term between pollination treatment and habitat was significant (Table 3), and fruit per flower and seeds per flower were both similar for natural pollination and supplemental pollination in both the forested and open patches (Fig. 2b, d). Moreover, a significant effect of habitat was detected (Table 3). Supplementally pollinated flowers in the open patch produced more fruit per flower ($F_{1,78} = 18.324$, $P < 0.001$; Fig. 2b) and seeds per flower ($F_{1,78} = 60.930$, $P < 0.001$; Fig. 2d) than supplementally pollinated flowers in the forested patch.

Resource reallocation

Control racemes on control plants in the forested and open patches in population BYS produced 0.56 ± 0.01 (SE) and 0.68 ± 0.01 fruit per flower, and 8.61 ± 0.27 and 13.41 ± 0.35 seeds per flower, respectively. Control racemes on control plants in the forested and open patches in population SNC produced 0.55 ± 0.01 and 0.62 ± 0.01 fruit per flower, and 7.38 ± 0.29 and 10.09 ± 0.24 seeds per flower, respectively. There was no difference observed in fruit production between control racemes on experimental plants and control racemes on control plants, as indicated by the lack of treatment effect ($F_{1,292} = 0.929$, $P = 0.336$), the interactive effects between treatment and habitat ($F_{1,292} = 0.002$, $P = 0.964$) and between treatment and population ($F_{1,292} = 0.118$, $P = 0.732$), and the three-way interactive effects between them ($F_{1,292} = 0.067$, $P = 0.795$). In addition, there was no difference in seed production between control racemes on experimental plants and control racemes on control plants, as indicated by the lack of treatment effect ($F_{1,292} = 0.840$, $P = 0.360$), the interactive effects between treatment and habitat ($F_{1,292} = 0.033$, $P = 0.856$) and between treatment and population ($F_{1,292} = 0.190$, $P = 0.663$), and the three-way interactive effects between them ($F_{1,292} = 0.024$, $P = 0.877$). These results indicated that there was no evidence of reallocation of resources from control racemes to experimental racemes within experimental plants.

Discussion

Effects of local habitat on pollinator visitation

Hosta ventricosa racemes in the open patch produced larger floral displays than racemes in the forested patch, a common result reflecting differences in overall plant resource status between patches (Kato and Hiura 1999; Meagher and Delph 2001). In population BYS, both light availability and floral display had positive effects on pollinator visitation rate per flower; in population SNC, however, only light availability had positive effects on pollinator visitation rate per flower. Therefore, indirect effects on visitation rate through display size in population BYS must be interpreted with some caution. For example, Kilkenny and Galloway (2008) reported that both light environment and floral display had positive effects on pollinator visitation rate per flower in *Campanulastrum americanum* plants growing in sunny and shady areas of a natural population, but visitation rate per flower responded positively to light levels and did not respond to display size in experimental arrays. Similar experimental arrays with different display sizes and light levels may verify whether floral display had an effect on visitation rate per flower in *H. ventricosa*.

Our results are consistent with those of Herrera (1997), who found that Hymenopterans preferred to forage in areas with high irradiance. Low light may be associated with lower visitation rate, if the flower color of *H. ventricosa* is less vibrant and therefore less visible in forested patches (Chittka and Raine 2006). Light availability might also correlate with visitation rate because of temperature (Corbet et al. 1993; Totland 2001). Since light availability has a positive effect on temperature (Herrera 1995; Kilkenny and Galloway 2008), it is likely that differences in visitation rate between forested patches and open patches could be due to both variations in light level and temperature.

Variations in pollen and resource limitation between local habitat

In population SNC, pollen supplementation did not enhance fruit and seed production at the whole population level, indicating a lack of pollen limitation. These results are in accordance with those of Niesenbaum (1993) and Hasegawa and Kudo (2005) in other systems, who found that female reproduction in both sunny and shady areas was not limited by pollen. In population BYS, pollen supplementation increased fruit and seed production in the forested patch but not in the open patch, and there was no evidence of reallocation of resources to supplementally pollinated racemes, indicating that pollen limitation occurred in the forested patch but not the open patch, therefore differences in visitation rate between habitat influenced female reproduction. These results are consistent with those of Kilkenny and Galloway (2008) and Kudo et al. (2008) in other systems, who reported that a decline in light availability was a mechanism causing pollen limitation. Our study suggested that whether female reproduction in shady areas was limited by pollen availability is population-specific.

Light availability appears to limit reproduction in *H. ventricosa*. Fruit and seed production per flower for supplementally pollinated racemes were higher in the open patch than in the forested patch in both populations, results consistent with the findings of many studies (Niesenbaum 1993; Kato and Hiura 1999; Setsuko and Tomaru 2011), supporting the hypothesis that light availability could affect female reproduction through resource availability.

Conclusion

The present study evaluated the effects of the local light environment on pollinator visitation, pollen and resource limitation of female reproduction in two populations of *H. ventricosa*. The results showed that light environment could directly influence pollinator behavior, although the relative importance of resource-mediated effect and pollinator-mediated effect on female reproduction may vary among populations.

Authors' contributions

CGX conceived and designed the experiments. WBX, XXJ, WX and YCP performed the experiments. CGX and WX analyzed the data. CGX wrote the manuscript. All authors read and approved the final manuscript.

Author details

[1] Department of Forestry, Sichuan Agricultural University, Chengdu 611130, China. [2] Xishuangbanna Tropical Botanical Garden, Chinese Academy of Sciences, Mengla 666303, China. [3] Soil and Fertilizer Research Institute, Sichuan Academy of Agricultural Sciences, Chengdu 610066, Sichuan, China.

Acknowledgements
This research was supported by the National Natural Science Foundation of China (Project Nos. 30870388, 31370436).

Competing interests
The authors declare that they have no competing interests.

References
Ashman TL, Knight TM, Steets JA, Amarasekare P, Burd M, Campbell DR, Dudash MR, Johnston MO, Mazer SJ, Mitchell RJ, Morgan MT, Wilson WG (2004) Pollen limitation of plant reproduction: ecological and evolutionary causes and consequences. Ecology 85:2408–2421

Burd M (1994) Bateman's principle and plant reproduction: the role of pollen limitation in fruit and seed set. Bot Rev 60:83–139

Cai ZQ (2011) Shade delayed flowering and decreased photosynthesis, growth and yield of Sacha Inchi (*Plukenetia volubilis*) plants. Ind Crops Prod 34:1235–1237

Cao GX, Kudo G (2008) Size-dependent sex allocation in a monocarpic perennial herb, *Cardiocrinum cordatum* (Liliaceae). Plant Ecol 194:99–107

Cao G, Xue L, Li Y, Pan K (2011) The relative importance of architecture and resource competition in allocation to pollen and ovule number within inflorescences of *Hosta ventricosa* varies with the resource pools. Ann Bot 107:1413–1419

Cao GX, Xie T, Wu BX, Yang CP (2015) Floral sex allocation and reproductive success within inflorescences of *Hosta ventricosa*, a pseudogamous apomict. J Plant Ecol 8:142–153

Chittka L, Raine NE (2006) Recognition of flowers by pollinators. Curr Opin Plant Biol 9:428–435

Corbet SA, Fussell M, Ake R, Fraser A, Gunson C, Savage A, Smith K (1993) Temperature and the pollinating activity of social bees. Ecol Entomol 18:17–30

Cunningham SA (1997) The effect of light environment, leaf area, and stored carbohydrates on inflorescence production by a rain forest understory palm. Oecologia 111:36–44

Griffin SR, Barrett SCH (2002) Factors affecting low seed: ovule ratios in a spring woodland herb, *Trillium grandiflorum* (Melanthiaceae). Int J Plant Sci 163:581–590

Grindeland JM, Sletvold N, Ims RA (2005) Effects of floral display size and plant density on pollinator visitation rate in a natural population of *Digitalis purpurea*. Funct Ecol 19:383–390

Haig D, Westoby M (1988) On limits to seed production. Am Nat 131:757–759

Hasegawa T, Kudo G (2005) Comparisons of growth schedule, reproductive property and allocation pattern among three *rhizomatous Polygonatum* species with reference to their habitat types. Plant species biol 20:23–32

Herrera CM (1995) Floral biology, microclimate, and pollination by ectothermic bees in an early-blooming herb. Ecology 76:218–228

Herrera CM (1997) Thermal biology and foraging responses of insect pollinators to the forest floor irradiance mosaic. Oikos 78:601–611

Kato E, Hiura T (1999) Fruit set in *Styrax obassia* (Styracaceae): the effect of light availability, display size, and local floral density. Am J Bot 86:495–501

Kilkenny FF, Galloway LF (2008) Reproductive success in varying light environments: direct and indirect effects of light on plants and pollinators. Oecologia 155:247–255

Kitajima K (1994) Relative importance of photosynthetic traits and allocation patterns as correlates of seedling shade tolerance of 13 tropical trees. Oecologia 98:419–428

Klinkhamer PGL, de Jong TJ, de Bruyn GJ (1989) Plant size and pollinator visitation in *Cynoglossum officinale*. Oikos 54:201–204

Knight TM, Steets JA, Vamosi JC, Mazer SJ, Burd M, Campbell DR, Dudash MR, Johnston MO, Mitchell RJ, Ashman TL (2005) Pollen limitation of plant reproduction: pattern and process. Annu Rev Ecol Syst 36:467–497

Kudo G, Ida TY, Tani T (2008) Linkages between phenology, pollination, photosynthesis, and reproduction in deciduous forest understory plants. Ecology 89:321–331

Larson BMH, Barrett SCH (2000) A comparative analysis of pollen limitation in flowering plants. Biol J Linn Soc 69:503–520

Lee TD (1988) Patterns of fruit and seed production. In: Doust JL, Doust LL (eds) Plant reproductive ecology: patterns and strategies. Oxford University Press, New York, pp 179–202

Lee TD, Bazzaz FA (1982) Regulation of fruit and seed production in an annual legume. Ecology 63:1363–1373

McConnaughay KDM, Coleman JS (1999) Biomass allocation in plants: ontogeny or optimality? A test along three resource gradients. Ecology 80:2581–2593

Meagher TR, Delph LF (2001) Individual flower demography, floral phenology and floral display size in *Silene latifolia*. Evol Ecol Res 3:845–860

Niesenbaum RA (1993) Light or pollen—seasonal limitations on female reproductive success in the understory shrub *Lindera benzoin*. J Ecol 81:315–323

R Development Core Team (2014) R: a language and environment for statistical computing. R Foundation for Statistical Computing, Vienna. http://www.r-project.org/

Schmid WG (ed) (2010) The genus *Hosta*. Timber Press, Portland

Setsuko S, Tomaru N (2011) The effects of plant size and light availability on male and female reproductive success and functional gender in a hermaphrodite tree species, *Magnolia stellata*. Botany 89:593–604

Stephenson AG (1981) Flower and fruit abortion: proximate causes and ultimate functions. Annu Rev Ecol Syst 12:253–279

Sutherland S (1986) Patterns of fruit-set: what controls fruit-flower ratios in plants? Evolution 40:117–128

Totland Ø (2001) Environment-dependent pollen limitation and selection on floral traits in an alpine species. Ecology 82:2233–2244

Wesselingh RA (2007) Pollen limitation meets resource allocation: towards a comprehensive methodology. New Phytol 174:26–37

Wiens D (1984) Ovule survivorship, brood size, life history, breeding systems, and reproductive success in plants. Oecologia 64:47–53

Zhao D, Hao Z, Tao J (2012) Effects of shade on plant growth and flower quality in the herbaceous peony (*Paeonia lactiflora* Pall.). Plant Physiol Biochem 61:187–196

5

Influence of bio-fertilizer containing beneficial fungi and rhizospheric bacteria on health promoting compounds and antioxidant activity of *Spinacia oleracea* L.

Muhammad Khalid[1†], Danial Hassani[1†], Muhammad Bilal[2], Fayaz Asad[3] and Danfeng Huang[1*]

Abstract

Background: This study evaluates the influences of bio fertilizers containing mycorrhizal fungi (*Glomus fasciculatum, Glomus mosseae*) individually or in combination with N-fixer (*Azotobacter chroococcum*), K solubilizer (*Bacillus mucilaginous*) and P solubilizer (*Bacillus megaterium*) on soil fertility and phytochemical levels of spinach.

Results: Root colonization by mycorrhizal fungi was increased in the presence of bacterial inoculation in comparison to individual inoculation treatments. Inoculation of bio fertilizer containing mycorrhizal fungi and bacterial species considerably augmented the concentration of total phenolic compounds, flavonoids and phenolic acid contents. The 1, 1-diphenyl-2-picrylhydrazyl (DPPH) scavenging capacity of spinach was found to be positively coincided with flavonoid contents, while partially correlated with total phenolic compounds and phenolic acids. Further, the HPLC analysis showed that significantly higher antioxidant activity of spinach was correlated with quercetin contents and chlorogenic acid. Chlorophyll contents were also increased following the bio fertilization treatments.

Conclusion: Results revealed that these microbes are useful tool for improving health promoting compounds in spinach.

Keywords: Bio fertilizer, Arbuscular mycorrhizal fungi, Rhizospheric bacteria, Health-promoting compounds, Spinach

Introduction

The consumption of fruits and vegetables could increase the human innate immunity against chronic diseases (Bagchi et al. 2003; Yochum et al. 1999). The phytoconstituents including polyphenols, quercetin and flavonoids are largely demonstrated as important antioxidants and exhibit profound radical scavenging capabilities (Bravo 1998; Chu et al. 2000; Duthie et al. 2000; Gil et al. 1999; Middleton and Kandaswami 1994). *Spinacia oleracea* L. is one of the most important and commonly consumed leafy vegetable. It is commercially known as spinach which is claimed to possess therapeutic properties and being a rich source of flavonoids as well as phenolic compounds besides its economical and ease of availability (Bunea et al. 2008; Ferreres et al. 1997; Metha and Belemkar 2014; Sultana and Anwar 2008). The pro-health properties of spinach are attributed to its low calorific value, and its large supply of vitamins, micro- and macronutrients and others phytochemicals, including polyphenols and fiber (Llorach et al. 2008). The quality of fresh vegetables could be assessed based on their nutritional value, growing conditions and usage of fertilizer. Despite the fact that the genetic modification and agronomic manipulation methods are widely used to improve the nutritional value of plants, the inadequate public acceptance and soil specificity of genetically modified food are still the challenges (Martínez-Ballesta et al. 2008).

*Correspondence: hdf@sjtu.edu.cn
†Muhammad Khalid and Danial Hassani contributed equally as the first author
[1] School of Agriculture and Biology, Shanghai Jiao Tong University, Shanghai 200240, People's Republic of China
Full list of author information is available at the end of the article

Started about 60 years ago, several studies have revealed the potentiality of beneficial microbes in increasing the plants resistance to biotic and abiotic stresses through the up-production of secondary metabolites (Shen 1997). Beneficial bacteria or fungi inhabit various sites such as plants rhizosphere, while others colonize on rhizoplane or even intercellular spaces (McCully 2001). Former studies revealed that phosphate and potassium solubilizing bacteria decompose the phosphate and potassium from their sources and make them available to the plants, assisting essential mineral uptake. Plant growth promoting rhizobacteria (PGPR) thrives in the rhizosphere of plants. It is worth mentioning that a substantial number of bacterial and fungal species entertain a functional relationship and establish an integrated system with the plants. They enhance the plant growth either by assisting in essential nutrients acquisition (minerals, nitrogen and phosphorus), eliciting pertinent hormones or acting as bio-control agents to reduce the inhibitory effects of various pathogens (Yang et al. 2009). Some strains such as *Azotobacter chroococcum* and *Azospirillum brasilense* have shown to possess properties of biological nitrogen fixation both in legume and non-legume, exerting a positive effect on overall physiology and development of the plants (Dobbelaere et al. 2001; Goldstein and Liu 1987). Former literature survey revealed 30% improvement in the yield of wheat by the *Azotobacter* inoculation (Zablotowicz et al. 1991). Likewise, root/shoot length and dry weight has been significantly increased in tomato, lettuce and canola by inoculation of *Pseudomonas putida* and *Pseudomonas fluorescens* (Glick et al. 1997; Hall et al. 1996).

Arbuscular mycorrhizal fungi (AMF) are associated with majority of the plants, growing under natural conditions and its contribution for micronutrients uptake is well-documented in the previous reports. Moreover, these beneficial microbes protect the plants from oxidative stress by synthesizing antioxidant enzymes including, peroxidase, catalase, superoxide and non-enzymatic antioxidants such as glutathione, ascorbate and α-tocopherol; hence, providing an suitable way to replace the hazardous agricultural chemical and agro-ecosystems destabilizing fertilizers.

The current study was appraised to evaluate the influence of beneficial bacteria (*A. chroococcum, Bacillus megaterium* and *Bacillus mucilaginous*) and fungi (*Glomus fasciculatum* and *Glomus mosseae*) on the antioxidant properties and physiological activities of *S. oleracea* L. and to develop an alternative method for improving the quality of health promoting phytochemicals and antiradical activity of spinach.

Materials and methods

Chemicals and reagents

Standard laboratory grade chemicals/reagents including, Folin–Ciocalteau reagent, 2,2-diphenyl-1-picrylhydrazyl radical (DPPH), gallic acid, ascorbic acid, Dithiothreitol (DTT), and (±)-6-Hydroxy-2, 5, 7, 8-tetramethylchromane-2-carboxylic acid (Trolox) were mainly procured from Sigma-Aldrich (USA) and DSL Chemicals (Shanghai) Co., Ltd. All the experimental works were carried out in the Joint Laboratory of Digital Horticulture, School of Agriculture and Biology, Shanghai Jiao Tong University, Shanghai China.

Soil

Soil sample, was collected from the botanical garden of School of agriculture and biology, Shanghai Jiao Tong University, China. The collected soil was air-dried, grinded, passed through a sieve (2 mm for chemical analysis and 8 mm for pot experiment) and mixed thoroughly. Soil was autoclaved three times and analyses were made prior to seeding. Basic properties of soil were; pH, 7.32; EC, 0.14 (dS/m); available N, 111.6 (ppm); available P, 181.7 (ppm), available K, 306.8 (ppm), cation exchange capacity (CEC), 13.2 (cmol(+)/kg); NH_4^+, 7.86 (ppm); NO_3^-, 2.67 (ppm); total C, 1.92 (%); total N, 0.19 (%); total K, 2063 (ppm).

Inocula development and bio-fertilizer

Sterilized peat moss was chosen as a carrier for the rhizobacteria including *A. chroococcum* (nitrogen fixer), *B. megaterium* (phosphate solubilizer) and *B. mucilaginous* (potassium solubilizer) inoculums. All the bacterial strains were cultured in Luria–bertani (LB) broth at 28 °C for 48 h in a rotary shaker at an agitation speed of 120 rpm. After designated time period, the culture density was measured by means of a haemocytometer (improved neubauer counting chamber) following the previously described method (Wu et al. 2005). The strains were centrifuged at 5000 rpm (at 4 °C) and resulting cells were thoroughly mixed with the sterilized peat moss. Mixture acting as the microbial inoculum contained a final population density of 1.33×10^8, 2.08×10^8, 2.66×10^8 cfu g^{-1} inoculum (wet weight) for K, P and N fixing bacteria, respectively. The sand-based two fungal inoculums consisting hyphae and spores were purchased from the Central bureau voor Schimmel cultures, Fungal Biodiversity Centre, Institute of the Royal Netherlands Academy of Arts and Sciences (KNAW).

Experimental set up

Greenhouse trial was conducted to investigate the efficiency of different combination of beneficial fungi and rhizobacteria on the growth and antioxidant profile of spinach. According to the treatment design, 15 g of sand-based mycorrhizal inoculum and 2 g of peat moss-based bacterial inoculum were inoculated into each pot according to treatment design. Treatments were set up for comparison, where 1st treatment (T1) considered as control with no fertilization, while the 2nd (T2) and 3rd (T3) treatment contained chemical fertilizers (KCl, KH_2PO_4 and urea) and autoclaved organic bio-fertilizer, respectively. The detailed information of experimental design has been listed in Table 1.

Seeds germination and analysis

The seeds of spinach were procured from Shanghai WELLS seed Co., LTD. Prior to sowing, the seeds were surface disinfected three times by soaking in 70% ethanol for 5 min and then in distilled water. After seed germination in each pot (height, 10 cm; bottom diameter 9 cm, top diameter 10 cm, soil, 800 g per each pot), the seedling were thinned and only two seedling per pot were allowed to continue their growth. The pots were placed randomly in a greenhouse with an average temperature of 21 ± 5.0 °C and watered (twice a week) with distilled water to maintain the appropriate soil humidity level. After 45 days, the plants were harvested, collected and used for analytical purposes. Each treatment was carried out in 15 replicate pots to maintain the reproducibility of the data.

Fungi and bacteria colonization assay

At harvesting, the root were washed, treated with KOH (10.0%) for 20 min at 90 °C, acidified with HCL 1.0% for 3 min and then stained with trypan blue 0.05%, and subjected to fungal root analysis in a manner as described previously (Giovannetti and Mosse 1980). Differentiating media with suspension dilution techniques from the soil samples were used to isolate and measure bacteria growth. For phosphate solubilizing bacteria:, NaCl 0.4 g, $(NH_4)_2SO_4$ 0.6 g, $Ca_3(PO_4)_3$ 9.0 g, KCl 0.3 g, $FeSO_4{\cdot}7H_2O$ 0.03 g, $MgSO_4{\cdot}7H_2O$ 0.5 g, $MnSO^4{\cdot}4H_2O$ 0.03 g, agar 21.0 g, glucose 10.0 g, sterilized water 1.0 L, pH 7.0; for Nitrogen fixer:, NaCl 0.3 g, 2% congo red solution 6 mL, K_2HPO^4 0.4 g, $MgSO^4{\cdot}7H_2O$ 0.2 g, 3 drops of 1% $FeCl_3$ and 1% $MnCl_2$ solution, agar 20.0 g, glucose 10.0 g, sterilized water 1.0 L, pH 7.0; for potassium solubilizing bacteria: Na_2HPO_4 2.0 g, $FeCl_3$ 0.005 g, $CaCO_3$ 0.1 g, $MgSO_4{\cdot}7H_2O$ 0.5 g, glass powder 1.0 g, agar 20.0 g, sucrose 5.0 g, 1.0 L distilled water, pH 7.0 (Wu et al. 2005).

Extract preparation of health-promoting compounds

The phenolic compounds were extracted following the method previously described (Khalid et al. 2017). The phonolic compounds were extracted following the method previously described Frozen leaf tissue (2.5 g) was grinded with a mortar and pestle using 15 mL of 50% (v/v) acidified methanol (0.1 M HCl) and the phenolic compounds were extracted for 20 min at room temperature, then centrifuged at 9000g for 30 min. This procedure was repeated three times and the supernatants were combined to produce a crude extract of polyphenols. The raw methanolic extract was then evaporated to dryness under a vacuum at a temperature of 40 °C and rinsed with 100% methanol to a final volume of 10 mL.

Study of health promoting phytochemical

HPLC–MS analysis

Quantitative assessment of phenolic compounds was carried out through HPLC–MS (LTQ XL, Thermo Fisher Scientific, San Jose, CA, USA) analysis (Świeca et al. 2012). The HPLC–MS system was equipped with a ternary pump, auto sampler, and thermostatic column compartment, diode array detector (Surveyor, Thermo Fisher), and a linear ion trap mass spectrometer (LTQ XL, Thermo Fisher Scientific, San Jose, CA, USA) equipped with an electrospray ionization (ESI) source. A CORTECS C18 column (2.1 mm × 100 mm, 2.6 μm; Waters) was used; the column temperature was maintained at 35 °C. The mobile phase A (0.1% formic acid/water) and B (acetonitrile) was used, the gradient program was as follows: 0–2 min 5.0% B; 4–11 min 15–35% B; 15–17 min, 100% B; 17.5–22 min, 5.0% B; flow rate was 0.25 mL min^{-1}, the injection volume was 4 μL. UV detection was performed at 270 and 370 nm, the wavelength was scanned from 200–600 nm. MS was scanned in ESI source in negative mode, mass range: *m/z* 92–1000;

Table 1 Experimental design

Treatments	T1	T2	T3	T4	T5	T6	T7
Bacteria inoculation	No	No	No	Yes	No	No	Yes
Mycorrhizal fungi	None	None	None	None	Yes	Yes	Yes
Fertilizer used	None	CF	OF^A	None	None	None	None
Code	Control	CF	OF	B	GF	GM	B + GF + GM

CF chemical fertilizer, *OF^A* organic fertilizer (autoclaved bio-fertilizer), *B Azotobacter chroococcum* + *Bacillus megaterium* + *Bacillus mucilaginous*, *GF Glomus fasciculatum*, *GM Glomus mosseae*

source voltage was 3.5 kV, capillary temperature was 350 °C, sheath gas flow was 35, aux gas flow was 15.0, sweep gas flow was 1.0, and capillary voltage was 43 V. Data acquisition, handling, and instrument control were performed using Xcalibur 2.3.1 software.

Determination of total phenolic contents, flavonoids and phenolic acid

Total phenolic contents (TPCs) were determined by the method as reported earlier (Singleton et al. 1999). Briefly, 0.5 mL H_2O in combination with 2.0 mL Folin–Ciocalteau reagent (1:5 H_2O) was mixed with 0.5 mL of the plant sample. After 3–5 min of incubation at room temperature, 10 mL of Na_2CO_3 (10%, *w/v*) was added to the mixture and incubated at room temperature for 30 min. Optical density of each sample was recorded at 725 nm in a UV–Vis spectrophotometer (HITACHI, U-2900) using Gallic acid as standard.

Total flavonoid contents (TFCs) were estimated in a manner described by (Lamaison and Carnat 1990). Shortly, 1.0 mL sample extract was allowed to react with 1.0 mL of aluminium chloride ($AlCl_3{\cdot}6H_2O$) solution (2.0%, w/v) for 10 min at room temperature and absorbance was monitored at 430 nm.

A previously reported method of (Szaufer-Hajdrych 2004) was followed for measuring the phenolic acid contents (PACs) in the extracted sample. To this end, 1.0 mL of sample was thoroughly mixed with a combination of 5.0 mL of distilled water, 1.0 mL HCl (0.5 M), Arnov reagent and NaOH (1 M) followed by OD measurement at 490 nm.

Chlorophyll measurement

Chlorophyll content was evaluated as reported (Lin et al. 2013). Freeze-dried leaves samples were grinded with acetone, centrifuged at 13,000 rpm for 5.0 min. Supernatants were collected and spectrophotometrically measured at 663 and 645 nm to analyze chlorophyll a, and chlorophyll b.

Total antioxidant activity

For antioxidant activity, free radical scavenging assay was carried out using 1, 1-diphenyl-2-picrylhydrazyl (DPPH) as described (Brand-Williams et al. 1995; Khanam et al. 2012). An 80 µL of methanolic extract was mixed with 1.92 mL DPPH solution, and absorbance was immediately noted at 515 nm. The quenching affinity of the sample was measured according to the relation given below.

$$\text{Scavenging } (\%) = [(AC - AA)/AC)] \times 100$$

where, AC represents the control absorbance at "0" min and AA denotes the absorbance of sample after "*t*" time.

Statistical analysis of data

All the analytic determinations were carried out in triplicates. Statistical analysis was performed using STATISTICA 7.0 software for mean comparison using Tukey's test at the significance level of $P < 0.05$.

Results and discussion

Colonization assay results in Fig. 1 manifest that the plant roots were found to be significantly colonized by Arbuscular mycorrhizal fungi (AMF) as compared to un-inoculated control treatments. Inoculation with beneficial bacteria and both *G. fasciculatum* and *G. mosseae* in T7 treatment led to augmented root as compared to T5 and T6. It has been demonstrated that the rhizobacteria can elicit the ability of mycorrhizal fungi in colonizing plant roots (Fitter and Garbaye 1994). Though, the exact mechanism underlying this triggering pathway is still unclear, but the up-production biosynthesis of plants hormones and vitamins as a result of bacterial infection may be involved in this process (Barea et al. 2002). The presence of beneficial rhizobacteria might cause the germination of a large number of spores by fungi, and as a consequence higher chances to infect/colonize plant roots (Tandon and Prakash 1998). It is not surprising that some beneficial bacteria actively secrete pectinolytic and cellulolytic enzymes that could further assist in mycorrhizal infection (Verma et al. 2001). In an earlier study, *A. chroococcum* has been isolated from the rhizospheric soil of *Spinacia oleracea* (Jiménez et al. 2011). Similarly, (Çakmakçı et al. 2007) reported that *B. megaterium* and *B. mucilaginous* successfully colonized rhizospheric zone in wheat and spinach. Recently (Song et al. 2015) observed a great synergy between the above mentioned strains and their aptitude to improve the quality of soil.

The present findings indicate that the level of bacteria (*A. chroococcum, B. megaterium, B. mucilaginous*) in the

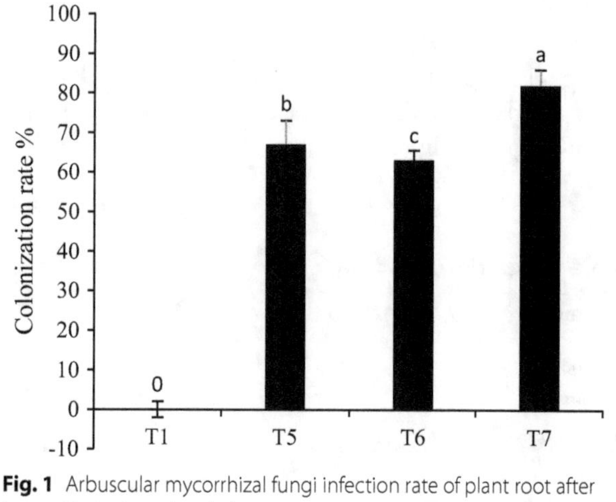

Fig. 1 Arbuscular mycorrhizal fungi infection rate of plant root after harvest in co-inoculated plants

rhizosphere as well as AMF integration in the roots was different in co-inoculated treatments (Fig. 1; Table 2). The numbers of bacteria and AMF infection were found to be higher in T7 compared to T4, 5 and 6 treatments (Tables 1, 2) suggesting the auxiliary property of micro-organism in combination with each other.

HPLC analysis results from spinach extract

The possible compounds were identified by MS1, MS2 fragments and compared with the reported literatures (Kim et al. 2008; Ribas-Agustí et al. 2011; Złotek et al. 2014). The retention time, m/z in negative mode, MS2 fragments and the possible chemical name are listed in Table 1. The major antioxidant and health benefiting compounds that were identified such as caffeic acid, ferulic acid, flavones (luteolin), flavonols (quercetin, kaempferol), isorhamnetin-3-gentiobioside-7-glucoside have been reported variously in previous studies in spinach (Alarcón-Flores et al. 2014; Justesen 2001; Nuutila et al. 2002). Chlorogenic acid, coumaric acid are also reported by Okazaki and coworkers while analyzing the effect of nitrogen concentration on the constituent's profile of spinach (Okazaki et al. 2008).

Influence of experimental treatments on health promoting compound in spinach leaves

In view of potential medicinal and preventive roles of secondary metabolites in chronic diseases such as cancer, cardiovascular and neurodegenerative disorders (Riedel et al. 2012), the health promoting metabolites of spinach were analyzed. Results showed that the total phenolic, flavonoids and phenolic acid contents were found to be significantly varied under different experimental treatments. The amount of total phenolic contents was 58.72 and 51.43% higher in T5 and T7 treatment, respectively, as compared to un-inoculated control plants (Fig. 2). Similar to our findings, the eliciting effect of *A. chroococcum* and *G. fasciculatum* on various other plants growth and total phenolic compounds have been observed (Baslam et al. 2011; Teixeira da Silva and Egamberdieva 2013). The flavonoid content was evaluated under different treatments and results are shown in Fig. 2. It was observed that experimental plants co-inoculated with *G. fasciculatum* and *G. mosseae* showed the highest content of flavonoids in T5 and T6, which were 48.02 and 40.46% greater than that of control plants. Investigation of antioxidant biosynthesis and induction of phytochemicals including flavonoids by AMF has earlier been documented in several reports (Carlsen et al. 2008; Eftekhari et al. 2012; Nisha and RajeshKumar 2010). Similarly, phenolic acids were also recorded to be significantly higher in all treatments compared with control. Nevertheless, the T4 treatment and T7 treatment led to the 27.68 and 28.07% higher values, respectively in comparison to un-inoculated plants. The results were in good agreement with (Taie et al. 2008) who have reported up to 75% improvement of phenolic acid biosynthesis in Rhizobacteria inoculated soybean seedling.

Antioxidant activity

The results of antioxidant activity analyzed through DPPH scavenging assay are portrayed in Fig. 2. It was observed that utmost antioxidant activity was determined in T7 (1.9 mM Trolox/g FW) elicited by bacterial and fungal combination followed by T6 (1.8 mM Trolox/g FW) treated with fungus *G. mosseae*. Observed improvements in antioxidant level were 70.17, 66.66 and 61.29% for T7, T6 and T4, respectively, as compared to control (without any bacterial and fungal inoculation). The DPPH scavenging capacity of spinach was found to be positively coincided with flavonoid contents, while partially correlated with total phenolic compounds and phenolic acids. Polyphenols are important class of biologically-active compounds with extensively reported antioxidant characteristics. However, other properties, such as, the capability to suppress enzymes [lipoxygenase or cyclooxygenase (COX)] involved in the inflammation process have recently taken more attention (Gawlik-Dziki et al. 2011; Mulabagal et al. 2010). Nevertheless, earlier reports highlighted that antioxidant capacity of any plant extract highly depends on the type and relative proportion of phenolics presence. In our study, the significantly higher DPPH scavenging potential of spinach might be positively correlated with quercetin contents and chlorogenic acid (Table 4). The results strongly corroborates with (Kim et al. 2007; Liu et al. 2007) who observed pronounced antioxidant activity of lettuce in the presence of quercetin and chlorogenic acid, while luteolin and caffeic acid negatively influences the antioxidant activity. In addition to phenolic compounds, several other bioactive

Table 2 Population of beneficial bacteria in the rhizosphere of co-inoculated spinach after 45 days of growth

Treatment	NFB (10^4 cfu/g dry soil)	PSB (10^6 cfu/g dry soil)	KSB (10^6 cfu/g dry soil)
T4	30.6 ± 3.30 b	44.1 ± 0.09 b	44.2 ± 1.50 b
T7	63.1 ± 11.5 a	79.8 ± 3.55 a	74.7 ± 3.91 a

NFB nitrogen fixing bacteria, *PSB* phosphate solubilizing bacteria, *KSB* potassium solubilizing bacteria

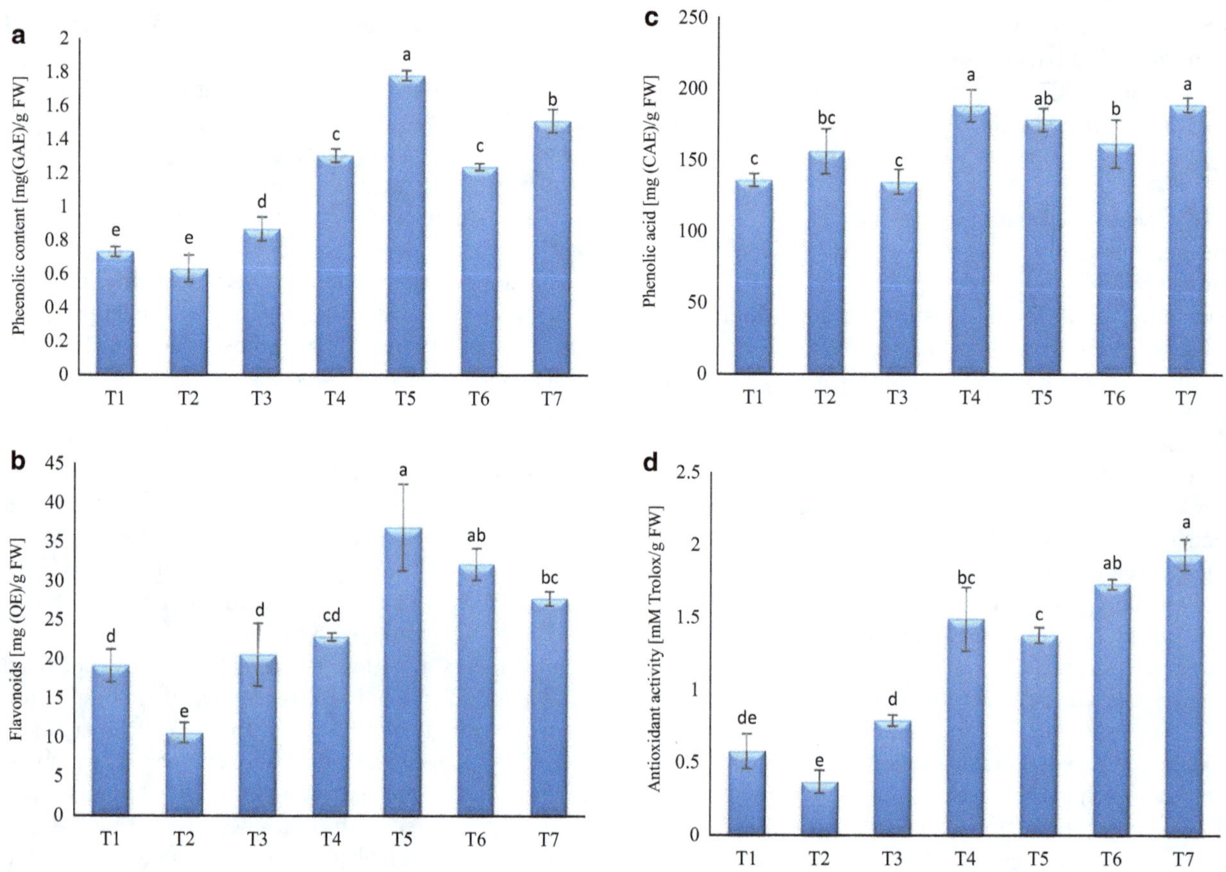

Fig. 2 Influence of selected treatments on **a** phenolic contents, **b** flavonoids, **c** phenolic acids, **d** antioxidant activity

constituents such as carotenoids and vitamins particularly vitamin C potentially contribute a key role in the elicitation of antioxidant potentialities of the plants (Sun et al. 2012).

Chlorophyll content

Green leafy vegetables are known to be the richest source of major dietary metabolites. Spinach possess a considerable amount of chlorophyll which play critical role in reducing the risk of cancer and cardiovascular diseases (CVDs) (Caldwell and Britz 2006). It has been demonstrated that antioxidant rich diets, particularly chlorophyll inhibit the functions of COX-1 and COX-2 (Kim et al. 2007; Mulabagal et al. 2010). There are a variety of previously reported studies concerning the stimulating effect of rhizobacteria combination in co-cultivation with *Ociumum basilicm* L. (Vivas et al. 2003). The responses of chlorophyll contents in spinach treated with different elicitors are illustrated in Table 4. Chlorophyll a content was recorded to be significantly higher in plants co-inoculated by *A. chroococcum*, *B. megaterium* and *B. mucilaginous* bacterial strains (T4) in contrast with the control. Whereas, no significant difference was observed in chlorophyll a content between the control and T3, T5 and T6 treatments (Table 4). The co-cultivation of plants with Mycorrhizal fungi as well as chemical and organic fertilizer had no important impact on chlorophyll b content whereas a significantly higher chl b content was recorded in experimental treatment inoculated with bacterial strains (T4) (Table 3). Bacterial inoculation had also an elevating role in total chlorophyll content (a + b). Chemical fertilizer (T2) as well as bacterial and fungal strains combination treatment (T7) displayed higher values 458.66 ± 7.27 b and 515.79 ± 13.15 c, respectively. No significant difference was observed in other treatments (Table 4).

Conclusions

In the light of above findings, it is concluded that the application of bio fertilizer containing beneficial microbes displayed a stimulating effect on the soil properties and health qualities of spinach. The rhizobacterial inoculation resulted in increase of root infection by arbuscular mycorrhizal fungi. Moreover, the total

Table 3 List of constituents based on the of HPLC analysis of *Spinacia oleracea* L extract

t_R	Negative mode (m/z)	Possible compound	Chemical name	MS² (−)
0.93	193	$C_{10}H_{10}O_4$	Ferulic acid	175
0.98	179	$C_9H_8O_4$	Caffeic acid	161, 143
1.58	353	$C_{16}H_{18}O_9$	Chlorogenic acid	335, 249
1.67	285	$C_{15}H_{10}O_6$	Kaempferol	267, 225
4.26	285	$C_{15}H_{10}O_6$	Luteolin	267, 225, 153
6.85	295	$C_{13}H_{12}O_8$	Caffeoylmalic acid	163, 149
7.10	163	$C_9H_8O_3$	Coumaric acid	119
7.60	787	$C_{33}H_{40}O_{22}$	Quercetin-3-gentiobioside-7-glucoside	772, 769, 655, 637,505, 373, 330, 313
8.20	801	$C_{34}H_{42}O_{22}$	Isorhamnetin-3-Gentiobioside-7-glucoside	786,769, 669, 651, 387, 345, 330, 329
8.51	465	$C_{20}H_{22}O_9$	Cassiaside	437, 407, 379, 259, 241
8.63	301	$C_{15}H_{10}O_7$	Quercetin	283, 265 255, 237
8.96	337	$C_{16}H_{18}O_8$	5-*p*-coumaroylquinic acid	277, 191
10.3	521	$C_{30}H_{34}O_8$	Benzoylgomisin H	345
12.1	519	$C_{29}H_{28}O_9$	Schisantherin D	343
12.5	503	$C_{29}H_{28}O_8$	Interiotherin A	327
12.8	533	$C_{27}H_{34}O_{11}$	Phillyrin	357
16.5	311	$C_{13}H_{12}O_9$	Caffeoyltartaric acid	149

Table 4 Influence of selected treatments on chlorophyll content in Spinach leaves

Treatment	Constituents (mg/100 g dw)		
	Chl a	Chl b	Chl a + b
T1	237.45 ± 3.7 d	128.31 ± 14.16 g	365.33 ± 8.67 f
T2	335.23 ± 6.45 b	123.20 ± 8.60 f	458.66 ± 7.27 b
T3	206.33 ± 7.61 e	140.74 ± 2.42 d	346.33 ± 6.15 g
T4	474.24 ± 10.41 a	312.62 ± 17.04 a	786.83 ± 20.79 a
T5	251.91 ± 18.22 c	149.80 ± 11.03 c	400.25 ± 10.83 d
T6	253.69 ± 3.05 c	100.80 ± 5.51 g	354.10 ± 22.17 f
T7	351.28 ± 13.40 c	163.45 ± 18.07 b	515.79 ± 13.15 c

Chl a chlorophyll A, *Chl b* chlorophyll B, *Chl a + b* chlorophyll A and B

phenolic compounds, antioxidant activity and chlorophyll content were also significantly enhanced by bio fertilizer treatments. These outcomes suggested that exploration of microbes display a high potential for use in the improvement of nutritious properties of fresh vegetables which could be a potential alternative to conventional approaches. However the mechanism underlying this phenomenon is not yet fully understood and will be remained for further investigations.

Authors' contributions

MK and DH carried out the experiment work. MB and FA helped in drafting the paper and interpreting the data. All authors read and approved the final manuscript.

Author details

[1] School of Agriculture and Biology, Shanghai Jiao Tong University, Shanghai 200240, People's Republic of China. [2] State Key Laboratory of Microbial Metabolism, School of Life Sciences and Biotechnology, Shanghai Jiao Tong University, Shanghai 200240, China. [3] Key Laboratory of Alpine Ecology and Biodiversity, Key Laboratory of Tibetan Environment, Changes and Land Surface Processes, Institute of Tibetan Plateau Research, Chinese Academy of Sciences, Beijing 100101, China.

Acknowledgements

This work was sponsored in part by the National High-tech R&D Program of China (863 Program) (Grant No. 2013AA103000) and Shanghai Agriculture Applied Technology Development Program, China (Grant No. T20140502).

Competing interests

The authors declare that they have no competing interests.

References

Alarcón-Flores MI, Romero-González R, Vidal JLM, Frenich AG (2014) Determination of phenolic compounds in artichoke, garlic and spinach by Ultra-High-Performance Liquid Chromatography coupled to Tandem Mass Spectrometry. Food Anal Methods 7:2095–2106

Bagchi D, Sen CK, Ray SD, Das DK, Bagchi M, Preuss HG, Vinson JA (2003) Molecular mechanisms of cardioprotection by a novel grape seed proanthocyanidin extract. Mutation Res Fundam Mol Mech Mutagen 523:87–97

Barea J, Azcón-aguilar C, Azcón R (2002) Interactions between mycorrhizal fungi and rhizosphere micro-organisms within the context of sustainable soil-plant systems. In: Gange AC, Brown VK (eds) Multitrophic interactions in terrestrial systems. pp 65–68

Baslam M, Garmendia I, Goicoechea N (2011) Arbuscular mycorrhizal fungi (AMF) improved growth and nutritional quality of greenhouse-grown lettuce. J Agric Food Chem 59:5504–5515

Brand-Williams W, Cuvelier M-E, Berset C (1995) Use of a free radical method to evaluate antioxidant activity. LWT-Food Sci Technol 28:25–30

Bravo L (1998) Polyphenols: chemistry, dietary sources, metabolism, and nutritional significance. Nutr Rev 56:317–333
Bunea A, Andjelkovic M, Socaciu C, Bobis O, Neacsu M, Verhé R, Van Camp J (2008) Total and individual carotenoids and phenolic acids content in fresh, refrigerated and processed spinach (*Spinacia oleracea* L.). Food Chem 108:649–656
Çakmakçı R, Erat M, Erdoğan Ü, Dönmez MF (2007) The influence of plant growth–promoting rhizobacteria on growth and enzyme activities in wheat and spinach plants. J Plant Nutr Soil Sci 170:288–295
Caldwell CR, Britz SJ (2006) Effect of supplemental ultraviolet radiation on the carotenoid and chlorophyll composition of green house-grown leaf lettuce (*Lactuca sativa* L.) cultivars. J Food Compos Anal 19:637–644
Carlsen S, Understrup A, Fomsgaard I, Mortensen A, Ravnskov S (2008) Flavonoids in roots of white clover: interaction of arbuscular mycorrhizal fungi and a pathogenic fungus. Plant Soil 302:33–43
Chu YH, Chang CL, Hsu HF (2000) Flavonoid content of several vegetables and their antioxidant activity. J Sci Food Agric 80:561–566
Dobbelaere S, Croonenborghs A, Thys A, Ptacek D, Vanderleyden J, Dutto P, Labandera-Gonzalez C, Caballero-Mellado J, Aguirre JF, Kapulnik Y (2001) Responses of agronomically important crops to inoculation with *Azospirillum*. Funct Plant Biol 28:871–879
Duthie GG, Duthie SJ, Kyle JA (2000) Plant polyphenols in cancer and heart disease: implications as nutritional antioxidants. Nutr Res Rev 13:79–106
Eftekhari M, Alizadeh M, Ebrahimi P (2012) Evaluation of the total phenolics and quercetin content of foliage in mycorrhizal grape (*Vitis vinifera* L.) varieties and effect of postharvest drying on quercetin yield. Ind Crops Prod 38:160–165
Ferreres F, Castañer M, Tomás-Barberán FA (1997) Acylated flavonol glycosides from spinach leaves (*Spinacia oleracea*). Phytochemistry 45:1701–1705
Fitter A, Garbaye J (1994) Interactions between mycorrhizal fungi and other soil organisms. Plant Soil 159:123–132
Gawlik-Dziki U, Swieca M, Sugier D, Cichocka J (2011) Comparison of in vitro lipoxygenase, xanthine oxidase inhibitory and antioxidant activity of *Arnica montana* and *Arnica chamissonis* tinctures. Acta Sci Pol Hortorum Cultus 10:15–27
Gil MI, Ferreres F, Tomás-Barberán FA (1999) Effect of postharvest storage and processing on the antioxidant constituents (flavonoids and vitamin C) of fresh-cut spinach. J Agric Food Chem 47:2213–2217
Giovannetti M, Mosse B (1980) An evaluation of techniques for measuring vesicular arbuscular mycorrhizal infection in roots. New Phytol 84:489–500
Glick BR, Liu C, Ghosh S, Dumbroff EB (1997) Early development of canola seedlings in the presence of the plant growth-promoting rhizobacterium *Pseudomonas putida* GR12-2. Soil Biol Biochem 29:1233–1239
Goldstein A, Liu S (1987) Molecular cloning and regulation of a mineral phosphate solubilizing gene from *Erwinia herbicola*. Nat Biotechnol 5:72–74
Hall JA, Peirson D, Ghosh S, Glick BR (1996) Root elongation in various agronomic crops by the plant growth promoting rhizobacterium *Pseudomonas putida* GR12–2. Isr J Plant Sci 44:37–42
Jiménez DJ, Montaña JS, Martínez MM (2011) Characterization of free nitrogen fixing bacteria of the genus *Azotobacter* in organic vegetable-grown Colombian soils. Braz J Microbiol 42:846–858
Justesen U (2001) Collision-induced fragmentation of deprotonated methoxylated flavonoids, obtained by electrospray ionization mass spectrometry. J Mass Spectrom 36:169–178
Khalid M, Hassani D, Bilal M, Liao J, Huang D (2017) Elevation of secondary metabolites synthesis in *Brassica campestris* ssp. chinensis L. via exogenous inoculation of *Piriformospora indica* with appropriate fertilizer. PLoS ONE 12:e0177185
Khanam UKS, Oba S, Yanase E, Murakami Y (2012) Phenolic acids, flavonoids and total antioxidant capacity of selected leafy vegetables. J Funct Foods 4:979–987
Kim H-J, Fonseca JM, Choi J-H, Kubota C (2007) Effect of methyl jasmonate on phenolic compounds and carotenoids of romaine lettuce (*Lactuca sativa* L.). J Agric Food Chem 55:10366–10372
Kim KH, Lee KH, Choi SU, Kim YH, Lee KR (2008) Terpene and phenolic constituents of *Lactuca indica* L. Arch Pharmacal Res 31:983–988
Lamaison J, Carnat A (1990) Levels of principal flavonoids in flowers and leaves of Crataegus-Monogyna Jacq and Crataegus-Laevigata (Poiret) Dc (Rosaceae). Pharm Acta Helv 65:315–320
Lin K-H, Huang M-Y, Huang W-D, Hsu M-H, Yang Z-W, Yang C-M (2013) The effects of red, blue, and white light-emitting diodes on the growth, development, and edible quality of hydroponically grown lettuce (*Lactuca sativa* L. var. capitata). Sci Hortic 150:86–91
Liu X, Ardo S, Bunning M, Parry J, Zhou K, Stushnoff C, Stoniker F, Yu L, Kendall P (2007) Total phenolic content and DPPH radical scavenging activity of lettuce (*Lactuca sativa* L.) grown in Colorado. LWT-Food Sci Technol 40:552–557
Llorach R, Martínez-Sánchez A, Tomás-Barberán FA, Gil MI, Ferreres F (2008) Characterisation of polyphenols and antioxidant properties of five lettuce varieties and escarole. Food Chem 108:1028–1038
Martínez-Ballesta MC, López-Pérez L, Hernández M, López-Berenguer C, Fernández-García N, Carvajal M (2008) Agricultural practices for enhanced human health. Phytochem Rev 7:251–260
McCully ME (2001) Niches for bacterial endophytes in crop plants: a plant biologist's view. Funct Plant Biol 28:983–990
Metha D, Belemkar S (2014) PHARMACOLOGICAL ACTIVITY OF SPINACIA OLERACEA LINN.-A COMPLETE OVERVIEW. Asian J Pharm Res Dev 2:83–93
Middleton E Jr, Kandaswami C (1994) The impact of plant flavonoids on mammalian biology: implications for immunity, inflammation and cancer. The flavonoids. Chapman and Hall, London
Mulabagal V, Ngouajio M, Nair A, Zhang Y, Gottumukkala AL, Nair MG (2010) In vitro evaluation of red and green lettuce (*Lactuca sativa*) for functional food properties. Food Chem 118:300–306
Nisha MC, RajeshKumar S (2010) Influence of arbuscular mycorrhizal fungi on biochemical changes in Wedilla chinensis (Osbeck) Merril. Anc Sci Life 29:26
Nuutila A, Kammiovirta K, Oksman-Caldentey K-M (2002) Comparison of methods for the hydrolysis of flavonoids and phenolic acids from onion and spinach for HPLC analysis. Food Chem 76:519–525
Okazaki K, Oka N, Shinano T, Osaki M, Takebe M (2008) Differences in the metabolite profiles of spinach (*Spinacia oleracea* L.) leaf in different concentrations of nitrate in the culture solution. Plant Cell Physiol 49:170–177
Ribas-Agustí A, Gratacós-Cubarsí M, Sárraga C, García-Regueiro JA, Castellari M (2011) Analysis of eleven phenolic compounds including novel p-coumaroyl derivatives in lettuce (*Lactuca sativa* L.) by Ultra-high-performance Liquid Chromatography with photodiode array and mass spectrometry detection. Phytochem Anal 22:555–563
Riedel H, Akumo DN, Saw NMMT, Kütük O, Neubauer P, Smetanska I (2012) Elicitation and precursor feeding influence phenolic acids composition in *Vitis vinifera* suspension culture. Afr J Biotechnol 11:3000–3008
Shen D (1997) Microbial diversity and application of microbial products for agricultural purposes in China. Agric Ecosyst Environ 62:237–245
Singleton VL, Orthofer R, Lamuela-Raventos RM (1999) [14] Analysis of total phenols and other oxidation substrates and antioxidants by means of folin-ciocalteu reagent. Methods Enzymol 299:152–178
Song X, Liu M, Wu D, Griffiths BS, Jiao J, Li H, Hu F (2015) Interaction matters: synergy between vermicompost and PGPR agents improves soil quality, crop quality and crop yield in the field. Appl Soil Ecol 89:25–34
Sultana B, Anwar F (2008) Flavonols (kaempeferol, quercetin, myricetin) contents of selected fruits, vegetables and medicinal plants. Food Chem 108:879–884
Sun B, Yan H, Zhang F, Wang Q (2012) Effects of plant hormones on main health-promoting compounds and antioxidant capacity of Chinese kale. Food Res Int 48:359–366
Świeca M, Gawlik-Dziki U, Kowalczyk D, Złotek U (2012) Impact of germination time and type of illumination on the antioxidant compounds and antioxidant capacity of Lens culinaris sprouts. Sci Hortic 140:87–95
Szaufer-Hajdrych M (2004) Phenolic acids in leaves of species of the *Aquilegia* L. genus. Herba Pol 50:50–54
Taie HA, El-Mergawi R, Radwan S (2008) Isoflavonoids, flavonoids, phenolic acids profiles and antioxidant activity of soybean seeds as affected by organic and bioorganic fertilization. Am Eurasian J Agric Environ Sci 4:207–213
Tandon V, Prakash A (1998) Influence of soil inoculation with VA mycorrhiza and phosphorus solubilising microorganisms on growth and phosphorus uptake in *Sesamum indicum*. Int J Trop Agric 16:201–209
Teixeira da Silva J, Egamberdieva D (2013) Plant-growth promoting rhizobacteria and medicinal plants. Recent Prog Med Plants 38:26–42
Verma SC, Ladha JK, Tripathi AK (2001) Evaluation of plant growth promoting

and colonization ability of endophytic diazotrophs from deep water rice. J Biotechnol 91:127–141

Vivas A, Marulanda A, Ruiz-Lozano JM, Barea JM, Azcón R (2003) Influence of a *Bacillus* sp. on physiological activities of two arbuscular mycorrhizal fungi and on plant responses to PEG-induced drought stress. Mycorrhiza 13:249–256

Wu S, Cao Z, Li Z, Cheung K, Wong M (2005) Effects of biofertilizer containing N-fixer, P and K solubilizers and AM fungi on maize growth: a greenhouse trial. Geoderma 125:155–166

Yang J, Kloepper JW, Ryu C-M (2009) Rhizosphere bacteria help plants tolerate abiotic stress. Trends Plant Sci 14:1–4

Yochum L, Kushi LH, Meyer K, Folsom AR (1999) Dietary flavonoid intake and risk of cardiovascular disease in postmenopausal women. Am J Epidemiol 149:943–949

Zablotowicz RM, Tipping EM, Lifshitz R, Kloepper JW (1991) Plant growth promotion mediated by bacterial rhizosphere colonizers. In: The rhizosphere and plant growth. Springer, pp 315–326

Złotek U, Świeca M, Jakubczyk A (2014) Effect of abiotic elicitation on main health-promoting compounds, antioxidant activity and commercial quality of butter lettuce (*Lactuca sativa* L.). Food Chem 148:253–260

6

A termite symbiotic mushroom maximizing sexual activity at growing tips of vegetative hyphae

Huei-Mei Hsieh[1], Mei-Chu Chung[1], Pao-Yang Chen[1], Fei-Man Hsu[1], Wen-Wei Liao[1], Ai-Ning Sung[1], Chun-Ru Lin[1], Chung-Ju Rachel Wang[1], Yu-Hsin Kao[1], Mei-Jane Fang[1], Chi-Yung Lai[2], Chieh-Chen Huang[3], Jyh-Ching Chou[4], Wen-Neng Chou[5], Bill Chia-Han Chang[6] and Yu-Ming Ju[1*]

Abstract

Background: *Termitomyces* mushrooms are mutualistically associated with fungus-growing termites, which are widely considered to cultivate a monogenotypic *Termitomyces* symbiont within a colony. *Termitomyces* cultures isolated directly from termite colonies are heterokaryotic, likely through mating between compatible homokaryons.

Results: After pairing homokaryons carrying different haplotypes at marker gene loci *MIP* and *RCB* from a *Termitomyces* fruiting body associated with *Odontotermes formosanus*, we observed nuclear fusion and division, which greatly resembled meiosis, during each hyphal cell division and conidial formation in the resulting heterokaryons. Surprisingly, nuclei in homokaryons also behaved similarly. To confirm if meiotic-like recombination occurred within mycelia, we constructed whole-genome sequencing libraries from mycelia of two homokaryons and a heterokaryon resulting from mating of the two homokaryons. Obtained reads were aligned to the reference genome of *Termitomyces* sp. *J132* for haplotype reconstruction. After removal of the recombinant haplotypes shared between the heterokaryon and either homokaryons, we inferred that 5.04% of the haplotypes from the heterokaryon were the recombinants resulting from homologous recombination distributed genome-wide. With RNA transcripts of four meiosis-specific genes, including *SPO11*, *DMC1*, *MSH4*, and *MLH1*, detected from a mycelial sample by real-time quantitative PCR, the nuclear behavior in mycelia was reconfirmed meiotic-like.

Conclusion: Unlike other basidiomycetes where sex is largely restricted to basidia, *Termitomyces* maximizes sexuality at somatic stage, resulting in an ever-changing genotype composed of a myriad of coexisting heterogeneous nuclei in a heterokaryon. Somatic meiotic-like recombination may endow *Termitomyces* with agility to cope with termite consumption by maximized genetic variability.

Keywords: Symbiosis, *Odontotermes formosanus*, *Termitomyces* mushroom, Meiotic-like, Basidiomycota

Background

Fungus-growing termites (Macrotermitinae, Isoptera) have practiced farming with their fungal symbionts of the basidiomycete genus *Termitomyces* R. Heim (Lyophyllaceae, Agaricales) as the cultivars since ca. 24–34 million years ago in the African rainforest (Aanen and Eggleton 2005; Roberts et al. 2016). Decomposing activity on plant matters of *Termitomyces* symbionts appears complemented by that of termite gut microbiota, which was suggested by analyzing annotated draft genomes from *Macrotermes natalensis* (Haviland) and its *Termitomyces* symbiont along with gut metagenomes from individuals of various termite castes (Poulsen et al. 2014). Fungus-growing termites, much as leaf-cutting ants, cultivate only one fungal symbiont in a colony (Aanen et al. 2009; Katoh et al. 2002; Mueller et al. 2005; Shinzato et al. 2005), which is widely accepted as a genotypic

*Correspondence: yumingju@gate.sinica.edu.tw
[1] Institute of Plant and Microbial Biology, Academia Sinica, Taipei 11529, Taiwan
Full list of author information is available at the end of the article

monoculture involving only one genotype of a particular *Termitomyces* species. This is in sharp contrast to genotypic polycultures and species polycultures, which are commonly adopted in human agriculture and have widely been perceived as better farming practices than monocultures, because diverse systems can be more productive, stable, and resistant to invasions than less diverse systems (Cook-Patton et al. 2011).

A *Termitomyces* mushroom (Fig. 1a–c) is cultivated by the only macrotermitine species known in Taiwan, black-winged subterranean termite *Odontotermes formosanus* Shiraki (Fig. 1e), at fungus gardens (Fig. 1c, d), which are built underground with fecal balls excreted by termite workers. Two types of spores are produced by the *Termitomyces* mushroom: while conidia are produced at fungus gardens, ingested by termite workers, and repeatedly mixed within fecal balls, basidiospores are only found on the gill surface of fruiting bodies, which emerge above-ground from fungus gardens in summer. Mutualistic relationship between *O. formosanus* and its *Termitomyces* symbiont can be demonstrated by rearing of *O. formosanus* in laboratory, which only becomes sustainable with incipient termite colonies artificially inoculated with spores of the symbiont (Fig. 1d). Associations between African macrotermitine termites and *Termitomyces* mushrooms have previously been established by artificially inoculating incipient termite colonies with *Termitomyces* spores (Johnson 1981; Johnson et al. 1981; Sands 1960).

A typical life cycle in the majority of basidiomycetes has a prolonged heterokaryotic (often dikaryotic) somatic stage, usually containing two genetically heterogeneous nuclei resulting from fusion of compatible homokaryons (Fig. 2a–f), which is governed by a self-incompatible mating system (Anderson and Kohn 2007; Brown and Casselton 2001; Raper 1966; Raudaskoski and Kothe 2010). Nuclei of heterokaryons stay unfused within every cell until formation of basidia at fruiting bodies, where the nuclei undergo fusion and meiosis (Brown and Casselton 2001). Four basidiospores are normally produced on the top of a basidium, each receiving a nucleus resulting from meiosis. A potentially great genetic variability thus exists among basidiospores, i.e., because they are products of meiosis, where genome-wide recombination can give rise to a great amount of genetic diversity in the offspring.

Here we show that the *Termitomyces* symbiont of *O. formosanus* possesses a life cycle different from other basidiomycetes in having a somatic sexual process involving karyogamy and meiotic-like nuclear division during cell division at active hyphal tips (Fig. 2g–m). The sexual process empowers its growing mycelium a hyperdynamic genotype composed of a myriad of nuclear haplotypes and renders the symbiont with immense genetic variability. The sexual process was confirmed by cytological studies, whole-genome sequencing, and real-time quantitative PCR of four meiosis-specific genes. Pairing tests indicated that mating system of the *Termitomyces* symbiont may not be governed by heterogenic compatibility as in the typical life cycle found in heterothallic basidiomycetes (Fig. 2a–f).

Methods

Fungal isolations

Termitomyces cultures were isolated from a fruiting body *T0984* collected from Nan-tou County, Taiwan, in 2009. The termite colony was not excavated to confirm its connection with *T0984*. However, by inoculating the conidia produced in culture by *T0984* into colonies of *O. formosanus* (see below), we were able to confirm the symbiotic relationship between the two partners. Two types of fungal cultures were obtained or derived from this fruiting body: a heterokaryon (*FB*) and homokaryons (*BS*). *FB* was isolated directly from the stipe tissue and served as the inoculum for laboratory-reared termite colonies, from which the fungus was reisolated by obtaining single conidia from guts of worker termites from the colonies and was confirmed to be *FB*. The *FB* culture was originally obtained by paring away the stipe surface with a heated razor blade, removing a piece of the stipe tissue from underneath by sterilized forceps, and placing the removed tissue on 2% Difco malt extract agar (MEA, 20 g/L Difco malt extract, 20 g/L Difco agar). Emerging hyphal tips were excised and transferred to fresh MEA. *BS* cultures were obtained from single basidiospores and were used for pairing tests. Those *BS* cultures were initiated from spreading onto MEA a large number of basidiospores, which were collected by placing the pileus of *T0984* on a sterilized plate for several hours, allowing basidiospores to fall from gills. Germination rates of the *Termitomyces* basidiospores were extremely low, and only one out of several hundred started to germinate after 2 days. A fine glass needle was used to gently guide germinating basidiospores bearing a short germ tube to clean areas on the agar, where agar blocks each overlain with a germinating basidiospore were excised and transferred to fresh MEA. We eventually obtained 43 *BS* isolates, whose growth rates varied greatly, 1–4 cm diam/month. Twenty of the *BS* cultures, i.e., *BS02*, *BS03*, *BS04*, *BS05*, *BS06*, *BS09*, *BS10*, *BS11*, *BS14*, *BS17*, *BS18*, *BS19*, *BS21*, *BS26*, *BS27*, *BS31*, *BS32*, *BS35*, *BS38*, and *BS39*, were randomly selected for subsequent pairing tests.

Termite rearing and fungal inoculation

To test if the *Termitomyces* culture (*FB*) that we isolated from the fruiting body *T0984* was indeed a symbiont of *O. formosanus*, laboratory-reared colonies of the termite

Fig. 1 *Termitomyces* mushroom and *Odontotermes formosanus*. **a** Surface of the pileus showing a central umbo. **b** Underside of the pileus showing white gills and a centrally attached stipe. **c** Soil profile to show the fruiting bodies emerging from a subterranean fungus garden. **d** Laboratory-reared termite colony at 3.5 years with a fungus garden ca. 7 cm diam. **e** Queen (*upper*) and king (*lower*) on the fungus garden shown in **d**. *Scale bars* in **a**– **d** = 5 cm, and **e** = 0.5 cm. **a**, **b** From the fruiting body *T0984*; **c** From National Museum of Natural Science, Taichung, Taiwan; and **d**, **e** from a termite colony inoculated with *T0984*

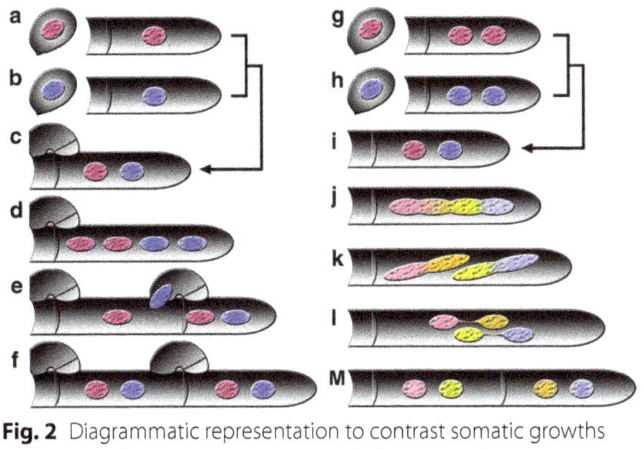

Fig. 2 Diagrammatic representation to contrast somatic growths between basidiomycetes with a typical life cycle and the *Termitomyces* symbiont. **a–f** Various stages of hyphal growth in a typical basidiomycete, whose heterokaryons (usually dikaryons), unlike those of *Termitomyces*, have a clamp connection at hyphal septa and dikaryotic mycelia, two commonly used indicators for successful matings. **a**, **b** Uninucleate homokaryons carrying different markers (*red* and *blue*) from single uninucleate basidiospores. **c** Binucleate heterokaryon resulting from pairing of two homokaryons. **d** Mitosis giving rise to four nuclei. **e** Formation of clamp connection, through which a blue nucleus migrates back to the penultimate cell, while two new septa are formed in the clamp cell and hyphal tip. **f** The final stage of hyphal cell division. **g–m** Various stages of *Termitomyces* hyphal growth. Note that a growing hyphal cell of a homokaryon or heterokaryon is binucleate or trinucleate. Only binucleate tip cells are shown here. **g**, **h** Binucleate homokaryons carrying different markers (*red* and *blue*) from single uninucleate basidiospores. **i** Binucleate heterokaryon resulting from pairing of two homokaryons. Homokaryons and heterokaryons cannot be discerned by nuclear numbers, both having two or three nuclei in a growing hyphal cell. **j** Karyogamy. **k** Stage resembling Meiosis I. **l** Stage resembling Meiosis II. **m** Two cells separated by a newly formed septum as the final stage

were initiated from alates captured during nuptial flight on the campus of Academia Sinica, Taiwan in June, 2009. Six dealates were placed in each 9-cm plastic Petri dish containing wet sifted soil to ensure the presence of both sexes. A chamber was created by the dealates at bottom of the soil within a few hours. Eggs were laid in 10 days. After the first brood of workers matured and emerged for foraging ca. 2.5 months after the nuptial flight, conidial suspension (ca. 10^5 conidia/mL) prepared from the culture *FB* (Table 1) was inoculated into 34 termite colonies, in which 21 formed fungus gardens within several days. They were survived by eight termite colonies after 6 years. The colonies were reared mainly with autoclaved twigs of Norfolk Island pine *Araucaria heterophylla* (Salisb.) Franco.

DNA extraction, PCR, cloning, and sequencing

All of the cultures that we used in this study were subjected to sequencing of their nuclear ribosomal internal transcribed spacers (*ITS*), mitochondrial intermediate peptidase (*MIP*), and pheromone receptor protein homologue (*RCB*). Mycelia were grown in 100 mL of malt extract broth (20 g/L Difco malt extract) on a rotary shaker at 120 rpm for 6–8 weeks. They were harvested by filtration through Whatman No. 1 filter papers, freeze-dried, and stored at −20 °C. Total DNA from a dried fungal mycelium was extracted by using a SLA-32 automatic nucleic acid extractor (Taiwan Advanced Nanotech Inc.) with a TANBead® Fungi DNA extraction kit. The three nuclear DNA loci, *ITS*, *MIP*, and *RCB* were amplified with the primer pairs *ITS1*/*ITS4* (White et al. 1990), *MAF*/*MAR*, and *MR8F*/*MR8R*, respectively, via polymerase chain reaction (PCR) in 96-well GeneAmpt® PCR System 9700 (Applied Biosystems, Foster City, California, USA) and sequenced with ABI Big-dye primer sequencing kit (Applied Biosystems). The primer pair and PCR condition of *ITS* followed the cited reference (White et al. 1990). See below for the primer pairs and the PCR condition of *MIP* and *RCB*. The resulting sequencing reactions were electrophoresed on an ABI 3730xl DNA sequencer. PCR products were cleaned with PCR-M™ clean-up system (Viogene-Biotek Corp., Hsi-chih, New Taipei City, Taiwan) following the manufacturer's protocol. While PCR products of homokaryons were sequenced directly, those of heterokaryons were subjected to cloning process using pGEM®-T vector System I (Promega, Madison, Wisconsin, USA) or Zero Blunt TOP10 plasmid vector (Invitrogen Inc., Carlsbad, California, USA). The plasmids were extracted with plasmid DNA extraction kit Mini Plus™ (Viogene-Biotek Corp.). Multiple clones of a PCR product were sequenced.

The PCR condition was the same for *MIP* and *RCB*: ca. 100 ng template DNA, 1.0 mM $MgCl_2$, 0.2 mM each dNTP, 0.2 μM each primer, and 1.25 U Easy-A High-Fidelity PCR Cloning Enzyme (Agilent Technologies, Santa Clara, California, USA) in a 25 μL reaction. Thermal cycling parameters were: an initial denaturation step at 95 °C for 2 min followed by 30 cycles of denaturing at 95 °C for 40 s, annealing at 57 °C for 30 s, extension at 72 °C for 1 min, and a 7 min final extension at 72 °C. The *MIP* locus used in the current study was a shorter version of that published in (James et al. 2004). The primer pair *MAF*/*MAR* (5′-AATGTGTCTGGTACCCGTTGTCCC-3′/5′-ATGTCCAAACTGGGTCTGGAAAGAT-3′) was designed to frame a 327 bp fragment, where 13 SNP sites were available. The *RCB* locus was amplified with the primer pair *MR8F*/*MR8R* (5′-AACCTGCATACCCTCTCTTCCCTAT-3′/5′-GTAATCCAACGGAAGCACCAAGCAT-3′), which framed a 289 bp fragment with seven SNP sites available. The reverse primer *MR8R* was located within *RCB1* (Aimi et al. 2005), whereas the forward primer *MR8F* was located at a 5′-untranslated region, which

Table 1 Heterokaryons resulting from pairings between 20 homokaryons, which were sorted into four genotypes, i.e., A_1B_1, A_2B_2, A_1B_2, and A_2B_1, by haplotypes at *MIP* and *RCB*

	A_1B_1 x A_2B_2	A_1B_2 x A_2B_1	A_2B_2 x A_2B_1 (common at *A*)	A_1B_2 x A_2B_2 (common at *B*)
Mating pairs	*BS11* × *BS09*, *BS11* × *BS32*, *BS18* × *BS09*, *BS18* × *BS27*, *BS18* × *BS32*, *BS39* × *BS09*, *BS39* × *BS32*	*BS02* × *BS03*, *BS17* × *BS03*, *BS17* × *BS05*, *BS17* × *BS10*	*BS04* × *BS31*, *BS06* × *BS35*, *BS09* × *BS31*	*BS14* × *BS19*, *BS14* × *BS38*, *BS21* × *BS19*, *BS21* × *BS32*, *BS21* × *BS38*, *BS26* × *BS19*, *BS26* × *BS32*, *BS26* × *BS38*
MIP (A)	+	+	ND	+
RCB (B)	+	+	+	ND

For each of the four pairing groups, we randomly selected three homokaryons of a genotype to pair with three homokaryons of another genotype. Note that there were nine mating pairs within each pairing group, but only confirmed heterokaryons are listed

+ Both haplotypes detected, *ND* not determined because of identical haplotype

was reached by chromosome walking from *RCB1*. The chromosome walking was done with TaKaRa LA PCR InVitro Cloning Kit (TaKaRa Bio Inc., Otsu, Shiga, Japan). Template DNAs were prepared according to the manufacturer's instructions and were subjected to digestion with *Sal*I. The digested DNA fragments were then ligated with nucleotide linkers and used as templates for two rounds of cassette amplification by PCR, which was conducted with two primer pairs *SU1* (5′-GCATCGTTTGCCCAGACGACAGAA-3′)/*SU2* (5′-ACGCCATGTAGAAACAAGTTCCTG-3′) and *SU7-1* (5′-GCGTCGTTCGCCCAGACGACAGAA-3′)/*SU7-2* (5′-ACGTCATGTAGAAACAAGTTCCTG-3′). After cassette amplification, all amplified DNA fragments were purified, cloned, and sequenced.

DNA marker loci for pairings and pairing tests

MIP and *RCB* were used as marker loci to confirm whether or not a pairing was successful, because *Termitomyces* species lack phenotypic markers. In other basidiomycetes, *MIP* is closely linked to mating locus *A* (James et al. 2004), whereas *RCB* is located within mating locus *B* (Aimi et al. 2005).

Pairings of the 20 homokaryotic *BS* cultures were conducted for pairs carrying different haplotypes at both *MIP* and *RCB* (A_1B_1 × A_2B_2 and A_1B_2 × A_2B_1), at *MIP* only (A_1B_2 × A_2B_2, i.e., common-B), and at *RCB* only (A_2B_2 × A_2B_1, i.e., common-A). Genotypes of the 20 homokaryons were determined and sorted into A_1B_1, A_2B_2, A_1B_2, and A_2B_1. For each of the four pairing groups shown in Table 1, we randomly selected three homokaryons of a genotype to pair with three homokaryons of another genotype, i.e., nine pairs in each pairing group. Homokaryons of a pair were placed 1 cm apart on a 6-cm petri plate containing MEA. Agar blocks containing hyphae at the confrontation line were isolated after 1 month for molecular typing. For homokaryons of those pairs with different haplotypes at both *MIP* and *RCB*, successful matings would result in heterokaryons within which both haplotypes of *MIP* and both of *RCB* coexist. In those cases where paired homokaryons carrying different haplotypes at only one of the two loci, resulting heterokaryons would be expected to carry either both haplotypes of *MIP* or both of *RCB*.

Nuclear staining

Nuclei of basidiospores, hyphae, and conidia were stained directly with DAPI (4′,6-diamidino-2-phenylindole in an antifade solution, Vector Laboratories, California, USA). Basidiospores were collected from the fruiting body *T0984*. Hyphae and conidia were obtained from culturing mycelia of *FB*, *BS*s, and paired *BS*s (Table 1) in 250 mL Erlenmeyer flasks, each of which contained a 100 mL liquid medium consisting of 3 g/L Bacto beef extract, 5 g/L Bacto peptone, and 15 g/L D-glucose, and incubated without shaking at 25 °C for 3 week. The stained fungal components were examined by fluorescence microscopy with a Zeiss LSM510 Meta confocal microscope equipped with C-Apochromat 63 ×/1.2 W corr objective. Two channels were used: DAPI channel (specifications: Two Photo Laser Excitation: 740 nm; Transmission 1.5%; Main Beam Splitter: KP650; Beam Splitter 2: mirror; BP 390–415IR; Detector Gain: 650; Amplifier Offset: −0.1) and DIC channel (specifications: Two Photo Laser Excitation: 740 nm; Detector Gain: 130; Amplifier Offset: −0.1).

Hyphae and conidia were also subjected to HCl-Giemsa stains (Giemsa's azur eosin methylene blue solution, Merck KGaA, Darmstadt, Germany) in accordance with published protocols (Knox-Davies and Dickson 1960; Ward and Ciurysek 1962) with slight modification and is summarized as follows. A harvested mycelium was washed with MQ water, fixed in a freshly prepared Farmer's fluid (95% ethanol: glacial acetic acid = 3:1) for 1 h, and successively washed in 95 and 70% ethanol, and immersed in MQ water for 10 min. It was then placed in 1 N HCl at room temperature for 10 min, hydrolyzed in 1 N HCl at 60 °C for 8 min, washed in MQ water with 5 changes, suspended in phosphate buffer (0.2% KH_2PO_4,

0.4% $Na_2HPO_4 \cdot 7H_2O$, pH 7.2) for 5 min, and stained in HCl-Giemsa for 15 min or longer if required. The stained mycelium was washed thoroughly with MQ water and phosphate buffer in series, and was mounted in MQ water. The stained fungal components were examined by light microscopy with a Leica DMRB microscope equipped with a PL Fluotar 100 ×/1.30 oil objective.

Slide culture technique was also used with small blocks (ca. 0.5 × 0.5 cm) of Difco oatmeal agar (OA) (BD, Sparks, Maryland, USA), each of which was inoculated with a culture, sandwiched between a cover slip and a microscope slide, and grown for 2 week. The hyphal mats and conidia that attached on the cover slip were then stained with DAPI and HCl-Giemsa as described above.

Fluorescence in situ hybridization (FISH)

Hyphae from cultures of the heterokaryon *BS05* × *BS17* and the homokaryon *BS05* were harvested and fixed in freshly prepared Farmer's fluid overnight. Hyphae were submerged within a drop of 45% acetic acid while being evenly spread onto a microscope slide prior to air-dried. Cell wall digestion was performed with 100 mg/mL lysing enzyme from *Trichoderma harzianum* (Sigma-Aldrich, St. Louis, Missouri, USA) dissolved in PME (50 mM PIPES, pH 6.7; 25 mM EGTA, pH 8; 5 mM $MgSO_4$) for 2 h at 37 °C. Probe DNA was prepared with the *ITS1-5.8S rDNA-ITS2* sequence of the *FB* culture amplified and labelled with digoxigenin-11-dUTP by using PCR DIG Labelling Mix (Roche Diagnostics GmbH, Penzberg, Germany). The PCR condition is as described above for *ITS*. The FISH procedure (Chung et al. 2008) is briefly described as follows. The nuclei on the microscope slide were denatured in 70% deionized formamide in 2× SSC (0.15 M NaCl, 0.015 M sodium citrate, pH 7.5) at 80 °C for 60 s. Hybridization mixture was prepared from 50% deionized formamide, 2× SSC, sheared salmon sperm DNA (1 μg/μL), 10% dextran sulfate, and probe DNA (10–100 ng in 20 μL per microscope slide). Hybridization was performed in a moisture chamber at 37 °C overnight. Hybridization signals were visualized using a rhodamine-conjugated anti-digoxigenin antibody (Roche Diagnostics GmbH). The nuclei were counterstained with DAPI.

For a fungal culture, FISH preparations were conducted for more than ten times, from each of which eight microscope slides were resulted. More than 20 images of a nuclear stage, i.e., prophase, metaphase, anaphase, or telophase, with clear FISH signals were taken from each FISH preparation.

PCR-free library construction and high throughput sequencing

Genomic DNAs were extracted from dried mycelia of two homokaryons (*BS05* and *BS17*) and a heterokaryon (*BS05* × *BS17*) by using a SLA-32 automatic nucleic acid extractor with a TANBead® Fungi DNA extraction kit. Approximately 1 μg of a genomic DNA was sonicated to ~350 bp with the Covaris M220 system using the following parameters: duty factor 20%, peak/display power 50 W, 200 cycles per burst, 65 s, 20 °C. The sonicated DNA was purified, end-repaired, A-tailed and ligated with adaptors according to the manufacture's guide (Illumina Truseq DNA PCR-Free LT library prep kit) followed by purification and sequencing. A paired-end DNA-seq library was sequenced using Illumina HiSeq 2500 platform per manufacturer's instructions. Sequencing was performed up to 250 cycles. Image analysis and base calling were performed with the standard Illumina pipeline. In total 17,621,957, 23,362,228, and 18,348,369 reads were aligned to the reference genome *Termitomyces* sp. *J132* (Poulsen et al. 2014), with the estimated depths of 32.3×, 42.8×, and 33.6× for *BS05* × *BS17*, *BS05*, and *BS17*, respectively.

Alignments, heterozygous SNP calling, detection of recombinants, and haplotype inference

After the adapter trimming by cutadapt 1.8.1 (Martin 2011), the reads were aligned to the reference genome of *Termitomyces* sp. *J132* (Poulsen et al. 2014), downloaded from http://gigadb.org/dataset/100056, by using Bowtie 2 with local alignment mode to optimize the mapability (Langmead and Salzberg 2012). Alignments of the homokaryon and heterokaryon samples are publicly available at NCBI as *Study SRP073090*.

To identify potential recombinants, we first searched for heterozygous SNPs with two alleles, as a proxy to the parental alleles. During the SNP calling, the alignments with low mapping quality (MAPQ score <10) were skipped. A heterozygous SNP was identified when the following conditions were met: the depth was ≥10 reads, the two major alleles accounted for ≥90% of the reads, and the difference between them was ≤20%. In total we found 71.5, 60.9, and 74.7 K heterozygous SNPs in *BS05* × *BS17*, *BS05*, and *BS17*, respectively. Due to the high variability of fungus genomes, a large number of the SNPs were likely due to the genetic difference between our strains and the reference strain. We further filtered out the SNPs from the heterokaryon that were also detected in either of the two homokaryons.

Reads that had haplotypes phased by at least five heterozygous SNPs were subjected to haplotype inference. The two most frequently observed haplotypes were considered parental haplotypes, whereas other haplotypes were recombinants. These haplotypes are summarized in Table 2 and are listed in the Additional file 1.

An example of haplotype inference is presented here to demonstrate how we inferred haplotypes from the

PCR-free libraries that we constructed for the heterokaryon ($N \times N$). The screenshot of Integrative Genomics Viewer (IGV) (Robinson et al. 2011) in Fig. 3 shows an alignment of 32 reads at scaffold 278 of the filtered heterokaryon, where 14 reads of one parental haplotype, 14 reads of the other parental haplotype, and two reads of two recombinant haplotypes have been aligned to the reference (indicated by a black arrow). The five heterozygous SNPs at positions 51,054, 51,057, 51,063, 51,069, and 51,071 are framed in black rectangles. Five incomplete reads are crossed out in black lines.

Reverse transcription PCR (RT-PCR) and real-time quantitative PCR (RT-qPCR)

Total RNA was isolated from a mycelial sample and a gill sample using TRIzol (Invitrogen Inc.) following the manufacturer's protocol. cDNA synthesis was performed using the First-Strand cDNA Synthesis Kit (Roche Diagnostics GmbH) and anchored-oligo(dT)18 primers. RT-qPCR reactions were performed using FastStart Universal SYBR Green Master (Roche Diagnostics GmbH). Translation elongation factor 1 alpha (*TEF1α*) was used as the internal normalization control with the primer pair GGTGTCAGGCAGCTCATCGT/CGGTCCTCACTCCACTTGGT. PCR was performed on the QuantStudio 12K Flex ABI 7500 real-time PCR system in 8-tube strips. The exon-spanning primer pairs used for RT-qPCR analyses of *SPO11*, *DMC1*, *MSH4*, and *MLH1* are: CCAGCCACGAAACGAGACAT/GCTCAAGAGTGGCAGCGATAT, AAATATTCTCTACGCCCGTGCAT/AAGTCCTTGTCCTCGGCAAA, GCGCCTTGCGAAGATAGAGA/CGATTGCGATGGGCCTTA, and TCTGCACTGAAAGAGCTGGTAGA/GGTCCGATTTGCGTATTCCA, respectively. The RT-qPCR reaction was performed in triplicate for each sample. Specificity of the amplifications and Ct values were analyzed using ABI PRISM dissociation curve analysis software.

Results

Pairing tests

The mating system of *Termitomyces* mushrooms has not been verified by pairing studies previously, being inferred heterothallic only in a study of the *Termitomyces* symbiont of *Macrotermes natalensis* from South Africa (de Fine Licht et al. 2005). We set out to pair 20 homokaryons (Table 1), which were obtained from the fruiting body *T0984* produced by a *Termitomyces* symbiont of *O. formosanus* in Taiwan. The molecular markers that we used had two haplotypes (alleles) at each of the two gene loci, mitochondrial intermediate peptidase (*MIP*), which is closely linked to mating type locus *A* in other studied mushrooms (James et al. 2004), and pheromone receptor (***RCB*), which is a region within mating type locus *B* in other mushrooms (Aimi et al. 2005; Brown and Casselton 2001). Successful matings would render both haplotypes of *MIP*, denoted as A_1 and A_2, and both of *RCB*, denoted as B_1 and B_2, coexisting within a resulting heterokaryon. Cloning and sequencing amplicons of the two loci from the resulting heterokaryons confirmed that the two original haplotypes had come together in a heterokaryon as we expected from successful matings (Table 1). In other basidiomycetes, mating behavior dictated by two loci is considered tetrapolar heterothallism, where compatibility between paired homokaryons requires different haplotypes at each locus, whereas incompatibility occurs when only one or both of the two loci has the same haplotype (Brown and Casselton 2001; Raper 1966). Interestingly, pairings between homokaryons of the *Termitomyces* symbiont that were heterozygous at either *MIP* only ($A_1B_2 \times A_2B_2$) or *RCB* only ($A_2B_2 \times A_2B_1$) also resulted in heterokaryons with a normal mycelial growth,

Table 2 Recombinants in the homokaryons and heterokaryon

	BS05		*BS17*		*BS05* x *BS17*[a]		Filtered *BS05* x *BS17*[a]	
	Number of reads	Percentage (%)	Number of reads	Percentage (%)	Number of reads	Percentage (%)	Number of reads	Percentage (%)
Non-recombinants (2 haplotype)	1,832,757	80.28	1213,986	80.71	1450,156	81.41	401,623	94.96
Recombinants (30 haplotypes)	450,166	19.72	290,056	19.29	331,170	18.59	21,336	5.04
Total	2282,923	100	1504,042	100	1781,326	100	422,959	100

The table is a summary for haplotypes phased by at least 5 heterozygous SNPs within a read. The 32 ($=2^5$) haplotypes, i.e., two non-recombinants and 30 recombinants, were inferred from every 5 heterozygous SNPs on the same reads

[a] The heterokaryon *BS05* × *BS17* resulted from a mating between the two homokaryons *BS05* and *BS17*. The filtered *BS05* × *BS17* was obtained by excluding those recombinant haplotypes of *BS05* × *BS17* that were shared by *BS05* or *BS17* in the haplotype inference (see "Methods" for detail)

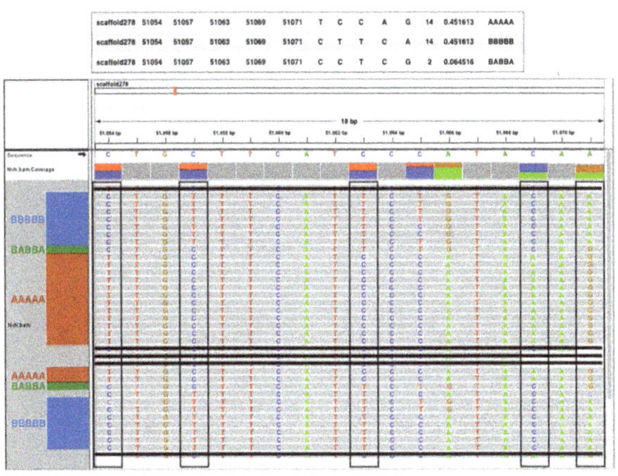

Fig. 3 An IGV screenshot showing inference of somatic meiotic-like recombination at scaffold 278

suggesting that self-incompatible mating is not strictly followed by or even lacked in the *Termitomyces* symbiont.

Nuclear behavior revealed by cytological studies

Curiously, cytological studies on heterokaryons resulting from successful pairings indicated that meiotic-like nuclear behavior occurred somatically in *Termitomyces*. Nuclei in a mycelium stained with HCl-Giemsa or DAPI were shown to be in a dikaryotic or trikaryotic arrangement in a newly formed hyphal tip cell (Fig. 4a–h). Nuclei in the cell fused into a single nucleus before dividing into four and six in the dikaryotic and trikaryotic cells, respectively. A septum was then formed to bisect the divided nuclei, resulting in two or three nuclei per cell depending on the nuclear number of their parent cells. The nuclear region distributed with nuclear ribosomal RNA genes (rDNA), which are organized in tandem arrays of multiple copies, was probed by fluorescence in situ hybridization (FISH) (Fig. 4m–x). A chromosome set had one FISH signal only as demonstrated in uninucleate basidiospores (Fig. 4m). Multiple FISH signals were frequently observed in close association within a nucleus, suggesting that pairing of homologous chromosomes indeed occurred (Fig. 4n–q, t). While single FISH signal was normally detected in postmeiotic nuclei at hyphal tips (Fig. 4v), nuclei with two FISH signals (Fig. 4x) or more were occasionally found. The same nuclear behavior was also observed in the culture isolated from stipe tissue of the fruiting body *T0984* (*FB*) and the cultures reisolated from termite colonies inoculated with *FB*.

Cytological studies were also extended to homokaryons derived from single basidiospores, which were uninucleate but germinated into mycelia each containing both dikaryotic and trikaryotic cells. Interestingly, the hyphal cells divided in exactly the same fashion as those of heterokaryons. The sexual process revealed in the homokaryons is similar to the same-sex mating reported from the human pathogenic fungus *Cryptococcus neoformans* (Lin et al. 2005).

Conidia are commonly perceived as asexual spores of fungi, but in the *Termitomyces* symbiont, regardless of whether conidia were produced from heterokaryons or homokaryons, their nuclei, two or three per conidium, underwent karyogamy and a meiotic-like process as those of active hyphal cells (Fig. 4i, k). *Termitomyces* conidia, unlike those of other fungi which serve as clonally propagating and dispersal agents, are produced within nodules at fungus gardens within sequestered termite colonies. Nodules, which contain thin-walled, inflated cells and hyphae in addition to conidia, are consumed by termite workers as an important part of the termite diets (Batra and Batra 1979). Conidia appear to be the only fungal elements surviving the alimentary tracts. Fecal balls excreted and deposited onto fungus gardens by worker termites thus mainly contain conidia and ingested foraged plant debris. Conidia germinate within hours after fecal ball deposition to resume the fungal growth at fungus gardens.

Occurrence of genetic recombination at somatic stage

Recombination of homologous nuclear DNA does not normally occur within a lone heterokaryon in other basidiomycetes. To detect if genetic recombination occurred during vegetative growth in the *Termitomyces* symbiont, we performed whole-genome sequencing of two homokaryons *BS05* and *BS17* and a heterokaryon *BS05* × *BS17*, which resulted from mating of the two homokaryons, with an Illumina HiSeq 2500 platform. The sequencing libraries were deliberately constructed without involving PCR amplifications to avoid PCR-mediated recombination (see Additional file 2: Text S1), and obtained reads were aligned to the reference genome *Termitomyces* sp. *J132* (Poulsen et al. 2014). To identify recombinants among the reads, we reconstructed haplotypes by phasing heterozygous SNPs in close proximity. Estimated recombinants from *BS05*, *BS17*, and *BS05* × *BS17* were 19.72, 19.29, and 18.59%, respectively (see Table 2; Additional file 1). The recombinants found in the homokaryons possibly resulted from long-term accumulation of gene duplication and/or repetitive sequences within a genome, which is known to be rich in fungal genomes (Raffaele and Kamoun 2012), or from chromosomal rearrangements (Heitman et al. 2013; Perkins 1997) (see *Possible account for recombination events occurring in the homokaryon* of "Discussion"). We thus excluded the recombinant haplotypes from alignments of *BS05* × *BS17* that were shared with *BS05* or *BS17*. Estimated recombinants in the resulting alignments of the total haplotypes were 5.04% (Table 2). Genetic recombination occurred genome-wide, with a slight, less than

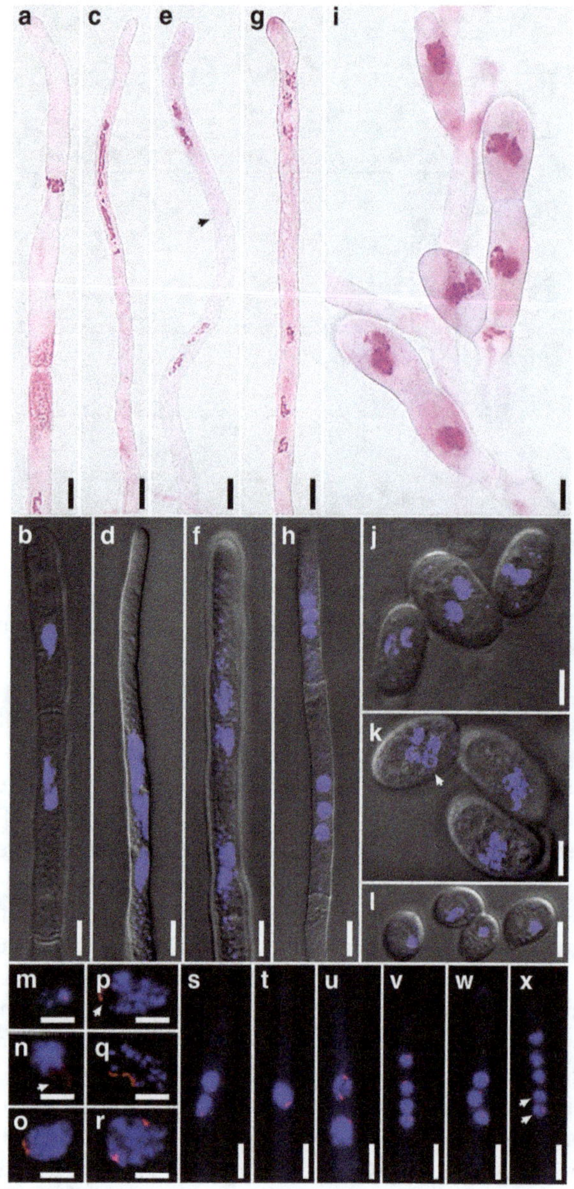

Fig. 4 Nuclei at various stages of the *Termitomyces* symbiont. **a–h** Vegetative hyphal tips. Nuclei in **a**, **c**, **e**, **g**, and **i** were stained with HCl-Giemsa, whereas those in **b**, **d**, **f**, and **h** were with DAPI. **a**, **b** A fused nucleus in the terminal cell. Nuclei of the penultimate cell in **b** are about to fuse. **c**, **d** Elongated nuclei at the stage resembling Meiosis I. **e**, **f** Nuclei resulting from a binucleate parent cell right after the stage resembling Meiosis II. Septum (*arrowhead*) is forming to bisect the four nuclei in **e**. **g**, **h** Nuclei resulting from a trinucleate parent cell right after the stage resembling meiosis II. **i** Conidiophore attached with developing conidia, each of which has a fused nucleus or two nuclei ready to fuse. **j**, **k** Conidia. Those in **j** are binucleate, whereas those in **k** are either uninucleate or quadrinucleate (*arrowhead*). **l** Uninucleate basidiospores. **m–x** Nuclei labelled with *red* FISH probes at *ITS1-5.8S rDNA-ITS2* region and counterstained with DAPI to render the nuclei *blue*. **m** Haploid nucleus of a basidiospore showing a FISH signal. **n–r** Nuclei at various stages of the meiotic-like process released from hyphae. **n** Nucleus at prophase showing FISH signals on four stringy chromatids attached to a nucleolus (the *dark region* pointed by an *arrowhead*). **o** Nucleus at prophase. **p** Nucleus at early metaphase showing FISH signals on four condensed chromatids attached to a nucleolus (the *dark region* pointed by an *arrow*). **q** Nucleus at metaphase, with chromosomes shown as condensed *blue spots*. **r** Nucleus at anaphase. **s–x** Nuclei at vegetative hyphal tips, which appear hazy owing to cell wall being partially digested by enzymes. **s** Two haploid nuclei before karyogamy. **t** Fused nucleus at prophase. **u** Two nuclei at a stage resembling anaphase of meiosis II. **v** Four nuclei after a meiotic-like division. **w** Three haploid nuclei before karyogamy. **x** Aberrant hyphal tip showing five postmeiotic nuclei with two (*arrowheads*) still active in division. *Scale bars* in **a–m**, **s–x** = 5 μm, and **n–r** = 2.5 μm

twofold enrichment and depletion at genic and intergenic regions, respectively (Fig. 5).

Expression of meiosis-specific genes

To further test if the genetic recombination occurring in mycelia is meiotic-like, four meiosis-specific genes, including *SPO11*, *DMC1*, *MSH4*, and *MLH1*, which are not normally expressed at somatic stage in other organisms (Schlecht and Primig 2003), were selected for examination in a mycelial sample. Fruiting bodies of a typical mushroom species have thin, vertically oriented, plate-like gills beneath the cap, which are entirely overlain with crowdedly arranged basidia. Meiosis is commonly considered to be restricted to basidia (Anderson and Kohn 2007; Kües 2000). A gill sample was thus included and compared with the mycelial sample for expression of the four meiosis-specific genes. RNA transcripts of the four genes from the two samples were first detected by reverse transcription PCR (RT-PCR) (Fig. 6a), and the confirmed RNA transcripts were cloned and used for designing primers for real-time quantitative PCR (RT-qPCR) analyses. Results of RT-qPCR showed that all four meiotic genes were expressed in the mycelial sample (Fig. 6b). Although higher expression of these four meiotic genes was detected in the gill sample, successful amplification and <30 threshold cycle value (Ct) in the mycelial sample confirmed their expression in mycelia. The difference in expression levels between mycelia and gills may be due to cells (i.e., basidia) undergoing meiosis in the gills being higher in percentage than those (i.e., only hyphal tips) in the mycelia. While gene expressions do not always correspond to gene functions, it should be noted that meiotic genes, especially genes involving recombination, are normally under tight expression control to prevent deleterious consequences of genomic instability.

Discussion

Sexual process in the *Termitomyces* mushroom

The novel life cycle that we present here for the *Termitomyces* symbiont represents a highly interesting case in evolution of sex, where the fungus maximizes the sexual process by vigorously reshuffling its genetic makeup in a meiotic-like fashion during each somatic cell division and conidial production irrespective of whether or not mating has occurred. It is to be found out whether or not this process continues in tissue of the fruiting body and terminates at basidia. Basidia are the only sites of sexual activity known in other basidiomycetes, where sex (karyogamy and meiosis) and reproduction (basidiospore formation) are tightly coupled (Aanen and Hoekstra 2007). Sex of the *Termitomyces* symbiont prevails at somatic stage and looks as if uncoupled with reproduction process except for the brief duration of basidiospore production. Nonetheless, growth of vegetative hyphae in fungus gardens can readily be diverted to conidial production after colonization of fecal balls, and conidia are sexual rather than asexual. The coupling between sex and reproduction in the *Termitomyces* symbiont is not tight, but more frequent than what it appears by way of conidial production.

Sexual processes have been widely regarded as being costlier than asexual ones (Barton and Charlesworth 1998; Bell 1982; Burt 2000; Maynard Smith 1971; Otto 2009; Williams 1975). The "cost of meiosis" or "cost of sex" remains a puzzling question in evolutionary biology despite the fact that sex in general has been accepted by many for generating genetically variable progeny, upon which natural selection can act to accelerate evolution for adaption and for eliminating deleterious mutations. Nests of fungus-growing termites are sequestered environments, where *Termitomyces* fungi are essentially sheltered from outside (Mueller et al. 2005). The nests are generally considered homeostatic with fine-tuned microclimates, which are characterized by high moistures, elevated temperatures ca. 30 °C, and high CO_2 concentrations, in favor of *Termitomyces* growth (Korb 2003; Thomas 1987; Turner 2004) but stressful to other microbes, which are relatively low in diversities (Cooke and Whipps 1987). If the nest environments are in homeostasis and if sex is for adaptation, what could be the driving force propelling the *Termitomyces* symbiont to keep sexuality over asexuality? Mycelia and nodules of the symbiont are frequently and constantly under termite consumption, which undoubtedly imposes an immediate, continuous pressure upon the symbiont, whose cellular components, notably nuclei and DNA-containing organelles, will be digested if they fail to enter indigestible conidia. Conidia are thus like Noah's ark in that only a certain number or percentage of cellular components can gain entrance into them to ultimately survive the termite gut passage. Strong competition in entering conidia among rivalling cellular components may exist and exert a direct driving force upon the symbiont to produce conidia high in genetic variability as opposed to clonality in order to adapt to the next round of termite consumption.

Choice of using FISH in detecting ploidy level of nuclei at hyphal tips

Flow cytometry has been employed extensively in detecting ploidy level of nuclei. However, using this method to study nuclei located at hyphal tips is extremely difficult, if not infeasible, unless hyphal tip cells could be excised and collected selectively to pass through the narrowing channel of a flow cytometer. FISH with rDNA (ITS1-5.8S rDNA-ITS2) as the probe DNA, on the other hand, is a useful method for visualizing and determining ploidy level of nuclei at hyphal tips, where the meiotic-like divisions occur. The images shown in Fig. 4m–x are representatives at various nuclear stages observed during our FISH experiments. For example, the nucleus at Fig. 4n was unmistakably identified as a diploid at prophase because it had four FISH signals on four stringy chromatins, which means a pair of 45S rDNA in a diploid genome undergoing a duplication process at this stage. Nuclei in Fig. 4v–x were haploid resulting from meiotic-like division because there was only one FISH signal detected in each of the nuclei.

Possible account for recombination events occurring in the homokaryon

Fungal genomes are highly dynamic, rich in chromosomal rearrangement (Heitman et al. 2013; Perkins 1997). In the *Termitomyces* symbiont, recombinants were detected not only in the heterokaryon *BS05* × *BS17* but

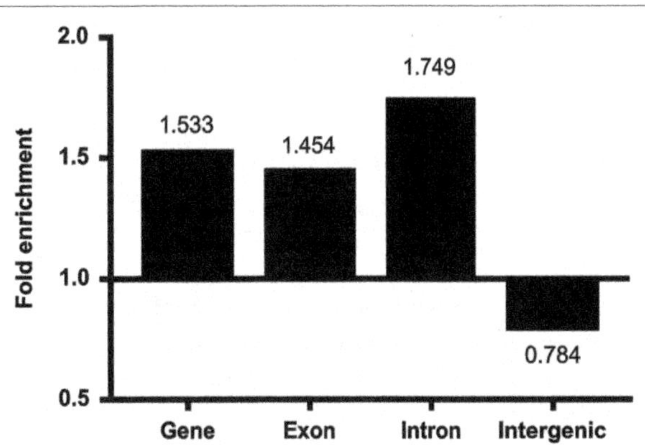

Fig. 5 Distribution of recombinants in different genomic regions. The fold enrichment was calculated by comparing the percentage of the recombinants in a specific genomic region against the percentage of the region in the genome

also in the homokaryon *BS05*. We excluded the possibility that the recombinants in the homokaryon sample originated from contaminants by checking reads at four protein-coding gene loci listed in Additional file 2: Table S2, where only one parental haplotype was detected at each locus.

To explain how recombinants could have arisen in the homokaryon but not in the heterokaryon and vice versa, we use a schematic diagram (Fig. 7) to demonstrate that one unique recombination was detected in each of the heterokaryon and the homokaryon and that one recombination was shared by both. At step 1, two parental haploid nuclei, P1 and P2, fused to form a diploid nucleus (*P1 × P2*), with a reciprocal recombination occurring at each of the two chromosome sets, i.e., one between loci c and d, and the other between loci y and z. At step 2, four daughter nuclei, D1, D2, D3, and D4, were produced after the meiotic-like process. At step 3, nuclei D1 and D2 fused, and the chromosome arm carrying alleles c_1, d_2, e_2, f_2 was translocated onto the chromosome carrying alleles w_1, x_1, y_1, and z_1. Among the meiotic products, nuclei N1 and N2 require further attention, because N1 acquired the chromosome arm that was lost in N2. At step 4, N1′, a nucleus derived from N1, was encapsulated within a basidiospore, which developed into homokaryon I. Within homokaryon I, F1 was developed from N1′ through the intermediate state N1′–F1, where a non-homologous recombination occurred between loci d and e of the two chromosomes. When homokaryon I mated with homokaryon II, only nucleus N1″, which derived from N1′, migrated to the heterokaryon (step 5), where it paired with nucleus D3″, which derived from nucleus D3' of homokaryon II (step 7). Nucleus D3' had its origin from nucleus D3 (step 6).

Recombination sites at the heterokaryon that resulted from plasmogamy between homokaryon I and homokaryon II were detected between loci c and d and between loci y and z, while those at homokaryon I were between loci c and d and between loci d and e. In the heterokaryon, the recombination site between loci y and z is considered reliable but the site between loci c and d was filtered out because it was shared by homokaryon I and the heterokaryon. The recombination site between loci d and e was unique to homokaryon I, becoming possible after the chromosome translocation event at step 3.

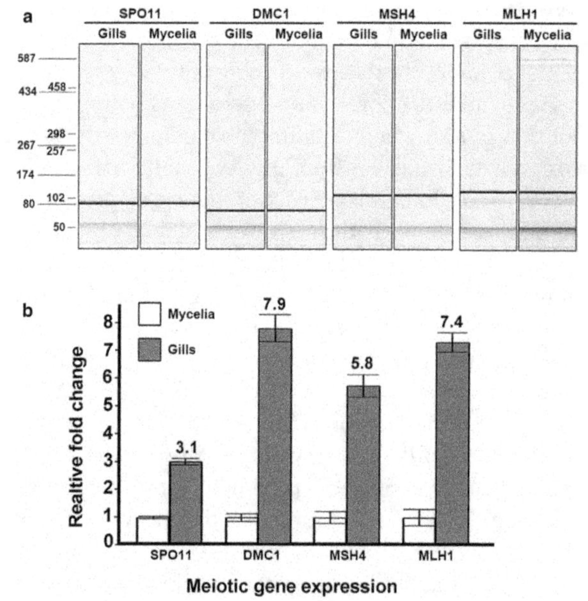

Fig. 6 Expression of four meiotic genes in both mycelia and gills. In basidiomycetes, meiosis is commonly considered to be restricted to basidia, which are crowdedly arranged as a palisade-like layer on the surface of gills. **a** The RT-PCR results detected by the high-resolution capillary electrophoresis. The size markers in the RNA ladder on the *left* are in bp. **b** Gene expression determined by RT-qPCR using exon-spanning primers. The expression of translational elongation factor 1 alpha (*TEF1α*) was used as a reference to normalize the data

Mating system of the *Termitomyces* symbiont not dictated by heterogenic compatibility

Two basic mating systems are known in mushrooms: homothallic (self-fertile) and heterothallic (self-sterile). Only 10% of the mushroom species have a homothallic mating system and can complete a life cycle without mating. The others have a heterothallic mating system and require fusion between mycelia with different mating types, which, depending on mushroom species, are determined by either one locus (bipolar heterothallism) or two loci (tetrapolar heterothallism) (Whitehouse 1949). Compatibility can only be achieved when the two mates have different alleles at each locus, i.e., heterogenic compatibility (Esser 2006).

MIP and *RCB*, as initially planned, were used as markers to suggest the mating system of *Termitomyces*. These markers were assumed to be associated with the mating type loci *A* and *B* (Aimi et al. 2005; James et al. 2004), with the mating system of the *Termitomyces* symbiont being assumed to be tetrapolar heterothallism. This assumption was thus made simply because there happened to be exactly two haplotypes for each of the two marker loci among the 45 homokaryons that we obtained from *T0984*. While capability of mating between some homokaryons has been verified in our study, it remains uncertain if mating is indeed a necessity for the *Termitomyces* symbiont to complete a life cycle. Definite answers will be possible only after a reliable fruiting system becomes available. If mating is eventually proven to be required, only heterokaryons will produce fruiting bodies, while homokaryons remain fruitless. Otherwise, both homokaryons and heterokaryons can fruit.

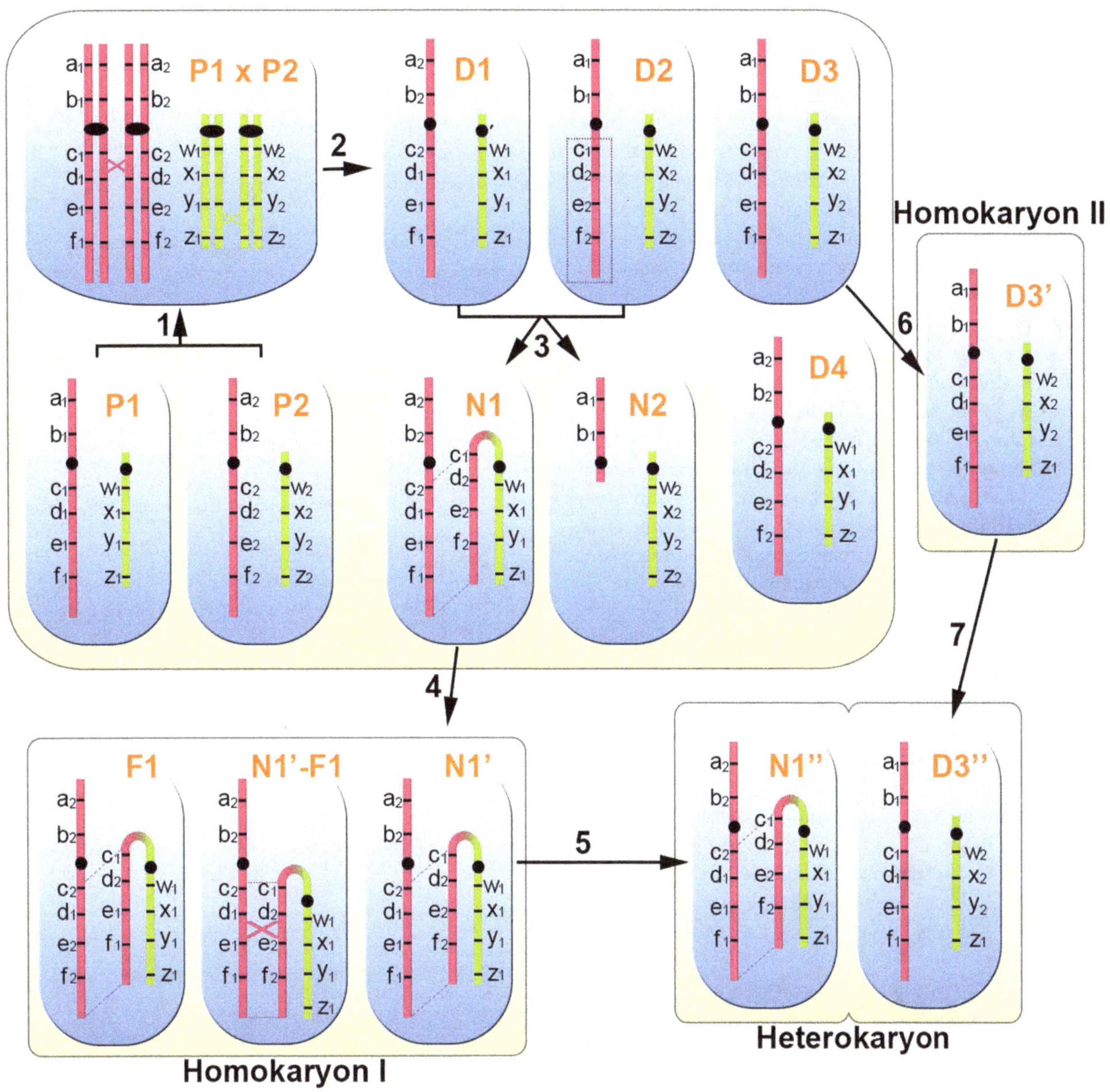

Fig. 7 Schematic diagram showing how certain recombinants could have arisen in the heterokaryon but not in the homokaryon and vice versa

The only previous study concerning the mating system of the *Termitomyces* symbiont of *Macrotermes natalensis* from South Africa (de Fine Licht et al. 2005) suggests that the fungal symbiont is heterothallic, requiring two compatible mating types to complete a life cycle. It is because two haplotypes of internal transcribed spacers (*ITS*) of nuclear ribosomal DNA were detected in cultures obtained from fungus gardens but only one of the two haplotypes was in each of the five single basidiospore cultures that they obtained. Despite the fact that their results agree with data showing that *Termitomyces* cultures isolated from termite colonies are heterokaryotic (Aanen et al. 2009), uncertainty remains because extensive pairings between single basidiospore isolates (homokaryons) were not conducted and, moreover, a reliable system for mushroom fruiting was not developed.

Somatic genetic recombination known or suspected to exist in other basidiomycetes

In many basidiomycetes, somatic genetic recombination does not normally occur but can be triggered in a heterokaryon upon its confrontation with a homokaryon (the Buller phenomenon) during an incompatible heterokaryon-homokaryon mating, $(A_1B_2 + A_2B_1) \times A_1B_1$. The resulting recombinant A_2B_2 nuclei then migrate from the heterokaryon into the homokaryon to pair with A_1B_1 nuclei (Anderson and Kohn 2007; Raper 1966). In addition to the Buller phenomenon, somatic genetic recombination in basidiomycetes is known to occur in three different ways: a meiotic-like process, parasexuality, and internuclear specific factor transfer (Anderson and Kohn 2007). A meiotic-like process is more likely for the *Termitomyces* symbiont, because specific factor transfer is highly unlikely to be genome-wide, and parasexuality occurs in extremely low frequencies.

We suspect that the novel life cycle that we describe in this study can also be found in other *Termitomyces* species and in basidiomycetes of other genera. Intragenic recombination has been reported from *EF1a* and other DNA regions among *Termitomyces* isolates from 31 colonies of *Macrotermes natalensis* in South Africa, which apparently belonged to a breeding population but curiously lacked apparent mushroom fruiting (de Fine Licht et al. 2006). Matings between compatible haploid monokaryons in two honey mushrooms, *Armillaria gallica* (Peabody et al. 2005) and *Armillaria tabescens* (Grillo et al. 2000), resulted in a haploid dikaryon and subsequently a persistent diploid monokaryon by karyogamy. Tissue of fruiting bodies, however, contained haploid dikaryotic cells and was shown to be genetically mosaic, implying somatic genetic recombination during haploidization. Somatic genetic recombination were occasionally detected within dikaryons of *Schizophyllum commune* (Frankel 1979; Parag 1962), and episodic somatic recombination was monitored and detected in culture of dikaryons for 18 months (Clark and Anderson 2004).

Conclusion

Despite the fact that the sexual process that we found in the *Termitomyces* mycelia is much like meiosis, we still have reservation in considering it typical meiosis mainly due to: (1) karyogamy within a hyphal tip cell involving fusion of two nuclei in some cases and three nuclei in other cases, and (2) further confirmation, in addition to our cytological studies, needed on ploidy changes, especially in those cases where three nuclei are involved. Somatic meiotic-like recombination in the *Termitomyces* symbiont is seemingly analogous with VDJ recombination known in human immune system in being able to recombine genetic material somatically. Further studies are needed to investigate whether or not meiotic-like recombination also confers to the *Termitomyces* symbiont similar vigilance and agility to the human immune system in quickly responding to invaders. Our study emphasizes the necessity of further understanding the genetics of *Termitomyces* symbionts, which are organisms centered on the stage of termite agriculture, in order to learn what and how meiotic-like recombination affects the termite agroecosystem. Our findings are likely to generate a profound influence on how we perceive termite agriculture as well as the endurance of *Termitomyces* monoculture with termites over tens of millions of years.

Authors' contributions

HMH, MCC, ANS, CRL, MJF, BCHC, and YMJ collected data; HMH, MCC, and YMJ analyzed data; YMJ wrote the manuscript; WNC collected the *Termitomyces* fruiting body; CYL, CCH, and JCC commented on the manuscript; MCC, CYL, CCH, and YMJ were project leaders. All authors read and approved the final manuscript.

Author details

[1] Institute of Plant and Microbial Biology, Academia Sinica, Taipei 11529, Taiwan. [2] Graduate Institute of Biotechnology, National Changhua University of Education, Changhua 50074, Taiwan. [3] Department of Life Sciences, National Chung Hsing University, Taichung 40227, Taiwan. [4] Department of Natural Resources and Environmental Studies, National Dong Hwa University, Hualien 97401, Taiwan. [5] National Museum of Natural Science, Taichung 40453, Taiwan. [6] Yourgene Bioscience, New Taipei 23863, Taiwan.

Acknowledgements

We thank Jack D. Rogers, Washington State University, USA, for reading the manuscript and Archana Allishe, Academia Sinica, for helping the submission of SRA data to NCBI.

Competing interests

The authors declare that they have no competing interests.

Funding

This work was supported by Ministry of Science and Technology (formerly National Science Council, Grant Number NSC 99-2321-B-001-030-MY3) of Taiwan, by Summit Project, Academia Sinica, and by Innovative Translational Agricultural Research, Academia Sinica to Y-MJ.

References

Aanen DK, Eggleton P (2005) Fungus-growing termites originated in African rain forest. Curr Biol 15:851–855

Aanen DK, Hoekstra RF (2007) Why sex is good: on fungi and beyond. In: Heitman J, Kronstad JW, Taylor JW, Casselton LA (eds) Sex in fungi: molecular determination and evolutionary implications. ASM Press, Washington, DC, pp 527–534

Aanen DK, de Fine Licht HH, Debets AJ, Kerstes NA, Hoekstra RF, Boomsma JJ (2009) High symbiont relatedness stabilizes mutualistic cooperation in fungus-growing termites. Science 326:1103–1106

Aimi T, Yoshida R, Ishikawa M, Bao D, Kitamoto Y (2005) Identification and linkage mapping of the genes for the putative homeodomain protein (*hox1*) and the putative pheromone receptor protein homologue (*rcb1*) in a bipolar basidiomycete, *Pholiota nameko*. Curr Genet 48:184–194

Anderson JB, Kohn LM (2007) Dikaryons, diploids, and evolution. In: Heitman J, Kronstad JW, Taylor JW, Casselton LA (eds) Sex in fungi: molecular determination and evolutionary implications. ASM Press, Washington, DC, pp 333–348

Barton NH, Charlesworth B (1998) Why sex and recombination? Science 281:1986–1990

Batra LR, Batra SWT (1979) Termite-fungus mutualism. In: Batra LR (ed) Insect-fungus symbiosis: nutrition, mutualism and commensalism. Montclair, Allanheld, Osmun & Co, pp 117–163

Bell G (1982) The masterpiece of nature: the evolution and genetics of sexuality. University of California Press, Berkeley

Brown AJ, Casselton LA (2001) Mating in mushrooms: increasing the chances but prolonging the affair. Trends Genet 17:393–400

Burt A (2000) Perspective: sex, recombination, and the efficacy of selection—was Weismann right? Evolution 54:337–351
Chung MC, Lee YI, Cheng YY, Chou YJ, Lu CF (2008) Chromosomal polymorphism of ribosomal genes in the genus *Oryza*. Theor Appl Genet 116:745–753
Clark TA, Anderson JB (2004) Dikaryons of the basidiomycete fungus *Schizophyllum commune*: evolution in long-term culture. Genetics 167:1663–1675
Cooke RC, Whipps JM (1987) Saprotrophy, stress and symbiosis. In: Rayner ADM, Brasier CM, Moore D (eds) Evolutionary biology of the fungi. Cambridge University Press, Cambridge, pp 137–148
Cook-Patton SC, McArt SH, Parachnowitsch AL, Thaler JS, Agrawal AA (2011) A direct comparison of the consequences of plant genotypic and species diversity on communities and ecosystem function. Ecology 92:915–923
de Fine Licht HH, Andersen A, Aanen DK (2005) *Termitomyces* sp. associated with the termite *Macrotermes natalensis* has a heterothallic mating system and multinucleate cells. Mycol Res 109:314–318
de Fine Licht HH, Boomsma JJ, Aanen DK (2006) Presumptive horizontal symbiont transmission in the fungus-growing termite *Macrotermes natalensis*. Mol Ecol 15:3131–3138
Esser K (2006) Heterogenic incompatibility in fungi. In: Kües U, Fischer R (eds) The mycota a comprehensive treatise on fungi as experimental systems for basic and applied research I growth, differentiation and sexuality, 2nd edn. Springer, Berlin, pp 141–165
Frankel C (1979) Meiotic-like recombination in vegetative dikaryons of *Schizophyllum commune*. Genetics 92:1121–1126
Grillo R, Korhonen K, Hantula J, Hietala AM (2000) Genetic evidence for somatic haploidization in developing fruit bodies of *Armillaria tabescens*. Fungal Genet Biol 30:135–145
Heitman J, Sun S, James TY (2013) Evolution of fungal sexual reproduction. Mycologia 105:1–27
James TY, Kües U, Rehner SA, Vilgalys R (2004) Evolution of the gene encoding mitochondrial intermediate peptidase and its cosegregation with the A mating-type locus of mushroom fungi. Fungal Genet Biol 41:381–390
Johnson RA (1981) Colony development and establishment of the fungus comb in *Microtermes* sp. nr *usambaricus* (Sjöstedt) (Isoptera: Macrotermitinae) from Nigeria. Insectes Sociaux 28:3–12
Johnson RA, Thomas RJ, Wood TG, Swift MJ (1981) The inoculation of the fungus comb in newly founded colonies of some species of the Macrotermitinae (Isoptera) from Nigeria. J Nat Hist 15:751–756
Katoh H, Miura T, Maekawa K, Shinzato N, Matsumoto T (2002) Genetic variation of symbiotic fungi cultivated by the macrotermitine termite *Odontotermes formosanus* (Isoptera: Termitidae) in the Ryukyu Archipelago. Mol Ecol 11:1565–1572
Knox-Davies PS, Dickson JG (1960) Cytology of *Helminthosporium turcicum* and its ascigerous stage, *Trichometasphaeria turcica*. Am J Bot 47:328–339
Korb J (2003) Thermoregulation and ventilation of termite mounds. Naturwissenschaften 90:212–219
Kües U (2000) Life history and developmental processes in the basidiomycete *Coprinus cinereus*. Microbiol Mol Biol Rev 64:316–353
Langmead B, Salzberg SL (2012) Fast gapped-read alignment with Bowtie 2. Nat Methods 9:357–359
Lin XR, Hull CM, Heitman J (2005) Sexual reproduction between partners of the same mating type in *Cryptococcus neoformans*. Nature 434:1017–1021
Martin M (2011) Cutadapt removes adapter sequences from high-throughput sequencing reads. EMBnet J 17:10–12
Maynard Smith J (1971) What use is sex. J Theor Biol 30:319–335
Mueller UG, Gerardo NM, Aanen DK, Six DL, Schultz TR (2005) The evolution of agriculture in insects. Annu Rev Ecol Evol Syst 36:563–595
Otto SP (2009) The evolutionary enigma of sex. Am Nat 174:S1–S14
Parag Y (1962) Studies on somatic recombination in dikaryons of *Schizophyllum commune*. Heredity 17:305–318
Peabody RB, Peabody DC, Tyrrell MG, Edenburn-MacQueen E, Howdy RP, Semelrath KM (2005) Haploid vegetative mycelia of *Armillaria gallica* show among-cell-line variation for growth and phenotypic plasticity. Mycologia 97:777–787
Perkins DD (1997) Chromosome rearrangements in *Neurospora* and other filamentous fungi. In: Hall J, Dunlap J, Friedmann T, Giannelli F (eds) Advances in Genetics, vol 36. Academic Press, San Diego, pp 239–398
Poulsen M, Hu HF, Li C, Chen ZS, Xu LH, Otani S, Nygaard S, Nobre T, Klaubauf S, Schindler PM et al (2014) Complementary symbiont contributions to plant decomposition in a fungus-farming termite. Proc Natl Acad Sci USA 111:14500–14505
Raffaele S, Kamoun S (2012) Genome evolution in filamentous plant pathogens: why bigger can be better. Nat Rev Microbiol 10:417–430
Raper JR (1966) Genetics of sexuality in higher fungi. Ronald Press Co., New York
Raudaskoski M, Kothe E (2010) Basidiomycete mating type genes and pheromone signaling. Eukaryot Cell 9:847–859
Roberts EM, Todd CN, Aanen DK, Nobre T, Hilbert-Wolf HL, O'Connor PM, Tapanila L, Mtelela C, Stevens NJ (2016) Oligocene termite nests with in situ fungus gardens from the Rukwa Rift Basin, Tanzania, support a Paleogene African origin for insect agriculture. PLoS ONE 11:e0156847
Robinson JT, Thorvaldsdottir H, Winckler W, Guttman M, Lander ES, Getz G, Mesirov JP (2011) Integrative genomics viewer. Nat Biotechnol 29:24–26
Sands WA (1960) The initiation of fungus comb construction in laboratory colonies of *Ancistrotermes guineensis* (Silvestri). Insectes Sociaux 7:251–263
Schlecht U, Primig M (2003) Mining meiosis and gametogenesis with DNA microarrays. Reproduction 125:447–456
Shinzato N, Muramatsu M, Watanabe Y, Matsui T (2005) Termite-regulated fungal monoculture in fungus combs of a macrotermitine termite *Odontotermes formosanus*. Zool Sci 22:917–922
Thomas RJ (1987) Factors affecting the distribution and activity of fungi in the nests of Macrotermitinae (Isoptera). Soil Biol Biochem 19:343–349
Turner JS (2004) Gaia, extended organisms, and emergent homeostasis. In: Schneider SH, Miller JR, Crist E, Boston PJ (eds) Scientists Debate Gaia: the next century. MIT Press, Cambridge, pp 57–71
Ward EWB, Ciurysek KW (1962) Somatic mitosis in *Neurospora crassa*. Am J Bot 49:393–399
White TJ, Bruns B, Lee S, Taylor J (1990) Amplification and direct sequencing of fungal ribosomal RNA genes for phylogenetics. In: Innis MA, Gelfand DH, Sninsky JJ, White TJ (eds) PCR protocols, a guide to methods and applications. Academic Press, San Diego, pp 315–322
Whitehouse HLK (1949) Multiple-allelomorph heterothallism in the fungi. New Phytol 48:212–244
Williams GC (1975) Sex and evolution. Princeton University Press, Princeton

7

Complete chloroplast genome sequences of two endangered *Phoebe* (Lauraceae) species

Yingang Li[1], Wuqin Xu[2], Wentao Zou[3], Dongyue Jiang[1] and Xinhong Liu[1*]

Abstract

Background: *Phoebe* (Lauraceae) comprises of evergreen trees or shrubs with approximately 100 species, distributed in tropical and subtropical Asia and Neotropical America. A total of 34 species and three varieties occur in China. Despite of economic and ecological value, only limited genomic resources are available for this genus.

Results: We sequenced the two complete chloroplast (cp) genomes of *Phoebe chekiangensis* and *P. bournei* using Illumina sequencing technology via a combined strategy of de novo and reference-guided assembly. We also performed comparative analyses with the cp genomes of *P. sheareri* and *P. sheareri* var. *oineiensis* previously reported. The chloroplast genomes of *P. chekiangensis* and *P. bournei* identically contain 112 genes consisting of 78 protein coding genes, 30 tRNA genes, and 4 rRNA genes, with the size of 152,849 and 152,853 bp, respectively. From the two chloroplast genomes, 131 SSRs were identified and 12 different SSRs located in five protein coding genes. The analysis showed the extremely conserved structure of chloroplast genomes with surprisingly little variations at the LSC/IR and SSC/IR boundaries. Moreover, the mean nucleotide diversity was found to be 0.162% for 77 regions, suggesting an extraordinarily low level of sequence divergence. Four highest divergent regions (*trnH-psbA, rps14-trnT, petA-psbJ, ccsA-ndhD*) with the percentage of nucleotide diversity higher than 0.50% were identified, which had potential use for species identification and phylogenetic studies.

Conclusion: This study will facilitate our understanding of population genetics, phylogenetic relationship and plant evolution of *Phoebe* species.

Keywords: *Phoebe chekiangensis, Phoebe bournei*, Chloroplast genomes, Repeat analysis, SSRs, Divergent regions

Background

Phoebe is a genus of evergreen trees or shrubs belonging to family, Lauraceae. *Phoebe* comprises of approximately 100 species, distributed in tropical and subtropical Asia and neotropical America. A total of 34 species and three varieties are endemic to China (Wu et al. 2008). *Phoebe* species, with their high-quality wood, were widely used to make column during palace construction in the Ming and Qing dynasties and built high valuable furniture which stood for the power and status of the noble (Ding et al. 2015). The most famous and valuable *Phoebe* wood are called 'wood with golden wire', which comes from several specific and sporadically distributed rare *Phoebe* species, endemics to China, including *Phoebe chekiangensis, Phoebe bournei, Phoebe sheareri, Phoebe zhennan*, and *Phoebe lichuanensis.* Our target species, *P. chekiangensis* is distributed in Zhejiang and its adjacent areas (including Fujian, Jiangxi, and Anhui Province), whereas *P. bournei* is distributed in Yangtze River Basin and its south region in China. The *Phoebe* species have a very high economic and ecological value but with extremely limited studies on biochemical compound and population diversity (Hegde et al. 1997; Zhang et al. 2012a, b).

The chloroplast (cp) is an important organelle that plays a key role in plant photosynthesis providing energy to green plants and carbon fixation (Douglas 1990). With

*Correspondence: lkyliuxh@163.com
[1] Zhejiang Academy of Forestry, Hangzhou 310023, China
Full list of author information is available at the end of the article

the rapid development of next-generation sequencing, it is now cheaper and faster to obtain genomes than by traditional Sanger sequencing. Therefore, cp genome-scale data have been increasingly used to infer phylogenetic relationships at high taxonomical levels, and even in lower taxa (Moore et al. 2007; Parks et al. 2009; Huang et al. 2014; Yang et al. 2017; Zeng et al. 2017). Although the cp genome is more conserved than the nuclear genome in plants, many mutation events in the chloroplast DNA sequence have been identified, including indels, substitutions, and inversions (Ingvarsson et al. 2003). Most angiosperm cp genomes have a quadripartite circular structure ranging from 115 to 165 kp in length, and are composed of two copies of inverted repeat (IR) regions that are separated by a large single copy (LSC) region and a small single copy (SSC) region (Wicke et al. 2011; Wang et al. 2015). Massive information for genetics, taxonomy, and phylogeny could be mined from cp genomes, because of their relatively conserved gene structure and sequence divergence between species and individuals (Parks et al. 2009; Huang et al. 2014). Furthermore, cp genomes can provide effective genetic markers for evolutionary studies from population level. In addition, analyzing the composition and structure of the cp genomes for such an important genus *Phoebe* can explore further genetic variations, which could improve quantitative and quality traits.

In this study, we reported the complete and annotated DNA sequences for the cp genomes of *P. chekiangensis* and *P. bournei* using next-generation sequencing platform, which is also the first comprehensive analysis on cp genomes for *Phoebe* combining the cp genomes of *P. sheareri* and *P. sheareri* var. *oineiensis* previously reported (Song et al. 2016). The specific aims of the present study were to: (1) present the complete chloroplast genome sequences and investigate global structural patterns of *P. chekiangensis* and *P. bournei*; (2) examine variations of repeat sequences and simple sequence repeats (SSRs) among the two *Phoebe* chloroplast genomes; (3) screen sequence divergence hotspot regions in the four *Phoebe* chloroplast genomes.

Methods

Plant material and DNA extraction

Young leaves of *P. chekiangensis* and *P. bournei* were sampled from single seedlings growing in the nursery located at 30.21N, 120.02E, of Zhejiang Academy of Forestry. The provenance was Hangzhou, Zhejiang Province and Mingxi, Fujian Province, PR China, which was the primary distribution area of *P. chekiangensis* and *P. bournei*, respectively. Total genomic DNA per species was extracted from 30 mg of the silica-dried leaf using the modified CTAB method (Porebski et al. 1997). The quality and concentration of the genomic DNA were assessed using agarose gel electrophoresis and an Agilent BioAnalyzer 2100 (Agilent Technologies).

Chloroplast genome illumina sequencing, assembly and annotation

Genomic DNA was used to generate short-insert (500 bp) paired-end sequencing libraries according to the Illumina standard protocol. Genomic DNA from each species was sequenced using a HiSeq™ 2000 analyzer (Illumina, San Diego, California, USA) at Beijing Genomics Institute (BGI, Shenzhen, China).The raw reads (20,777,674 and 20,787,108 bp for *P. chekiangensis* and *P. bournei*) were generated with 125 bp length and assembled into whole chloroplast genomes in a multi-step approach employing a modified pipeline that involved a combination of both reference guided and de novo assembly approaches. First, paired-end sequence reads were trimmed to remove low-quality bases ($Q < 20$, 0.01 probability error) and adapter sequences using CLC-quality trim tool (quality_trim software included in CLC ASSEMBLY CELL package, http://www.clcbio.com/products/clc-assembly-cell/) before undertaking sequence assembly. Second, the contigs were assembled using CLC de novo assembler with the following optimized parameters: bubble size of 98, minimum contig length of 200, mismatch cost of two, deletion and insertion costs of three, length fraction of 0.9, and similarity fraction of 0.8. Third, all the contigs were aligned to the reference chloroplast genome of *Machilus yunnanensis* (NC028073) using BLAST (http://blast.ncbi.nlm.nih.gov/), and aligned contigs (≥90% similarity and query coverage) were ordered according to the reference chloroplast genome. Then, contigs were aligned with the reference genome to construct the draft chloroplast genome of each species in Geneious 9.0.5 software (http://www.geneious.com). Finally, clean reads were re-mapped to the draft cp genomes of two *Phoebe* species, and the mapping ratio were 3.41% for *P. chekiangensis* and 2.15% for *P. bournei*, and average coverage depth were 578.2 and 367.1 for *P. chekiangensis* and *P. bournei*, respectively. Using the Dual Organellar Genome Annotator (DOGMA) program (Wyman et al. 2004), the two *Phoebe* chloroplast genomes were annotated. Protein-coding genes were identified by using the plastid/bacterial genetic code. Intron/exon boundaries were further determined using MAFFT v7 with those of the *M. yunnanensis* chloroplast genomes as a reference (Katoh and Standley 2013). Using the program tRNAscan-SEwith default settings (Schattner et al. 2005), tRNA boundaries were verified. The circular chloroplast genome maps of the *Phoebe* were drawn using the Organellar Genome DRAW (OGDRAW) software, with subsequent manual editing (Lohse et al. 2007).

Characterization of repeat sequences and SSRs

REPuter was used to visualize both forward, palindrome, reverse and complement repeats, with a minimum repeat size of 30 bp and a sequence identity greater than 90% with hamming distance equal to 3 in *P. chekiangensis* and *P. bournei* (Kurtz and Schleiermacher 1999). The SSRs, which usually have a higher mutation rate, are easily genotyped using PCR, used as markers for phylogenetic analysis and in marker-assisted breeding when located on nuclear chromosomes. Microsatellites, or simple sequence repeats (SSRs) were detected using MISA perl script with thresholds of ten repeat units for mononucleotide SSRs, five repeat units for dinucleotide SSRs, four repeat units for trinucleotide SSRs, and three repeat units for tetra-, penta-, and hexanucleotide SSRs (Thiel et al. 2003).

Divergence hotspot identification

The four complete chloroplast genome sequences (*P. chekiangensis, P. bournei, P. sheareri*, and *P. sheareri* var. *oineiensis*) were aligned,using the chloroplast genome of *M. yunnanensis* as a reference by mVISTA program (Frazer et al. 2004). Default parameters were utilized to align the chloroplast genomes in Shuffle-LAGAN mode and a sequence conservation profile was visualized in an mVISTA plot. To discover the divergence hotspot regions in *Phoebe*, protein coding gene, intron, and intergenic spacer region were evaluated with DnaSP 5.10 (Librado and Rozas 2009). All the regions were sequentially extracted under the following criteria: (1) total number of mutation (Eta) >0; (2) an aligned length >200 bp. Any large structural events, such as gene order rearrangements and IR expansions/contractions, were recorded.

Results

Chloroplast genomes organization of *P. chekiangensis* and *P. bournei*

For these two *Phoebe* species, just three contigs which were found to be significantly homologous to the reference genome were combined to generate each chloroplast genome, with no gaps or missing nucleotides (Ns) found. The total genome size of *P. chekiangensis*, with a length of 152,849 bp (deposited in GenBank, Accession No. KY346511), was in close proximity to those of other *Phoebe* species and only 4, 6, and 27 bp smaller than that of *P. bournei* (152,853 bp, deposited in GenBank, Accession No. KY346512), *P. sheareri* var. *oineiensis* (152,855 bp, GenBank Accession No. KX437772), and *P. sheareri* (152,876 bp, GenBank Accession No. KX437773), respectively. The size of complete chloroplast genome of the four *Phoebe* species was within the range of angiosperms (Yang et al. 2010). Similar to the vast majority of angiosperms, showing a typical quadripartite structure, the two *Phoebe* chloroplast genomes consist of a pair of inverted repeats (IRs) of 18,927 bp in *P. chekiangensis* and 18,928 bp in *P.bournei*, a large single copy (LSC) region of 93,772 bp in *P. chekiangensis* and 93,777 bp in *P. bournei* and a small single copy (SSC) region of 20,775 bp in *P. chekiangensis* and 20,774 bp in *P. bournei* (Fig. 1; Table 1). Protein-coding regions accounted for 46.23% of the whole genome, while tRNA and rRNA regions accounted for 1.79 and 5.91%, respectively, and the remaining 46.07% was non-coding regions (Table 1). The overall GC content was 39.1%, whereas the GC content in the LSC, SSC and IR regions were 38.0, 34.4, and 44.4%, respectively, indicating identical levels among the two *Phoebe* chloroplast genomes.

SSR and repeat sequences analysis

Repeat sequences played a vital role in phylogenetic analysis and genome rearrangement (Nie et al. 2012). We used REPuter to analyze the repeat sequence of *Phoebe* cp genomes and found forward repeats, palindrome repeats and reverse repeats of at least 30 bp per repeat unit with a sequence identity of ≥90%. *P. chekiangensis* contained 36 repeats comprising of 13 forward repeats, 15 palindromic repeats, 6 reverse repeats and 2 complement repeats. The only quantity difference in repeat type between *P. chekiangensis* and *P. bournei* was that the latter contained 14 palindromic repeats, one less than the former (Fig. 2a; Additional file 1). The quantity of repeats with 30–40 bp length was 33 and 32 for *P. chekiangensis* and *P. bournei*, respectively. In addition, *P. chekiangensis* and *P. bournei* both contained one repeat with 41, 42, and 48 bp length (Fig. 2b; Additional file 1).

With MISA analysis, *Phoebe* chloroplast genome was found to contain 66 (*P. chekiangensis*) and 65 (*P. bournei*) SSRs longer than 10 bp, of which 49 SSRs were the same for the two chloroplast genomes (similar repeat units located in similar genomic regions) (Fig. 3a; Additional file 2). Among the total 131 SSRs, most loci were located in intergenic spacer (IGS) regions (61.07%), followed by introns (23.37%) and CDS (17.56%) (Fig. 3b). We observed that 12 different SSRs were located in five protein-coding genes [*ycf1* (×5), *cemA*(×2), *rpoC2* (×2), *ycf2* (×2), and *matK*] of the two *Phoebe* chloroplast genomes.

Inverted repeats (IRs) contraction and expansion of four *Phoebe* species

The IR region expanded into the *ycf2* gene, creating a pseudogene fragment *ψycf2* at the IRa/LSC border with length of 3,161 bp (*P. bournei* and *P. sheareri* var. *oineiensis*) and 3,162 bp (*P. chekiangensi* and *P. sheareri*). The *ycf1* gene crossed the SSC/IRa region and the pseudogene fragment *ψycf1* was located at the IRb region with 1381 and 1399 bp. The *trnH* was the unique gene with

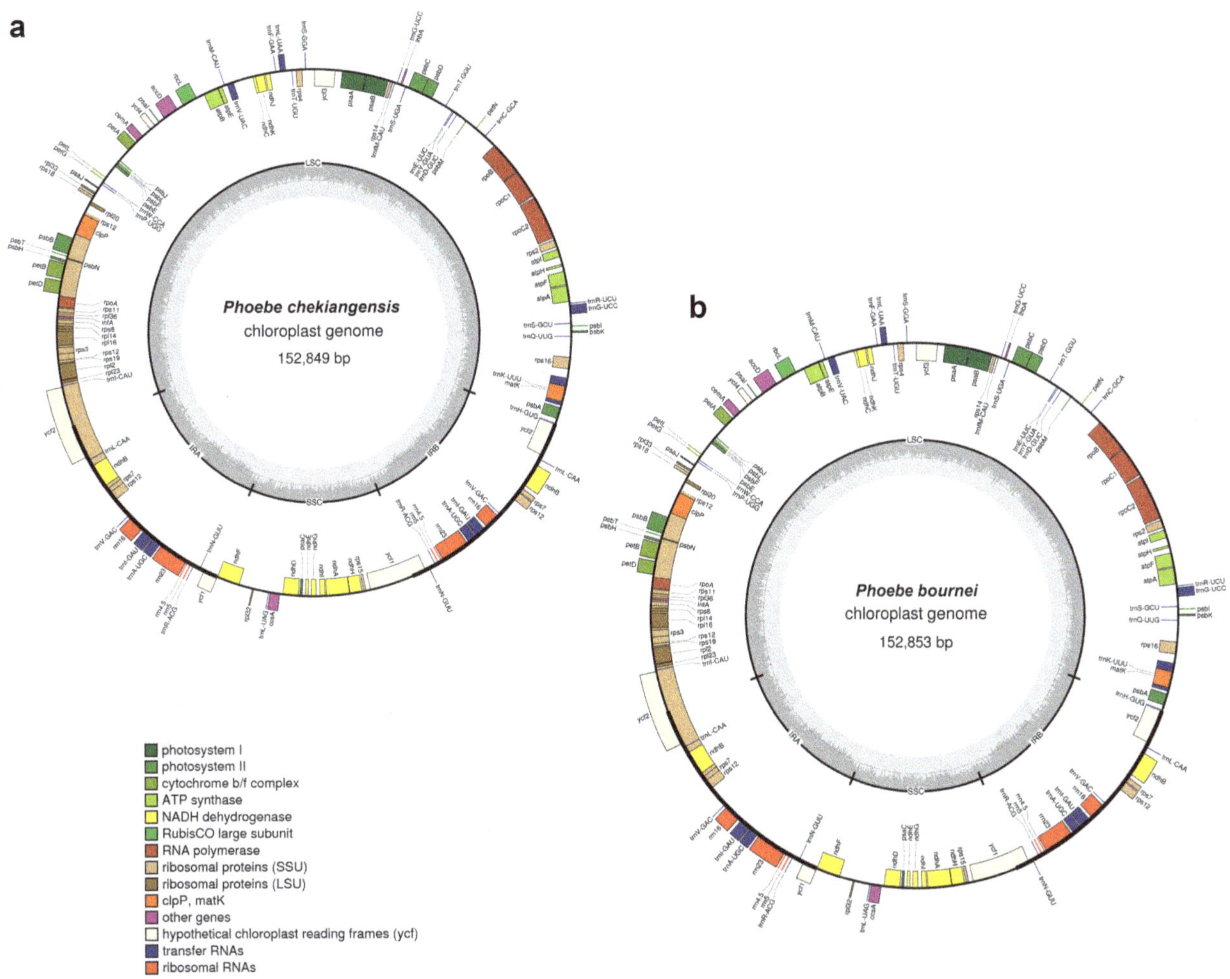

Fig. 1 Chloroplast genome of *Phoebe chekiangensis* and *Phoebe bournei*. **a** *Phoebe chekiangensis*, **b** *Phoebe bournei*. Genes shown on the outside of the *circle* are transcribed clockwise, and genes inside are transcribed counter-clockwise. Large single copy (LSC), small single copy (SSC), and inverted repeats (IRa, IRb) are indicated. Genes belonging to different functional groups are *color-coded*. The *darker gray* in the *inner* corresponds to GC content, and the *lighter gray* corresponds to AT content

large difference among *Phoebe* species. The *trnH* genes of *P. sheareri* and *P. sheareri* var. *oineiensis* were separated by 21 bp at the IRa/LSC border, whereas those of *P. bournei* and *P. chekiangensis* were separated by 43 bp (Fig. 4).

Divergence sequence hotspots in *Phoebe* species

The overall sequence identity of the four *Phoebe* chloroplast genomes was compared and plotted using the mVISTA program (Frazer et al. 2004), with the annotation of *M. yunnanensis* as a reference to elucidate the level of sequence divergence (Fig. 5). Being largely consistent with recent studies (Yao et al. 2015; Zhang et al. 2016), most of the sequence variations were found to be located in the LSC and SSC regions, while the IR regions exhibited comparatively lower sequence diversity. The lower sequence divergence observed in the IRs rather than LSC and SSC regions for *Phoebe* species is likely due to copy correction between IR sequences by gene conversion (Khakhlova and Bock 2006).

Seventy-seven regions (30 coding regions, 38 intergenic spacers, eight introns, and one rRNA) with more than 200 bp in length were eventually identified. Of these 77 regions, nucleotide diversity (*Pi*) ranged from 0.00018 (*rrn23*) to 0.01389 (*ccsA-ndhD*) among four *Phoebe* species (Fig. 6; Additional file 3). As found in most angiosperms (Choi et al. 2016), sequence divergence in intergenic regions was higher than that in genic regions of these four *Phoebe* chloroplast genomes. The mean value of *Pi* in non-coding regions was 0.221%, which was nearly more than twice as much as that (0.123% on average) in the coding regions. Intergenic regions with a percentage of *Pi* exceeding 0.5% were *trnH-psbA* (0.506%),

Table 1 The basic characteristics of the sequencing data for *P. chekiangensis* and *P. bournei*

Characteristics	*P. chekiangensis*	*P. bournei*
Clean reads	20,777,674	20,787,108
Average read length	125	125
Total cpDNA size (bp)	152,849	152,853
Length of large single copy (LSC) region	93,772	93,777
Length of inverted repeat (IR) region	18,927	18,928
Length of small single copy (SSC) region	20,075	20,074
Total CDS length	70,668	70,668
Total tRNA length	2740	2740
Total rRNA length	9038	9038
Total GC content (%)	39.1	39.1
LSC	38.0	38.0
IR	44.4	44.4
SSC	34.4	34.4
Total number of genes	113	113
Protein-coding genes	79	79
rRNAs genes	4	4
tRNAs genes	30	30

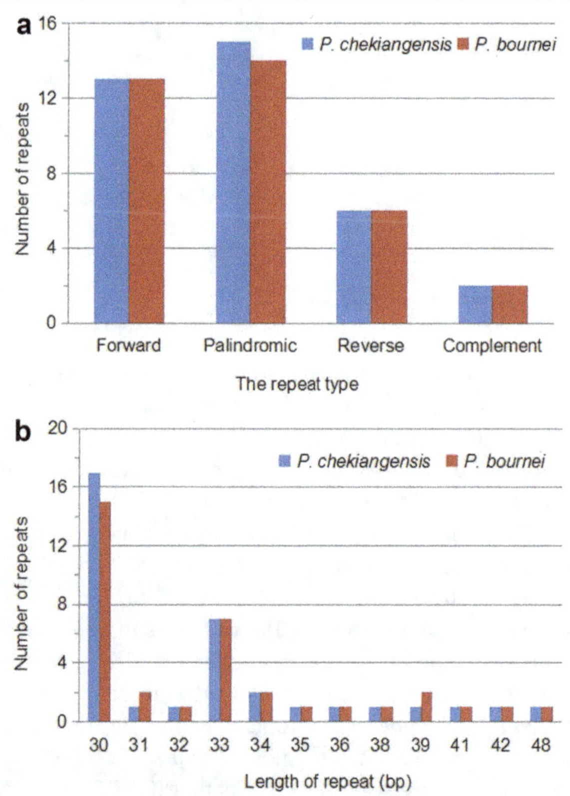

Fig. 2 Analysis of repeated sequences in *P. chekiangensis* and *P. bournei* chloroplast genomes. **a** Frequency of repeats by length; **b** frequency of repeat types

rps4-trnT (0.716%), *petA-psbJ* (0.887%), and *ccsA-ndhD* (1.389%). However, the highest proportion of variability in genic regions was 0.251% (*rps8*) (Fig. 6; Additional file 3).

Discussion

Results from this study showed the GC content of *P. chekiangensis* and *P. bournei* chloroplast genomes is close to that reported in other Lauraceae chloroplast genomes (Song et al. 2015, 2016). Although the GC percentage in the IR regions (44.4%) of the two *Pheobe* was higher than that of *Nicotiana otophora* (43%) (Asaf et al. 2016), the presence of rRNA in *Pheobe* (four) was lower than that of *N. otophora* (eight). The results in this study were incompatible to previous report which suggested that a high GC percentage in the IR regions could be due to the presence of rRNA (Qian et al. 2013). Similarly, 112 different genes, including 78 protein-coding genes (47 genes encoding photosynthesis-related proteins, four DNA dependent RNA polymerases, 20 ribosomal proteins, one translation initiation factor, four genes encoding other proteins, and two genes of unknown function), 30 tRNA genes, and four rRNA genes were annotated in *P. chekiangensis* and *P. bournei* chloroplast genomes (Fig. 1; Additional file 4). The pattern of protein coding genes was similar to that of *Persea americana* (Song et al. 2016). Among all the protein-coding genes, ten genes possessed a single intron, two genes (*ycf3* and *clpP*) contain two introns, whereas six tRNA genes contain a single intron (Additional file 4).

Overall, a total of 71 repeats, including 36.6% forward repeats (26), 40.8% palindromic repeats (29), 16.9% reverse repeats (12), and 5.7% complement repeats (4), were detected in *P. chekiangensis* and *P. bournei* chloroplast genomes. About 61.07% of these repeats were distributed in intergenic spacer. The result was comparable to chloroplast genomes of most angiosperm plant (Uthaipaisanwong et al. 2012; Yao et al. 2015). Previous studies suggested that the presence of these repeats indicates that the region is a crucial hotspot for genome reconfiguration (Gao et al. 2009). One 30 bp forward repeat occurred in the *ndhC-trnV* intergenic spacer and one 30 bp palindromic repeat in *Ψycf1* were unique to *P. chekiangensis*. In contrast, one 31 bp forward repeat occurred in the *ndhC-trnV* intergenic spacer were only contained by *P. bournei* (Additional file 1). Apart from the above three repeats, the others were shared between two *Phoebe* species. SSRs in the chloroplast with high polymorphism in copy numbers have been recognized as one of the main sources of molecular markers, and extensively used for population genetics and phylogenetic investigation (Pauwels et al. 2012; Zhang et al. 2012a, b; Zhao et al. 2015).

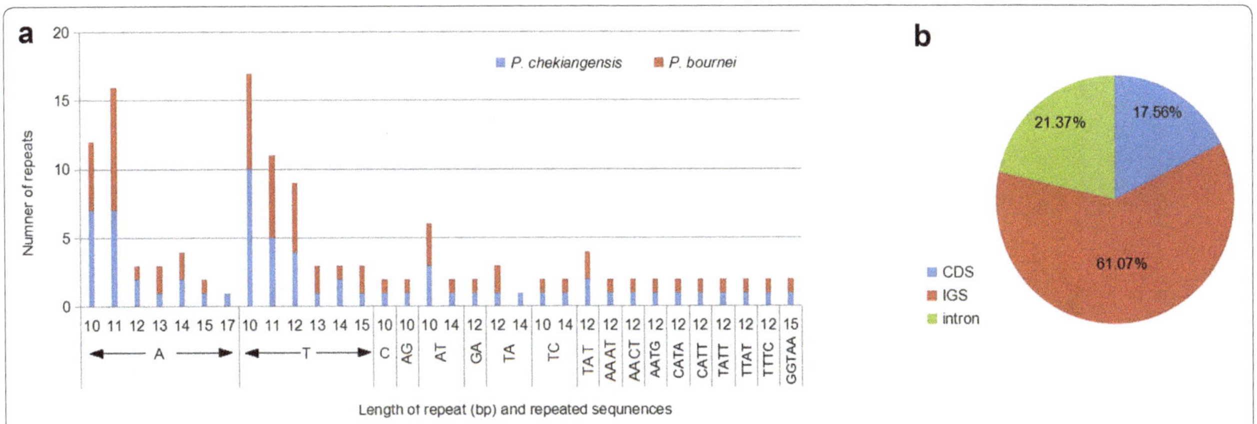

Fig. 3 Simple sequence repeats (SSRs) in the two *Pheobe* chloroplast genomes. **a** Numbers of SSRs by length; **b** distribution of SSR loci. *IGS* intergenic spacer region

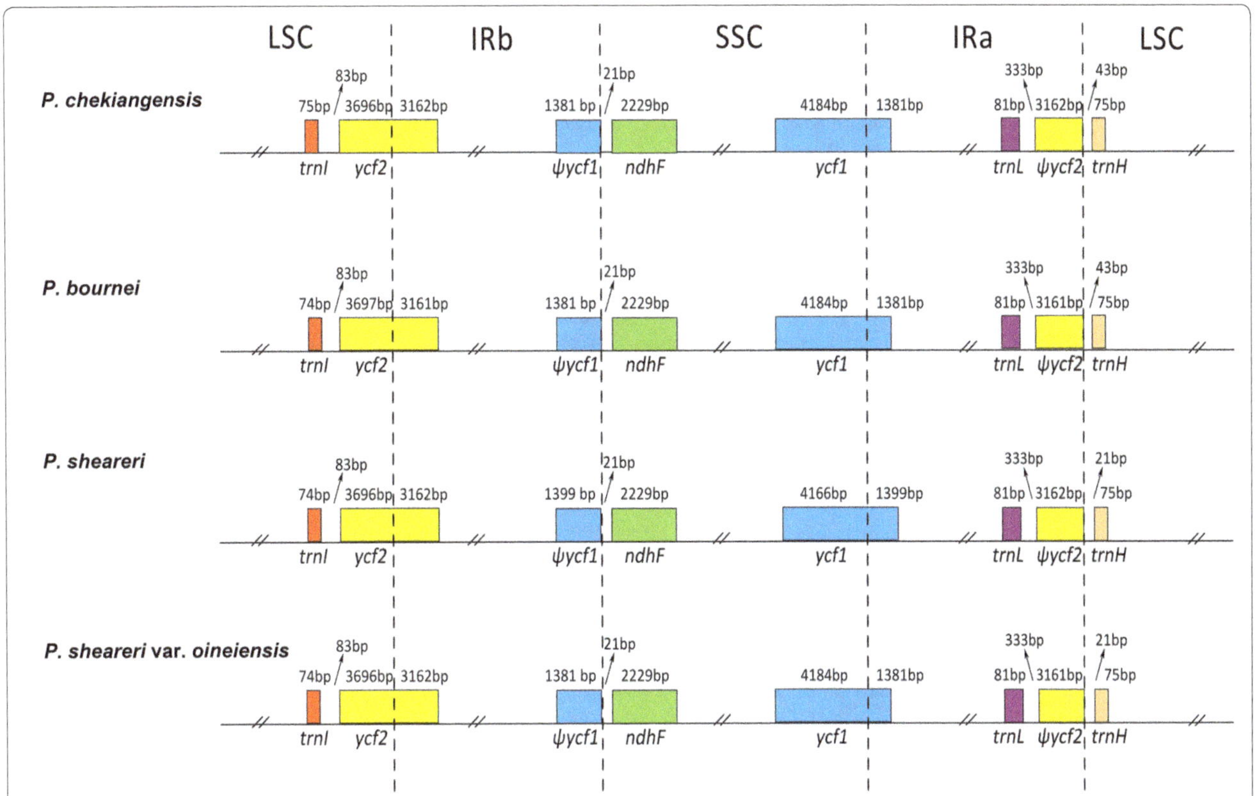

Fig. 4 Comparison of LSC, IR, and SSC junction positions among chloroplast genomes of *P. chekiangensis*, *P. bournei*, *P. sheareri*, and *P. shearerivar. oineiensis*

Among 131 SSRs longer than 10 bp, almost all mononucleotide was composed of A/T (97.7%), and majority of dinucleotides was composed of AT (60.0%). The AT richness in SSRs of the two *Pheobe* genome was similar to previous reports suggesting that SSRs found in the chloroplast genome were generally composed of polythymine (T) or polyadenine (A) repeats, and infrequently contained tandem cytosine (C) and guanine (G) repeats (Kuang et al. 2011; Qian et al. 2013; Chen et al. 2015). The SSRs identified in the two *Phoebe* species might be used in future population genetic studies as well as similar studies of other species, such as *Panax ginseng* (Kim and

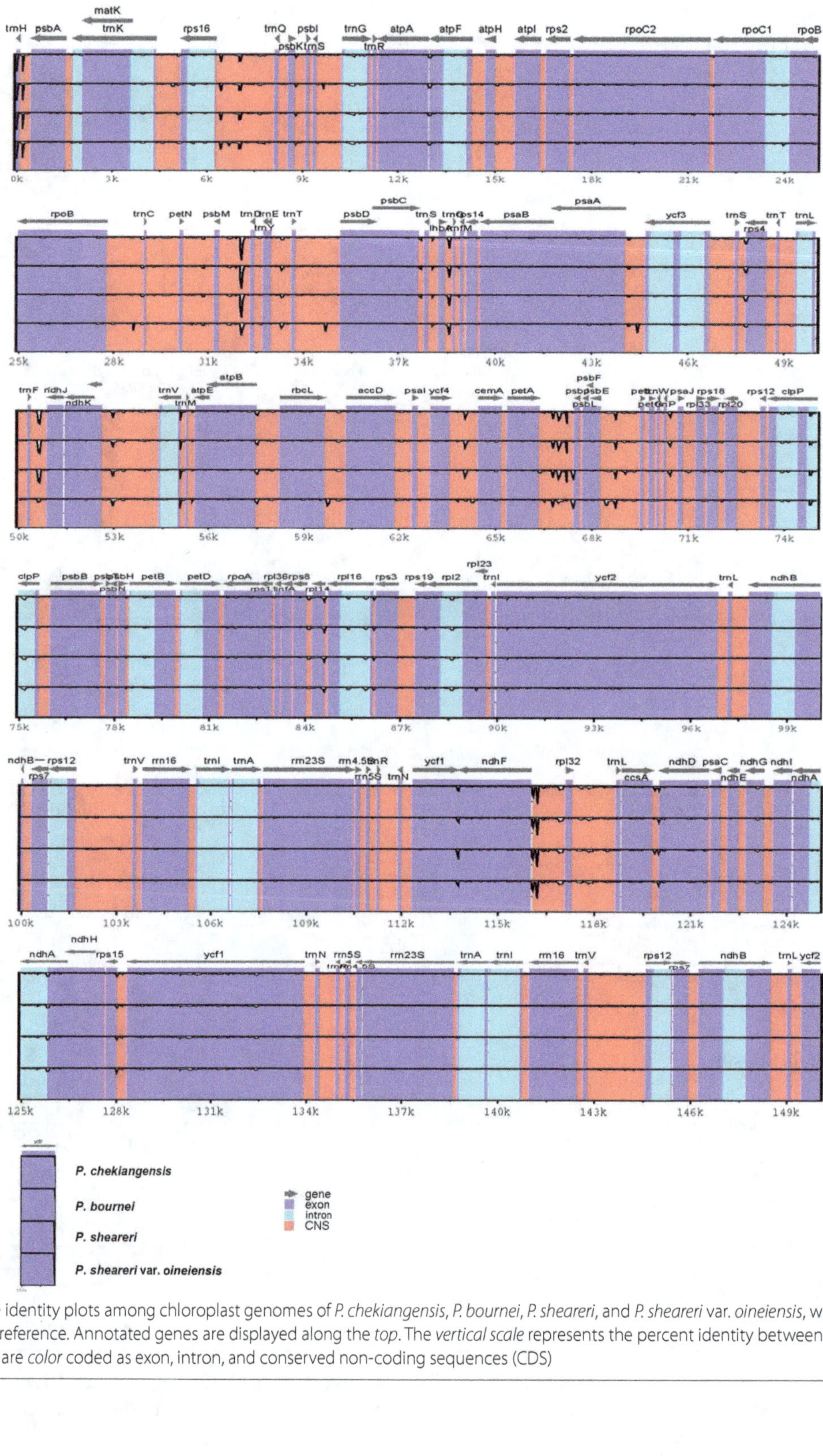

Fig. 5 Sequence identity plots among chloroplast genomes of *P. chekiangensis*, *P. bournei*, *P. sheareri*, and *P. sheareri* var. *oineiensis*, with *Machilus yunnanensis* as a reference. Annotated genes are displayed along the *top*. The *vertical scale* represents the percent identity between 50 and 100%. Genome regions are *color* coded as exon, intron, and conserved non-coding sequences (CDS)

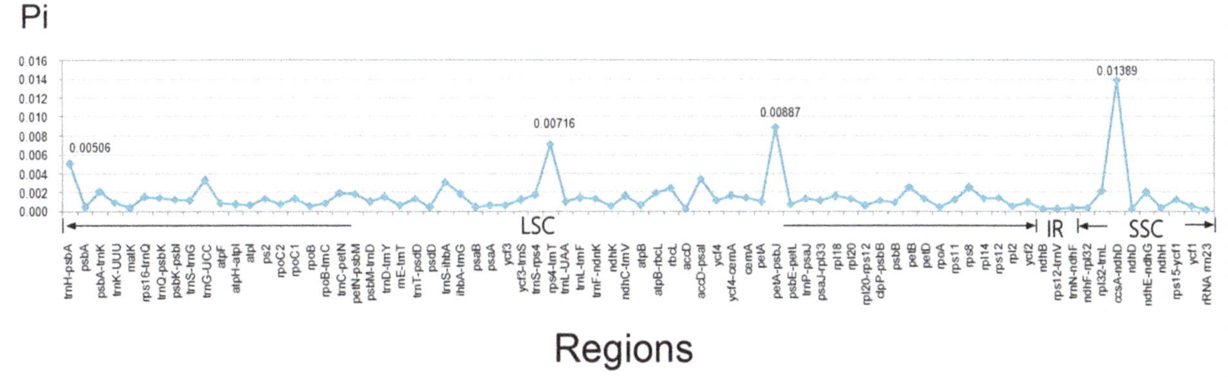

Fig. 6 The nucleotide diversity (Pi) values were compared among four *Phoebe* species

Lee 2004), *Cucumis sativus* (Kim et al. 2006), *Vigna radiate* (Tangphatsornruang et al. 2010), and *Pyrus pyrifolia* (Terakami et al. 2012).

Based on the analysis of inverted repeats contraction and expansion, a conclusion could be drawn that chloroplast genomes of four *Phoebe* species exhibited comparatively little difference at the IR/LSC and IR/SSC boundary regions, which was similar in Veroniceae (Choi et al. 2016). The genetic divergence within Lauraceaeis surprisingly low (Rohwer 2000; Song et al. 2016). The mean nucleotide diversity of the complete chloroplast genome of the four *Phoebe* species was only 0.162%, lower than that of two *Panax* species (0.40%) (Dong et al. 2014), three *Veroniceae* species (0.40%) (Choi et al. 2016), and nine *Gossypium* species (0.62%) (Xu et al. 2012), and extremely lower than that of five *Epimedium* species (3.97%) (Zhang et al. 2016). Molecular markers with nucleotide diversity over 1.5% have been reported before as highly variable regions and could be used to promote the further phylogenetic analysis and species identification in other seed plants (Särkinen and George 2013; Korotkova et al. 2014; Huang et al. 2014). However, due to the relative conservative of chloroplast genomes of *Phoebe* species, the quantity of the regions with nucleotide diversity exceeding 0.5% was only four, whereas in *Nicotiana* (Asaf et al. 2016), the quantity of the regions with nucleotide diversity over 0.7% was 15, and in *Machilus* (Song et al. 2015), the quantity of the regions with nucleotide diversity over 0.8% was 7. In this study, the highest nucleotide diversity was 1.389% (*ccsA-ndhD*), followed by 0.887% (*petA-psbJ*), 0.716% (*rps4-trnT*), and 0.506% (*trnH-psbA*). Therefore, only the above four regions with nucleotide diversity higher than 0.5% probably had the potential use for phylogeographic analyses and plant identification of *Phoebe* species.

Conclusions

The 77 repeat sequences were identified in the *Phoebe* chloroplast genomes. The analysis showed the extremely conserved structure of chloroplast genomes, the mean nucleotide diversity was found to be 0.162% for 77 regions, suggesting an extraordinarily low level of sequence divergence. Four highest divergent regions (*trnH-psbA, rps14-trnT, petA-psbJ, ccsA-ndhD*) were also identified, which might be useful in species identification and phylogenetic studies. Overall, this study will facilitate our understanding of population genetics, phylogenetic relationship and plant evolution of *Phoebe* species.

Authors' contributions

LYG, XWQ and LXH conceived and designed the experiment; ZWT and JDY contributed to the sampling; XWQ and LYG performed the experiment and analyzed the data. The manuscript was written by LYG. All authors read and approved the final manuscript.

Author details

[1] Zhejiang Academy of Forestry, Hangzhou 310023, China. [2] Key Laboratory of Conservation Biology for Endangered Wildlife of Ministry of Education, College of Life Sciences, Zhejiang University, Hangzhou 310058, China. [3] Institute of Tropical Forestry, Chinese Academy of Forestry, Guangzhou 510520, China.

Acknowledgements

We thank Yingxiong Qiu, Ruisen Lu, Shanshan Zhu of Zhejiang University for providing experiment design and laboratory support.

Competing interests

The authors declare that they have no competing interests.

Funding

This work was supported by the Major Science and Technology Program (No. 2010C12009), Key Research & Development Program (No. 2017C02028) of Zhejiang Province, and National Key Research & Development Program (No. 2016YFD0600603) of China.

References

Asaf S, Khan AL, Khan AR, Waqas M, Kang SM, Khan MA, Lee SM, Lee IJ (2016) Complete chloroplast genome of *Nicotianaotophora* and its comparison with related species. Front Plant Sci 7:843–854

Chen JH, Hao ZD, Xu HB, Yang LM, Liu GX, Sheng Y, Zheng C, Zheng WW, Chen TL, Shi JS (2015) The complete chloroplast genome sequence of the relict woody plant *Metasequoia glyptostroboides* Hu et Cheng. Front Plant Sci 6:447

Choi KS, Chung MG, Park SJ (2016) The complete chloroplast genome sequences of three Veroniceae species (Plantaginaceae): comparative analysis and highly divergent regions. Front Plant Sci 7:342–355

Ding YJ, Zhang JH, Lu YF, Lin EP, Lou LH, Tong ZK (2015) Development of EST-SSR markers and analysis of genetic diversity in natural populations of endemic and endangered plant *Phoebe chekiangensis*. Biochem Syst Ecol 63:183–189

Dong WP, Liu H, Xu C, Zuo YJ, Chen Z, Zhou SL (2014) A chloroplast genomic strategy for designing taxon specific DNA mini-barcodes: a case study on ginsengs. BMC Genet 15:138–145

Douglas SE (1990) Plastid evolution: origins, diversity, trends. Curr Opin Genet Dev 8:655–661

Frazer KA, Pachter L, Poliakov A, Rubin EM, Dubchak I (2004) VISTA: computational tools for comparative genomics. Nucleic Acids Res 32:W273–W279

Gao L, Yi X, Yang YX, Su YJ, Wang T (2009) Complete chloroplast genome sequence of a tree fern *Alsophila spinulosa*: insights into evolutionary changes in fern chloroplast genomes. BMC Evol Biol 9:130–142

Hegde VR, Dai P, Ladislaw C, Patel MG, Puar MS, Pachter JA (1997) D_4 dopamine receptor-selective compounds from the Chinese plant *Phoebe chekiangensis*. Bioorg Med Chem Lett 9:1207–1212

Huang H, Shi C, Liu Y, Mao SY, Gao LZ (2014) Thirteen *Camellia* chloroplast genome sequences determined by high-throughput sequencing: genome structure and phylogenetic relationships. BMC Evol Biol 14:151–167

Ingvarsson PK, Ribstein S, Taylor DR (2003) Molecular evolution of insertions and deletion in the chloroplast genome of *Silene*. Mol Biol Evol 20:1737–1740

Katoh K, Standley DM (2013) MAFFT multiple sequence alignment software version 7: improvements in performance and usability. Mol Biol Evol 30:772–780

Khakhlova O, Bock R (2006) Elimination of deleterious mutations in plastid genomes by gene conversion. Plant J 46:85–94

Kim KJ, Lee HL (2004) Complete chloroplast genome sequences from Korean ginseng (*Panax schinseng* Nees) and comparative analysis of sequence evolution among 17 vascular plants. DNA Res 11:247–261

Kim JS, Jung JD, Lee JA, Park HW, Oh KH, Jeong WJ, Choi DW, Liu JR (2006) Complete sequence and organization of the cucumber (*Cucumis sativus* L cv. Baekmibaekdadagi) chloroplast genome. Plant Cell Rep 25:334–340

Korotkova N, Nauheimer L, Ter-Voskanyan H, Allgaier M, Borsch T (2014) Variability among the most rapidly evolving plastid genomic regions is lineage-specific: implications of pairwise genome comparisons in *Pyrus* (Rosaceae) and other angiosperms for marker choice. PLoS ONE 11:e112998

Kuang DY, Wu H, Wang YL, Gao LM, Zhang SZ, Lu L, Bonen L (2011) Complete chloroplast genome sequence of *Magnolia kwangsiensis* (Magnoliaceae): implication for DNA barcoding and population genetics. Genome 54:663–673

Kurtz S, Schleiermacher C (1999) REPuter: fast computation of maximal repeats in complete genomes. Bioinformatics 15:426–427

Librado P, Rozas J (2009) DnaSP v5: a software for comprehensive analysis of DNA polymorphism data. Bioinformatics 25:1451–1452

Lohse M, Drechsel O, Bock R (2007) OrganellarGenomeDRAW (OGDRAW): a tool for the easy generation of high-quality custom graphical maps of plastid and mitochondrial genomes. Curr Genet 52:267–274

Moore MJ, Bell CD, Soltis PS, Soltis DE (2007) Using plastid genome-scale data to resolve enigmatic relationships among basal angiosperms. Proc Natl Acad Sci USA 104:19363–19368

Nie XJ, Lv SZ, Zhang YX (2012) Complete chloroplast genome sequence of a major invasive species Crofton Weed (*Ageratina adenophora*). PLoS ONE 7:e36869

Parks M, Cronn R, Liston A (2009) Increasing phylogenetic resolution at low taxonomic levels using massively parallel sequencing of chloroplast genomes. BMC Biol 7:84–100

Pauwels M, Vekemans X, Gode C, Frerot H, Castric V, Saumitou-Laprade P (2012) Nuclear and chloroplast DNA phylogeography reveals vicariance among European populations of the model species for the study of metal tolerance *Arabidopsis halleri* (Brassicaceae). New Phytol 193:916–928

Porebski S, Bailey LG, Baum BR (1997) Modification of a CTAB DNA extraction protocol for plants containing high polysaccharide and polyphenol components. Plant Mol Biol Rep 15:8–15

Qian J, Song JY, Gao HH (2013) The complete chloroplast genome sequence of the medicinal plant *Salvia miltiorrhiza*. PLoS ONE 8:e57607

Rohwer JG (2000) Toward a phylogenetic classification of the Lauraceae: evidence from *matK* sequences. Syst Bot 25:60–71

Särkinen T, George M (2013) Predicting plastid marker variation: can complete plastid genomes from closely related species help? PLoS ONE 11:e82266

Schattner P, Brooks AN, Lowe TM (2005) The tRNAscan-SE, snoscan and snoGPS web servers for the detection of tRNAs and snoRNAs. Nucleic Acids Res 33:686–689

Song Y, Dong WP, Liu B, Xu C, Yao X, Gao J, Corlett RT (2015) Comparative analysis of complete chloroplast genome sequences of two tropical trees *Machilus yunnanensis* and *Machilus balansae* in the family Lauraceae. Front Plant Sci 6:662–669

Song Y, Yao X, Tan YH, Gan Y, Richard TC (2016) Complete chloroplast genome sequence of the avocado: gene organization, comparative analysis, and phylogenetic relationships with other Lauraceae. Can J Forest Res 46:1293–1301

Tangphatsornruang S, Sangsrakru D, Chanprasert J, Uthaipaisanwong P, Yoocha T, Jomchai N (2010) The chloroplast genome sequence of mungbean (*Vigna radiata*) determined by high-throughput pyrosequencing: structural organization and phylogenetic relationships. DNA Res 17:11–22

Terakami S, Matsumura Y, Kurita K, Kanamori H, Katayose Y, Yamamoto T (2012) Complete sequence of the chloroplast genome from pear (*Pyrus pyrifolia*): genome structure and comparative analysis. Tree Genet Genomes 8:1–14

Thiel T, Michalek W, Varshney RK, Graner A (2003) Exploiting EST databases for the development and characterization of gene-derived SSR-markers in barley (*Hordeum vulgare* L). Theor Appl Genet 106:411–422

Uthaipaisanwong P, Chanprasert J, Shearman J (2012) Characterization of the chloroplast genome sequence of oil palm (*Elaeis guineensis* Jacq). Gene 500:172–180

Wang MX, Cui LC, Feng KW, Deng PC, Du XH, Wan FH, Song WN, Nie XJ (2015) Comparative analysis of Asteraceae chloroplast genomes: structural organization, RNA editing and evolution. Plant Mol Biol Rep 33:1526–1538

Wicke S, Schneeweiss GM, Müller KF, Quandt D (2011) The evolution of the plastid chromosome in land plants: gene content, gene order, gene function. Plant Mol Biol 76:273–297

Wu ZY, Peter HR, Hong DY (2008) Flora of China. Science press, Beijing

Wyman SK, Jansen RK, Boore JL (2004) Automatic annotation of organellar genomes with DOGMA. Bioinformatics 20:3252–3255

Xu Q, Xiong G, Li P, He F, Huang Y, Wang K, Li ZH, Hua JP (2012) Analysis of complete nucleotide sequences of 12 gossypium chloroplast genomes: origin and evolution of allotetraploids. PLoS ONE 8:e37128

Yang M, Zhang XW, Liu GM, Yin YX, Chen KF, Yun QZ, Zhao DJ, Al-Mssallem IS, Yu J (2010) The complete chloroplast genome sequence of date palm (*Phoenix dactylifera L*). PLoS ONE 5:e12762

Yang J, Yue M, Niu C, Ma XF, Li ZH (2017) Comparative analysis of the complete chloroplast genome of four endangered herbals of *Notopterygium*. Genes 8:124

Yao X, Tang P, Li Z, Li D, Liu Y, Huang H (2015) The first complete chloroplast genome sequences in Actinidiaceae: genome structure and comparative analysis. PLoS ONE 10:e0129347

Zeng SY, Zhou T, Han K, Yang YC, Zhao JH, Liu ZL (2017) The complete chloroplast genome sequences of six *Rehmannia* species. Genes 8:103

Zhang Q, Li J, Zhao Y, Korba SS, Han Y (2012a) Evaluation of genetic diversity in Chinese wild apple species along with apple cultivars using SSR markers. Plant Mol Biol Rep 30:539–546

Zhang R, Zhou ZC, Jin GQ, Wang SH, Wang XH (2012b) Genetic diversity and differentiation within three species of the family Lauraceae in southeast China. Biochem Syst Ecol 44:317–324

Zhang YJ, Du LW, Liu A, Chen JJ, Wu L, Hu WM, Zhang W, Kim KH, Lee SC, Yang TJ, Wang Y (2016) The complete chloroplast genome sequences of five *Epimedium* species: lights into phylogenetic and taxonomic analyses. Front Plant Sci 7:306–317

Zhao Y, Yin J, Guo H (2015) The complete chloroplast genome provides insight into the evolution and polymorphism of *Panax ginseng*. Front Plant Sci 5:696–707

8

Synergistic effects of plant defense elicitors and *Trichoderma harzianum* on enhanced induction of antioxidant defense system in tomato against Fusarium wilt disease

Andleeb Zehra[1,2*], Mukesh Meena[1], Manish Kumar Dubey[1], Mohd. Aamir[1] and R. S. Upadhyay[1]

Abstract

Plant defense against their pathogens can be induced by a complex network of different inducers. The present study investigates the synergistic effect of *Trichoderma harzianum*, exogenous salicylic acid (SA) and methyl jasmonate (MeJA) over the response and regulation of the antioxidant defense mechanisms and lipid peroxidation in tomato plants against Fusarium wilt disease. In the present work, tomato plants were infected by *Fusarium oxysporum* f. sp. *lycopersici* 3 days after inoculated with *T. harzianum* and/or sprayed daily for 3 days with chemical inducers (SA and MeJA). Plants were analysed at 0, 24, 48, 72 and 96 h after inoculation with *Fusarium oxysporum* f. sp. *lycopersici*. Infection of tomato plants by pathogen led to strong reduction in the dry weight of roots and shoots with the enhanced concentration of H_2O_2 and varying degree of lipid peroxidation. Concurrently, exogenous SA, when applied with pathogen greatly enhanced H_2O_2 content as well as activities of antioxidant enzymes except catalase (CAT) and ascorbate peroxidase (APx). The pathogen challenged plants pretreated with *T. harzianum* and MeJA together exhibited less lipid peroxidation and as well as the elevated level of ascorbic acid and enhanced activities of antioxidant enzymes. All applied treatments protected tomato seedlings against Fusarium wilt disease but the percentage of protection was found higher in plants pretreated with the combination of *T. harzianum* and chemical inducers.

Keywords: *Trichoderma harzianum*, Chemical inducers, Reactive oxygen species, Antioxidant system, Lipid peroxidation

Introduction

Tomato (*Lycopersicon esculentum* Mill.) is a major contributor to the fruits and vegetable diet of humans throughout the world (Kapsiya et al. 2015). However, its growth, yield and economic productivity are significantly reduced by several pathogens. Fusarium wilt disease of tomato caused by *Fusarium oxysporum* f. sp. *lycopersici* is one of the most devastating disease worldwide which causes yield reduction and affects tomato production (Hanaa et al. 2011). The fungal pathogen resides in the soil and causes infection to susceptible tomato plants by the roots throughout their course of development causing necrosis and inducing wilting of the plants at the later stages of their infection and reduces the yield and hence limits the overall economic productivity (El-khallal 2007a, b).

Many biotic and abiotic agents can effectively induce plant resistance against various pathogens (Ramamoorthy et al. 2001; Edreva 2004). Bio-agents can induce resistance against various diseases caused by many pathogens such as viruses (Maurhofer et al. 1994), bacteria (Park and Kloepper 2000) and fungi (Bokhari and Perveen 2012). In spite of these, some bioagents such as *Trichoderma* spp. have been reported to promote plant growth, nutrient uptake, and induction of plant defense responses against different biotic and abiotic stresses (Harman et al. 2004; Shoresh et al. 2010; Hermosa et al. 2012). There are several evidences which support that

*Correspondence: zandleeb142@gmail.com
[2] Department of Botany, Banaras Hindu University, Varanasi 221 005, India
Full list of author information is available at the end of the article

Trichoderma spp. are able to induce the defense mechanisms in several plants (Vinale et al. 2008; Brotman et al. 2012). The results obtained through 2D electrophoresis and high-density oligonucleotide microarrays confirmed the induction of proteome and differential gene expression in plant system when inoculated with *Trichoderma* spp. (Marra et al. 2006; Alfano et al. 2007; Segarra et al. 2007). Different chemical inducers such as salicylic acid (SA), jasmonic acid (JA) and methyl jasmonate (MeJA) are signal molecules that play a key role in plant growth and development, and in the induction of plant defense responses to various abiotic and biotic stress factors (Sticher et al. 1997; War et al. 2011a, b). Various physiological, biochemical and molecular processes in plants including antioxidative enzyme activities can be manipulated by exogenous application of SA and JA (Idrees et al. 2011; War et al. 2011a, b). SA and MeJA play a role in various signal transduction mechanisms associated with plant defense and also elicit the expression of various important enzymes catalyzing a wide array of biosynthetic reactions whose products, through a complex network results in formation of various types of defense compounds including polyphenols, alkaloids and pathogenesis-related (PR) proteins (Hahlbrock and Scheel 1989; Creelman and Mullet 1995), conferring plant protection from pathogen-attack (Delaney et al. 1994; Kozlowski et al. 1999). The transcriptional reprogramming associated with plant defense responses are mediated by SA against different biotic and abiotic stresses (Herrera-Vásquez et al. 2015). Both endogenous and exogenous SA was evidenced to play roles in antioxidant metabolism and have a tight control over cellular ROS (Kang et al. 2014). In the current scenario, the plant protection provided by induction of systemic resistance is an effective and simple approach to disease management. The resistance of plants to pathogens can be enhanced by the application of various biotic and abiotic agents. However, combining different biocontrol agents with chemical inducers can significantly enhance the resistance of a plant and contribute to better control of pathogens (El-khallal 2007a, b).

Plants have developed effective defense mechanisms to survive after pathogen attack. The generation of reactive oxygen species (ROS) including superoxide anions (O_2^-), hydroxyl radicals (OH^-) and hydrogen peroxide (H_2O_2) is the principal indication of early plant defense response encountered with the pathogen and recognized by the same (Torres et al. 2006; Lehmann et al. 2015). One of the key functions associated with ROS system is the regulation of several biological and physiological processes including growth and development of plants and mitigation of both biotic and/or abiotic stresses by acting as a key signaling intermediates (Mittler et al. 2011; Baxter et al. 2013). Moreover, these ROS intermediates have been found to be implicated in defense response of plant through a plethora of mechanistic actions which is exemplified by cross-linking reactions of lignin and proteins causing plant cell-wall reinforcement, exposure of hypersensitive response (HR), or development of SAR against the targeted pathogen and least by acting as toxic agents against either the host plant cells, or the with development of SAR, or against the pathogen through killing them or ceasing their microbial activity. These ROS molecules have also been implicated in defense signaling pathway by acting as secondary messengers in signaling mechanism leading into the expression and activation of plant defense-related genes (Shimizu et al. 2006). Fortunately, plants possess effective mechanisms for detoxification of ROS in order to protect themselves from the toxic effects. Induction of antioxidant enzymes is a major defense mechanism of plants. Different antioxidant enzymes such as superoxide dismutase (SOD), catalase (CAT), ascorbate peroxidase (APx) and guaiacol peroxidase (GPx) involved in ROS metabolism during pathogen infection. SOD catalyses the dismutation reaction of O_2^- radicals to O_2 and H_2O_2. H_2O_2 is rapidly removed by CAT and peroxidases. CAT effectively eliminates the most of the H_2O_2, whereas APx can scavenge H_2O_2 that is inaccessible for CAT because of their higher affinity towards H_2O_2 and their occurrence in distinctive sub cellular locations (Creissen et al. 1994; Scandalios et al. 1997).

Here, the objective of this study was to find out the integrated effect of the biocontrol agent, *T. harzianum* and chemical inducers, SA and MeJA on biochemical changes in tomato plants against *Fusarium oxysporum* f. sp. *lycopersici* which leads to elicitation of plant defense response.

Materials and methods

Isolation and identification of pathogen

Fusarium oxysporum f. sp. *lycopersici* (Fol) was isolated from diseased tomato plants showing typical wilt symptoms on potato dextrose agar (PDA) media and incubated at 25–27 °C for 5–7 days. Stock cultures of Fol were prepared and stored at 4 °C. Pathogenicity test for all the isolates of Fol was done by following Koch's postulate. A highly virulent isolate was selected for further experiments.

Collection of *Trichoderma* species

Trichoderma harzianum BHU BOT RYRL4 (NCBI Accession No—KR 856210) and *T. asperellum* BHU BOT RYRL1 (NCBI Accession No—KR 856207) were provided from RY Roy Laboratory of Mycopathology, Department of Botany, Banaras Hindu University, Varanasi, India.

T. harzianum MTCC 936, *T. viride* MTCC 793 and *T. asperellum* MTCC 4347 were procured from Microbial Type Culture Collection (MTCC), IMTECH, Chandigarh, India. Some spp. of *Trichoderma* was also isolated from soil samples of the rhizospheric soil of tomato plants from different places of India. All the fungal cultures were maintained in tubes of Potato Dextrose Agar (PDA) slant medium supplemented with streptomycin antibiotic and stored at 4 °C for 6 months and revived thereafter.

Evaluation of the antagonistic activity of *Trichoderma* spp.

Different strains of *Trichoderma* spp. were screened for the evaluation of antagonistic activity against the pathogen by inoculating a 5 mm mycelial plug in the Petri plates. In the same Petri dishes perpendicular to the *Trichoderma* fungal disc, a 5 mm mycelial disc of 5-day-old pathogen culture was placed 6 cm away at the opposite side of the Petri dish. Plates were incubated at 27 ± 2 °C and growth of the pathogen mycelia towards the fungal disc was observed, and percent inhibition were measured 5 days after incubation. Data were obtained for the percentage inhibition of radial growth $[100 \times (C - T)/C]$ where C = radial growth of the pathogen in control and T = radial growth of the pathogen in dual culture with antagonist (Garrett 1956). The experiment was repeated thrice.

Inoculum preparation

For the preparation of the inoculum, spore suspensions from 7 days old cultures of *Trichoderma* spp. were flooded and mixed with glass rod using sterilized distilled water. The suspension was then filtered through cheese cloth and further diluted with distilled water to obtain a final concentration of 1×10^5 spores/ml using haemocytometer. Similarly, the inoculum preparation for pathogen was also done and the final concentration was maintained to 2×10^5 spores/ml.

Plant material and growth

Tomato (*Lycopersicon esculentum* cv. Punjab chhuara) seeds were obtained from Indian Institute of Vegetable Research (Adalpura, Varanasi) and were surface-sterilized in 0.5% sodium hypochlorite solution for 20 min and grown in autoclaved soil in greenhouse under alternate 14 h light and 10 h dark cycle at 27 and 22 °C, respectively. After attaining a height of 15–20 cm, these plants were used for testing pathogenicity against pathogen and also for further experiments.

Pathogenicity test

Pathogenicity test was carried out by inoculating healthy 15–20 cm tall plants with the spore suspension of pathogen containing 2×10^5 spores/ml followed by 48 h of incubation in the moist chamber after which the plants were returned to the green house chamber. Greenhouse temperature ranged from 25 to 29 °C with alternate 14 h light and 10 h dark cycles.

Evaluation of tomato plant growth characteristics of different *Trichoderma* strains

An in vivo assay was used to evaluate the abilities of different *Trichoderma* strains to promote tomato plant growth. Tomato seeds were coated with an aqueous suspension containing 1×10^5 spores of *Trichoderma* per ml (1 ml of spore suspension/30 seeds) and then air dried in an open petri-dish overnight under a laminar flow hood. Treated tomato seeds were sown in pots containing soil autoclaved at 121 °C for 1 h on 2 successive days. Pots with untreated tomato seeds were used as controls. The pots were incubated in a greenhouse at temperature ranged from 25 to 29 °C and watered as needed. Measurements of shoot length, stem diameter, and main root length were taken after 4 weeks.

In vitro antifungal activity assay of SA and MeJA against *F. oxysporum* f. sp. *lycopersici* and *Trichoderma* species

The effect of SA and MeJA was tested on radial growth of Fol and different strains of *Trichoderma* spp. cultured on potato dextrose agar (PDA) medium. The PDA was amended with SA and MeJA at 50, 100, 150, 200, 250 and 300 μM concentrations. Diameters of the fungal colony were measured at 7 days after inoculation with a 5 mm diameter plug of Fol and *Trichoderma* species.

Phytotoxic assay of SA and MeJA against tomato plants

The tomato plants were sprayed with 50, 100, 150, 200, 250 and 300 μM concentrations of SA and MeJA to detect the phytotoxicity of these elicitors against tomato plants. After 7 days, the plants were assayed for their phytotoxicity.

Treatment of plant material

When the tomato seedlings were 3–4 weeks old and 15–20 cm tall, tomato seedlings were inoculated with *T. harzianum* at 1×10^5 spores/ml and/or sprayed daily with chemical inducers, SA and MeJA (200 μM) on the leaf surface every 24 h for 3 consecutive days. Then, plants were inoculated with the spore suspension of Fol (2×10^5 spores/ml).

Plant harvest and analysis

The plants were harvested and the leaves from treated and control plants at five different stages at 0, 24, 48, 72 and 96 h post inoculation with pathogen, were collected for the assessment of active concentration of

H_2O_2, ascorbic acid, malondialdehyde (MDA). The enzymatic activities of different antioxidative enzymes including (SOD, APx, GPx and CAT) were also determined. In addition, we have also analyzed the effect of all the treatments given over the dry weight of roots and shoots. The treatments given were provided in the following combinations, *Fusarium oxysporum* f. sp. *lycopersici* + SA + *Trichoderma harzianum*, *F. oxysporum* f. sp. *lycopersici* + MeJA + *T. harzianum*, *F. oxysporum* f. sp. *lycopersici* + SA, *F. oxysporum* f. sp. *lycopersici* + MeJA, *F. oxysporum* f. sp. *lycopersici* + *T. harzianum*, *F. oxysporum* f. sp. *lycopersici* challenged and healthy control. For each treatment, three replicates having five seedlings per replicate were maintained.

3,3′-Diaminobenzidine (DAB) staining to detect H_2O_2 accumulation

Histochemical detection of H_2O_2 was done by 3,3′-diaminobenzidine (DAB) staining as described by Thordal-Christensen et al. (1997). Leaves of control and treated plants were cut with the razor blade and were immediately placed in a beaker containing 1 mg/ml 3,3′-diaminobenzidine-hydrochloric acid (DAB-HCL), adjusted to pH 5.6 with NaOH, and were incubated in a growth chamber for 8 h in the dark. This was followed by further washing and destaining in 96% boiling ethanol. Microscopic observation was made to find out the location of cells with H_2O_2 accumulation. The red–brown coloration was observed in the sample that shows the sites of H_2O_2 deposition.

H_2O_2 estimation

The H_2O_2 produced was determined as described by Sagisaka (1976). Leaf tissues from control and treated plants were homogenised in 5% cold trichloro acetic acid (TCA) and the homogenate was centrifuged at 17,000*g* for 10 min at 0 °C. The reaction mixture consisted of 1.6 ml of supernatant, 0.4 ml of 50% TCA, 0.4 ml of ferrous ammonium sulphate and 0.2 ml of potassium thiocyanate. The amount of H_2O_2 was measured by a calibration curve prepared with known concentrations of H_2O_2 by measuring the absorbance at 480 nm after 15 min of incubation.

Measurement of cell death

Histochemical analysis was done to check the degree of cell death using Evans blue staining. For this, the control and treated leaves were boiled for 1 min in a freshly prepared solution of phenol, lactic acid, glycerol and distilled water (1:1:1:1) containing 20 mg/ml Evans blue. The tissues after boiling were cleared overnight in a solution of 2.5 g/ml chloral hydrate in water. This staining easily demarcates the dead cells having intense blue stain as compared to the control plants that remained unstained.

The extent of cell death was measured by the method by Baker and Mock (1994). In this method, the samples from both control and treated leaf tissues were treated separately with 1 ml 0.25% solution of Evans blue and the samples were further incubated on a platform shaker at 27 °C and 80 rpm for 20 min. The samples were washed with deionized water in order to remove the traces of the dye from leaf tissues. For the removal of tissues bound dye, the samples were transferred to a 1.5 µl centrifuge tube and treated with 1.5% aqueous sodium dodecyl sulphate (SDS). The samples were crushed with mortar and pestle and the homogenates thus obtained were then diluted with 0.5% deionized water and were vortexed for 30 s. The samples were further transferred to the fresh tubes and centrifuged at 9000*g* for 3 min. The supernatant was taken in 0.8 ml aliquot and optical density was measured spectrophotometrically at 600 nm.

Ascorbic acid (AsA) content

Ascorbic acid was estimated by the method described by Keller and Schwager (1977). 0.1 g of fresh leaves from each of the treatments were homogenised with 2.0 ml of extracting solution (5 g oxalic acid + 0.75 g EDTA in 1000 ml of distilled water), centrifuged at 10,000*g* for 10 min and the supernatant was collected. To 1 ml of supernatant, 5 ml of 20 µg/ml of 2,6-dichlorophenol-indophenol (DCPIP) dye was added to develop color. The absorbance of the mixture was noted at 520 nm against the blank. The standard curve was prepared by using the different concentration of ascorbic acid by following the same method. The concentration of ascorbic acid is calculated by using the method of Keller and Schwager (1977).

Lipid peroxidation (LPO)

The extent of LPO was determined by using the method as described by Ohkawa et al. (1979). Estimation of total malondialdehyde (MDA), content produced as a secondary byproduct of polyunsaturated fatty acid peroxidation by the thiobarbituric acid (TBA) reaction is a major determinant of the extent of LPO content. The leaf samples from each of the treatments were homogenized in a buffer, centrifuged and the supernatant was used as extract for the assay. The reaction mixture consisted of 0.1 ml of extract, 0.2 ml of 8.1% SDS, 1.5 ml of 20% acetic acid and 1.5 ml of 0.8% aqueous solution of TBA. The volume of the reaction mixture was maintained up to 4.0 ml with distilled water and incubated at 95 °C for 1 h in water bath. After cooling, 1.0 ml of distilled water and 5.0 ml of mixture of *n*-butanol and pyridine (15:1 by vol) was added and centrifuged at 10,000*g* for 15 min. The spectrophotometric absorbance was measured at 532 nm and the amount of MDA was calculated based on an extinction coefficient of 155 mM^{-1} cm^{-1} and expressed as nmol MDA g^{-1} FW.

Antioxidant enzyme assay

Superoxide dismutase (SOD) activity

SOD (EC 1.15.1.1) activity was determined using a method described by Fridovich (1974). Leaf samples (0.1 g) from control and treated plants were homogenised in 2.0 ml of extraction buffer (0.1 M phosphate buffer containing 0.5 mM EDTA at pH 7.5) and centrifuged at 12,000*g* for 15 min at 4 °C. The supernatants thus obtained were used as enzyme extract. The reaction mixture consisted of 200 mM methionine, 2.25 mM NBT, 3 mM EDTA, 100 mM phosphate buffer (pH 7.8), 1.5 M sodium carbonate and enzyme extract. The final volume of the reaction mixture was maintained up to 3 ml. The reaction was initiated by adding 400 µl of riboflavin (2 µM) and the tubes were illuminated for 15 min. The reaction was stopped by keeping the tubes in dark. The absorbance was measured at 560 nm by measuring the ability of enzyme extract to inhibit photochemical reduction of nitroblue tetrazolium (NBT) chloride.

Ascorbate peroxidase (APx) activity

APx (E.C. 1.11.1.11) activity was estimated according to the method given by Nakano and Asada (1981) by measuring the oxidation of ascorbic acid at 290 nm with slight modifications. About 0.1 g leaf tissue were homogenised in 2 ml of extraction buffer (90 mM Na_2HPO_4 buffer, at pH 7.8, 8% glycerol, 1 mM EDTA, and 5 mM ascorbate). PVP (0.3 g/g tissue) was added and the homogenate was centrifuged at 15,000*g* for 10 min at 4 °C. The resulting supernatants were used for the enzymatic assay. Enzyme extract (200 µl) was added to the reaction mixture of 25 mM phosphate buffer (pH 7.0), 0.1 mM EDTA, 0.25 mM ascorbic acid and 1.0 mM H_2O_2 and the decrease in absorbance was recorded 30 s after addition of the enzyme extract. APx activity was calculated based on an extinction coefficient 2.8 mM^{-1} cm^{-1} and the enzymatic activity was expressed as nmol ascorbate oxidized min^{-1} mg^{-1} protein.

Catalase (CAT) activity

CAT (EC 1.11.1.6) activity was determined by using the method described by Aebi (1984). Plant tissue (0.1 g) from control and treated plants were homogenised in 50 mM Tris HCl buffer (pH 8.0) containing 0.5 mM EDTA, 2% w/v polyvinylpyrrolidone, and 0.5% (v/v) Triton X100. The homogenate was centrifuged at 12,000*g* for 15 min at 4 °C, and the resulting supernatant was used as enzyme extract. The reaction mixture consisted of 300 µM phosphate buffer (pH 7.2) and 100 µM H_2O_2 in 1 ml enzyme extract. Activity was determined by recording O_2 released from enzymatic dissociation of H_2O_2 in darkness for 1 min. O_2 produced by the enzymatic reaction was estimated by measuring the decrease in H_2O_2 absorption at 240 nm (extinction coefficient of H_2O_2 is 0.036 mM^{-1} cm^{-1}) and enzyme activity was expressed as µM H_2O_2 oxidized min^{-1} g^{-1} FW.

Guaiacol peroxidase (GPx) activity

GPx (EC 1.11.1.7) activity was spectrophotometrically determined as described by Zheng and Van Huystee (1992). The oxidation of guaiacol to tetra guaiacol was estimated by the increase in absorbance at 470 nm in a reaction mixture that contained 10 mM sodium phosphate (pH 6.0), 0.3% (v/v) H_2O_2, 1% (v/v) tetraguaiacol, and 0.3 ml enzyme extract. The reaction was started by the addition of H_2O_2. The enzymatic activity was determined in terms of U/mg protein as can be extrapolated by the linear portion of the activity curve whose absorbance was measured at 470 nm. The amount of the enzyme catalysing the oxidation of 1 µmol of guaiacol min^1 represents one unit of enzyme activity. Protein was estimated following the method of Lowry and others (1951).

Statistical analysis

All statistical analysis was done by using SPSS ver. 16. The results are the mean of the three replicates of each of the experiment. The data were statistically analysed by using one-way analysis of variance (ANOVA) and mean separations were compared with Duncan's multiple range tests at the $p \leq 0.05$ significance level. Differences at $p \leq 0.05$ were considered to be significant.

Results

Analysis of the antagonistic activity of *Trichoderma* species against Fol

Among the different species of *Trichoderma*, *T. harzianum* BHU-BOT-RYRL4 showed the highest percentage of inhibition of pathogen growth (83.17%). Whereas, *T. asperellum* BHU-BOT-RYRL1 showed the lowest inhibition (39.98%). *T. harzianum* MTCC 936, *T. viride* MTCC 793 and *T. asperellum* MTCC 4347 inhibited the growth of pathogen by 72.13, 68.12 and 50.27%, respectively.

Effect of the different *Trichoderma* strains on the growth characteristics of tomato plants

All the tested strains of *Trichoderma* had a significant effect on the shoot length, root length and stem diameter of tomato plants in comparison to control (Table 1). However, among the different strains of *Trichoderma*, *T. harzianum* BHU BOT RYRL 4 showed the highest tomato plant growth characteristics viz. shoot length (147.89%) root length (192.09%) and stem diameter (147.5%) followed by *T. harzianum* MTCC 936 (shoot

Table 1 Effect of the different strains of *Trichoderma* on the plant growth characteristics of tomato

Trichoderma isolates	Plant growth characteristics of tomato		
	Shoot length (cm)	Stem diameter (mm)	Root length (cm)
Control	19.23[a]	0.80[a]	8.26[a]
T. harzianum BHU BOT RYRL 4 (NCBI Accession No—KR 856210)	28.44[e]	1.18[c]	15.86[c]
T. asperellum BHU BOT RYRL 1 (NCBI Accession No—KR 856207)	22.22[ab]	0.89[ab]	10.22[ab]
T. harzianum MTCC 936	26.30[de]	1.13[b]	15.23[b]
T. viride MTCC 793	25.40[cd]	1.06[b]	13.60[b]
T. asperellum MTCC 4347	23.34[bc]	0.96[ab]	11.67[ab]

Values followed by the same letters are not significantly different at $p \leq 0.05$

length; 136.76%, root length; 184.38%, stem diameter; 141.25%) in comparison to control (Table 1). Whereas, *T. asperellum* BHU BOT RYRL 1 showed the lowest plant growth characteristics (shoot length; 115.54%, root length; 123.72%, stem diameter; 111.25%).

Based on these results, *T. harzianum* BHU-BOT-RYRL 4 was selected for further experiments.

In vitro antifungal activity of SA and MeJA against Fol and *T. harzianum*

The mycelial growth of Fol and *T. harzianum* was not significantly affected by SA and MeJA amendment in PDA medium. Four concentrations of SA and MeJA tested, viz., 50, 100, 150 and 200 μM, were not found to inhibit mycelial growth of Fol as well as *T. harzianum* significantly as compared to control. However, beyond 200 μM, significant reductions in the growth of both the fungi were recorded.

Phytotoxic assay of SA and MeJA against tomato plants

Four concentrations of SA and MeJA did not show any symptoms of phytotoxicity against tomato plants. Beyond 200 μM, SA and MeJA showed phytotoxicity against the tomato plants. Hence, the concentration of 200 μM for SA and MeJA was selected for spraying on tomato plants.

Total biomass

Significant increase in growth of tomato plants in terms of dry weight of shoot and root were observed in Fol + SA + Th and Fol + MeJA + Th treated plants in comparison to pathogen challenged plants. But the maximum dry weight of shoot and root by 137.19 and 218.75% was shown by Fol + SA + Th treated plants. Plants challenged with pathogen had decreased plant root and shoot dry weight when compared with control (Table 2).

Localisation of H_2O_2 production

For the detection of potent sites generating H_2O_2 due to the oxidative burst, DAB staining was performed using DAB as a substrate. The principle sites of H_2O_2 generation among the dead cell lesions emerged with having dark brown coloration of polymerized DAB and these sites were found to be more prevalent in pathogen challenged plants followed by those plants which were pretreated with SA along with pathogen. The microscopic observation reveals that the pathogen challenged leaves showed the maximum number of cells undergoing intense dark brown coloration exposing the affected sites for H_2O_2 accumulation (Fig. 1d) followed by Fol + SA treated plants (Fig. 1f). Moreover, it was observed that the vascular tissues were also targeted and engaged for generating H_2O_2 incited by oxidative burst for those samples which were pretreated with SA along with pathogen. These dark brown coloration precipitates were totally absent from controlled untreated samples (Fig. 1a, b). In comparison between treatments, minimum induction was observed in Fol + MeJA + Th treated plants (Fig. 1g, h).

Table 2 Effect of the pre-treatment of *T. harzianum* and/or chemical inducers (SA and MeJA) on the changes in dry weight of shoot and root of tomato plants challenged with *Fusarium oxysporum* f. sp. *lycopersici* (Fol)

Treatments	Dry weight of shoot (g/plant)	Dry weight of root (g/plant)
Control	4.92[c]	2.37[c]
Fol	3.28[a]	0.96[a]
Fol + Th	4.13[bc]	1.61[ab]
Fol + SA	3.91[ab]	1.32[ab]
Fol + MeJA	3.72[a]	1.21[ab]
Fol + Th + SA	4.56[c]	2.10[c]
Fol + Th + MeJA	4.31[c]	1.93[bc]

Values followed by the same letters are not significantly different at $p \leq 0.05$

H_2O_2 estimation

Treatment of tomato leaves with the pathogen, SA, MeJA and *T. harzianum* produced an oxidative burst in the form of H_2O_2 production. Treatment with SA especially

Fig. 1 Localisation of H_2O_2 in different treatments of tomato leaves as revealed by reddish brown stain. **a, b** Unchallenged control (control), **c, d** challenged with pathogen (Fol). **e, f** Pre-treated with SA and challenge inoculated with pathogen (Fol + SA). **g, h** Pre-treated with *T. harzianum* (Th), MeJA and challenge inoculated with pathogen (Fol + Th + MeJA)

when applied with pathogen markedly increased ROS production. Pathogen stress, SA, MeJA and Th treatment caused a sharp change in H_2O_2 content at 24 and 48 h. After 48 h of treatment, H_2O_2 decreased gradually in all treatments. Minimum H_2O_2 production was shown by Fol + Th + MeJA treated plants (Fig. 3a).

Measurement of cell death

Microscopic examination was done after staining the cells with Evans blue dye which clearly reveals the localized region of cell death and extent of damage incited by the oxidative burst. This was easily observed and determined by examining both the treated tissues and control samples under the microscope. The variation in color intensity from light to dark blue clearly pointed out the sites of damaged tissues and the extent of their destruction profile lying in between the live cells. This observed variation in colour intensity is directly proportional to the degree of damages as the fully dead tissues take the intense blue coloration when compared to active tissues from control samples having no coloration. In this study, maximum cell death was shown in pathogen inoculated leaves followed by Fol + SA treated leaves as compared with other treatments (Figs. 2c–f, 3b). Evans blue staining showed that death did not occur in untreated leaf tissue (Fig. 2a, b) while Fol + Th + MeJA treated plants showed minimum cell death (Fig. 2g, h).

Ascorbic acid (AsA) content

The level of AsA started to rise 24 h after inoculation reaching an optimal level up to 48 h and then started to decline after 72 h. The maximum AsA content was more prevalent in leaves treated with Fol + Th + SA when compared to single treatments. The total ascorbic acid was found to be highest in Fol + Th + SA treated samples whereas the control plants showed no variation in their AsA profile. AsA content was 345.6 and 176.32% higher in Fol + Th + SA treated plants as compared to control and pathogen challenged plants, respectively (Fig. 4).

Lipid peroxidation

The Malondialdehyde (MDA) content was significantly higher in pathogen challenged plants and was greatly reduced in the plants pretreated with Th and MeJA together as compared to the other treatments. Levels of MDA slowly increased in leaves of all the treatments but at a particularly lower level in Fol + Th + MeJA treated tomato plants as compared with the other treatments. The level of MDA gradually increased up to 48 h after pathogen infection and declined thereafter in all treatments. MDA content was 321.17% less and 188.36% higher in the Fol + Th + MeJA treated plants when compared with pathogen challenged and control plants, respectively. In Fol + Th + SA treated plants, it was 132.92% lesser than pathogen inoculated plants (Fig. 5a).

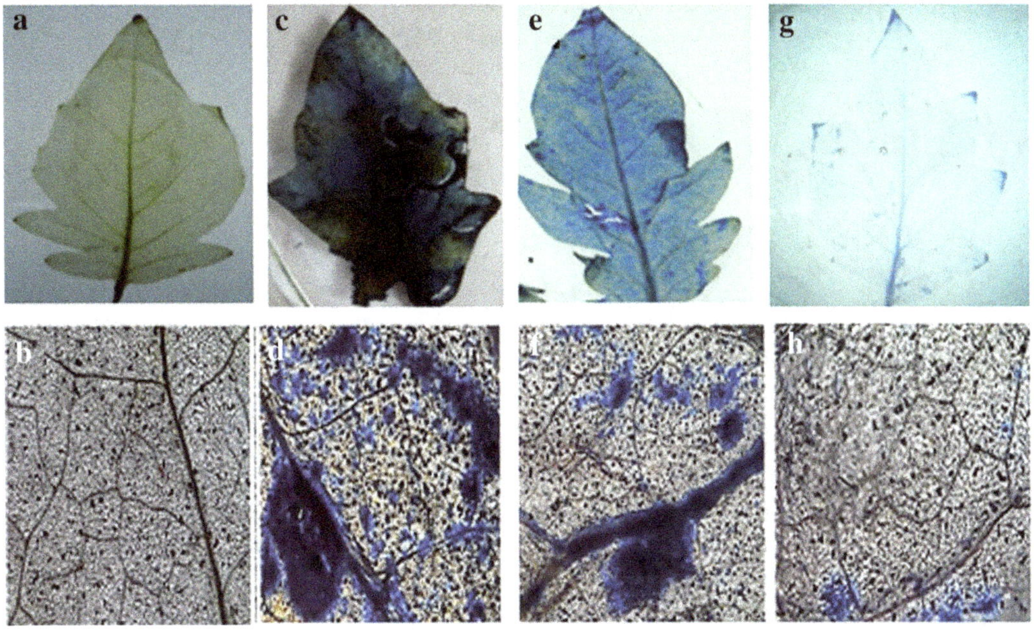

Fig. 2 Localisation of cell death (evans blue uptake) in different treatments of tomato leaves as revealed by blue stain. **a**, **b** Unchallenged control (control), **c**, **d** challenged with pathogen (Fol). **e**, **f** Pre-treated with SA and challenge inoculated with pathogen (Fol + SA). **g**, **h** Pre-treated with *T. harzianum* (Th), MeJA and challenge inoculated with pathogen (Fol + Th + MeJA)

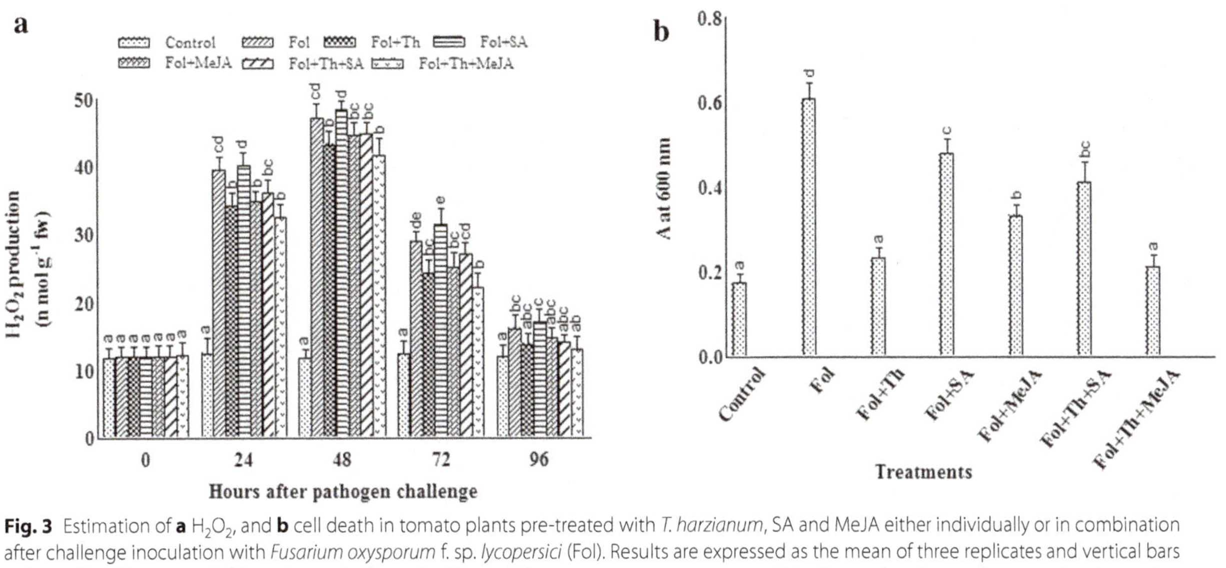

Fig. 3 Estimation of **a** H_2O_2, and **b** cell death in tomato plants pre-treated with *T. harzianum*, SA and MeJA either individually or in combination after challenge inoculation with *Fusarium oxysporum* f. sp. *lycopersici* (Fol). Results are expressed as the mean of three replicates and vertical bars indicate SD of the mean. Different letters indicate significant differences among treatments within the results taken at the same time interval according to Duncan's multiple range test at $p \leq 0.05$

Antioxidant enzyme activities

Infection with the pathogen resulted in the general increase in activity of antioxidant enzymes as compared to the control plants. The activities of APx, GPx and CAT increased at 24 h, peaking at 72 h and decreased slightly thereafter in all applied treatments (Fig. 6a–c). However, the SOD activity was found maximum at 48 h and declined thereafter (Fig. 5b). Treatment with chemical inducer especially SA when applied with pathogen markedly decreased H_2O_2 scavenging enzymes (APx and CAT) and greatly increased SOD activity in comparison with other treatments. Pre-treatment of tomato

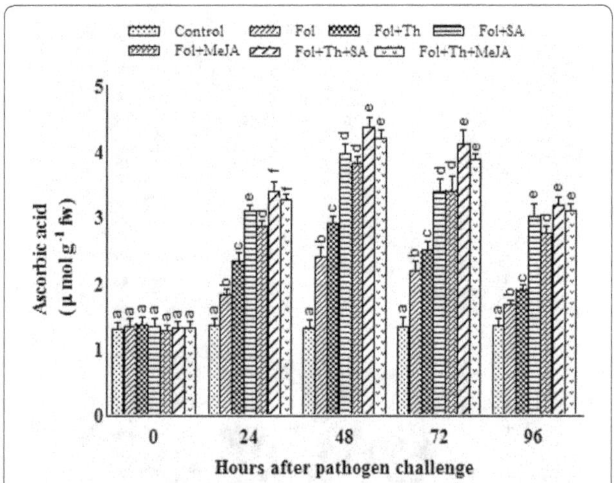

Fig. 4 Estimation of ascorbic acid in tomato plants pre-treated with *T. harzianum*, SA and MeJA either individually or in combination after challenge inoculation with *Fusarium oxysporum* f. sp. *lycopersici* (Fol). Results are expressed as the mean of three replicates and vertical bars indicate SD of the mean. Different letters indicate significant differences among treatments within the results taken at the same time interval according to Duncan's multiple range test at $p \leq 0.05$

plants with Th and MeJA markedly induced H_2O_2 scavenging enzymes and the highest induction was recorded when applied together. The activities of APx and CAT in Fol + Th + MeJA treated plants were 173.6 and 130.87% greater than those of the pathogen inoculated plants whereas in the case of Fol + Th + SA treated plants, the activities of GPx and SOD were 168.09 and 145.13% higher than pathogen challenged plants respectively which declined thereafter.

Discussion

ROS is speculated as the first line of active plant defense against various invading pathogens and may directly act as an antimicrobial agent (Malolepsza and Rozalska 2005; Torres et al. 2006; Shetty et al. 2007). Many reports have shown that H_2O_2 is a signaling molecule that plays a significant role in plant defense response (Neuenschwander et al. 1995; Orozco-Cárdenas et al. 2001). ROS generation leads to cellular damage and ultimately cell death. The estimation of cell death and degree of damages due to ROS system can be easily visualized by Evans blue dye whereas the DAB staining predicts the potent sites of H_2O_2 generation (Faoro and Iriti 2005). In our study, staining with DAB and Evans blue, which can visualise H_2O_2 accumulation and cell death demonstrates that the pathogen challenged followed by Fol + SA treated plants induced maximum H_2O_2 accumulation and cell death in leaf tissues in comparison to other treatments. In the present study, it was evident from the results that there was a more or less steady increase or decrease in the level of the H_2O_2 in all the treatments. Maximum H_2O_2 production was shown by the plants which were treated with SA along with pathogen or in combination with *T. harzianum*, while minimum production was shown by the Fol + Th + MeJA treated plants. Suppression of H_2O_2 production in leaves of Fol + Th + MeJA treatments could be associated with the increased activity of antioxidant enzymes which decreased levels of H_2O_2 (either by direct decomposition or oxidation). SA has been shown to increase accumulation of H_2O_2 in plants after pathogen infection which activates defense related genes by acting as a secondary messenger (Kauss and

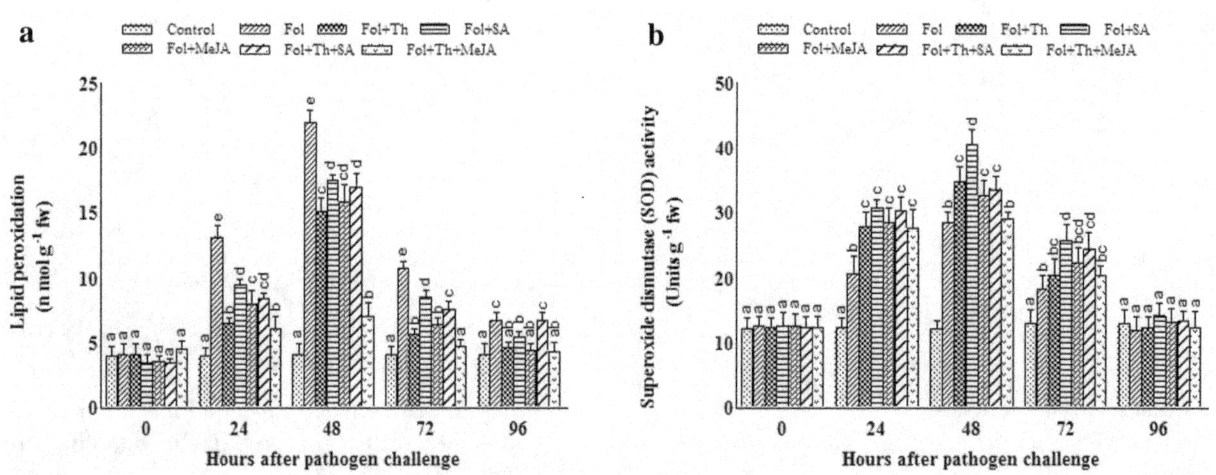

Fig. 5 Estimation of changes in the **a** concentration of MDA **b** activity of SOD, in tomato plants pre-treated with *T. harzianum*, SA and MeJA either individually or in combination after challenge inoculation with *Fusarium oxysporum* f. sp. *lycopersici* (Fol). Results are expressed as the mean of three replicates and vertical bars indicate SD of the mean. Different letters indicate significant differences among treatments within the results taken at the same time interval according to Duncan's multiple range test at $p \leq 0.05$

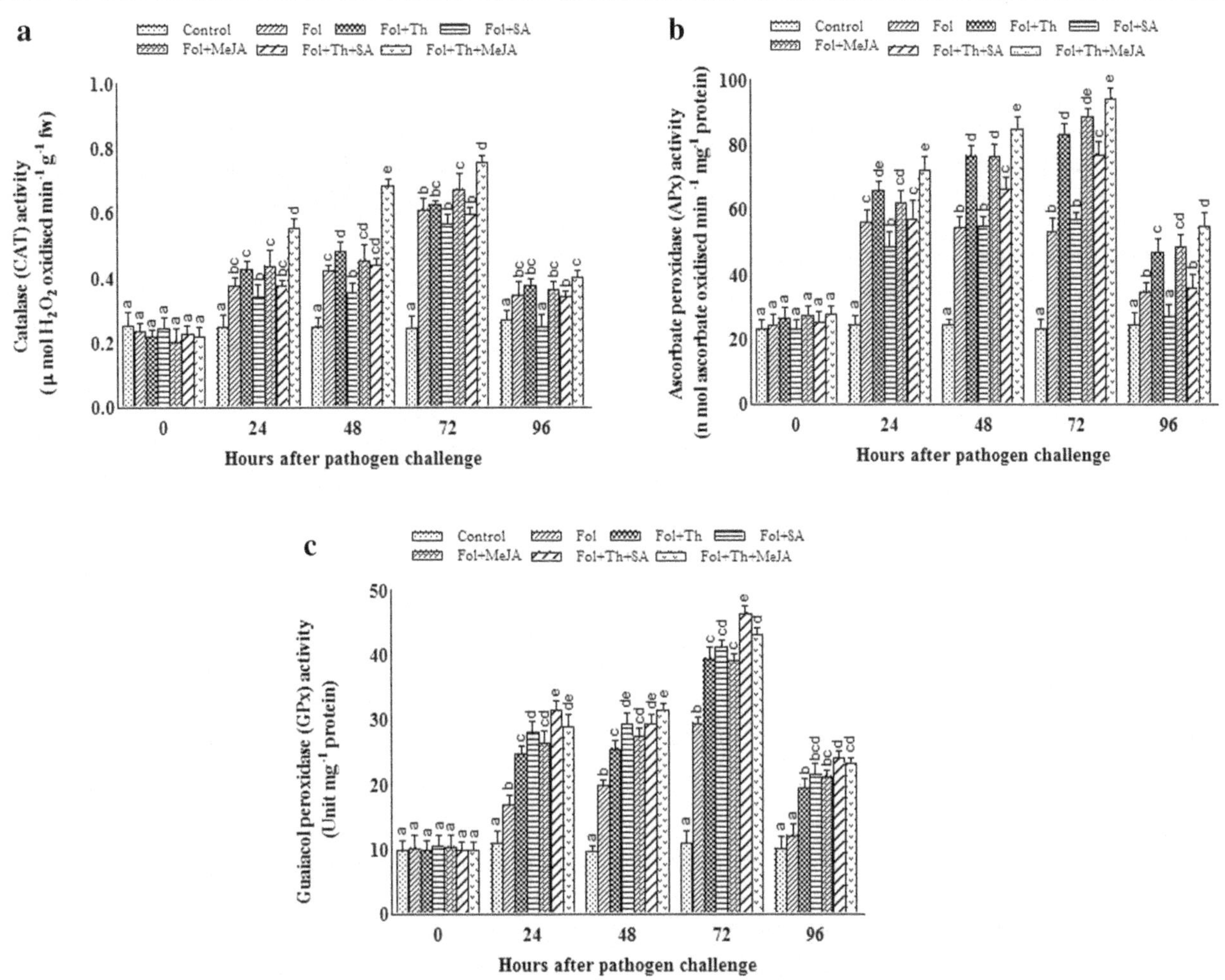

Fig. 6 Estimation of changes in the activities of **a** CAT, **b** APx, and **c** GPx in leaves of tomato plants pre-treated with *T. harzianum*, SA and MeJA either individually or in combination after challenge inoculation with *Fusarium oxysporum* f. sp. *lycopersici* (Fol). Results are expressed as the mean of three replicates and vertical bars indicate SD of the mean. Different letters indicate significant differences among treatments within the results taken at the same time interval according to Duncan's multiple range test at $p \leq 0.05$

Jeblick 1996; Garreton et al. 2002; Anand et al. 2008; Lee et al. 2010; Khokon et al. 2011; Miura and Tada 2014). The remarkable role of SA in plants is found to be associated with the activation of many genes whose expression terminates into plant defense activities including generation of H_2O_2 and induction of cell death in response to fungal elicitors and wounds (Kauss et al. 1992a, b; Shirasu et al. 1997; Mur et al. 2006). On the contrary, SA is considered to play an important role in preventing the damage to plants against oxidative stress by reducing the accumulation of ROS (Yang et al. 2004). An ambivalent outcome of SA as a prooxidant and antioxidant has been shown in many stress models against various pathogens and several abiotic stresses such as high light intensity, drought, salinity, and cold (Mou et al. 2003; Mateo et al. 2006; Miura and Tada 2014).

The antioxidant enzymes play a defensive role in plants against many oxidative stresses (Saed-Moucheshi et al. 2014). It has been shown that H_2O_2 enhances the antioxidant capacity of cells by increasing the activity of various antioxidant enzymes such as CAT, APx and SOD (Xia et al. 2009). The results showed that in the plants infected with *F. oxysporum* f. sp. *lycopersici*, the activities of APx, CAT and GPx significantly increased at 48 and 72 h in comparison to control, but the highest activity of these enzymes were found in plants treated with MeJA and *T. harzianum* together. However, the plants treated with the chemical inducer SA especially when applied with pathogen, noticeably decreased H_2O_2 scavenging enzymes (APx and CAT) and greatly increased SOD activity. The increased SOD activity due to SA might protect biomolecules from being attacked by superoxide radicals

(Belkadhi et al. 2013). The increased antioxidant enzyme activities in the MeJA and *T. harzianum* treated plants were more pronounced and higher levels were observed, indicating that the MeJA and Th pretreated plants were highly efficient in scavenging from the deleterious effects of ROS. The SOD enzyme of defense system dismutate the superoxide radicals generated after oxidative metabolism into H_2O_2 and O_2 which acts as a first line of defense response in plants (Gratão et al. 2012; Radwan 2012). SA plays an important role in generation of H_2O_2 (by inhibiting the expression of genes encoding for APx and catalases) which can be correlated with the activation of defense related genes and other enzymes such as SA-insensitive guaiacol peroxidases having an important role in other defense related responses including lignification and crosslinking of cell wall proteins (Gillham and Dodge 1987; Rajan and Murugan 2010). Chen et al. (1993) reported that SA inhibits the catalytic activity of enzyme catalase by binding to it. It has also been proved in tobacco where SA inhibits catalase activity (Conrath et al. 1995). Similarly, the in vitro inhibitory action of SA catalase-inhibiting effect has also been demonstrated in many other plant species such as *Arabidopsis*, tomato, cucumber (Sanchez-Casas and Klessig 1994), and maize (Horváth et al. 2002).

An initial increase in ascorbic acid content was observed in our study, which was significantly higher in Fol + Th + SA treated plants when compared with pathogen challenged and healthy controls. Thereafter, a marked decline in ascorbic acid content was observed in all treatments. The initial increase in ascorbic acid content in the present study can be explained by the fact that ascorbic acid along with glutathione are present at high concentrations in chloroplast and other cellular organelles which play a promising role in redox buffering of plant cells where they play a key role in plant defense.

The free fatty acids generated after degradation of membrane lipids act as substrate for enzyme lipoxygenase (LOX), causing oxidative deterioration resulting into membrane peroxidation. The membrane peroxidation finally produces various sorts of alkoxy and peroxy radicals including singlet oxygen. These biochemical reactions occurring inside the membrane lead into the major source of ROS generation for plant cells (Farmer and Mueller 2013). In our study, the levels of MDA gradually increased in leaves of infected tomato plants with increasing the time of infection. Tomato plants which were pretreated with Th and chemical inducers (SA and MeJA), MDA levels markedly decreased as compared with pathogen challenged plants. Among these treatments, pathogen challenged plants had the highest MDA level followed by Fol + SA treated plants. The catalytic activity of SOD is associated with the extent of lipid peroxidation as evidenced from the fact that SA treated plants results into the lower degree of membrane damage when compared to plants those were inoculated with the pathogen. One of the critical function of SA involves the activation of SOD that helps in the deactivation of lipid peroxidation thus facilitates in maintaining the integrity of membrane structures of root cells (Zenkov et al. 2001; Belkadhi et al. 2013). Minimum lipid peroxidation was found in Fol + Th + MeJA treated plants. The reduction in lipid peroxidation in tomato plants treated with Th along with MeJA might be related to the enhanced activity of antioxidant enzymes preventing formation and accumulation of free radicals, and subsequently membrane damage.

The individual treatment with chemical inducers like SA or MeJA or Fol elicitors did not have major effect on defense gene expression. However, pre-treatment of plants with these chemical inducers followed by pathogenic inoculation leads into aggravated defense response and characterized by increased expression of defense related genes such as Phenylalanine ammonia lyase (*PAL*) (in case of SA treatment) or increased expression of *LOX* genes (MeJA treatment). The two different chemical inducers have different effect on plant defense gene expression in which SA works through systemic acquired resistance (SAR) pathway. The SA in this SAR mediated defense may go through direct activation of PR genes expression or in low doses do not activate defense genes directly but prime the tissue for potentiated defense-gene expression upon subsequent Fol infection. SA-mediated transcriptional responses control the temporal patterns of gene expression in response to stress. Moreover, it has been well demonstrated that SA plays crucial role in *Trichoderma* root early colonization as revealed through expression studies (Alonso-Ramírez et al. 2014). In contrast, the MeJA as well as *Trichoderma* inoculated plants mediated defense response through ISR pathways involving ethylene and other hormones. This was confirmed through mutant studies where *Trichoderma harzianum* was inoculated with wild type as well mutants affected in the biosynthesis of specific defense-related hormones were selected, including the JA-deficient *defenseless1* (*def1*), the SA-deficient *NahG*, the ABA-deficient *sitiens* and the ET-under producing ACC deaminase *ACD* (Martínez-Medina et al. 2013). In all cases the mutant plants were found to be susceptible for disease development even after *Trichoderma* inoculation, indicating that JA-regulated pathway is required for TISR against pathogen challenged conditions. Recently, Rubio et al. (2017) have evaluated the defense gene expression following the combined treatment of *Trichoderma harzianum* T34 with NPK supplemented and under salt stressed conditions and demonstrated the expression of expression of eight genes out of nine analyzed: *EIN2*, encoding a central component of the ET signaling pathway; *NPR1*, which encodes a key transducer of SA signaling involved in plant defense responses; *AREB2*, encoding a transcription factor (TF) of ABA signaling; *LERBOH1*, involved in ROS production; *APX1*, encoding an ascorbate peroxidase; *SOS1*, which encodes a salt tolerance marker; *ARF1*, encoding a TF that binds to auxin response elements; and *DREB3*, a marker gene involved in tolerance to drought. AsA is a very important antioxidant, which protects the plants under oxidative stress (Shafiq et al. 2014). It has been reported that exogenous application of JA increases AsA in different plants suggesting that JA may regulate AsA metabolism (Ayala-Zavala et al. 2010; Chen and Gu 2014).

Trichoderma induced systemic resistance (TISR) in tomato against Fol challenged conditions is mainly based on boosted JA-dependent responses, the pathways regulated by the plant hormones-SA is also required for successful TISR development (Martínez-Medina et al. 2013). Priming for enhanced resistance to biotic and abiotic stress obviously is operating via various pathways involved in different metabolic processes. This suggests that the mode of action of priming and the resulting potentiation of cellular defense responses rather than the direct upregulation of defense signaling cascades might be of great advantage for living organisms. Plants pretreated with bioagent and chemical inducers together become primed to respond faster and show stronger activation of cellular defense responses after pathogen challenge compared with control plants. Higher induction of antioxidant enzymes in the plants which are pretreated with biological and chemical inducers in the present study can be correlated as a defense response triggered against *F. oxysporum* f. sp. *lycopersici* invasion in tomato plants.

Conclusion

The present study showed that the pre-treatment of the tomato plants with chemical inducers SA, MeJA and biocontrol agent *T. harzianum* induced the defense system against *F. oxysporum* f. sp. *lycopersici* infection with the increased expression of various antioxidant enzymes leading to enhanced defense response. In conclusion, it can be said that combinations of biocontrol agent and chemical inducers can be a useful and better promising measures for suppressing Fusarium wilt disease more efficiently than using alone due to the involvement of a number of defense mechanisms.

Abbreviations

Fol: *Fusarium oxysporum* f. sp. *lycopersici*; Th: *Trichoderma harzianum*; SA: salicylic acid; MeJA: methyl jasmonate; MDA: malondialdehyde; CAT: catalase; SOD: superoxide dismutase; APx: ascorbate peroxidase; GPx: guaiacol peroxidase; ROS: reactive oxygen species; H_2O_2: hydrogen peroxide; SAR: systemic acquired resistance; ISR: induced systemic resistance.

Authors' contributions

AZ sample collected, executed most of the experiments and designed the manuscript. MM, MKD and MA executed some of the experiments, RSU Resolution of the critical questions related to the accuracy of the data. All authors read and approved the final manuscript.

Author details

[1] Laboratory of Mycopathology and Microbial Technology, Centre of Advanced Study in Botany, Banaras Hindu University, Varanasi, Uttar Pradesh 221 005, India. [2] Department of Botany, Banaras Hindu University, Varanasi 221 005, India.

Acknowledgements

Financial support for this work by the University Grants Commission, (UGC), Government of India is gratefully acknowledged.

Competing interest

The authors declare that they have no competing interests.

References

Aebi H (1984) Catalase in vitro. Methods Enzymol 105:12–16

Alfano G, Ivey MLL, Cakir C, Bos JIB, Miller SA, Madden LV, Kamoun S, Hoitink HAJ (2007) Systemic modulation of gene expression in tomato by *Trichoderma hamatum* 382. Phytopathology 97:429–437

Alonso-Ramírez A, Poveda J, Martín I, Hermosa R, Monte E, Nicolás C (2014) Salicylic acid prevents *Trichoderma harzianum* from entering the vascular system of roots. Mol Plant Pathol 15(8):823–831

Anand A, Uppalapati SR, Ryu CM, Allen SN, Kang L, Tang Y, Mysore KS (2008) Salicylic acid and systemic acquired resistance play a role in attenuating crown gall disease caused by *Agrobacterium tumefaciens*. Plant Physiol 146:703–715

Ayala-Zavala J, Rosas-Domínguez C, Vega-Vega V, González-Aguilar G (2010) Antioxidant enrichment and antimicrobial protection of fresh-cut fruits using their own byproducts: looking for integral exploitation. J Food Sci 75(8):R175–R181

Baker CJ, Mock NM (1994) An improved method for monitoring cell death in cell suspension and leaf disc assays using evans blue. Plant Cell Tissue Organ Cult 39:7–12

Baxter A, Mittler R, Suzuki N (2013) ROS as key players in plant stress signalling. J Exp Bot. https://doi.org/10.1093/jxb/ert375

Belkadhi A, De Haro A, Soengas P, Obregon S, Cartea ME, Djebali W, Chaïbi W (2013) Salicylic acid improves root antioxidant defense system and total antioxidant capacities of flax subjected to cadmium. OMICS J Integr Biol 17:398–406

Bokhari NA, Perveen K (2012) Antagonistic action of *Trichoderma harzianum* and *Trichoderma viride* against *Fusarium solani* causing root rot of tomato. Afr J Microbiol Res 6:7193–7197

Brotman Y, Lisec J, Meret M, Chet I, Willmitzer L, Viterbo A (2012) Transcript and metabolite analysis of the *Trichoderma*-induced systemic resistance response to *Pseudomonas syringae* in *Arabidopsis thaliana*. Microbiology 158:139–146

Chen H, Gu Z (2014) Effect of ascorbic acid on the properties of ammonia caramel colorant additives and acrylamide formation. J Food Sci 79:C1678–C1682

Chen Z, Silva H, Klessig RF (1993) Active oxygen species in the induction of plant systemic acquired resistance by SA. Science 262:1883–1886

Conrath U, Chen ZX, Ricigliano JR, Klessig DF (1995) Two inducers of plant defense responses, 2,6-dichloroisonicotinec acid and salicylic acid, inhibit catalase activity in tobacco. Proc Natl Acad Sci USA 92:7143–7147

Creelman RA, Mullet JE (1995) Jasmonic acid distribution and action in plants:

regulation during development and response to biotic and abiotic stress. Proc Natl Acad Sci USA 92:4114–4119

Creissen G, Edwards A, Mullineaux P (1994) Glutathione reductase and ascorbate peroxidase. In: Foyer CH, Mullineaux PM (eds) Causes of photooxidative stress and amelioration of defense system in plants. CRC Press, Boca Raton, pp 343–364

Delaney TP, Uknes S, Vernooij B, Friedrich L, Weymann K, Negrotto D, Gaffney T, Gut-Rella M, Kessmann H, Ward E et al (1994) A central role of salicylic acid in plant disease resistance. Science 266:1247–1249

Edreva A (2004) A novel strategy for plant protection: induced resistance. J Cell Mol Biol 3:61–69

El-Khallal SM (2007a) Induction and modulation of resistance in tomato plants against Fusarium wilt disease by bioagent fungi (arbuscular mycorrhiza) and/or hormonal elicitors (jasmonic acid & salicylic acid): 1-changes in growth, some metabolic activities and endogenous hormones related to defence mechanism. Aust J Basic Appl Sci 1:691–705

El-Khallal SM (2007b) Induction and modulation of resistance in tomato plants against *Fusarium* wilt disease by bioagent fungi (arbuscular mycorrhiza) and/or hormonal elicitors (jasmonic acid & salicylic acid): 2-changes in the antioxidant enzymes, phenolic compounds and pathogen related-proteins. Aust J Basic Appl Sci 1:717–732

Faoro F, Iriti M (2005) Cell death behind invisible symptoms: early diagnosis of ozone injury. Biol Plant 49:585–592

Farmer EE, Mueller MJ (2013) ROS-mediated lipid peroxidation and RES-activated signaling. Annu Rev Plant Biol 64:429–450

Fridovich I (1974) Superoxide dismutases. Adv Enzymol Relat Areas Mol Biol 41:35–97

Garreton V, Carpinelli J, Jordana X, Holuigue L (2002) The as-1 promoter element is an oxidative stress-responsive element and salicylic acid activates it via oxidative species. Plant Physiol 130:1516–1526

Garrett SD (1956) Biology of root-infecting fungi. Cambridge University Press, New York, p 293

Gillham DJ, Dodge AD (1987) Chloroplast superoxide and hydrogen peroxide scavenging systems from pea leaves: seasonal variation. Plant Sci 50:105–109

Gratão PL, Monteiro CC, Carvalho RF, Tezotto T, Piotto FA, Peres LEP, Azevedo RA (2012) Biochemical dissection of diageotropica and never ripe tomato mutants to Cd-stressful conditions. Plant Physiol Biochem 56:79–96

Hahlbrock K, Scheel D (1989) Physiology and molecular biology of phenylpropanoid metabolism. Annu Rev Plant Physiol Plant Mol Biol 40:347–369

Hanaa RMF, Abdou ZA, Salama DA, Ibrahim MAR, Sror HAM (2011) Effect of neem and willow aqueous extracts on fusarium wilt disease in tomato seedlings: Induction of antioxidant defensive enzymes. Ann Agric Sci 56:1–7

Harman GE, Howell CR, Viterbo A, Chet I, Lorito M (2004) *Trichoderma* species-opportunistic, avirulent plant symbionts. Nat Rev Microbiol 2:43–56

Hermosa R, Viterbo A, Chet I, Monte E (2012) Plant-beneficial effects of *Trichoderma* and of its genes. Microbiology 158:17–25

Herrera-Vásquez A, Salinas P, Holuigue L (2015) Salicylic acid and reactive oxygen species interplay in the transcriptional control of defense genes expression. Front Plant Sci 6:171

Horváth E, Janda T, Szalai G, Páldi E (2002) In vitro salicylic acid inhibition of catalase activity in maize: differences between the isoenzymes and a possible role in the induction of chilling tolerance. Plant Sci 163:1129–1135

Idress M, Naeem N, Aftab T, Khan MMA (2011) Salicylic acid mitigates salinity stress by improving antioxidant defense system and enhances vincristine and vinblastine alkaloids production in periwrinkle (*Catharanthus roseus* (L.) G. Don). Acta Physiol Plant 33:987–999

Kang G, Li G, Guo T (2014) Molecular mechanism of salicylic acid-induced abiotic stress tolerance in higher plants. Acta Physiol Plant 36:2287–2297

Kapsiya J, Gungula DT, Tame VT, Bukar N (2015) Effects of storage chemicals and packaging systems on physicochemical characteristics of tomato (*Solanum lycopersicum* L.) fruits. AASCIT J Biosci 1:41–46

Kauss H, Jeblick W (1996) Influence of salicylic acid on the induction of competence for H_2O_2 elicitation. Plant Physiol 11:755–763

Kauss H, Krause K, Jeblick W (1992a) Methyl jasmonate conditions parsley suspension cells for increased elicitation of phenylpropanoid defense responses. Biochem Biophys Res Commun 189:304–308

Kauss H, Theisinger-Hinkel E, Mindermann R, Conrath U (1992b) Dichloroisonicotinic and salicylic acid, inducers of systemic acquired resistance, enhance fungal elicitor responses in parsley cells. Plant J 2:655–660

Keller T, Schwager H (1977) Air pollution and ascorbic acid. Eur J For Pathol 7:338–350

Khokon AR, Okuma E, Hossain MA, Munemasa S, Uraji M, Nakamura Y et al (2011) Involvement of extracellular oxidative burst in salicylic acid induced stomatal closure in *Arabidopsis*. Plant Cell Environ 34:434–443

Kozlowski G, Buchala A, Metraux JP (1999) Methyl jasmonate protects Norway spruce [*Picea abies* (L.) Karst.] seedlings against *Pythium ultimum* Trow. Physiol Mol Plant Pathol 55:53–58

Lee S, Kim SG, Park CM (2010) Salicylic acid promotes seed germination under high salinity by modulating antioxidant activity in Arabidopsis. New Phytol 188:626–637

Lehmann S, Serrano M, L'Haridon F, Tjamos SE, Metraux JP (2015) Reactive oxygen species and plant resistance to fungal pathogens. Phytochemistry 112:54–62

Lowry OH, Rosebrough NJ, Farr AL, Randall RJ (1951) Protein measurement with the Folin phenol reagent. J Biol Chem 193:265–275

Malolepsza U, Rozalska S (2005) Nitric oxide and hydrogen peroxide in tomato resistance, nitric oxide modulates hydrogen peroxide level in o-hydroxyethylorutin-induced resistance to *Botrytis cinerea* in tomato. Plant Physiol Biochem 43:623–635

Marra R, Ambrosino P, Carbone V, Vinale F, Woo SL, Ruocco M, Ciliento R, Lanzuise S, Ferraioli S, Soriente I, Gigante S, Turra D, Fogliano V, Scala F, Lorito M (2006) Study of the three-way interaction between *Trichoderma atroviride*, plant and fungal pathogens by using a proteomic approach. Curr Genet 50:307–321

Martínez-Medina A, Fernández I, Sánchez-Guzmán MJ, Jung SC, Pascual JA, Pozo MJ (2013) Deciphering the hormonal signalling network behind the systemic resistance induced by *Trichoderma harzianum* in tomato. Front Plant Sci 4:206

Mateo A, Funck D, Mühlenbock P, Kular B, Mullineaux PM, Karpinski S (2006) Controlled levels of salicylic acid are required for optimal photosynthesis and redox homeostasis. J Exp Bot 57:1795–1807

Maurhofer M, Hase C, Meuwly P, Metraux JP, Defago G (1994) Induction of systemic resistance of tobacco to tobacco necrosis virus by the root-colonizing *Pseudomonas fluorescens* strain CHAO: influence of the gacA gene and of pyoverdine production. Phytopathology 84:139–146

Mittler R, Vanderauwera S, Suzuki N, Miller G, Tognetti VB, Vandepoele K, Gollery M, Shulaev V, Van Breusegem F (2011) ROS signaling: the new wave? Trends Plant Sci 16:300–309

Miura K, Tada Y (2014) Regulation of water, salinity, and cold stress responses by salicylic acid. Front Plant Sci 5:4

Mou Z, Fan W, Dong X (2003) Inducers of plant systemic acquired resistance regulate NPR1 function through redox changes. Cell 113:935–944

Mur LA, Kenton P, Atzorn R, Miersch O, Wasternack C (2006) The outcomes of concentration-specific interactions between salicylate and jasmonate signaling include synergy, antagonism, and oxidative stress leading to cell death. Plant Physiol 140:249–262

Nakano Y, Asada K (1981) Hydrogen peroxide is scavenged by ascorbate-specific peroxidase in spinach chloroplasts. Plant Cell Physiol 22:867–880

Neuenschwander U, Vernooij B, Fridrich L, Uknes S, Kessmann H, Ryals J (1995) Is hydrogen peroxide a second messenger of salicylic acid in systemic acquired resistance? Plant J 8:227–233

Ohkawa H, Ohishi N, Yagi K (1979) Assay for lipid peroxides in animal tissues by thiobarbituric acid reaction. Anal Biochem 95:351–358

Orozco-Cárdenas ML, Narváez-Vásquez J, Ryan CA (2001) Hydrogen peroxide acts as a second messenger for the induction of defense genes in tomato plants in response to wounding, systemin, and methyl jasmonate. Plant Cell 13:179–191

Park KS, Kloepper JW (2000) Activation of PR-1 a promoter by rhizobacteria that induce systemic resistance in tobacco against *Pseudomonas syringae* pv. *tabaci*. Biol Control 18:2–9

Radwan DEM (2012) Salicylic acid induced alleviation of oxidative stress caused by clethodim in maize (*Zea mays* L.) leaves. Pest Biochem Physiol 102:182–188

Rajan SS, Murugan K (2010) Purification and kinetic characterization of the liverwort *Pallavicinia lyelli* (Hook.) S. Gray. cytosolic ascorbate peroxidase. Plant Physiol Biochem 48:758–763

Ramamoorthy V, Viswanathan R, Raguchandar J, Prakasham T, Samiyappan R (2001) Induction of systemic resistance by plant growth promoting rhizobacteria in crop plants against pest and diseases. Crop Prot 20:1–11

Rubio MB, Hermosa R, Vicente R, Gómez-Acosta FA, Morcuende R, Monte E, Bettiol W (2017) The combination of *Trichoderma harzianum* and chemical fertilization leads to the deregulation of phytohormone networking, preventing the adaptive responses of tomato plants to salt stress. Front Plant Sci. https://doi.org/10.3389/fpls.2017.00294

Saed-Moucheshi A, Pakniyat H, Pirasteh-Anosheh H, Azooz MM (2014) Role of ROS as signaling molecules in plants. Elsevier Inc, Amsterdam

Sagisaka S (1976) The occurrence of peroxide in a perennial plant, *Populus gelrica*. Plant Physiol 57:308–309

Sanchez-Casas P, Klessig DF (1994) A salicylic acid-binding activity and a salicylic acid-inhibitable catalase activity are present in a variety of plant species. Plant Physiol 106:1675–1679

Scandalios JG, Guan L, Polidoros AN (1997) Catalases in plants: gene structure, properties, regulation, and expression. In: Scandalios JG (ed) Oxidative stress and the molecular biology of antioxidant defenses. Cold Spring Harbor Laboratory Press, Cold Spring Harbor, pp 343–406

Segarra G, Casanova E, Bellido D, Odena MA, Oliveira E, Trillas I (2007) Proteome, salicylic acid and jasmonic acid changes in cucumber plants inoculated with *Trichoderma asperellum* strain T34. Proteomics 7:3943–3952

Shafiq S, Akram NA, Ashraf M, Arshad A (2014) Synergistic effects of drought and ascorbic acid on growth, mineral nutrients and oxidative defense system in canola (*Brassica napus* L.) plants. Acta Physiol Plant 36:1539–1553

Shetty NP, Mehrabi R, Lutken H, Haldrup A, Kema GH, Collenge DP, Jorgenson HJ (2007) Role of hydrogen peroxide during the interaction between the hemibiotrophic fungal pathogen *Septoria tritici* and wheat. New Phytol 174:637

Shimizu N, Hosogi N, Hyon G-S, Jiang S, Inoue K, Park P (2006) Reactive oxygen species (ROS) generation and ROS-induced lipid peroxidation are associated with plasma membrane modifications in host cells in response to AK-toxin I from *Alternaria alternata* Japanese pear pathotype. J Gen Plant Pathol 72:6–15

Shirasu K, Nakajima H, Rajasekhar VK, Dixon RA, Lamb C (1997) Salicylic acid potentiates an agonist-dependent gain control that amplifies pathogen signals in the activation of defence mechanisms. Plant Cell 9:261–270

Shoresh M, Harman GE, Mastouri F (2010) Induced systemic resistance and plant responses to fungal biocontrol agents. Annu Rev Phytopathol 48:21–43

Sticher L, Mauch-Mani B, Metraux JP (1997) Systemic acquired resistance. Annu Rev Phytopathol 35:235–270

Thordal-Christensen H, Zhang Z, Wei Y, Collinge DB (1997) Subcellular localization of H_2O_2 in plants. H_2O_2 accumulation in papillae and hypersensitive response during the barley—powdery mildew interaction. Plant J 11:1187–1194

Torres MA, Jonathan DG, Dangl JL (2006) Reactive oxygen species signaling in response to pathogen. Plant Physiol 141:373–378

Vinale F, Sivasithamparam K, Ghisalberti EL, Marra R, Woo SL, Lorito M (2008) *Trichoderma*—plant-pathogen interactions. Soil Biol Biochem 40:1–10

War AR, Paulraj MG, War MY, Ignacimuthu S (2011a) Role of salicylic acid in induction of plant defense system in chickpea (*Cicer arietinum* L.). Plant Signal Behav 6:1787–1792

War AR, Paulraj MG, War MY, Ignacimuthu S (2011b) Jasmonic acid-mediated induced resistance in groundnut (*Arachis hypogaea* L.) against *Helicoverpa armigera* (Hubner) (Lepidoptera: Noctuidae). J Plant Growth Regul 30:512–523

Xia XJ, Wang YJ, Zhou YH, Tao Y, Mao WH, Shi K, Asami T, Chen Z, Yu JQ (2009) Reactive oxygen species are involved in brassinosteroid-induced stress tolerance in cucumber. Plant Physiol 150:801–814

Yang YN, Qi M, Mei CS (2004) Endogenous salicylic acid protects rice plants from oxidative damage caused by aging as well as biotic and abiotic stress. Plant J 40:909–919

Zenkov NK, Lankin VE, Menshchikova EV (2001) Okislitel'nyi stress (oxidative stress). Nauka, Moscow, p 280

Zheng X, Van Huystee RB (1992) Peroxidase-regulated elongation of segments from peanut hypocotyls. Plant Sci 81:47–56

Screening transferable microsatellite markers across genus *Phalaenopsis* (Orchidaceae)

Ya-Zhu Ko[1], Huei-Chuan Shih[2], Chi-Chu Tsai[3,4], Hsing-Hua Ho[1], Pei-Chun Liao[5*] and Yu-Chung Chiang[1,6*]

Abstract

Background: Molecular identification based on microsatellite loci is an important technology to improve the commercial breeding of the moth orchid. There are more than 30,000 cultivars have been enrolled at the Royal Horticultural Society (RHS). In this study, genomic microsatellite primer sets were developed from *Phalaenopsis aphrodite* subsp. *formosana* to further examine the transferability of across 21 *Phalaenopsis* species.

Methods and results: Twenty-eight polymorphic microsatellite markers were obtained using the magnetic bead enrichment method, with high transferability of the 21 species of the genus *Phalaenopsis*, especially in the subgenus *Phalaenopsis*. The 28 newly developed polymorphic microsatellite markers with high polymorphism information content values. The best and second fit grouping (*K*) are inferred as two and four by the ΔK evaluation in the assignment test. This result indicates that these microsatellite markers are discernible to subgenus *Phalaenopsis*.

Conclusions: Our results indicate that these new microsatellite markers are useful for delimiting species within genus *Phalaenopsis*. As expected, the genetic relationships between species of subgenus *Phalaenopsis* can be well distinguished based on the assignment test. These molecular markers could apply to assess the paternity of *Phalaenopsis* as well as investigating hybridization among species of genus *Phalaenopsis*.

Keywords: *Phalaenopsis*, Microsatellites, Polymorphism, Transferability

Background

The subtropical Taiwan Island that is situated off the southeastern Asian continent has well-suited climate conditions for the growth of orchids. Since the high quality of breeding and micropropagation technology coupled with market demands of the orchid genus *Phalaenopsis* Blume (Orchidaceae), Taiwan has become one of the important exporting countries of orchids in the world (Chen and Chen 2007, 2011; Tang and Chen 2007). The genus *Phalaenopsis* belongs to the family Orchidaceae, subfamily Epidendroideae, tribe Vandeae and subtribe Aeridinae (Dressler 1993), which is often known as moth orchid and comprises approximately 66 species (Christenson 2001). *Phalaenopsis* species is broadly distributed across Himalayas of northern India, South India, Sri Lanka, Southeast China, Taiwan, Indonesia, Thailand, Myanmar, Malaysia, the Philippines, Papua New Guinea and northeastern Australia (Chen and Chen 2011; Christenson 2001). According to the pollinia numbers (Christenson 2001) and molecular evidences (Tsai et al. 2010), *Phalaenopsis* can be divided into five subgenera: the four pollinia clades of subgenera *Proboscidioides*, *Aphyllae*, and *Parishianae* and the two pollinium clades of subgenera *Polychilos* and *Phalaenopsis*. Among these subgenera, the *Polychilos* and *Phalaenopsis* was each subdivided into four sections *Polychilos*, *Fuscatae*, *Amboinenses*, *Zebrinae* and *Phalaenopsis*, *Deliciosae*, *Esmeralda*, *Stauroglottis*, respectively (Dressler 1993; Christenson 2001).

The species of genus *Phalaenopsis* are most popular epiphytic monopodial orchids for their distinctive and

*Correspondence: pcliao@ntnu.edu.tw; yuchung@mail.nsysu.edu.tw
[1] Department of Biological Sciences, National Sun Yat-sen University, Kaohsiung 804, Taiwan
[5] Department of Life Science, National Taiwan Normal University, Taipei, Taiwan
Full list of author information is available at the end of the article

varied flowers with the unique structure. Horticultural breeding by hybridization remixed floral characters, such as the colors, shapes, and sizes, to create diversified varieties and cultivars. Based on the high breeding and cultivation techniques for the regulation of light and feeding and the development in interspecific and intergeneric hybrids breeding and polyploidy, improvement of the long-lasting quality of the floral traits made *Phalaenopsis* as one of an important orchid source for cut-flower crop.

There are two indigenous species of *Phalaenopsis* native to Taiwan, the *P. aphrodite* subsp. *formosana* and *P. equestris* (Chen and Wang 1996). Both species were classified as the section *Phalaenopsis* (Christenson 2001). *Phalaenopsis aphrodite* subsp. *formosana*, commonly known as the Taiwan moth orchid, has been widely used as an important breeding hybrids parent, and it is one of the most important progenitors for the traits of modern large and white of floral organs commercial hybrids breeding (Tanaka et al. 2005). *Phalaenopsis equestris* is another important breeding parent for the miniature type of multi-flowers and artificial hybrids with white petals and sepals and a red lip (Men et al. 2003; Tang and Chen 2007).

Recently, intergeneric hybrids between *Phalaenopsis* and *Ascocenda* cultivars were developed to introduce orange color into hybrid cultivars (Liu et al. 2016). However, complex phenotype and long stage of juvenile make the identification of varieties and cultivars of *Phalaenopsis* plants difficult and time consuming. In addition, traditional horticultural breeding technique for new cultivars of *Phalaenopsis* by integrating the morphology, physiological development, and environmental factors as well as their complex interactions makes the breeding consequence unpredictable and uncertain. Molecular markers can provide sensitive and accurate tools for identifying species and cultivars. Therefore, development of highly reliable, rapid, and cheap technique for differentiating and identifying seedlings of species and cultivars of *Phalaenopsis* is necessary and useful for enhancing the efficiency of the breeding. Furthermore, development of molecular markers could apply to paternity analysis, phylogenetic reconstruction, and resolving long-standing issues on *Phalaenopsis* breeding. Microsatellite markers with characteristics of high level of polymorphism, codominant inheritance and reproducibility (Powell et al. 1996) are useful tools for application in plant genetics and crop breeding, including fruit tree (Chiang et al. 2012; Chiou et al. 2012; Tsai et al. 2013; Lai et al. 2015) and orchid (Tsai et al. 2014, 2015). Compared to previous studies (Sukma 2011; Tsai et al. 2015), we intend to use more microsatellite loci as well as more extensive species testing in this study to enhance the discriminatory power between *Phalaenopsis* genus.

The genome size is small for *P. aphrodite* subsp. *formosana* (Hsiao et al. 2011) and roughly 2.81 pg in diploid genome (Chen et al. 2013), which is suitable for the development of microsatellite markers. Here, the objective of this study was to develop transferable microsatellite markers from *P. aphrodite* subsp. *formosana* using the modified magnetic bead enrichment method. Based on these transferable markers, the molecular identification systems is able to be established for accessing the hybridization and introgression among species of the genus *Phalaenopsis* in future work.

Materials and methods

Plant materials

There are 21 species of the genus *Phalaenopsis* comprised of five subgenera used in this study. The taxonomy and nomenclature are followed (Christenson 2001), and specimens information are listed in Table 1. All samples were collected from the plants planted in the greenhouse at the Kaohsiung District Agricultural Improvement Station (KDAIS) in Taiwan by C. C. Tsai. Voucher specimens were deposited in herbarium of the National Museum of Natural Science, Taiwan (TNM).

Screening, sequencing microsatellite loci, and primer designation

Total DNA was extracted from tissue culture seedlings or young leaves following the procedure by a Plant Genomic DNA Extraction Kit (RBC Bioscience, Taipei, Taiwan). The DNA sample from *P. aphrodite* subsp. *formosana* was screened for microsatellites by digested with the restriction enzyme *Mse*I (Promega, Madison, Wisconsin, USA) and confirmed with 1.5% agarose gel electrophoresis. The digested fragment sizes with a range from 400 to 1000 bps were extracted using agarose gel and then ligated with *Mse*I-adapter pair (5′-TACTCAGGACTC AT-3′ and 5′-GACGATGAGTCCTGAG-3′) using DNA T4 ligase. As the template DNA for the enrichment of the partial genomic library, the ligated products were then used to perform 20 cycles of pre-hybridization PCR amplification in a 20 μL reaction mixture using the adapter specific primer (*Mse*I-N: 5′-GATGAGTCCT GAGTAAN-3′). The PCR mixture contained 20 ng template DNA, 10 pmol *Mse*I-N, 2 μL 10× reaction buffer, 2 mM dNTP mix, 2 mM $MgCl_2$, 0.5 U *Taq* DNA polymerase (Promega), and sterile water was added to total volume of 20 μL, with the PCR program of initial denaturation of 94 °C for 5 min, followed by 18 cycles of 30 s at 94 °C, 1 min at 53 °C, 1 min at 72 °C, and a final extension at 72° C for 10 min using a Labnet MultiGene 96-well Gradient Thermal Cycler (Labnet, Edison, New Jersey, USA). The biotinylated oligonucleotide repeat probes $(AG)_{15}$, $(AC)_{15}$, $(TCC)_{10}$, and $(TTG)_{10}$ were used

Table 1 Information on geographic distribution, species code and voucher specimens of the genus *Phalaenopsis* used in this study

Classification	Geographical distribution	Code	Source[a]
Subgenus *Proboscidioides* (Rolfe) E. A. Christ.			
P. lowii Rchb.f.	Myanmar, and adjacent western Thailand	P4	KDAIS-KC88
Subgenus *Aphyllae* (Sweet) E. A. Christ.			
P. minus (Seidenf.) E. A. Christ.	Endemic to Thailand	P11	KDAIS-KC227
P. braceana (J. D. Hook.) E. A. Christ.	Bhutan and China	P13	KDAIS-KC289
Subgenus *Parishianae* (Sweet) E. A. Christ.			
P. parishii Rchb.f.	Eastern Himalayas, India, Myanmar, and Thailand	P15	KDAIS-KC316
Subgenus *Polychilos* (Breda) E. A. Christ.			
P. mannii Rchb.f.	Northeast India, Nepal, and China to Vietnam	P18	KDAIS-KC22
P. cornu-cervi (Breda) Bl. and Rchb.f.	Northeast India and the Nicobar Islands to Java and Borneo	P2	KDAIS-KC23
P. kunstleri J. D. Hook.	Myanmar and Malay Peninsula	P8	KDAIS KC-139
P. pulchra (Rchb.f.) Sweet	Endemic to the Philippines (Luzon and Leyte)	P1	KDAIS-KC17
P. violacea Witte	Indonesia (Sumatra) and Malaysia (Malay Peninsula)	P9	KDAIS-KC153
P. micholitzii Rolfe	Philippines (Mindanao)	P19	KDAIS-KC382
P. maculata Rchb.f.	Malaysia (Pahang), East Malaysia (Sabah and Sarawak), and Indonesia (Kalimantan Timur)	P3	KDAIS-KC49
P. amboinensis J. J. Sm.	Indonesia (Molucca Archipelago and Sulawesi)	P17	KDAIS-KC157
P. inscriptiosinensis Fowlie	Endemic to Indonesia (Sumatra)	P14	KDAIS-KC298
P. corningiana Rchb.f.	Borneo (Sarawak and elsewhere on the island)	P16	KDAIS-KC346
Subgenus *Phalaenopsis*			
P. amabilis (L.) Blume	Widespread from Sumatra and Java to the southern Philippines, east to New Guinea and Queensland, Australia	P5	KDAIS-KC23
P. aphrodite Rchb.f.	Northern Philippines and southeastern Taiwan	PN	KDAIS-KC96
P. schilleriana Rchb.f.	Endemic to the Philippines	P10	KDAIS-KC429
P. chibae Yukawa	Yukawa endemic to Vietnam	P20	KDAIS-KC488
P. pulcherrima (Lindl.) J. J. Sm.	Widespread from northeast India and southern China throughout Indochina to Malaysia (Malay Peninsula), Indonesia (Sumatra), and East Malaysia (Sabah)	P12	KDAIS-KC256
P. equestris (Schauer) Rchb.f.	Philippines and Taiwan	P7	KDAIS-KC203
P. lindenii Loher	Endemic to the Philippines	P6	KDAIS-KC119

[a] Plant materials were cultivated at the Kaohsiung District Agricultural Improvement Station, Taiwan and voucher specimens were deposited at the herbarium of the National Museum of Natural Science, Taiwan

to hybridize with the amplicons at 68 °C for 1 h. The hybridization mixture was then enriched using 1 mg of streptavidin magnesphere paramagnetic particles (Promega) at 42 °C for 2 h and then eluted. Subsequently, DNA fragments containing microsatellites were purified and then amplified by 25-cycle-PCR using purified captured DNA fragments as templates (5 μL), *Mse*I-N (10 pmol), 10× reaction buffer (2 μL), dNTP mix (2 mM), $MgCl_2$ (2 mM), 0.5 U *Taq* DNA polymerase (Promega), and supplement sterile water to 20 μL under the amplification conditions described above. The PCR products were purified by the HiYield™ Gel PCR DNA Fragments Extraction Kit (RBC Bioscience) and used for cloning. The purified DNA was ligated into the pGEM®-T Easy Vector System (Promega), and used to transformed into *E. coli* DH5α competent cells. The positive clones were randomly selected and used for sequencing. In total, 321 positive colonies were collected and amplified with T7 and SP6 primers and sequenced on an ABI PRISM 3700 DNA Sequencer (Applied Biosystems, Foster City, California, USA). Sequences containing microsatellites were detected using Tandem Repeats Finder version 4.09 (Benson 1999), and primer pairs were designed for microsatellite loci with suitable flanking regions to amplify using FastPCR software version 6.5.94 (Kalendar et al. 2009). Each primer pairs were designed to amplify with a fragment in the range of 100–400 bp.

Microsatellites PCR amplification

To verify the effectiveness and polymorphisms of 28 microsatellite loci, all primer pairs designed for amplifying these microsatellites were tested using the *P.*

aphrodite subsp. *formosana* DNA samples together with the other 20 *Phalaenopsis* species. The optimal annealing temperature was determined using gradient PCR on a Labnet MultiGene 96-well Gradient Thermal Cycler (Labnet). The PCR was carried out in a total reaction volume of 20 μL, in which the PCR reaction mixtures contained 20 ng template DNA, 0.2 μM forward and reverse primers, 2 μL 10× reaction buffer, 2 mM dNTP mix, 2 mM $MgCl_2$, and 0.5 U of *Taq* polymerase (Promega). The gradient PCR protocol was set at 94 °C for 5 min, followed by 35 cycles of 94 °C for 40 s, a temperature gradient 50–60 °C for 60 s, 72 °C for 60 s, and a final step of 72 °C for 10 min. Then, the PCR products were assessed using 10% polyacrylamide gel electrophoresis and stained with ethidium bromide and visualized (EtBr) under UV light exposure. The patterns and length of alleles were recorded digitally by Quantity One ver. 4.62 (Bio-Rad Laboratories, Hercules, California, USA).

Data analyses

In total, 146 repeatable amplicons with length variation were screened from 28 microsatellite primer pairs (Table 2) in 21 species (Table 1). The number and average of amplicons (alleles) per primer pairs and the polymorphism information content (PIC) value of each loci were estimated using Power-Marker version 3.25 (Liu and Muse 2005). The Bayesian clustering method was used to estimate genotyping group information and genetic components for 21 *Phalaenopsis* taxa with the assistance of STRUCTURE ver. 2.3.4 (Pritchard et al. 2000; Falush et al. 2003, 2007). The admixture model (Hubisz et al. 2009) was selected in the Bayesian clustering analysis. The posterior probability of the genetic grouping number (K = 1–21) was estimated using the Markov chain Monte Carlo (MCMC) approach and 10 independent runs with a first 10% discarding (burnin) followed by 5,000,000 MCMC steps for each grouping number. The first-two best grouping numbers were evaluated using ΔK process (Evanno et al. 2005) by STRUCTURE HARVESTER ver. 0.6.8 (Earl and vonHoldt 2012). The graphical display of the results was drawn by DISTRUCT program (Rosenberg 2004).

Results and discussion

All of 21 *Phalaenopsis* species reveal either zero, one or two PCR amplicons in each of 28 microsatellite loci. One or two PCR amplicons per locus represent homozygotes or heterozygotes, and no amplicon indicate lacking this homologous microsatellite locus (Table 2). The genome size of *Phalaenopsis aphrodite* subsp. *formosana* detected by flow cytometry reveals roughly 2.81 pg in diploid genome (Chen et al. 2013) and all diploid species of *Phalaenopsis* have 38 chromosome number (Christenson 2001). These related studies and our current results indicate that 21 *Phalaenopsis* taxa studied are diploid plants, except the *P. lowii* and *P. minus* are not listed in the study of Chen et al. (Christenson 2001; Chen et al. 2013).

In total, 146 amplicons (alleles) were identified by 28 microsatellite primer pairs across 21 native *Phalaenopsis* species, and the number of amplicons per primer pairs ranged from 2 to 12, with an average of 5.21 (Table 2). The cross-species amplification test for the 20 other species was conducted using 28 microsatellite primers developed by *P. aphrodite* subsp. *formosana*, and the species of *P. amabilis* (L.) Blume, *P. schilleriana* Rchb.f, *P. chibae*, *P. equestris* (Schauer) Rchb.f. and *P. lindenii* Loher have higher transferable loci. The above mentioned four species with *P. aphrodite* subsp. *formosana* are all classified under the genus *Phalaenopsis*. The microsatellite primers could be successfully transferable to an average of 6.21 species [range from two (PA7, PA11 and PA41) to 20 (PA101) species] (Table 3). Due to the high transferability to species of the subgenus *Phalaenopsis*, these newly developed microsatellite primers are able to apply to establish a standard molecular identification operating system in *Phalaenopsis*.

The allelic polymorphism information content (PIC) values reflect the extent of allele diversity among the species, the PIC values in the present study ranged from 0.38 to 0.87, with an average of 0.63 (Table 4). Previous studies showed that the PIC values ranged from 0.1754 to 0.6740 (Sukma 2011) and 0 to 0.682 (Tsai et al. 2015) for the genomic microsatellite loci and EST-SSR of *Phalaenopsis* species, respectively. Thus, the PIC value in our study is greater than previous studies on *Phalaenopsis*. This PIC result is consistent with genomic microsatellite studies in *Scutellaria austrotaiwanensis* (Hsu et al. 2009), mango (Chiang et al. 2012), and Indian jujube (Chiou et al. 2012).

For genetically delimiting 21 species of the genus *Phalaenopsis*, a model-based Bayesian clustering algorithm was performed in STRUCTURE 2.3.4. The result showed that the first two best clustering numbers are $K = 2$ and $K = 4$ (Table 4). The ΔK was 96.55 and 2.31 when $K = 2$ and $K = 4$ in the Bayesian clustering analysis, respectively. Under $K = 2$, most species of the subgenus *Phalaenopsis* were assigned to the same cluster with high percent of Component 1 (pink segment in Fig. 1A) except *P. pulcherrima* that is genetically assigned to sections *Esmeralda* (subgenus *Polychilos*). The subgenus *Proboscidioides*, *Aphyllae*, and *Parishianae*, and *Polychilos* were consigned to the cluster with high percent of Component 2 (green segment in Fig. 1A) except *P. kunstleri* belonging to subgenus *Polychilos* which revealed an admixture genetic composition (56.8% of Component 1 and 43.2% of Component

Table 2 Characteristics of the 28 polymorphic microsatellite primers derived from *Phalaenopsis aphrodite* subsp. *formosana*

Locus	Primer sequences (5′–3′)	Repeat motif	Allelic size (bp)	Annealing temperature (°C)	No. of alleles	PIC
PA5-1	F: TCCCATTATCACTCCCTCAC R: GGTTAGAGATATAGGGAGAG	$(TC)_{14}$	140–164	59	4	0.67
PA5-2	F: CTCTCTTTCCTTCTCACCTC R: AAGATAGAGGGAGAGAGTGG	$(TC)_{10}$	98–104	58	4	0.61
PA7	F: CTCTGCTTCTCACCTTTCAC R: GGACAGAAAGTGAGAGAGAG	$(TC)_{12}$	116–264	56	3	0.55
PA10	F: TCTTCAGTCCCTCACTCATC R: ACAAAGCGGTGGAGAATATG	$(CT)_{14}$	132–152	58	7	0.75
PA11	F: ATCTATTGCTCTTTGTCCTC R: TAGCAAAGAGATGCTGAAGG	$(CT)_{42}$	214–216	55	2	0.38
PA14	F: TTTTCACTCTCCCTCCATCC R: GATGTAGAGAATGAGGGAGC	$(CT)_{21}$	182–186	52	3	0.55
PA15	F: TCTCCTACTCCCTCTATCTCA R: CTTGAAAGGCAGAGAGATAG	$(CT)_{25}$	306–310	56	3	0.55
PA19	F: TCTCCCTATATCTCTGCATC R: TGGAAAGAGAAAGGTTCAGG	$(CTCC)_{4}$	154–158	51	3	0.58
PA21	F: TCTCTCACTTTGTCACTCGC R: AAAGGGAAGTAGGGAAGGAG	$(CT)_{14}$	134–146	57	6	0.77
PA24	F: TTGATCTCTCTGGCACCCAC R: AAGAGAGAGTTAGTTGGAGAT	$(TC)_{36}$	216–224	55	3	0.59
PA25-1	F: ACCCACTTTCTCCTATCTCC R: GATGAAAGAGAGTGAGAGCG	$(CT)_{20}$	176–202	58	5	0.64
PA25-2	F: TCTCCCTCTCTTTACCACTC R: GTGAGAGAGATAGAGTGAGC	$(CT)_{12}$	92–267	58	6	0.73
PA32-1	F: CTCTTCCTGCTTTTCCTAGG R: AAGAGGGTGTGAGGAAGAGG	$(CT)_{25}$	148–222	57	8	0.83
PA32-2	F: TCTCTCACTACTCTATCTTG R: GAGAAGATAGAAAGAGTGAG	$(CT)_{18}$	140–152	54	5	0.73
PA36	F: CTCCACTTTATCTCTCTACC R: ATTGAGCGAGATAAAACTAG	$(TC)_{39}$	220–250	55	4	0.67
PA37	F: TTTACCTCTTTTGCTAGCTC R: AAGAGAAAGGGAAGGAGAGC	$(TC)_{23}$	226–234	50	4	0.67
PA38	F: CTCTCTCACTCTATTACTCC R: AGCTAGATAGAGGGAGAAAG	$(CT)_{32}$	224–384	54	9	0.84
PA40	F: GAGCAACATTCACTAGAGAG R: CTGGCAAAGCTTTGAGAAGG	$(CA)_{14}$	258–320	56	6	0.79
PA41	F: GAGGAGAAATAATGATTCCG R: AGACACTCTCACACACTTTC	$(AG)_{12}$	138–140	50	2	0.38
PA63	F: TTCATTCCATCTACCCCATC R: GATAGAAAGACTAGAGTAGG	$(CT)_{8}$	130–136	55	3	0.59
PA64	F: CTCTCCTTTTTCTTATCTTTCAC R: TAGAGAGATAGAGGGCAAGC	$(CT)_{94}$	248–296	55	3	0.55
PA74-1	F: AATGACCTCTCTGCTCTCTC R: GCAAGAGAAGTTGTGGGATGG	$(TC)_{28}$	172–306	50	3	0.59
PA74-2	F: CATCCCACAACTTCTCTTGC R: AGTGCTCAAGCGAGTTAGAGAC	$(CT)_{13}$	122–134	55	5	0.58
PA83-1	F: CCCTCTTTCTCTCATTGTCC R: GGGACAGAGTGCATAAGATG	$(TC)_{9}$	190–198	54	2	0.35

Table 2 continued

Locus	Primer sequences (5′–3′)	Repeat motif	Allelic size (bp)	Annealing temperature (°C)	No. of alleles	PIC
PA83-2	F: CCTTATCTCTTCTCTCTACC R: AGAAAGGAAGGGTAGGAGAG	$(TC)_{36}$	168–210	50	12	0.87
PA100-1	F: TCCCTCTATTTTAGACACCC R: GGAGAAAGAGCAAGACAGTG	$(TC)_{11}$	132–136	52	2	0.35
PA100-2	F: TCTCCATCCGTTAGCCTCTC R: GGGTAGGCAGAGAGAGTGAT	$(CT)_{16}$	128–136	59	5	0.73
PA101	F: CCCACTCACACTCTATCTTC R: AGGGTCAAACAGAATGAAGG	$(TC)_{11}$	126–138	55	7	0.63
				Average	4.57	0.63

Table 3 The result of the 28 polymorphic microsatellite loci isolated from *Phalaenopsis aphrodite* subsp. *formosana* test on 21 samples

Locus	PN	P1	P2	P3	P4	P5	P6	P7	P8	P9	P10	P11	P12	P13	P14	P15	P16	P17	P18	P19	P20	IS
PA5-1	O	–	–	O	–	O	O	O	O	–	–	–	–	–	–	–	–	–	–	–	–	6
PA5-2	O	–	–	–	–	O	O	–	O	–	E	–	–	–	–	–	–	–	–	–	–	5
PA7	E	–	–	–	–	–	O	–	–	–	–	–	–	–	–	–	–	–	–	–	–	2
PA10	O	O	O	E	–	E	O	E	–	–	O	–	O	–	–	–	O	–	O	–	O	12
PA11	O	–	–	–	–	–	–	–	–	–	O	–	–	–	–	–	–	–	–	–	–	2
PA14	O	–	–	–	–	–	O	O	–	–	O	–	–	–	–	–	–	–	–	–	–	4
PA15	O	–	–	–	–	–	–	O	–	–	O	–	–	–	–	–	–	–	–	–	O	4
PA19	O	–	–	–	–	O	O	O	O	–	O	–	–	–	O	–	–	–	–	–	O	8
PA21	O	–	O	O	O	O	O	O	O	–	E	O	–	–	O	–	–	–	–	–	–	11
PA24	O	–	–	–	–	O	O	–	–	–	–	–	–	–	–	–	–	–	–	–	–	3
PA25-1	O	–	–	–	–	O	O	O	–	–	–	–	–	–	–	O	O	–	E	–	O	8
PA25-2	O	–	–	–	–	–	–	E	–	–	O	E	–	–	–	–	–	–	E	–	–	5
PA32-1	E	–	–	–	–	–	O	O	O	–	O	–	–	–	–	–	O	–	–	O	O	8
PA32-2	O	–	–	–	–	–	O	E	–	–	–	–	–	–	–	O	–	–	O	–	E	6
PA36	O	–	–	–	–	–	–	–	–	–	O	–	–	–	O	–	–	–	–	O	O	5
PA37	O	–	–	–	–	–	E	O	–	–	–	–	–	–	–	–	–	–	–	–	–	3
PA38	O	–	–	–	O	–	O	O	O	–	E	O	O	–	–	–	–	–	–	–	E	9
PA40	E	–	–	–	O	–	–	O	E	–	–	–	–	–	–	–	–	–	–	–	–	4
PA41	O	–	–	–	–	–	–	–	–	–	O	–	–	–	–	–	–	–	–	–	–	2
PA63	O	–	–	–	–	O	–	O	–	–	–	–	–	–	–	–	–	–	–	–	–	3
PA64	O	–	–	–	–	O	–	O	–	–	–	–	–	–	–	E	–	–	–	–	–	4
PA74-1	O	O	–	–	–	–	O	–	–	–	–	–	–	–	–	–	–	–	–	–	–	2
PA74-2	O	–	–	–	–	O	O	O	–	–	E	–	E	–	–	–	–	–	–	–	–	6
PA83-1	O	–	–	–	–	–	–	–	–	–	–	–	–	–	–	–	–	–	–	O	O	3
PA83-2	E	O	O	–	E	–	O	O	O	–	E	O	O	–	–	E	–	–	O	E	E	14
PA100-1	O	–	–	–	–	O	–	O	O	–	–	–	–	–	–	–	–	–	O	–	O	6
PA100-2	O	–	–	–	–	O	O	O	O	O	O	O	–	–	–	–	–	–	O	–	–	9
PA101	O	O	O	O	O	O	E	E	O	O	O	O	O	O	O	O	O	O	O	–	O	20
IL	28	4	4	4	6	13	18	20	11	2	16	5	5	1	4	5	4	1	8	4	12	

IS the number of successful amplified species, *IL* the number of successful amplification primer, *O* homozygote, *E* heterozygote

Table 4 Proportion of individuals of each pre-defined population in each of the 2 and 4 clusters

	K = 2		K = 4				Subgenus	Section
	Composition 1	Composition 2	Composition 1	Composition 2	Composition 3	Composition 4		
P. aphrodite	*0.995*	0.005	*0.625*	0.371	0.001	0.002	*Phalaenopsis*	*Phalaenopsis*
P. amabilis	*0.689*	0.311	*0.572*	0.164	0.238	0.027	*Phalaenopsis*	*Phalaenopsis*
P. schilleriana	*0.904*	0.096	*0.800*	0.155	0.028	0.017	*Phalaenopsis*	*Phalaenopsis*
P. equestris	*0.988*	0.012	*0.824*	0.145	0.003	0.029	*Phalaenopsis*	*Stauroglottis*
P. lindenii	*0.974*	0.026	0.231	*0.478*	0.002	0.288	*Phalaenopsis*	*Stauroglottis*
P. chibae	*0.529*	0.471	0.106	0.133	0.116	*0.645*	*Phalaenopsis*	*Deliciosae*
P. pulcherrima	0.117	*0.883*	0.047	0.028	*0.894*	0.031	*Phalaenopsis*	*Esmeralda*
P. cornu-cervi	0.009	*0.991*	0.002	0.002	*0.992*	0.004	*Polychilos*	*Polychilos*
P. mannii	0.313	*0.687*	0.183	0.054	*0.643*	0.121	*Polychilos*	*Polychilos*
P. kunsteri	*0.568*	0.432	*0.488*	0.111	0.391	0.01	*Polychilos*	*Fuscatae*
P. violacea	0.016	*0.984*	0.003	0.002	*0.992*	0.003	*Polychilos*	*Amboinenses*
P. maculata	0.047	*0.953*	0.01	0.012	*0.958*	0.02	*Polychilos*	*Amboinenses*
P. pulchra	0.032	*0.968*	0.007	0.004	*0.983*	0.006	*Polychilos*	*Amboinenses*
P. micholitzii	0.239	*0.761*	0.036	0.058	*0.643*	0.263	*Polychilos*	*Amboinenses*
P. amboinensis	0.007	*0.993*	0.002	0.002	*0.994*	0.002	*Polychilos*	*Amboinenses*
P. inscriptiosinensis	0.054	*0.946*	0.01	0.013	*0.922*	0.055	*Polychilos*	*Zebrinae*
P. corningiana	0.028	*0.972*	0.003	0.024	*0.859*	0.114	*Polychilos*	*Zebrinae*
P. lowii	0.214	*0.786*	0.063	0.078	*0.815*	0.043	*Proboscidioides*	*Proboscidioides*
P. parishii	0.147	*0.853*	0.072	0.019	*0.887*	0.022	*Parishianae*	*Parishianae*
P. minnii	0.081	*0.919*	0.029	0.008	*0.957*	0.006	*Aphyllae*	*Aphyllae*
P. braceana	0.007	*0.993*	0.002	0.002	*0.994*	0.002	*Aphyllae*	*Aphyllae*

Italic values indicate major component of the species

2) (Table 4). When $K = 4$, Component 1 of $K = 2$ was divided into three components, 1a (pink segment in Fig. 1B), 1b (orange segment in Fig. 1B), and 1c (purple segment in Fig. 1B) (Table 4). Under $K = 4$, sections *Deliciosae* and *Esmeralda* can be divided into different clusters, which are grouped together when $K = 4$. Two sections *Phalaenopsis* and *Stauroglottis* of subgenus *Phalaenopsis* were grouped together with high genetic similarity (Table 4 and Fig. 1B). In addition, section *Fuscatae* of subgenus *Polychilos* was genetically assigned to the subgenus *Phalaenopsis* cluster based on both section *Fuscatae* of subgenus *Polychilos* belong to pink segment group with more than 50% proportion of Component 1 (see Fig. 1A, B and Table 4). The assignment test by Bayesian clustering analysis reveals similar result with molecular phylogeny patterns described by Tsai et al. (2005). The Bayesian clustering analysis based on EST-SSR loci could not get high resolution between either subgenus or sections within subgenus (Tsai et al. 2015). Compare to EST-SSR results published by Tsai et al. (2015), these newly developed genomic microsatellite loci have higher resolution than EST-SSR loci when study on native moth orchids.

Conclusions

The *Phalaenopsis* species are important genetic resources for the breeding of hybrids in the horticultural market. The molecular identification markers are an important technology for breeder to improve the commercial cultivars. In this study, we developed 28 primer sets for the polymorphic microsatellite loci of *Phalaenopsis aphrodite* subsp. *formosana*, which are highly transferable among related species of the genus *Phalaenopsis*. Based on these transferable markers, delimitations between subgenera and between sections inferred by the Bayesian clustering analysis indicate that these SSR markers reveal

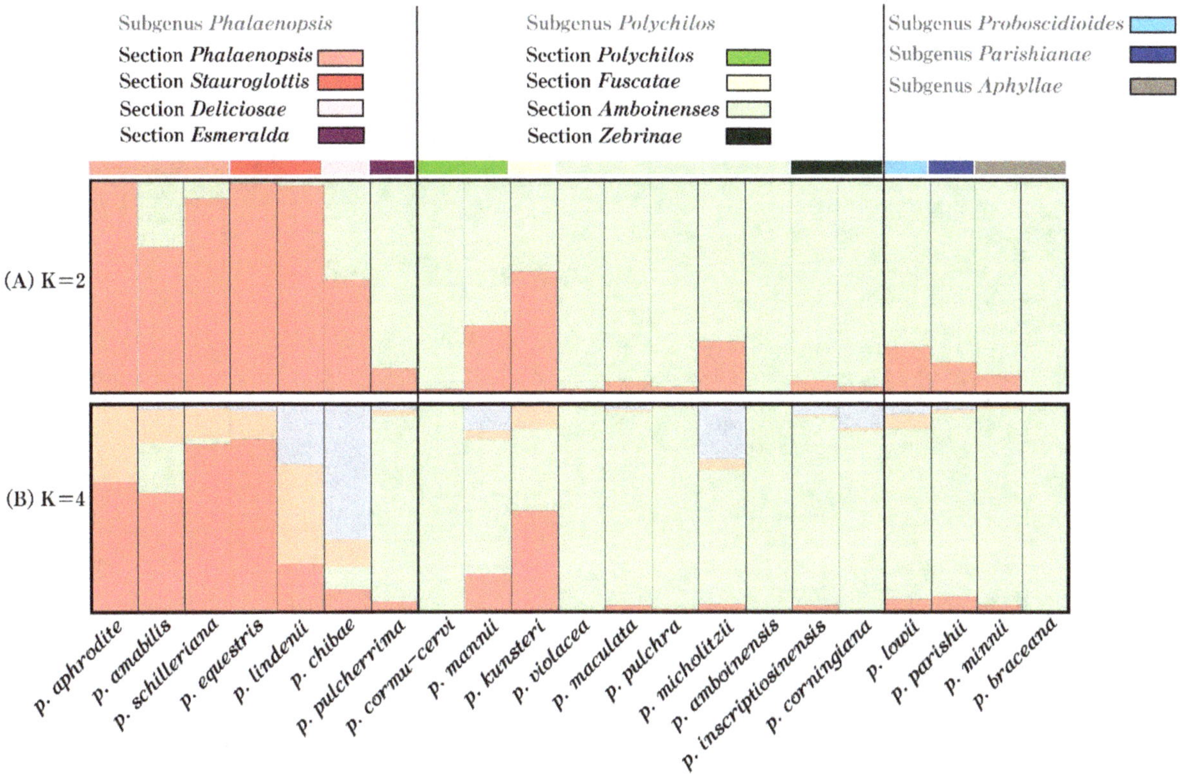

Fig. 1 Genotypic group structure of the genus *Phalaenopsis* used in this study in $K = 2$ (**A**) and 4 (**B**)

high taxonomic resolution for paternity and hybridization application among genus *Phalaenopsis*. In this study, we provided useful and cheap DNA barcoding markers for molecular breeding.

Authors' contributions

Conceived and designed the experiments: YZK and YCC. Performed the experiments: YZK, HCS, HHH, and YCC. Analyzed the data: YZK, HHH, and YCC. Contributed reagents/materials/analysis tools: HCS, CCT, HHH, PCL and YCC. Wrote the paper: YZK, PCL and YCC. Conceived of the study, edited the manuscript, and approved the final manuscript: YZK, HCS, CCT, HHH, PCL and YCC. All authors read and approved the final manuscript.

Author details

[1] Department of Biological Sciences, National Sun Yat-sen University, Kaohsiung 804, Taiwan. [2] Department of Nursing, Meiho University, Pingtung 912, Taiwan. [3] Kaohsiung District Agricultural Research and Extension Station, Pingtung 900, Taiwan. [4] National Pingtung University of Science and Technology, Pingtung 912, Taiwan. [5] Department of Life Science, National Taiwan Normal University, Taipei, Taiwan. [6] Department of Biomedical Science and Environment Biology, Kaohsiung Medical University, Kaohsiung, Taiwan.

Acknowledgements

Achievement of this study is dedicated to the memory of the scientific contributions of Dr. C. C. Tsai who is the third author. Dr. C.C. Tsai dead on November 05, 2015 in an unexpected stroke. Dr. C.C. This research was supported by funding from the Ministry of Science and Technology, Taiwan, MOST 105-2621-B-110-003-MY3 and MOST 105-2621-B-110-001 to YCC and the Multidisciplinary and Data Science Research Center (MDSRC), NSYSU.

Competing interests

The authors declare that they have no competing interests.

References

Benson G (1999) Tandem repeats finder: a program to analyze DNA sequences. Nucleic Acids Res 27(2):573–580

Chen WH, Chen HH (2007) Orchid biotechnology. World scientific Publishing Co. Ltd, Singapore

Chen WH, Chen HH (2011) Orchid biotechnology II. World scientific Publishing Co. Ltd, Singapore

Chen WH, Wang YT (1996) Phalaenopsis orchid culture. Taiwan Sugar 43:11–16

Chen WH, Kao YL, Tang CY, Tsai CC, Lin TY (2013) Estimating nuclear DNA content within 50 species of the genus *Phalaenopsis* Blume (Orchidaceae). Sci Hortic 161:70–75

Chiang YC, Tsai CM, Chen YKH, Lee SR, Chen CH, Lin YS, Tsai CC (2012) Development and characterization of 20 new polymorphic microsatellite markers from *Mangifera indica* L. (Anacardiaceae). Am J Bot 99(5):e117–e119

Chiou CY, Chiang YC, Chen CH, Yen CR, Lee SR, Lin YS, Tsai CC (2012) Development and characterization of 38 polymorphic microsatellite markers from an economical fruit tree, the Indian jujube. Am J Bot 99(5):e199–e202

Christenson EA (2001) *Phalaenopsis*: a monograph. Timber Press, Portland

Dressler RL (1993) Phylogeny and classification of the orchid family. Cambridge University Press, Cambridge

Earl DA, vonHoldt BM (2012) STRUCTURE HARVESTER: a website and program for visualizing STRUCTURE output and implementing the Evanno method. Conserv Genet Resour 4:359–361

Evanno G, Regnaut S, Goudet J (2005) Detecting the number of clusters of individuals using the software structure: a simulation study. Mol Ecol 14:2611–2620

Falush D, Stephens M, Pritchard JK (2003) Inference of population structure using multilocus genotype data: linked loci and correlated allele frequencies. Genetics 164:1567–1587

Falush D, Stephens M, Pritchard JK (2007) Inference of population structure using multilocus genotype data: dominant markers and null alleles. Mol Ecol Notes 7:574–578

Hsiao YY, Pan ZJ, Hsu CC, Yang YP, Hsu YC, Chuang YC, Shih HH, Chen WH, Tsai WC, Chen HH (2011) Research on orchid biology and biotechnology. Plant Cell Physiol 52(9):467–1486

Hsu TW, Kuo CC, Tsai CC, Chiang YC (2009) Isolation and characterization of 16 microsatellite markers from a rare and endemic species, *Scutellaria austrotaiwanensis* (Lamiaceae). Conserv Genet Resour 1(1):85–88

Hubisz MJ, Falush D, Stephens M, Pritchard JK (2009) Inferring weak population structure with the assistance of sample group information. Mol Ecol Resour 9:1322–1332

Kalendar R, Lee D, Schulman AH (2009) FastPCR software for PCR primer and probe design and repeat search. Genes Genomes Genom 3(1):1–14

Lai JM, Tsai CC, Yen CR, Ko YZ, Chen SR, Weng IS, Lin YS, Chiang YC (2015) Molecular characterization of twenty polymorphic microsatellite markers in the polyploid fruit tree species *Syzygium samarangense* (Myrtaceae). Genet Mol Res 14(4):13013–13021

Liu K, Muse SV (2005) PowerMarker: an integrated analysis environment for genetic marker analysis. Bioinformatics 21(9):2128–2129

Liu YC, Lin BY, Lin JY, Wu WL, Chang CC (2016) Evaluation of chloroplast DNA markers for intraspecific identification of *Phalaenopsis equestris* cultivars. Sci Hortic 203:86–94

Men S, Ming X, Wang Y, Liu R, Wei C, Li Y (2003) Genetic transformation of two species of orchid by biolistic bombardment. Plant Cell Rep 21(6):592–598

Powell W, Machray GC, Provan J (1996) Polymorphism revealed by simple sequence repeats. Trends Plant Sci 1(7):215–222

Pritchard JK, Stephens M, Donnelly P (2000) Inference of population structure using multilocus genotype data. Genetics 155:945–959

Rosenberg NA (2004) DISTRUCT: a program for the graphical display of population structure. Mol Ecol Notes 4:137–138

Sukma D (2011) Development of sequence-based microsatellite marker for *Phalaenopsis* orchid. Hayati J Biosci 18:71–76

Tanaka Y, Katsumoto Y, Brugliera F, Mason J (2005) Genetic engineering in floriculture. Plant Cell Tissue Organ Cult 80(1):1–24

Tang CY, Chen WH (2007) Breeding and development of new varieties in *Phalaenopsis*. In: Chen WH, Chen HH (eds) Orchids biotechnology. World Scientific Publishing Co. Pvt. Ltd, Singapore, p 1–22

Tsai CC, Huang SC, Chou CH (2005) Molecular phylogeny of *Phalaenopsis* Blume (Orchidaceae) based on the internal transcribed spacer of the nuclear ribosomal DNA. Plant Syst Evol 256:1–16

Tsai CC, Chiang YC, Huang SC, Chen CH, Chou CH (2010) Molecular phylogeny of *Phalaenopsis* Blume (Orchidaceae) on the basis of plastid and nuclear DNA. Plant Syst Evol 288(1–2):77–98

Tsai CC, Chen YKH, Chen CH, Weng IS, Tsai CM, Lee SR, Lin YS, Chiang YC (2013) Cultivar identification and genetic relationship of mango (*Mangifera indica*) in Taiwan using 37 SSR markers. Sci Hortic 164:196–201

Tsai CC, Wu PY, Kuo CC, Huang MC, Yu SK, Hsu TW, Chiang TY, Chiang YC (2014) Analysis of microsatellites in the vulnerable orchid *Gastrodia flavilabella*: the development of microsatellite markers and cross-species amplification in *Gastrodia*. Bot Stud 55:72

Tsai CC, Shih HC, Wang HV, Lin YS, Chang CH, Chiang YC, Chou CH (2015) RNA-seq SSRs of moth orchid and screening for molecular markers across genus *Phalaenopsis* (Orchidaceae). PLoS ONE 10(11):e0141761

10

Techniques for the collection, transportation, and isolation of orchid endophytes from afar: a case study from Madagascar

Lawrence W. Zettler[1*], Landy Rajaovelona[2], Kazutomo Yokoya[3], Jonathan P. Kendon[3], Andrew L. Stice[1], Amanda E. Wood[1,4] and Viswambharan Sarasan[3]

Abstract

Background: Tropical orchids need more study with respect to their mycorrhizal associations. For researchers in distant countries who aspire to study these orchids augmenting their conservation, the great distances involved, coupled with limited funds, pose formidable challenges. These challenges are sometimes exacerbated by political unrest, delays in securing permits, unexpected hardships, and the risk that the biological samples collected (e.g., roots harboring mycorrhizal fungi) will not survive long-distance transport.

Results: We describe a protocol for the collection and transport of root samples from Madagascar orchids to labs in the United Kingdom (Kew) and the United States (Illinois) where *Rhizoctonia*-like fungi were subsequently isolated. Three separate trips were made spanning 4 years (2012–2015), with emphasis on the collection of roots from epiphytic, lithophytic, and terrestrial orchids inhabiting the Itremo Massif of the Central Highlands. Collectively, the trips to Madagascar resulted in the isolation of all major groups of *Rhizoctonia*-like fungi (*Ceratobasidium*, *Tulasnella*, *Sebacina*) from all three orchid growth forms (terrestrials, epiphytes and lithophytes). Sampling of terrestrial and epiphytes during the rainy season (January) yielded best results.

Conclusions: Our study demonstrates that peloton-forming fungi in root samples can retain viability up to 3 weeks after collection.

Keywords: Orchidaceae, Conservation, Epiphytes, *Tulasnella*, *Ceratobasidium*

Background

Orchids have the unparalleled distinction of being the most diverse plant family on earth, but also the most vulnerable to extinction (Swarts and Dixon 2009; Merritt et al. 2014). Part of their vulnerability stems from the family's susceptibility to acute changes in their environment exacerbated by climate change, as well as their extreme dependence on pollinators and mycorrhizal fungi to complete their life cycles. Of the approximately 25,000 species of orchids worldwide (Dressler 1993; Cribb et al. 2003), about two-thirds are represented by epiphytes and lithophytes (Swarts and Dixon 2009), the majority of which are confined to tropical latitudes in areas prone to deforestation. The remaining one-third consist of terrestrials, many of which occupy cooler temperate zones that are undergoing rapid warming. Collectively, orchids face a conservation crisis of epic proportions, and understanding their biotic and abiotic needs is crucial to their survival.

Mycorrhizal fungi are needed by orchids as a carbon source to initiate germination of their dust-like seeds (Rasmussen 1995). Most of these fungi fall under the category of basidiomycetes in the *Rhizoctonia* complex

*Correspondence: lwzettle@ic.edu
[1] Department of Biology, Illinois College, 1101 West College Ave., Jacksonville, IL 62650, USA
Full list of author information is available at the end of the article

(Currah et al. 1997)—a group that persists mostly as free-living saprophytes (Moore-Landecker 1996), but also pathogens, microparasites, and orchid symbionts (Swarts and Dixon 2017). Unlike other mycorrhizal associations throughout the plant kingdom, orchids exploit their fungi as a food source (mycotrophy) while providing no substantial benefits in return (Rasmussen and Rasmussen 2009). At maturity, terrestrial and epiphytic orchids alike are thought to maintain a lifelong association with fungi, at least to some extent, providing these plants with 'trophic versatility', i.e., a dual mechanism that utilizes both photosynthesis and mycotrophy. The importance of mycorrhizal fungi, therefore, spans the life of the orchid, and conserving these unique plants requires that we also understand and conserve their fungal associates. To effectively do so, we must understand how orchids interact with fungi in the natural setting, particularly at the germination site, exemplified by recent studies (e.g., Jacquemyn et al. 2014; McCormick et al. 2012; Swarts et al. 2010). Mycorrhizal fungi must also be isolated from living orchid tissues, identified, screened for their ability to germinate seeds in vitro verifying their functionality, and safeguarded in cryopreservation.

During the past 30+ years, much has been published on orchid endophytes recovered from temperate terrestrials (e.g., Currah et al. 1987; Warcup 1981; Zelmer et al. 1996), and more recently the tropical epiphytes (e.g., Pereira et al. 2003; Nontachaiyapoom et al. 2010; Chen et al. 2012; Hoang et al. 2017), but the lithophytes remain in need of more study. The length of time that fungi can remain viable in orchid tissues from field collection to the lab is also unclear. Swarts and Dixon (2017) recommended that peloton extraction take place the same day of root collection, and indeed many studies have adopted this protocol (e.g., Aggarwal et al. 2012; Chutima et al. 2011; Otero et al. 2002; Pereira et al. 2005), as it is generally assumed that fungal pelotons lose their viability shortly after root detachment. Suárez et al. (2006), for example, reported that hyphae in Andean orchid tissues lose viability even after one night of storage in the laboratory regardless of chilling. Nevertheless, samples may be processed 3–4 days after collection if necessary (Swarts and Dixon 2017), or longer in some cases. Zettler et al. 2011, for example, recovered orchid endophytes from root samples that were collected 1-week prior. Similarly, Richardson et al. (1993) had success in isolating a diverse assemblage of *Rhizoctonia*-like fungi from epiphytes in Costa Rica up to 3 weeks after collection.

In Madagascar, where 90% of the island's 1000 orchid species are endemic (Tyson 2000), deforestation on a massive scale has resulted in patches of fragmented orchid populations that persist from year to year with little natural regeneration (Whitman et al. 2011). Many orchid species in Madagascar and elsewhere are in dire need of study to keep pace with projected extinction rates this century. For researchers in distant countries who aspire to study these orchids for conservation purposes, the great distances involved, coupled with limited funds, pose formidable challenges. These challenges are often exacerbated by political unrest, delays in securing permits, unexpected hardships, and the risk that the biological samples collected (e.g., roots harboring mycorrhizal fungi) will not survive long-distance transport. As this study has shown, endophytes of orchids persist in root samples at least 3 weeks after they are collected in the field, as Richardson et al. (1993) reported earlier for epiphytes collected in Costa Rica.

In 2012, we were presented with a unique opportunity to collect and study endophytes of rare orchids native to the Itremo Massif Protected Area in the Central Highlands of Madagascar—one of the top five biologically diverse "hotspots" (Tyson 2000). Our primary goal was to isolate, identify, and safeguard *Rhizoctonia*-like fungi to facilitate orchid conservation in the region (e.g., symbiotic germination). The orchids targeted included endemic species (e.g., *Angraecum protensum*, Fig. 1), epiphytes (e.g., *Bulbophyllum*, *Polystachya*), terrestrials (e.g., *Cynorkis flexuosa*, Fig. 2), and lithophytes (e.g., *Angraecum*, *Aerangis*). One prevailing concern, however, was the great distances between our two labs and the Indian Ocean country (Illinois = 15,300 km, London = 9000 km), namely if orchid endophytes would remain viable in living tissues during lengthy, long-distance transport over rugged terrain and by air.

In this paper, we describe our protocol for the collection and long-distance transport of orchid root samples from Madagascar to labs in the United Kingdom (Kew) and the United States (Illinois) leading to isolation and provisional identification of the endophytes. The goal of this paper is to provide other researchers with a workable framework for recovering *Rhizoctonia*-like fungi from aged (2–3 week-old) tissue samples acquired from terrestrial, epiphytic, and lithophytic orchids in remote areas.

Methods

Permits

Logistic and taxonomic support for this joint study between Royal Botanic Gardens Kew (Kew) and Illinois College was facilitated by Kew Madagascar Conservation Centre (KMCC) and Parc Botanique et Zoologique de Tsimbazaza (PBZT). During the 5-year project, more than 40 taxa spanning 24 genera were selected for study involving collection and movement of live material (roots, seeds) from Madagascar to labs in the United Kingdom (UK) and the United States (USA). Root samples were shared between the two partners to ensure

Fig. 1 (Left)—*Angraecum protensum*—one of several lithophytic orchids endemic to the Itremo Massif of the Central Highlands in Madagascar, shown in flower during the dry season (June 2012). The long nectar spur is visible to the lower left, and two young roots are seen on the lower right

Fig. 2 (Right)—The terrestrial orchid, *Cynorkis flexuosa*, shown against the backdrop of open rocky grassland of the Itremo Massif. There are ca 135 species of *Cynorkis* species in Madagascar alone. The lateral branch roots on this specimen were detached and transported to labs in the United States (Illinois) and United Kingdom (Kew) for fungal isolation

that live material could be processed in a timely manner leading to the isolation of *Rhizoctonia*-like fungi from root pelotons. To facilitate the legal collection and international transport of orchid material from Madagascar to the UK (Kew) and USA (Illinois), a CITES permit was obtained which allowed three tubes, each containing seedlings and mature roots per species, to be collected. The rarity of plants in the wild necessitated further restrictions, namely that collections be limited to three each of juvenile and mature plants. Depending on the species and its availability, 1–5 roots per specimen were collected. These were accompanied by a phytosanitary certificate which was secured prior to departure from Madagascar. The CITES permit and phytosanitary certificate were delivered by the Direction Generale des Forets (DGF Nanisana Antananarivo) and Service de la Quarantaine et de I'Inspection Nanisana, respectively. For import of root samples into the USA, an additional permit (PPQ 526) was obtained from the US Department of Agriculture (USDA), as the US Government regards all orchid endophytes in the *Rhizoctonia* complex to be plant pathogens. Samples entering the UK were accompanied by a Phytosanitary Certificate issued in Madagascar, and a UK Letter of Authority.

Dates and study sites in Madagascar

Three separate trips to Madagascar took place spanning 4 years (2012–2015), with emphasis on the collection of roots from mature plants and spontaneous seedlings of lithophytes, epiphytes and terrestrials inhabiting the Itremo Massif Protected Area of the Central Highlands. The first two trips (June 2012, April/May 2013) were planned after the rainy season (December to March) in an effort to collect spontaneous seedlings on orchid-rich substrates that may have germinated under the wetter conditions. The third trip (January 2015) took place during the rainy season to isolate additional fungal strains that may have been inactive during the previous two trips.

Collection and long-distance transport

The collection procedure reported by Yokoya et al. (2015) was employed for all three trips to Madagascar, and is summarized below. During the first trip, roots of mature orchids (= those that achieved anthesis or large enough to do so) were primarily targeted for collection to document the location of fungal pelotons in roots, and to isolate fungi in pure culture via hyphal tips. In the two trips that followed, emphasis was placed on collecting roots of seedlings and mature plants alike. Seedlings (= smallest leaf bearing stage, < 2 cm in length) and juveniles (plants > 2 cm in length, no anthesis) were provisionally identified on site by KMCC staff (Rajaovelona and Gardiner 2016) using subtle morphological features (e.g., presence of pseudobulbs) as well as proximity to mature plants on or near the same substrate. The identities of the seedlings were later confirmed by DNA analysis following the procedures described by Yokoya et al. (2015). To maximize our chances for isolating viable pelotons, younger-appearing roots were collected whenever possible. For epiphytic and lithophytic orchids, these roots appeared translucent to white in color, often with slight greenish pigmentation near the apex (Figs. 1 and 3). For terrestrials, roots that exhibited orange-yellowish patches of color were selected, as well as seedlings in close proximity to mature plants (Fig. 4). After detachment in the field, each root was placed over a small, pre-moistened cotton ball within pre-sterilized glass vial with screw cap (Fig. 5). To permit gas exchange leading up to departure from Madagascar (7–10 days after collection), the caps placed on each vial were not securely tightened. The vials were then placed within a 50 ml capacity centrifuge tube with screw cap (VWR International, LLC, Radnor, PA, USA). Tubes were then stored vertically within an insulated handbag for transport from field to shelter. Care was taken to keep the handbag out of direct sunlight so that the root samples would remain as cool as possible (15–25 °C).

Fig. 4 (Left) A strap leaf of the terrestrial orchid, *Cynorkis gibbosa*, showing characteristic mottled pigmentation, rooted in moss on a granite outcrop/seepage slope on the Itremo Massif. Small seedlings are shown at the lower left, presumably from the same species. Seedling stages are often detected in close proximity to mature plants sharing the same substrate

Fig. 3 (Bottom)—Actively-growing root tip of a lithophytic *Angraecum* species. Note the green pigmented tip, and root growth on bare rock substrate devoid of visible organic debris, often typical of lithophytes of the Itremo Massif

For terrestrial orchids, the root collection procedure differed slightly in that soil containing intact root systems (root ball) was also collected. This permitted the roots to remain in a semi-natural (moist) state leading up to departure from Madagascar. A trowel or small shovel was used to gently excavate the soil around individual plants, and to lift the root ball with minimal disturbance to the brittle root systems. Each root ball was then placed into its own separate plastic bag, and the bags were then carefully packed into an insulated handbag for transport. A wet bath towel was then placed

Fig. 5 (Right) A pre-sterilized glass sampling vial containing root segments of a terrestrial orchid. The screw cap lid remained untightened to permit gas exchange leading up to air transport out of Madagascar. Each glass vial was placed into a plastic, shatterproof centrifuge vial during transport from Madagascar to the lab

through the zip to facilitate wicking and evaporation to the outside air serving to cool the inside of the bag. Upon arrival at the KMCC base in Antananarivo, 2–7 days after field collection, root samples and root balls were placed into a refrigerator (ca. 6 °C). Approximately 24 h before departure from the country, roots of terrestrial orchids were carefully lifted from soil and rinsed off with UV-irradiated and/or bottled water to remove organic debris to comply with US and UK important regulations. Lateral branch roots were detached and placed over a pre-moistened cotton ball in a pre-sterilized glass vial (8 ml capacity). The screw cap was then tightened firmly and wrapped with a strip of Parafilm "M" (Pechiney Plastic Packaging, Menasha, WI, USA. Caps on glass vials containing roots of lithophytes and epiphytes were also tightened and wrapped with Parafilm "M" at that same time (ca. 24 h prior to departure by air). All sealed glass vials were then housed in 50 ml plastic (shatter-proof) centrifuge vials which were also firmly tightened and sealed with Parafilm "M". Vials were re-packed into insulated handbags and transported back to labs in the USA and UK as cabin luggage to ensure that samples were maintained at ambient temperature during the duration of each flight.

Fungal isolation and provisional identification

Immediately upon arrival into the UK and USA, within 24 and 48 h after departure from Madagascar, respectively, all root samples were placed in refrigeration (4–6 °C) for a period lasting 1–2 weeks. Fungi were isolated from root cortical regions using the method by Zettler et al. (2003), but our procedure differed in that Fungal Isolation Medium (FIM) substituted for Modified Melin-Norkrans' agar (MMN). Clumps of macerated cortical cells containing pelotons were immersed in FIM containing streptomycin sulfate [(10 ml/l of stock solution = 1 g dissolved in 70 ml); Clements and Ellyard 1979] and incubated at ambient temperature until actively-growing hyphae could be observed under a dissection microscope (typically 1–4 days). Hyphal tips from cortical cells and/or pelotons were then subcultured to Potato Dextrose Agar (PDA, Difco™, Becton, Dickinson and Co., Sparks, MD, USA) using a sterile scalpel. Provisional identification of *Rhizoctonia*-like fungi to genus level (*Ceratobasidium, Sebacina, Tulasnella*) reported herein was based on cultural descriptions reported by Currah et al. (1997).

Results

Rhizoctonia-like fungi from Madagascar samples

Pelotons were observed in mature roots in half of the lithophytes, half of the epiphytes, and both terrestrials (Table 1). Most of the pelotons were observed in the apex of the roots, in the 1–5 cm region from the tip (Table 1). Four of the seven lithophytic *Angraecum* species harbored pelotons (*A. calceolus, A. longicalcar, A. magdalenae, A. rutenbergianum*; Table 1). Once plated on agar, all pelotons yielded common conidial fungi (*Fusarium* sp., *Trichoderma* sp.), and none resulted in isolates that were assignable to the *Rhizoctonia* complex.

Following the second trip to Madagascar (April–May 2013), root samples returned to Illinois yielded *Rhizoctonia*-like fungi in two of the nine epiphytes (*Aerangis* sp., *Polystachya concreta*), both of which were seedlings (Table 2). Of the 14 terrestrial taxa sampled, roots from six species yielded endophytes assignable to the *Rhizoctonia* complex (*Benthamia rostratum, Cynorkis purpurea, Eulophia macra, Graphorkis concolor, Habenaria ambositrana*, and *Tylostigma nigrescens*), but none were recovered from the seven lithophytes (Table 2). Collectively, less than half (40%) of the roots of terrestrial orchids spanning all three growth stages (seedlings, juveniles, mature plants) yielded *Rhizoctonia*-like fungi, and 20%

Table 1 Peloton location by root region for orchid samples acquired in the Itremo Massif within the Central Highlands of Madagascar during the first of three trips (June 2012, dry season)

Growth habit	Orchid species	*n*	1	2	3	4	5	6	7	8	9	10
Lithophytic	*Angraecum calceolus*	2/2			X	X						X
	Angraecum longicalcar	1/3			X							
	Angraecum magdalenae	1/2		X								
	Angraecum obesum	0/1	–	–	–	–	–	–	–	–	–	–
	Angraecum protensum	0/1	–	–	–	–	–	–	–	–	–	–
	Angraecum sororium	0/2	–	–	–	–	–	–	–	–	–	–
	Angraecum rutenbergianum	1/1	X	X								
	Jumellea ibityana	0/2	–	–	–	–	–	–	–	–	–	–
Epiphytic	*Bulbophyllum* sp.	3/3	X		X				X			
	Bulbophyllum bicoloratum	0/2	–	–	–	–	–	–	–	–	–	–
	Jumellea sp.	1/4	X									
	Jumellea arborescens	0/2	–	–	–	–	–	–	–	–	–	–
	Jumellea intricata	1/2				X						
	Polystachya sp.	0/1	–	–	–	–	–	–	–	–	–	–
	Polystachya cultriformis	1/2					X					
	Polystachya fusiformis	0/1	–	–	–	–	–	–	–	–	–	–
Terrestrial	*Cynorkis* sp.	1/1		X								
	Eulophia sp.	1/2				X						

Numbers 1–10 reflect the distance (in cm) from the root tip (e.g., *1* terminal end of actively-growing tip, *2* second cm region from root tip), and *n* the number of roots harboring pelotons/number of roots collected

for the epiphytes (Table 2). Seedlings of *C. purpurea* harbored the most diverse assemblage of orchid endophytes with of all three major genera being represented in the samples (*Ceratobasidium, Tulasnella, Sebacina*).

Following the third and final trip to Madagascar coinciding with the rainy season (January 2015), a greater number of *Rhizoctonia*-like fungi were recovered from root samples collected 2–3 weeks prior, including a lithophyte, *Eulophia* sp. (Table 3). Among the six epiphytic taxa sampled, seedlings from *Aerangis* and *Polystachya* yielded *Ceratobasidium* and *Tulasnella* strains, respectively (Table 3). The majority of fungal isolates were acquired from terrestrials, in particular *Cynorkis* species (Table 3).

Discussion

As to why none of the samples from the first trip yielded *Rhizoctonia*-like fungi is perplexing, but might be attributed, in part, to the time of year of the collecting. For example, sampling during the first trip took place in June, whereas the second took place earlier in the year (April/May) closer to the end of the rainy season. Thus, it is conceivable that the samples collected in June were devoid of *Rhizoctonia*-like fungi because of drier conditions. The two subsequent trips that ensued yielded endophytes assignable to all three major genera of orchid mycorrhizal fungi (*Ceratobasidium, Tulasnella, Sebacina*). For roots of epiphytic and lithophytic orchids, placing detached roots into pre-sterilized vials that remain unsealed possibly served to allow gas exchange (oxygen) for roots and endophytes alike. Moist sterile cotton balls within the vials also may have maintained higher relative humidity levels needed for root and fungus longevity. For terrestrial orchids, keeping roots intact within the moist soil/root ball may have contributed to the high number of *Rhizoctonia*-like fungi acquired from the samples. Placing the vials in an insulated bag during field work, and temporary storage in refrigeration whenever possible, may have also benefited endophyte survival by maintaining cooler temperatures, slowing down metabolic rates in living cells.

Root colonization by endophytic fungi is thought to be influenced by two important factors—the growing season, and the growth stage of the plant (Swarts and Dixon 2017). For temperate terrestrial orchids, Harvais and Raitsakas (1975) and Warcup (1973) found that the most effective fungi for the purposes of seed germination were acquired earlier in the growing season. Huynh et al. (2004) reported that fungi isolated from pelotons in early growth stages were most effective at facilitating seed germination. A prevailing concern we had for collecting smaller roots of seedlings and juvenile stages during the dry season (April/May 2013) was dehydration of the samples given their high surface-to-volume ratio. This potential problem was

Table 2 A summary of the frequency of *Rhizoctonia*-like fungi acquired from roots of orchids inhabiting the Itremo Massif of the Central Highlands of Madagascar during April–May 2013 (dry season)

Growth habit	Orchid	Site	Sample	Fungus (# strains)
Lithophytic	*Angraecum coutrixii*	1, 3	Seedling	None
	Angraecum longicalcar	2	Mature	None
	Angraecum magdalenae	3	Seedling	None
	Angraecum protensum	1	Seedling Mature	None None
	Angraecum rutenbergianum	1	Seedling	None
	Angraecum sororium	3	Seedling	None
	Oeceoclades sp.	2	Mature	None
Epiphytic	*Aerangis* sp.	7	Seedling	*Ceratobasidium* (3)
	Aerangus citrata	5	Seedling	None
	Angraecum sp.	5	Seedling	None
	Angraecum protensum	1	Seedling	None
	Angraecum rutenbergianum	3	Seedling	None
	Bulbophyllum sp.	3	Seedling	None
	Jumellea denisfolliata	5	Seedling	None
	Polystachya concreta	1	Seedling 1 Seedling 2	None *Tulasnella* (7), *Sebacina* (6)
	Polystachya culturiformis	7	Seedling	None
Terrestrial	*Benthamia* sp.	1	Mature	None
	Benthamia glaberrima	3	Mature	None
	Benthamia rostratum	4	Juvenile	*Tulasnella* (1)
	Calanthe sp.	7	Mature	None
	Cynorkis gibbosa	7	Mature	None
	Cynorkis purpurea	7	Seedling	*Ceratobasidium* (7) *Tulasnella* (3), *Sebacina* (1)
	Disa incarnata	3	Mature	None
	Eulophia macra	2	Mature	*Tulasnella callospora* (1)
	Graphorkis concolor	7	Mature	*Ceratobasidium* (1)
	Habenaria sp.	1	Mature	None
	Habenaria ambositrana	1	Juvenile	None
		3	Seedling	*Tulasnella* (4), *sebacina* (1)
	Satyrium trinerve	4	Mature	None
	Tylostigma sp.	4	Mature	None
	Tylostigma nigrescens	4	Seedling	*Tulasnella* (5)

Fungal genera listed represent provisional identifications carried out at the time of isolation, based on cultural characteristics described by Currah et al. (1997). Growth habit reflects the substrate where the individual orchid was actually rooted at the time of collection. Collection sites: *1* exposed rocks, occasional tapia trees, *2* exposed marble outcrop, *3* exposed rocks, sandy stream bed, gnarled small trees, *4* open grassland, moist soil, occasional rocks, *5* reduced forest (canopy ca. 20 m), *6* exposed ridges, montane vegetation, *7* dense shaded forest, downhill stream. With one exception (2), all sites were within 5 km of one another

Terrestrial seedlings = 3/3, epiphytic seedlings = 2/10, lithophytic seedlings = 0/5

Terrestrial juveniles = 1/2, epiphytic juveniles = NA, lithophytic juveniles = NA

Terrestrial mature = 2/10, epiphytic mature = NA, lithophytic mature = 0/3

Total terrestrial = 6/15 (40%), total epiphytic = 2/10 (20%), total lithophytic = 0/8 (0%)

apparently avoided by the addition of a moist (sterile) cotton ball placed inside the vial with the sample, as several of *Rhizoctonia*-like isolates were later recovered. Some of these isolates were later tested for their ability to germinate seeds in vitro, with positive results. For example, one of the six strains of *Sebacina*, isolated from a *P. concreta* seedling, was most effective among 14 endophytes tested at inducing rapid in vitro seedling development of *C. purpurea* in symbiotic germination studies that ensued (Rafter et al. 2016). In another experiment, seeds of *H. ambositrana* and *T. nigrescens*, yielded leaf-bearing seedlings in vitro, 49 days after inoculation with *Sebacina* and *Tulasnella* endophytes acquired from seedlings of the same species, respectively (A. Wood, unpub. data).

Table 3 A summary of the frequency of *Rhizoctonia*-like fungi acquired from roots of orchids inhabiting the Itremo Massif of the Central Highlands of Madagascar during January 2015 (rainy season)

Growth habit	Orchid	Site	Sample	# Strains and fungus
Lithophytic	*Eulophia* sp. 1	1	Seedling	None
	Eulophia sp. 1	1	Mature	*Tulasnella* (1)
	Jumellea densifolia	2	Seedling	None
Epiphytic	*Aerangis* sp. 1	5	Seedling	*Ceratobasidium* (10)
	Angraecum sp. 1	2	Seedling, juvenile	None
	Angraecum sp. 3	6	Mature	None
	Aeranthes sp. 1	6	Juvenile, mature	None
	Bulbophyllum baronii	2	Mature	None
	Polystachya sp. 1	6	Seedling	*Tulasnella* (4)
	Polystachya concreta	3	Seedling	None
Terrestrial	*Angraecum* sp. 2	2	Mature	None
	Benthamia sp. 1	3	Mature	None
	Calanthe sylvatica	5	Mature	None
	Cynorkis sp. 1	3	Mature	*Ceratobasidium* (2)
	Cynorkis sp. 1	4	Mature	*Tulasnella, sebacina* (2)
	Cynorkis sp. 1	7	Mature	None
	Cynorkis fastigiata	4	Mature	*Ceratobasidium* (1)
	Cynorkis fastigiata	6	Mature	*Tulasnella* (3)
	Cynorkis flexuosa	1	Mature	None
	Cynorkis flexuosa	1	Seedling	*Tulasnella* (2)
	Cynorkis gibbosa	4	Mature	*Tulasnella* (4)
	Cynorkis gibbosa	4	Juvenile	None
	Cynorkis gibbosa	7	Seedling	None
	Cynorkis uniflora	3	Seedling	*Ceratobasidium* (1)
	Cynorkis uniflora	3	Mature	None
	Cynorkis uniflora	7	Mature	None
	Disa sp. 1	4	Mature	None
	Eulophia sp. 1	1	Seedling	None
	Eulophia plataginea	1	Juvenile	None
	Habenaria sp.1	1	Juvenile, mature	None
	Habenaria sp. 2	2	Seedling	*Tulasnella* (1)
	Habenaria sp. 2	4	Mature	*Tulasnella* (1)
	Jumellea sp.	1	Mature	None
	Liparis sp. 1	5	Seedling	*Tulasnella* (1)
	Polystachya sp. 1	2	Mature	None
	Satyrium sp.	5	Seedling	*Tulasnella* (1)
	Satyrium trinerve	3	Mature	None
	Satyrium trinerve	4	Seedling	*Tulasnella* (4)

Fungal genera listed represent provisional identifications carried out at the time of isolation, based on cultural characteristics described by Currah et al. (1997). Growth habit reflects the substrate where the individual orchid was actually rooted at the time of collection. Collection sites: *1* abandoned mine, rocky grassland, *2* tapia forest, *3* seepage slope, *4* grassland, seepage slope, *5* forest preserve, *6* forest, *7* rocky elevated grassland

Terrestrial seedlings = 6/8, epiphytic seedlings = 2/4, lithophytic seedlings = 0/2

Terrestrial juveniles = 0/3, epiphytic juveniles = 0/2, lithophytic juveniles = NA

Terrestrial mature = 6/18, epiphytic mature = 1/4, lithophytic mature = 1/1

Total terrestrial = 12/29 (41%), total epiphytic = 3/10 (30%), total lithophytic = 1/3 (33%)

To what extent mature epiphytic orchids harbor and utilize *Rhizoctonia*-like fungi in Madagascar needs further study. Due to permit restrictions, only a select few taxa could be collected the third and final year of the study, and the decision was made to target orchid tissues that were most likely to harbor *Rhizoctonia*-like fungi, namely seedlings. Roots collected from the lone mature epiphyte (*Bulbophyllum baronii*) were devoid of pelotons, and therefore no endophytes were isolated (Table 3). For future work on the Itremo Massif and the Central Highlands in general, we recommend that roots of mature epiphytes be collected during the rainy season (January). For epiphytic orchids occupying Madagascar's eastern and northern forests where rainfall is more plentiful throughout the year (> 2000 mm annually; Cribb and Hermans 2009), time of collection may be less critical, as higher moisture levels would be expected to favor fungal activity. To our knowledge, no studies have been reported that document *Rhizoctonia*-like fungi from Orchidaceae inhabiting the NE portion of the island. Given Madagascar's considerable biodiversity, securing fungi from both areas and comparing the orchid mycoflora between the two regions seems like a logical next step.

As a group, roots of mature lithophytic orchids yielded relatively few *Rhizoctonia*-like fungi with one exception (*Tulasnella* from *Eulophia* sp. 1, Table 3). Mature roots from six lithophytes collected the first year harbored pelotons primarily in the first 4 cm of the root tip (Table 1), yet these pelotons yielded conidial saprophytes (*Fusarium, Trichoderma*), not typical genera in the *Rhizoctonia* complex. As to why this occurred is puzzling and deserves further inquiry. It is conceivable that roots of these lithophytes were subjected to rapid dehydration given their dry placement on sun-exposed rocks with little or no associated (moist) organic debris (Fig. 3). Thus, the pelotons we observed in the tissues may have been formed by *Rhizoctonia*-like fungi during the rainy season, but quickly dried out before the pelotons could be completely digested by the orchid (lysis). Opportunistic saprophytes may then have gained entry as secondary invaders, which may explain why we isolated *Fusarium* in the present study, as well as slow-growing, dark-pigmented endophytes (*Toxicocladosporium, Cladophialophora, Lophiostoma*) recovered in samples at Kew (Yokoya et al. 2015). Studies are needed to explore the true nature of the peloton-forming, non-*Rhizoctonia* fungi to determine if these endophytes are potentially harmful and/or benign inhabitants, or if they serve a physiological purpose.

Moisture availability (retention) linked with seasonality may also explain why more *Rhizoctonia*-like fungi were isolated from epiphytes compared to lithophytes. For example, roots of many of the epiphytes we sampled were tightly affixed to crevices of tree bark often in association with lichens and mosses. All three of these substrates would be expected to retain water more effectively than bare rock alone, and could also serve as a carbon source for associated fungi, including the *Rhizoctonia*-like fungi present within the orchid. Members of *Ceratobasidium*, in particular, are known to produce polyphenoloxidases that are involved with lignin breakdown (Rasmussen 1995), and this may explain why roots of some epiphytes, namely *Aerangis*, harbored mostly *Ceratobasidium* (Tables 2 and 3). Thus, the utilization of *Ceratobasidium* strains may afford epiphytes like *Aerangis* with an additional source of organic carbon, and therefore a selective advantage for life in the canopy. For lithophytes, utilization of *Ceratobasidium* at an early (protocorm) stage of development may be critical to life on the rocks in pockets where organic debris and moisture accumulate.

Conclusions

Despite the distance, rugged terrain, and length of time between field collection and transport to the laboratory, roots of epiphytic, lithophytic and terrestrial orchids yielded all major groups of fungi in the *Rhizoctonia* complex (*Ceratobasidium, Tulasnella, Sebacina*). These fungi were present in roots of seedlings, juveniles and mature plants, especially terrestrials. Despite their small size, root pieces from seedlings stages of epiphytes (e.g., *P. concreta*) yielded fungi with our method despite being detached in the field 3 weeks prior. In cases where peloton extraction cannot be accomplished in a timely manner (1–4 days of collection), our study demonstrates that samples retain viability up to 3 weeks after collection.

Authors' contributions

LWZ, KY, JPK and ALS devised the techniques for sample collection and transport. LWZ, LR, KY, JPK, ALS and AEW performed field research in Madagascar, and LR identified Orchidaceae collected. VS headed the team effort. LWZ wrote the manuscript. All authors read and approved the final manuscript.

Author details

[1] Department of Biology, Illinois College, 1101 West College Ave., Jacksonville, IL 62650, USA. [2] Kew Madagascar Conservation Centre, Lot #J131B, Ambodivaonjo-Ivandry, 101, Antananarivo, Madagascar. [3] Royal Botanic Gardens Kew, Richmond, Surrey TW9 3AB, UK. [4] Environmental Studies Program, University of Pennsylvania, Philadelphia, PA 19104, USA.

Acknowledgements

In Madagascar, the invaluable contributions of Gaëtan Ratovonirina (KMCC), Jacky Andriantiana, (PBZT), Solo Rapanarivo (PBZT), and Edward Jones (Kew) are gratefully acknowledged. The helpful suggestions provided by Phillip Cribb (Kew), Rachel Helmich (IC), and Connor Melton (IC) are also appreciated.

Competing interests

The authors declare that they have no competing interests.

An explanation of why your manuscript should be published in Botanical Studies

This manuscript provides details for collection and long-distance transport of root samples from orchids for fungal isolation. The special issue on Orchid Conservation by Botanical Studies is a suitable place for it to be published because the journal has a wide readership for those interested in orchids and their conservation.

An explanation of any issues relating to journal policies

Not applicable.

Confirmation that all authors have approved the manuscript for submission

All authors have approved the manuscript for submission to Botanical Studies.

Confirmation that the content of the manuscript has not been published, or submitted for publication elsewhere

The whole or part of the manuscript has not been published or under consideration for publication anywhere else.

Funding

Financial support for this project was provided by the Threatened Plants Appeal and Sainsbury Orchid Endowment, and Bentham-Moxon Trust. Funds for the Illinois College (IC) team were provided by the Margaret A. Cargill Foundation.

References

Aggarwal S, Nirmala C, Beri S, Rastogi S, Adholeya A (2012) In vitro symbiotic seed germination and molecular characterization of associated endophytic fungi in a commercially important and endangered Indian orchid, *Vanda coerulea* Griff. ex Lindl. Eur J Environ Sci 2(1):33042

Chen J, Wang H, Guo SX (2012) Isolation and identification of endophytic and mycorrhizal fungi from seeds and roots of *Dendrobium* (Orchidaceae). Mycorrhiza 22:297–307

Chutima R, Dell B, Vessabutr S, Bussaban B, Lumyong S (2011) Endophytic fungi from *Pecteilis susannae* (L.) Rafin (Orchidaceae), a threatened terrestrial orchid in Thailand. Mycorrhiza 21(3):221–229

Clements MA, Ellyard RK (1979) The symbiotic germination of Australian terrestrial orchids. Am Orchid Soc Bull 48:810–816

Cribb PJ, Hermans J (2009) Field guide to the orchids of Madagascar. Kew Publishing, Royal Botanic Gardens, Kew

Cribb PJ, Kell SP, Dixon KW, Barrett RL (2003) Orchid conservation: a global perspective. (Chapter 1, pp 1–24). Mycorrhizal diversity (Chapter 11, pp 185–203). In: Dixon KW, Kell SP, Barrett RL, Cribb PJ (eds) Orchid conservation. Natural History Publications, Kota Kinabalu, pp 205–226 (ISBN 983-812-078-2)

Currah RS, Sigler L, Hambleton S (1987) New records and new taxa of fungi from the mycorrhizae of terrestrial orchids of Alberta. Can J Bot 65:2473–2482

Currah RS, Zelmer CD, Hambleton S, Richardson KA (1997) Fungi from orchid mycorrhizas. In: Arditti J, Pridgeon AM (eds) Orchid biology: reviews and perspectives, VII. Springer, Netherlands, pp 117–170

Dressler RL (1993) Phylogeny and classification of the orchid family. Dioscorides Press, Portland

Harvais G, Raitsakas A (1975) On the physiology of a fungus symbiotic with orchids. Can J Bot 53:144–155

Hoang NH, Kane ME, Radcliffe EN, Zettler LW, Richardson LW (2017) Comparative seed germination and seedling development of the Ghost Orchid, *Dendrophylax lindenii* (Orchidaceae), and molecular identification of its mycorrhizal fungus from south Florida. Ann Bot 119(3):379–393

Huynh TT, McLean CB, Coates F, Lawrie AC (2004) Effect of developmental stage and peloton morphology on success in isolation of mycorrhizal fungi in *Caladenia formosa* (Orchidaceae). Aust J Bot 52(2):231–241

Jacquemyn H, Brys R, Merckx VS, Waud M, Lievens B, Wiegand T (2014) Coexisting orchids species have distinct mycorrhizal communities and display strong spatial segregation. New Phytol 202(2):616–627

McCormick MK, Taylor DL, Juhaszova K, Burnett RK, Whigham DF, O'Neill JP (2012) Limitations on orchid recruitment: not a simple picture. Mol Ecol 21(6):511–523

Merritt DJ, Hay FR, Swarts ND, Sommerville KD, Dixon KW (2014) Ex situ conservation and cryopreservation of orchid germplasm. Int J Plant Sci 175:46–58

Moore-Landecker E (1996) Fundamentals of the fungi. Prentice Hall, Upper Saddle River

Nontachaiyapoom S, Sasirat S, Manoch L (2010) Isolation and identification of *Rhizoctonia*-like fungi from roots of three orchid genera, *Paphiopedilum*, *Dendrobium*, and *Cymbidium*, collected in Chaing Rai and Chiang Mai provinces of Thailand. Mycorrhiza 20:459–471

Otero JT, Ackerman JD, Bayman P (2002) Diversity and host specificity of endophytic *Rhizoctonia*-like fungi from tropical orchids. Am J Bot 89(11):1852–1858

Pereira OL, Rollemberg CL, Borges AC, Matsouka K, Kasuya MCM (2003) *Epulorhiza epiphytica* sp. nov. isolated from mycorrhizal roots in Brazil. Mycoscience 44:153–155

Pereira OL, Kasuya MCM, Borges AC, de Araújo EF (2005) Morphological and molecular characterization of mycorrhizal fungi isolated from neotropical orchids in Brazil. Can J Bot 83:54–65

Rafter M, Yokoya K, Schofield EJ, Zettler LW, Sarasan V (2016) Non-specific symbiotic germination of *Cynorkis purpurea* (Thouars) Kraezl., a habitat-specific terrestrial orchid from the Central Highlands of Madagascar. Mycorrhiza 26(6):541–552. https://doi.org/10.1007/s00572-016-0691-6

Rajaovelona L, Gardiner LM (2016) *Angraecum longicalcar*: saving a critically endangered orchid. Orchid Rev 124(1314):95–99

Rasmussen HN (1995) Terrestrial orchids, from seed to mycotrophic plant. Cambridge University Press, Cambridge

Rasmussen HN, Rasmussen FN (2009) Orchid mycorrhiza: implications of a mycophagous life style. Oikos 118(3):334–345

Richardson KA, Currah RS, Hambleton S (1993) Basidiomycetous endophytes from the roots of neotropical epiphytic Orchidaceae. Lindleyana 8:127–137

Suárez JP, Weiß M, Abele A, Garnica S, Oberwinkler F, Kottke I (2006) Diverse tulasnelloid fungi form mycorrhizas with epiphytic orchids in the Andean cloud forest. Mycol Res 110:1257–1270

Swarts ND, Dixon KW (2009) Terrestrial orchid conservation in the age of extinction. Ann Bot 104:543–556

Swarts ND, Dixon KW (2017) Conservation methods for terrestrial orchids. J. Ross Publishing, Plantation

Swarts ND, Sinclair EA, Francis A, Dixon KW (2010) Ecological specialization in mycorrhizal symbiosis leads to rarity in an endangered orchid. Mol Ecol 19(15):3226–3242

Tyson P (2000) The eighth continent: life, death and discovery in the lost world of Madagascar. William Morrow (Harper Collins) Publishers, New York, p 374

Warcup JH (1973) Symbiotic germination of some Australian terrestrial orchids. New Phytol 72:387–392

Warcup JH (1981) The mycorrhizal relationships of Australian orchids. New Phytol 87:371–381

Whitman M, Medler M, Randriamanindry JJ, Rabakonandrianina E (2011) Conservation of Madagascar's granite outcrop orchids: the influence of fire and moisture. Lankesteriana Int J Orchidol 11:55–67

Yokoya K, Zettler LW, Kendon JP, Bidartondo M, Stice AL, Skarha S, Corey LL, Knight A, Sarasan V (2015) Preliminary findings on identification of mycorrhizal fungi from diverse orchids in the Central Highlands of Madagascar. Mycorrhiza 25:611–625. https://doi.org/10.1007/s00572-015-0635-6

Zelmer CD, Cuthbertson L, Currah RS (1996) Fungi associated with terrestrial orchid mycorrhizas, seeds and protocorms. Mycoscience 37(4):439–448

Zettler LW, Sharma J, Rasmussen F (2003) Mycorrhizal diversity (Chapter 11, pp 185–203). In: Dixon KW, Kell SP, Barrett RL, Cribb PJ (eds) Orchid conservation. Natural History Publications, Kota Kinabalu, pp 205–226. (ISBN 983-812-078-2)

Zettler LW, Wood EM, Johnson JAN, Kirk AK, Perlman SP (2011) Seed propagation and reintroduction of the US federally endangered Hawaiian endemic, *Platanthera holochila* (Hbd.) Krzl. (Orchidaceae). Eur J Environ Sci 1(2):69–70

11

Actinidia callosa var. *callosa* suppresses metastatic potential of human hepatoma cell SK-Hep1 by inhibiting matrix metalloproteinase-2 through PI3K/Akt and MAPK signaling pathways

Jeng-Shyan Deng[1†], Jui-Shu Chang[2†], Jung-Chun Liao[3], Wei Chao[4], Ming-Ming Lee[1], Chien-Hua Cheng[5] and Guan-Jhong Huang[4*]

Abstract

Background: Cancer cell metastasis involving multi-step procedures and cytophysiological property changes may make difficult in the clinical management and death rate increasing.

Results: In this study, we first observed that ethyl acetate fraction of *Actinidia callosa* var. *callosa* (EAAC) carry out a dose-dependent inhibitory effect without cytotoxicity on the mobility and invasion of highly metastatic SK-Hep1 cells. To investigate the EAAC in cancer metastasis, SK-Hep1 cells were treated with EAAC at various concentrations and then subjected to gelatin zymography, casein zymography and western blot to study the impacts of EAAC on metalloproteinase-2 (MMP-2) and tissue inhibitor of metalloproteinase-1/2 (TIMP-1/2), respectively. Our results showed that EAAC treatment may decrease the expressions of MMP-2 and enhance the expression of TIMP-1/2 in a concentration-dependent manner. EAAC also inhibited effect on the phosphorylation of mitogen-activated protein kinase (MAPK) and phosphatidylinositol-3-kinase/serine/threonine protein kinase [or protein kinase B (PI3K/Akt)] and focal adhesion kinase (FAK).

Conclusions: These results indicate that EAAC inhibited SK-Hep1 cell of metastasis by reduced protein level of MMP-2 through the suppression of MAPK and FAK signaling pathway and of the activity of PI3K/Akt. These findings suggest that EAAC may be used as an antimetastatic agent.

Keywords: *Actinidia callosa* var. *callosa*, Migration, Invasion

Background

Hepatocellular carcinoma (HCC) is the most common type of cancer in the Southeast Asia, including Taiwan. Metastasis is a characteristic of highly malignant cancers associated with a poor clinical outcome. Malignant tumor progression relies on the capacity of cancer cells to invade and promoting angiogenic effects in tumor areas. One characteristic that metastatic cancer cells were induced the dissolution of basement membrane proteins and the extracellular matrix (ECM). The activity of the degradative process is mediated by matrix metalloproteinases (MMPs) (Forrbes et al. 2003), which belong to zinc-dependent neutral endopeptidases that can break down the ECM (Parmo-Cabanas et al. 2006). The MMPs are found as mediators of tumor migration and invasion by breaking down connective tissue barriers (Chao et al. 2017). Thus, the inhibition of MMP activity is important for the prevention of cancer metastasis. MMP-2 protein

*Correspondence: gjhuang@mail.cmu.edu.tw
†Jeng-Shyan Deng and Jui-Shu Chang contributed equally to this work
[4] School of Chinese Pharmaceutical Sciences and Chinese Medicine Resources, College of Chinese Medicine, China Medical University, Taichung, Taiwan
Full list of author information is available at the end of the article

expressions are high level in hepatoma cells, such as SK-Hep1 cells (Ho et al. 2012). In addition, urokinase-type plasminogen activator (uPA) is a serine protease that activation of pro-uPA occurs after binding to its receptor uPAR (uPA receptor). Plasminogen activator inhibitors (PAI-1 and PAI-2) were important component of the coagulation system that inhibits both receptor-bound and free uPA (Liao et al. 2009). Meanwhile, Tissue inhibitors of metalloproteinases (TIMPs) control MMP activities and the imbalance between MMPs and TIMPS may contribute to degradation or depositions of ECM involved in tissue remodeling and regulate tumor cell progression including tumor angiogenesis (Shih et al. 2009).

Actinidia callosa var. *callosa* (*Actinidiaceae*; AC) is a liana plant, which belongs to Actinidiaceae, the roots of AC have been used in traditional Chinese medicine for the treatment of anti-pyretic, anti-inflammatory diseases and various cancers, including gastric carcinoma, liver carcinoma, and breast carcinoma (Liao et al. 2013; Lee et al. 2014). Actinidia species had showed some pharmacological effects such as anti-inflammatory activity from the fruit of *Actinidia polygama* (Ren et al. 2007), antitumor, and immunomodulatory activity from the roots of *Actinidia eriantha* (Xu et al. 2009). In the present study, we used ethyl acetate fraction of AC (EAAC) to investigate the antimetastatic effects on a human hepatocarcinoma SK-Hep1 cell line in vitro.

Methods

Chemicals

Trypsin, EDTA, fetal bovine serum (FBS) and penicillin/streptomycin were from Gibco Life Technologies, Inc. (Paisley, UK). Cell culture supplies were purchased from Costar (Corning, Inc., Cypress, CA, USA). Dulbecco's modified Eagle's medium (DMEM), 3-(4,5-dimethylthiazolyl-2)-2,5-diphenyltetrazolium bromide (MTT), RNase A, and other chemicals were obtained from Sigma Chemical Co. (St. Louis, MO, USA). Plant materials were collected from Taichung Country in Taiwan. They were identified and authenticated by Dr. Shyh-Shyun Huang, Professor, School of Pharmacy, College of Pharmacy, China Medical University, Taichung, Taiwan.

Extraction and fractionation

The powder of *A. callosa* var. *callosa* (AC) (2 kg) was extracted with methanol (MAC) three times. The extract was evaporated under reduced pressure using a rotavapor, and then stored under light protection. A yield equivalent to 6.51% of the original weight was obtained. Next, MAC (130.2 g) was dissolved and suspended in 100 mL of water in a separatory funnel prior to being partitioned in sequence with *n*-hexane, ethyl acetate and *n*-butanol (800 mL each for three times). Under reduced pressure, fractions were yielded and collected: *n*-hexane fraction (17.79 g, 13.66%), ethyl acetate fraction (78.50 g, 60.73%), *n*-butanol fraction (17.18 g, 13.21%) and aqueous fraction (16.73 g, 12.76%). All extracts were stored in the refrigerator before the use.

Fingerprint analysis by HPLC

The analysis will be performed on a HITACHI HPLC L-5000 system equipped with a degasser, pumps, and a photodiode array detector linked to a PC computer running the software program HPLC LACHROM. The analytical column (250 × 4.6 mm i.d.) used is Mightysil 5 μm C18 (Japan). For HPLC analysis, an aliquot (10 μL) is injected into the columns and eluted at 30 °C. Mobile phase eluent (A) is 0.6‰ aqueous phosphoric acid, and eluent (B) MeOH, and the flow rate is kept constant throughout the analysis at 1 mL/min. Injections are accomplished with a 10 μL fixed loop. The elution programme used is as follows 13% A to 87% B. For photodiode array detection, the wavelengths of triterpenic acids at their respective maximum absorbance–wavelength can monitored at the same time. Identification is based on retention times and on-line spectral data in comparison with authentic standards. Quantification is performed by establishing calibration curves for each compound determined, using the standards.

Cell culture

The hepatocarcinoma SK-Hep1 cell was purchased from the Bioresources Collection and Research Center (BCRC) of the Food Industry Research and Development Institute (Hsinchu, Taiwan). Cells were cultured in plastic dishes containing DMEM supplemented with 10% fetal bovine serum (FBS) in a CO_2 incubator (5% CO_2 in air) at 37 °C and subcultured every 2 days at a dilution of 1:5 using 0.05% trypsin–0.02% EDTA in Ca^{2+}-, Mg^{2+}-free phosphate-buffered saline (DPBS).

Cell viability

The cells (2×10^5) were cultured in 96-well plate containing DMEM supplemented with 10% FBS for 1 day to become nearly confluent. Then cells were cultured with EAAC for 24 h. After that, the cells were washed twice with DPBS and incubated with 100 μL of 0.5 mg/mL MTT for 2 h at 37 °C testing for cell viability. The medium was then discarded and 100 μL dimethylsulfoxide (DMSO) was added. After 30-min incubation, absorbance at 570 nm was read by a microplate reader.

Determination of MMP by zymography

MMP released from SK-Hep1 cells into the medium was assayed by gelatin zymography (7.5% zymogram gelatin gels) (Hung et al. 2009). Briefly, the culture medium was electrophoresed in a 10% SDS-PAGE gel containing 0.1% gelatin. The gel was then washed at room temperature in a solution containing 2.5% (*v/v*) Triton X-100 with two changes and subsequently transferred to a reaction buffer for enzymatic reaction containing 1% NaN_3, 10 mM $CaCl_2$ and 40 mM Tris–HCl, pH 8.0, at 37 °C with shaking overnight (for 12–15 h). Finally, the MMP gel was stained for 30 min with 0.25% (*w/v*) Coomassie blue in 10% acetic acid (*v/v*) and 20% methanol (*v/v*) and destained in 10% acetic acid (*v/v*) and 20% methanol (*v/v*).

Cell migration assay

Tumor cell migration was assayed in transwell chambers (Millipore) (Hung et al. 2009). Transwell chambers with 6.5 mm PVPP-free polycarbonate filters of 8 µm pore size were used. SK-Hep1 cells (5×10^5/mL) and 0–200 µg/mL of EAAC were suspended in DMEM (100 µL, serum free), placed in the upper transwell chamber, and incubated for 24 h at 37 °C. Then, the cells on the upper surface of the filter were completely wiped away with a cotton swab, and the lower surface of the filter was fixed in methanol, stained with Giemsa, and counted under a microscope at a magnification of 200×. For each replicate, the tumor cells in 10 randomly selected fields were determined, and the counts were averaged.

Cell invasion assay

The invasion of tumor cells was assessed in transwell chambers containing a 6.5 mm PVPP-free polycarbonate filter with 8 µm pore size, as described in the cell migration assay (Ho et al. 2012) except that each filter was coated with 100 µL of a 1:20 diluted Matrigel in cold DMEM to form a thin continuous film on top of the filter. The number of cells was adjusted to 5×10^5/mL and 100 µL (containing 5×10^4 cells) was transferred to each of triplicate wells in DMEM containing 10% FBS. After incubating for 24 h, the cells were stained and counted as described above, and the number of cells invading the lower side of the filter was measured.

Preparation of whole-cell lysates

SK-Hep1 cells (1×10^5 cells) were plated in a 100 mL tissue culture flask and were treated with various concentrations of EAAC and 0.2% DMSO. SK-Hep1 cells were washed twice with PBS and were scraped into a microcentrifuge tube. The cells were centrifuged at 1250*g* for 5 min, and the pellet was lysed with iced-cold RIPA buffer (1% NP-40, 50 mM Tris-base, 0.1% SDS, 0.5% deoxycholic acid, 150 mM NaCl, pH 7.5), to which was added freshly prepared phenylmethylsulfonyl fluoride (10 mg/mL), leupeptin (17 mg/mL), and sodium orthovanadate (10 mg/mL). After incubation for 5 min on ice, the samples were centrifuged at 10,000*g* for 10 min, and then the supernatants were collected as whole-cell lysates. The lysates were denatured and subjected to SDS-PAGE and western blotting. The protein content was determined with Bio-Rad protein assay reagent using BSA as a standard.

Western blotting analysis

Whole-cell lysate proteins were mixed with an equal volume of electrophoresis sample buffer, and the mixture was then boiled for 10 min. Then, an equal protein content of total cell lysate from control, 0.2% DMSO, and EAAC-treated sample were resolved on 10–12% SDS-PAGE gels. Proteins were then transferred onto nitrocellulose membranes (Millipore, Bedford, MA) by electroblotting using an electroblotting apparatus (Bio-Rad). Nonspecific binding of the membranes was blocked with Tris-buffered saline (TBS) containing 1% (*w/v*) nonfat dry milk and 0.1% (*v/v*) Tween-20 (TBST) for more than 2 h. Membranes were washed with TBST three times each for 10 min and then incubated with an appropriate dilution of specific primary antibodies in TBST overnight at 4 °C. The membranes were washed with TBST and then incubated with an appropriate secondary antibody (horseradish peroxidase-conjugated, goat antimouse, or antirabbit IgG) for 1 h. After washing the membrane three times each for 10 min in TBST, the bands were visualized using ECL reagents (Millipore, Billerica, MA). Band intensity on scanned films was quantified using Kodak molecular imaging (MI) software and expressed as relative intensity compared with control.

Statistical analysis

Experimental results were presented as the mean ± standard deviation (SD) of three parallel measurements. A difference was considered to be statistically significant when $p < 0.05$, $p < 0.01$, and $p < 0.001$.

Results

Compositional analysis of EAAC by HPLC

In this study, we set up the fingerprint chromatogram in the EAAC. The three compounds of betulinic acid, ursolic acid, and oleanolic acid were used as markers. As shown in Fig. 1, these triperpenic acids have been identified as betulinic acid (retention time, 15.16 min), ursolic acid (17.43 min), and oleanolic acid (18.42 min) by their retention time and UV absorbance of purified standards. According to the plot of peak–area ratio (y) vs. concentration (x, µg/mL), the regression equations of the three constituents and their correlation coefficients (r) were determined. The relative amounts of the three triperpenic acids found in EAAC was in the order of ursolic

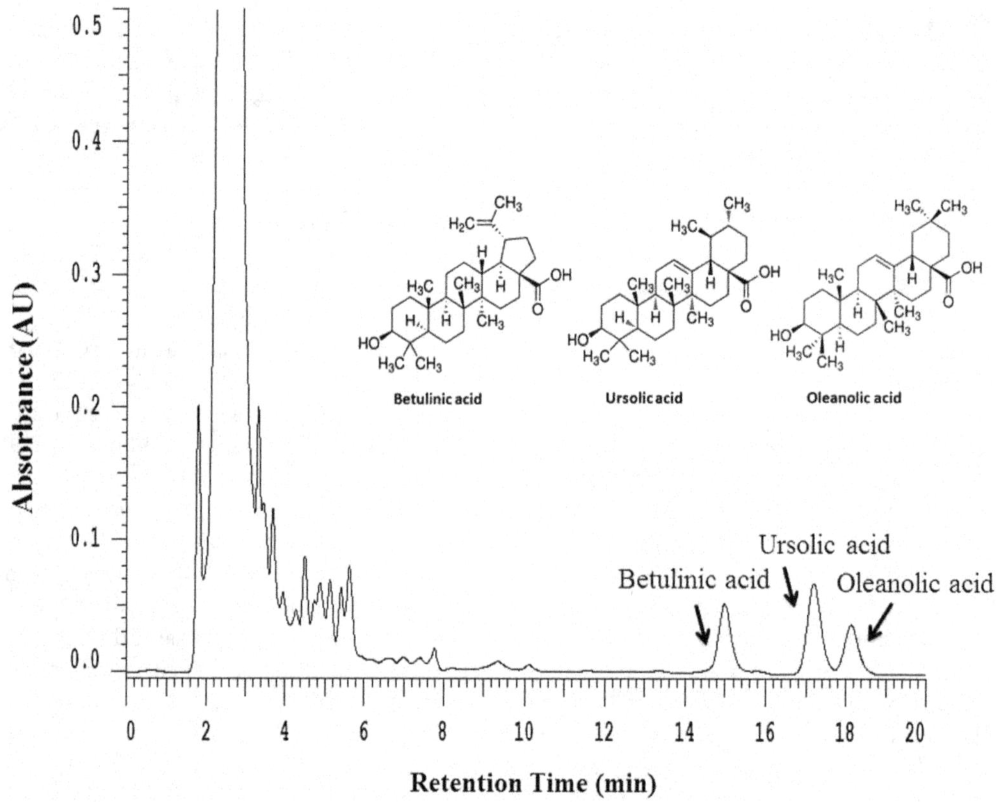

Fig. 1 HPLC chromatograms (280 nm) of ethyl-acetate fraction of the stem of *A. callosa* var. *callosa* (EAAC)

acid (6.34 mg/g) > oleanolic acid (5.00 mg/g) > betulinic acid (1.25 mg/g), respectively.

Effect of EAAC on cell proliferation

In this study, the cytotoxicity of EAAC toward SK-Hep1 human hepatoma cells was evaluated using the MTT assay. The effects of 0–1000 μg/mL EAAC on cell growth after 24 and 48 h are shown in Fig. 2a. After 24 and 48 h of incubation, EAAC inhibited cell proliferation in a dose-dependent fashion, with cell numbers significantly reduced by 4.6–50.5 and 7.5–69.4% compared to the control. EAAC has been shown to have anti-tumor effects in vitro, but the underlying mechanism is unclear. In this study, we demonstrated that EAAC inhibited the proliferation of SK-Hep1 cells in a concentration-dependent manner. It was therefore clear that a 24 h treatment of EAAC, at a concentration ranging from 0 to 200 μg/mL, has no cytotoxicity to SK-Hep1 cells, highly metastatic cancer cells.

EAAC inhibits the activation of MMP in SK-Hep1 cells

To check the possible antimetastatic mechanisms of EAAC, we determined the effect of these compounds on the expression of MMP. First, cultured conditioned media of SK-Hep1 cells were subjected to zymographic analysis. After 24 h, treatment with 0–200 μg/mL EAAC shows a dose–response relationship in MMP-2 (Fig. 2b). Treatment of EAAC at 50, 100, 150, and 200 μg/mL decreased MMP-2 expression by 11.5, 31.6, 44.5 and 54.5%, respectively. Thus, in the SK-Hep1 cell system, MMP-2 was down-regulated by EAAC treatments. In addition, we did not find MMP-9 activity in the zymography assay.

Effect of EAAC on migration and invasion of SK-Hep1 cells in vitro

The transwell assay was used to study the migration and invasion of SK-Hep1 cells 24 h after EAAC treatment. We found that EAAC at 0–200 μg/mL significantly decreased both the migration and invasion (Fig. 3a, b) of SK-Hep1 cells in dose-dependent manners. The IC_{50} values for EAAC on migration and invasion of SK-Hep1 was approximately 73.3 and 92.7 μg/mL, respectively.

Effect of EAAC on MMPs, TIMPs, and FAK expression in SK-Hep1 cells

To further search the modulation of pro-MMP activation mediated by EAAC, we determined MMP-2/-9 and TIMP-1/-2 protein expression levels. A variety of

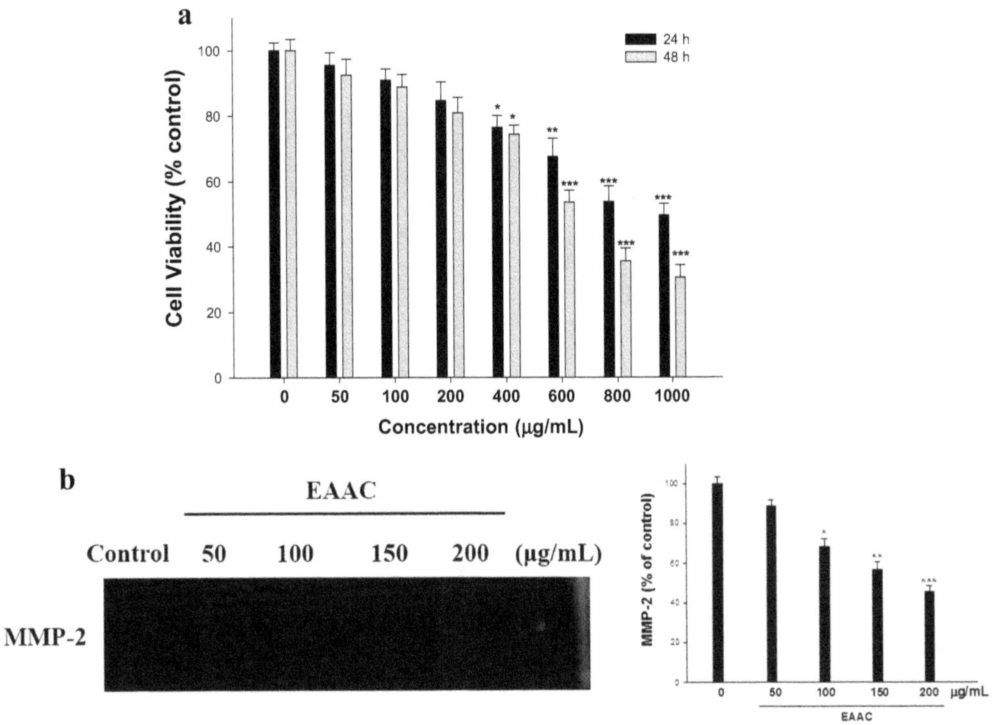

Fig. 2 Effects of EAAC on SK-Hep1 cell viability (**a**) and MMP-2 (**b**) activity. The cells were incubated for 24 and 48 h in the absence or present of EAAC (0, 50, 100, 150, and 200 μg/mL). After treatment, cell viability assay was performed using MTT assay. Optical density was determined at 570 nm and is expressed as cell survival relative to control. Cells were treated with various concentrations of EAAC for 24 h. The conditioned media were collected, and MMP-2 activity was determined by gelatin zymography. MMP-2 activity was quantified by densitomeric analysis. The data were presented as mean ± SD for three different experiments performed in triplicate. *$p < 0.05$, **$p < 0.01$, and ***$p < 0.001$ was compared with untreated control group

scientific papers have revealed that over-expression of TIMPs, especially TIMP-1 and TIMP-2, can inhibit tumor growth, invasion and metastasis. As results shown in Fig. 4a, EAAC especially decreased MMP-2 and MMP-9 activity in a concentration-dependent manner while activity was also increased TIMP-1 and TIMP-2 by SK-Hep1 (Fig. 4b).

Focal adhesion kinase (FAK) is a potentially important therapeutic target because it is a key point of convergence for the relevant growth factor pathways required for survival and metastatic functions of cancer cells. FAK was recognized as a protein tyrosine kinase that place to cell focal adhesion contacts and adhesion sites (Shi et al. 2017). To evaluate the effect of EAAC on FAK protein expression, SK-Hep1 was treated with EAAC at 0–200 μg/mL for 24 h. EAAC suppressed FAK expression (Fig. 4c). We also illustrated this suppressive effect on the phosphorylation of FAK by EAAC in SK-Hep1 cell.

Inhibition by EAAC of MAPK phosphorylation

To evaluate signaling pathway induced by EAAC in SK-Hep1 cells, we investigated the responses of MAPKs, which are signaling molecules receptive to extracellular stimuli. We dissected the phosphorylation of MAPKs in SK-Hep1 cells after treatment with EAAC (0–200 μg/mL) for 24 h. As shown in Fig. 5a, EAAC inhibited the phosphorylation activities of MAPKs in SK-Hep1 cells. Treatment with 200 μg/mL EAAC reduced the phosphorylation activity of ERK1/2, p38 and JNK1/2 about 64.5, 65.4 and 74.5%, respectively.

Effects of EAAC on the PI3K/Akt signaling in SK-Hep1 cells

PI3K/Akt signaling pathway plays a role in facilitating tumor cell invasion and metastasis. The constitutive activation of PI3K/Akt signaling is frequently found in cancer (Wang et al. 2016). Incubation of SK-Hep1 cells with EAAC (200 μg/mL) decreased the PI3K and the phosphorylation of Akt protein levels (Fig. 5b). EAAC (200 μg/mL) produced suppression on PI3K and the phosphorylated of Akt level about 85.5 and 68.4% compared to control intensity at 24 h. EAAC inhibited the expression of PI3K and the phosphorylated of Akt was coincided with the kinetics of cell migration and invasion.

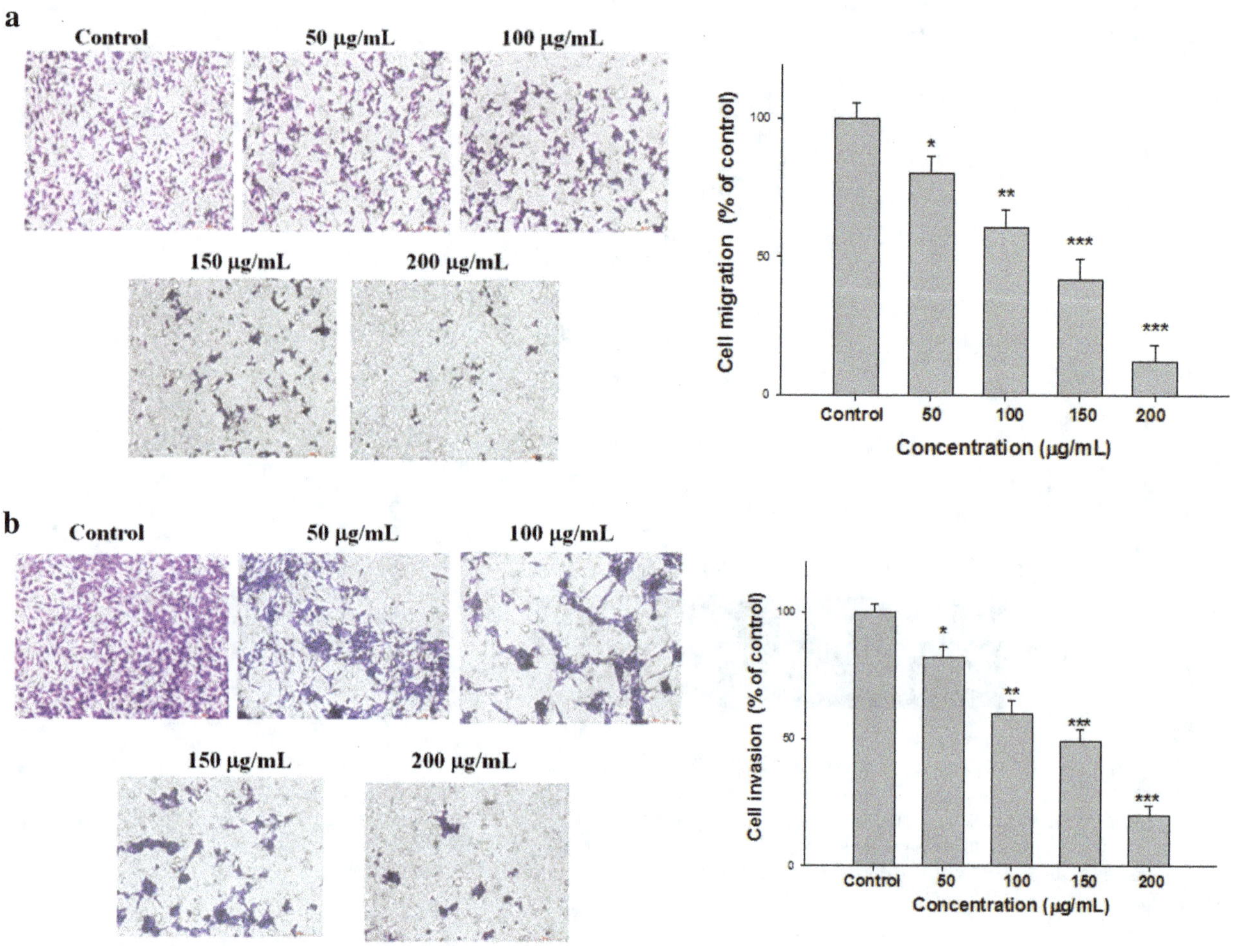

Fig. 3 Effects of EAAC on transwell migration (**a**) and invasion (**b**) of SK-Hep1 cells. SK-hep1 cells were incubated with EAAC (0, 50, 100, 150, and 200 µg/mL) for 24 h, and the transwell migration cells and invasion cells were counted. The data were presented as mean ± SD for three different experiments performed in triplicate. $^{*}p < 0.05$, $^{**}p < 0.01$, and $^{***}p < 0.001$ were compared with untreated control group

Effect of betulinic acid, ursolic acid and oleanolic acid on MMP expression

To investigate the possible anti-metastatic mechanisms of EAAC, we determined the effect of three compounds (betulinic acid, ursolic acid and oleanolic acid) on the activities of MMP-2. It was clear that a 24 h treatment of betulinic acid, ursolic acid and oleanolic acid, at a concentration ranging from 0 to 50 µM, has no cytotoxicity to SK-Hep1 cells (Fig. 6a). In addition, cultured conditioned media of SK-Hep1 cells were subjected to zymographic analysis. Treatment with 50 µM betulinic acid, ursolic acid and oleanolic acid after 24 h suppressed MMP-2 activity by 64.4, 59.4, and 68.4%, respectively. The value of half-maximal effective inhibitory concentration (IC_{50}) for MMP-2 by betulinic acid, ursolic acid and oleanolic acid was approximately 23.5, 45.6, and 19.3 µM (Fig. 6b). Thus, in the SK-Hep1 cell system, MMP-2 was down-regulated by betulinic acid, ursolic acid and oleanolic acid treatments.

Discussion

SK-Hep1 is an immortal, human cell line derived from the ascetic fluid of a patient with liver adenocarcinoma. High metastatic capacities of SK-Hep1 cells are suitable for cancer metastasis research and have been used wildly in the literature (Park et al. 2015). In this study we examined the antimetastatic effects of EAAC in human hepatoma SK-Hep1 cells. We found that EAAC inhibited the cell migration and cell invasion of SK-Hep1 cells without cytotoxicity by targeting the MAPK and PI3K/AKT signaling pathway. We showed that EAAC notably inhibited the activities of MMP-2. These results revealed that the antimetastatic effect of EAAC was associated with the inhibition of enzymatically degradative processes (MMP-2 and -9) of tumor metastasis. To our knowledge, this is the first study to demonstrate the biochemical mechanisms by which EAAC reduces the metastasis in SK-Hep1 cells.

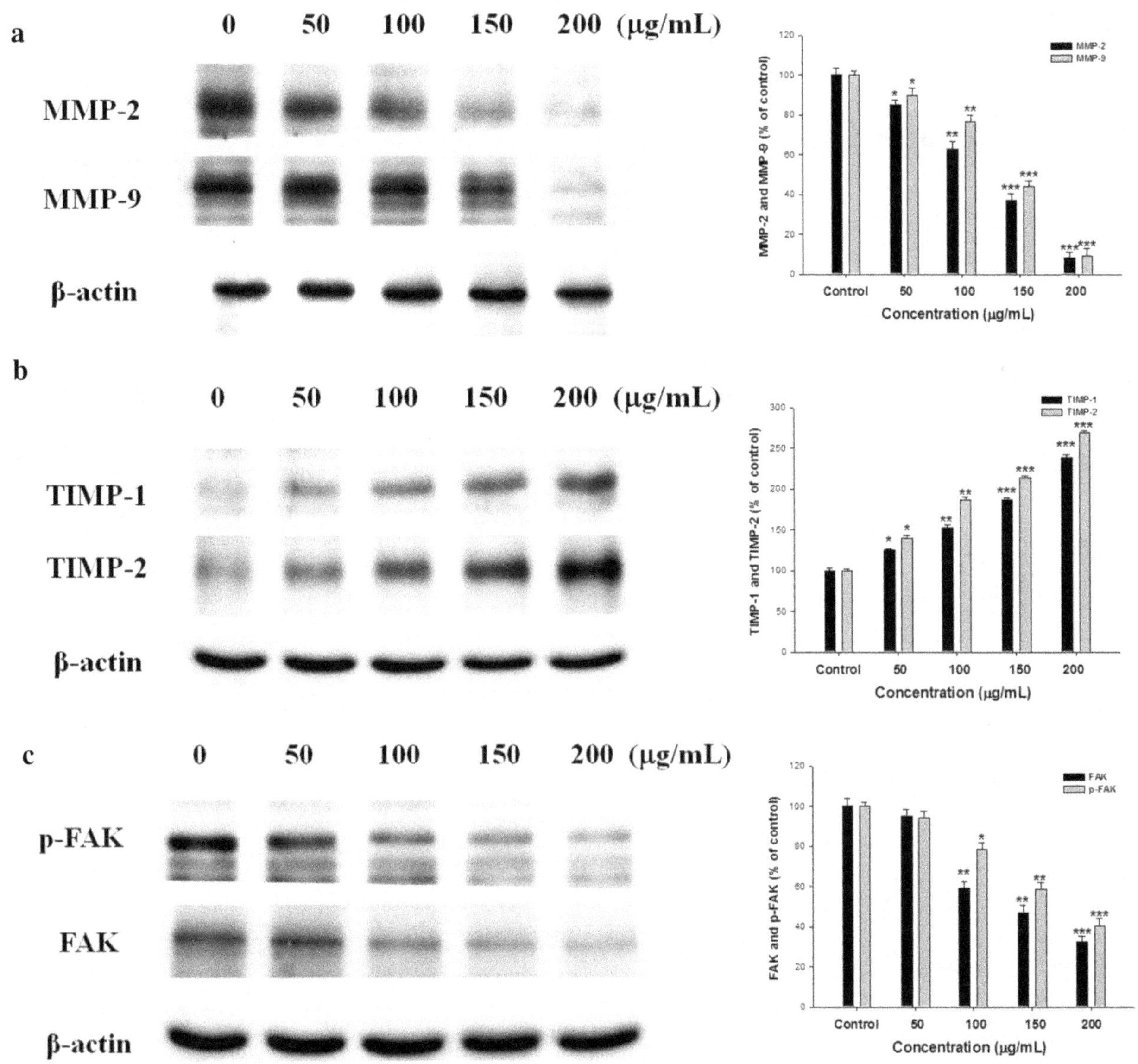

Fig. 4 Effects of EAAC on MMP (**a**), TIMP (**b**), and FAK (**c**) protein expressions. Analyzed by western blot, SK-Hep1 cells were treated with 0, 50, 100, 150, and 200 μg/mL for 24 h and cell lysates were subjected to SDS-PAGE followed by western blotting. Activities of these proteins were subsequently quantified by densitometric analyses, with the control set to 100%. The data were presented as mean ± SD for three different experiments performed in triplicate. *$p < 0.05$, **$p < 0.01$ and ***$p < 0.001$ 0.001 were compared with untreated control group

MMP-2 and -9 are type IV collagenase/gelatinase, and their expression is associated with aggressive tumor phenotypes and poor prognosis (Chen et al. 2015). These enzymes are markers associated with the tumor invasion and metastasis and degrade the environmental ECM and the basement membrane. MMP-2 activity is the highest expressions in the tissue of hepatocellular carcinoma with metastasis and plays a key role in the degradation of the basement membrane and ECM, thereby enhancing migration of endothelial cells (Byambaragchaa et al. 2013). Our results showed that EAAC at 0–200 μg/mL significantly inhibited MMP-2 activity of SK-Hep1 cells, significantly. Thus, it is likely that inhibition of MMP-2 activities by EAAC is responsible for EAAC inhibition of migration and invasion of the SK-Hep1 cells.

FAK is involved in ECM via expression and release of MMP-2 and MMP-9 and suggested to have an essential role in metastasis (Huang et al. 2010). Activated FAK binds to PI3K domain 2 (SH2) where it catalyzes the phosphorylation of inositol lipids (Casar et al. 2014). Thus, FAK also regulates phosphorylation of ERK and Akt, and the FAK-ERK and FAK-Akt signaling pathway

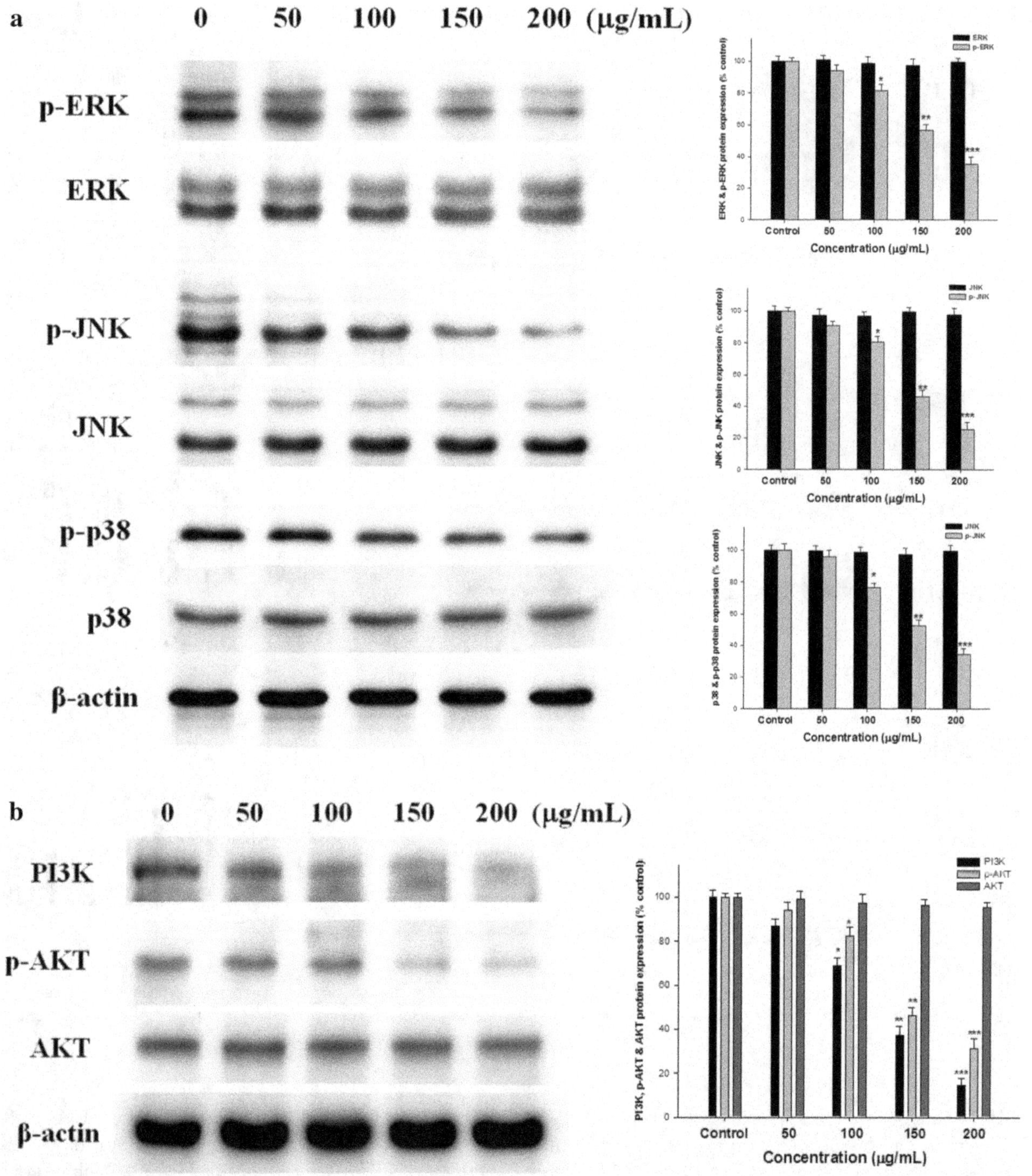

Fig. 5 Inhibitory effect of EAAC on MAPK (**a**), and PI3K, AKT (**b**) signaling pathway. SK-Hep1 cells were cultured in serum-free media containing 0, 50, 100, 150, and 200 mg/mL EAAC for 24 h, and the cell lysates were subjected to SDS-PAGE followed by western blotting. Activities of these proteins were subsequently quantified by densitometric analyses, with the control set to 100%. The data were presented as mean ± SD for three different experiments performed in triplicate. *$p < 0.05$, **$p < 0.01$ and ***$p < 0.001$ were compared with untreated control group

induces MMP-9 expression (Lu et al. 2016). In this study, we found that EAAC inhibited the activation of FAK, as evidenced by reduced phosphorylation of FAK. We also demonstrated that treatment with EAAC inhibited phosphorylation of ERK1/2, JNK1/2, and p38 activity. The results indicated that EAAC inhibited the

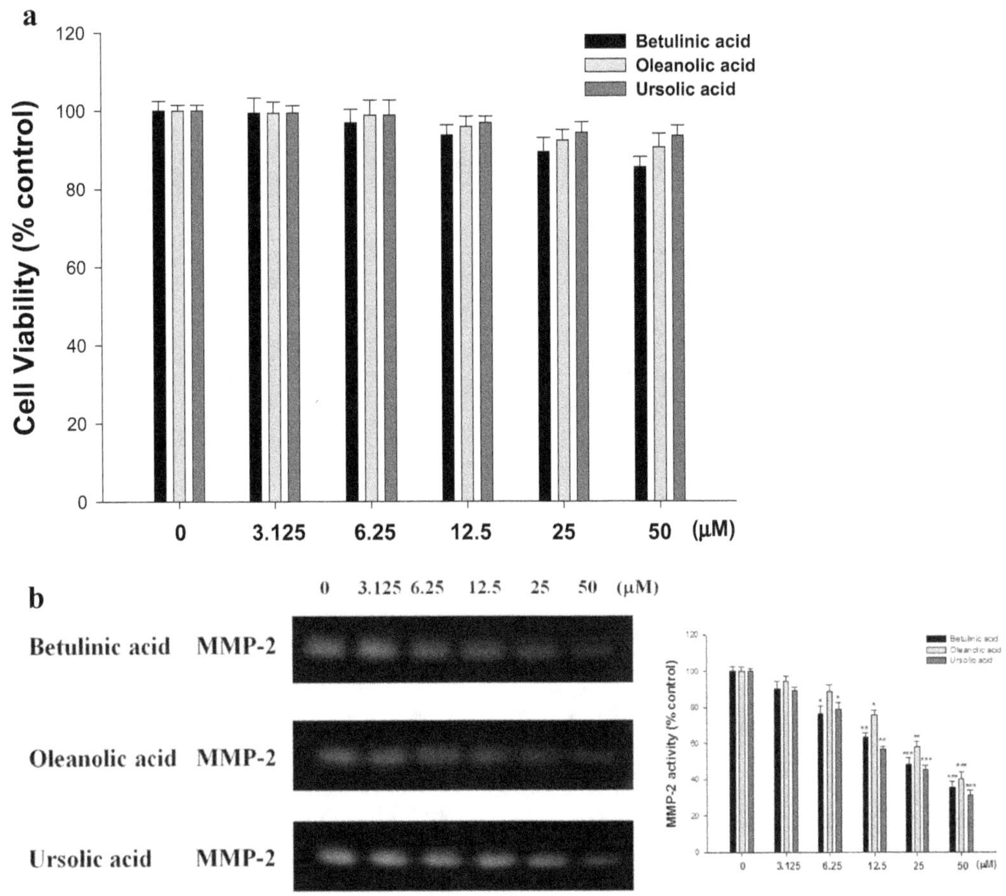

Fig. 6 Effects of betulinic acid, ursolic acid, and oleanolic acid on SK-Hep1 cell MMP-2 activity. The cells were incubated for 24 h in the absence or present of betulinic acid, ursolic acid and oleanolic acid (0, 3.125, 6.25, 12.5, 25, and 50 μM). After treatment, cell viability assay was performed using MTT assay. Optical density was determined at 570 nm and is expressed as cell survival relative to control. The conditioned media were collected, and MMP-2 activity was determined by gelatin zymography. MMP-2 activity was quantified by densitomeric analysis. The data were presented as mean ± SD for three different experiments performed in triplicate. *$p < 0.05$, **$p < 0.01$, and ***$p < 0.001$ was compared with untreated control group

phosphorylation of ERK1/2, p38, and JNK expressions. MAPK kinases associated with MMP-2 and MMP-9 expression in protein levels.

PI3K/Akt signaling pathways is the chief mechanism for controlling cell survival, metabolism, angiogenesis and malignant transformation in response to extracellular cues (Hseu et al. 2017). Akt is a protein kinase, as an important target gene of PI3K, which phosphorylates phosphatidylinositol-4, 5-bisphosphate (PIP2) to form phosphatidylinositol 3, 4, 5-trisphosphate (PIP3) (He et al. 2017). Dysregulation of the PI3K/Akt pathway is implicated in a number of human diseases including cancer, diabetes, cardiovascular disease and neurological diseases. Thus, we evidenced that FAK promotes SK-Hep1 cancer cell migration and invasion in concert with the activation of the MAPK and PI3K/Akt signaling pathways, which modulate the MMP expression. In addition, we showed that EAAC inhibited PI3K/Akt in SK-Hep1 cells.

Plants provide a vigorous source for food health and drug discovery. One important class of bioactive phytochemicals is triterpenoids, which represent a large family of compounds classified according to the number of isoprene units (Chen et al. 2016). Triterpenoids are widely spread out and 20,000 triterpenoids have been identified from the various parts of medicinal plants (Gopalakrishnan and Thomas 2014). In this study, three triperpenic acids (ursolic acid, oleanolic acid and betulinic acid) found in EAAC and there biosynthesized by the acetate/mevalonate pathway and (3*S*)-2,3-oxidosqualene cyclization. Ursolic acid, oleanolic acid and betulinic acid possess interesting pharmacological properties, including the antioxidant, microbicide, antidiabetic, anti-inflammatory, hypolipidemic, anticancer and antiatherosclerotic actions (Silva et al. 2016).

Betulinic acid, ursolic acid and oleanolic acid are pentacyclic triterpene acids naturally occurring in many medicinal herbs. These have been demonstrated to exert anticancer effects in various cancer cell systems (Yamai et al. 2009). It has been reported that treatment with a combination of anticancer drugs (5-fluorouracil) and triterpenes (betulinic acid, ursolic acid and oleanolic acid) resulted in additive or synergistic suppression of cell growth in vitro and in the reduction of experimental esophageal squamous carcinoma cells cell metastasis by direct suppression of cell growth in vivo (Yamai et al. 2009). Betulinic acid also resulted in suppression of cell growth in vitro and in the prevention of B16F10 melanoma experimental metastasis by suppression of cell growth in vivo (Sawada et al. 2004). In addition, ursolic acid also suppressed expression of MMP-2, as well as declined invasion and migration in these lung and gastric cancer cells (Hussain et al. 2017). Oleanolic acid decreased the expression of the angiogenic vascular endothelial growth factor (VEGF) and decreased the development of melanoma-induced lung metastasis (Žiberna et al. 2017).

In conclusion, we have demonstrated that EAAC inhibits the migration and invasion of carcinoma cancer SK-Hep1 cells. Mechanistically, we show that this effect of EAAC may occur through inactivation of the MAPK signaling pathway, exerting inhibitory effects on FAK and pFAK protein expressions and inhibiting PI3K, Akt, and phospho-Akt levels, thereby decreasing the activities of MMP-2 and MMP-9 leading to inhibition of metastasis of SK-Hep1 cells. These results also demonstrate the inhibition of both MMP-2 and MMP-9 activities process by EAAC may be helpful in developing new chemotherapeutic strategies for SK-Hep1 cell related human hepatoma.

Abbreviations

EAAC: ethyl acetate fraction of *Actinidia callosa* var. *callosa*; MMP-2/9: metalloproteinase-2/9; TIMP-1/2: metalloproteinase-1/2; MAPK: mitogen-activated protein kinase; PI3K: phosphatidylinositol-3-kinase; AKT (PKB): protein kinase B; FAK: focal adhesion kinase.

Authors' contributions

WC carried out the cell culture and western analysis. J-CL, M-ML, C-HC, J-SD, J-SC and G-JH design an experiment and writing a research paper. All authors read and approved the final manuscript.

Author details

[1] Department of Food Nutrition and Health Biotechnology, Asia University, Taichung, Taiwan. [2] School of Chinese Medicine, Graduate Institute of Integrated Medicine College of Chinese Medicine, China Medical University, Taichung, Taiwan. [3] School of Pharmacy, College of Pharmacy, China Medical University, Taichung, Taiwan. [4] School of Chinese Pharmaceutical Sciences and Chinese Medicine Resources, College of Chinese Medicine, China Medical University, Taichung, Taiwan. [5] Department of Visual Communication Design, Asia University, Taichung, Taiwan.

Acknowledgements

The authors want to thank the financial supports from the National Science Council (MOST 105-2320-B-039-046- and MOST 106-2320-B-039-045-), China Medical University (CMU) (ASIA104-CMUH-06 and CMU105-ASIA-23) and Asia University (105-ASIA-05, 105-ASIA-06 and 105-ASIA-07).

Competing interests

The authors declare that they have no competing interests. And there have neither financial competing interests nor other competing interests.

References

Byambaragchaa M, de la Cruz J, Yang SH, Hwang SG (2013) Anti-metastatic potential of ethanol extract of *Saussurea involucrata* against hepatic cancer in vitro. Asian Pac J Cancer Prev 14:5397–5402

Casar B, Rimann I, Kato H, Shattil SJ, Quigley JP, Deryugina EI (2014) In vivo cleaved CDCP1 promotes early tumor dissemination via complexing with activated β1 integrin and induction of FAK/PI3K/Akt motility signaling. Oncogene 33:255–268

Chao W, Deng JS, Li PY, Liang YC, Huang GJ (2017) 3,4-Dihydroxybenzalactone suppresses human non-small cell lung carcinoma cells metastasis via suppression of epithelial to mesenchymal transition, ROS-mediated PI3K/AKT/MAPK/MMP and NFκB signaling pathways. Molecules 22:537

Chen YY, Lu HF, Hsu SC, Kuo CL, Chang SJ, Lin JJ, Wu PP, Liu JY, Lee CH, Chung JG, Chang JB (2015) Bufalin inhibits migration and invasion in human hepatocellular carcinoma SK-Hep1 cells through the inhibitions of NF-κB and matrix metalloproteinase-2/-9-signaling pathways. Environ Toxicol 30:74–82

Chen WL, Ho YP, Chou JC (2016) Phenologic variation of major triterpenoids in regular and white *Antrodia cinnamomea*. Bot Stud 57:33

Forrbes K, Gillette K, Sehgal I (2003) Lycopene increases urokinase receptor and fails to inhibit growth or connexin expression in a metastatically passaged prostate cancer cell line: a brief communication. Exp Biol Med 228:967–971

Gopalakrishnan KK, Thomas TD (2014) Reproductive biology of *Pittosporum dasycaulon* Miq., (Family Pittosporaceae) a rare medicinal tree endemic to Western Ghats. Bot Stud 55:15

He LF, Xu HW, Chen M, Xian ZR, Wen XF, Chen MN, Du CW, Huang WH, Wu JD, Zhang GJ (2017) Activated-PAK4 predicts worse prognosis in breast cancer and promotes tumorigenesis through activation of PI3K/AKT signaling. Oncotarget 8:17573–17585

Ho YL, Li KC, Chao W, Chang YS, Huang GJ (2012) Korean red ginseng suppresses metastasis of human hepatoma SK-Hep1 cells by inhibiting matrix metalloproteinase-2/-9 and urokinase plasminogen activator. Evid Based Complement Alternat Med 2012:965846

Hseu YC, Chao YH, Lin KY, Way TD, Lin HY, Thiyagarajan V, Yang HL (2017) *Antrodia camphorata* inhibits metastasis and epithelial-to-mesenchymal transition via the modulation of claudin-1 and Wnt/β-catenin signaling pathways in human colon cancer cells. J Ethnopharmacol 208:72–83

Huang GJ, Yang CM, Chang YS, Amagaya S, Wang HC, Hou WC, Huang SS, Hu ML (2010) Hispolon suppresses SK-Hep1 human hepatoma cell metastasis by inhibiting matrix metalloproteinase-2/9 and urokinase-plasminogen activator through the PI3K/Akt and ERK signaling pathways. J Agric Food Chem 58:9468–9475

Hung SH, Shen KH, Wu CH, Liu CL, Shih YW (2009) α-Mangostin suppresses PC-3 human prostate carcinoma cell metastasis by inhibiting matrix metalloproteinase-2/9 and urokinase-plasminogen expression through the JNK signaling Pathway. J Agric Food Chem 57:1291–1298

Hussain H, Green IR, Ali I, Khan IA, Ali Z, Al-Sadi AM, Ahmed I (2017) Ursolic acid derivatives for pharmaceutical use: a patent review (2012–2016). Expert Opin Ther Pat 27:1061–1072

Lee CC, Lee BH, Wu SC (2014) *Actinidia callosa* peel (kiwi fruit) ethanol extracts protected neural cells apoptosis induced by methylglyoxal through Nrf2 activation. Pharm Biol 52:628–636

Liao YC, Shih YW, Chao CH, Lee XY, Chiang TA (2009) Involvement of the ERK signaling pathway in fisetin reduces invasion and migration in the human lung cancer cell line A549. J Agric Food Chem 57:8933–8941

Liao JC, Huang SS, Deng JS, Lee CY, Lin YC, Huang GJ (2013) Chemical characterization and in vivo anti-Inflammatory activities of *Actinidia callosa* var. ephippioides via suppression of proinflammatory cytokines. Am J Chin Med 41:405–423

Lu KH, Chen PN, Hsieh YH, Lin CY, Cheng FY, Chiu PC, Chu SC, Hsieh YS (2016) 3-Hydroxyflavone inhibits human osteosarcoma U2OS and 143B cells metastasis by affecting EMT and repressing u-PA/MMP-2 via FAK-Src to MEK/ERK and RhoA/MLC2 pathways and reduces 143B tumor growth in vivo. Food Chem Toxicol 97:177–186

Park MS, Kim NH, Kang CW, Oh CW, Kim GD (2015) Antimetastatic effects of gambogic acid are mediated via the actin cytoskeleton and NF-κB pathways in SK-Hep1 cells. Drug Dev Res 76:132–142

Parmo-Cabanas M, Molina-Ortiz I, Matias-Roman S, Garcia-Bernal D, Carvajal-Vergara X, Valle I, Pandiella A, Arroyo A, Teixido GJ (2006) Role of metalloproteinases MMP-9 and MT1-MMP in CXCL12-promoted myeloma cell invasion across basement membranes. J Pathol 208:108–118

Ren J, Han EJ, Chung SH (2007) In vivo and in vitro anti-inflammatory activities of alpha-linolenic acid isolated from *Actinidia polygama* fruits. Arch Pharm Res 30:708–714

Sawada N, Kataoka K, Kondo K, Arimochi H, Fujino H, Takahashi Y, Miyoshi T, Kuwahara T, Monden Y, Ohnishi Y (2004) Betulinic acid augments the inhibitory effects of vincristine on growth and lung metastasis of B16F10 melanoma cells in mice. Br J Cancer 90:1672–1678

Shi J, Guo B, Hui Q, Chang P, Tao K (2017) Fangchinoline suppresses growth and metastasis of melanoma cells by inhibiting the phosphorylation of FAK. Oncol Rep 38:63–70

Shih YW, Wu PF, Lee YC, Shi MD, Chiang TA (2009) Myricetin suppresses invasion and migration of human lung adenocarcinoma A549 cells: possible mediation by blocking the ERK signaling pathway. J Agric Food Chem 57:3490–3499

Silva FS, Oliveira PJ, Duarte MF (2016) Oleanolic, ursolic, and betulinic acids as food supplements or pharmaceutical agents for type 2 diabetes: promise or illusion? J Agric Food Chem 64:2991–3008

Wang H, Zhang C, Xu L, Zang K, Ning Z, Jiang F, Chi H, Zhu X, Meng Z (2016) Bufalin suppresses hepatocellular carcinoma invasion and metastasis by targeting HIF-1α via the PI3K/AKT/mTOR pathway. Oncotarget 7:20193–22208

Xu HS, Wu YW, Xu SF, Sun HX, Chen FY, Yao L (2009) Antitumor and immunomodulatory activity of polysaccharides from the roots of *Actinidia eriantha*. J Ethnopharmacol 125:310–317

Yamai H, Sawada N, Yoshida T, Seike J, Takizawa H, Kenzaki K, Miyoshi T, Kondo K, Bando Y, Ohnishi Y, Tangoku A (2009) Triterpenes augment the inhibitory effects of anticancer drugs on growth of human esophageal carcinoma cells in vitro and suppress experimental metastasis in vivo. Int J Cancer 125:952–960

Žiberna L, Šamec D, Mocan A, Nabavi SF, Bishayee A, Farooqi AA, Sureda A, Nabavi SM (2017) Oleanolic acid alters multiple cell signaling pathways: implication in cancer prevention and therapy. Int J Mol Sci 18:643

12

Sky-canopy border length, exposure and thresholding influence accuracy of hemispherical photography for complex plant canopies

Guo-Zhang M. Song[1*], Kuo-Jung Chao[2], David Doley[3] and David Yates[3]

Abstract

Background: Hemispherical photography (HP) is a popular method to estimate canopy structure and understorey light environment, which analyses photographs acquired with wide view-angle lens (i.e. fisheye lens). To increase HP accuracy, the approaches of most previous studies were to increase the preciseness of exposure and thresholding of photographs, while ours quantified effects of canopy properties (gap fraction and length of sky-canopy border (SCB)) and errors of exposure and thresholding on the accuracy of HP.

Results: Through analysing photographs of real and model canopies, it was showed that HP inaccuracy resulted from the mismatch between exposure and thresholding rather than exposure or thresholding errors alone. HP inaccuracy was a function of the SCB length and the extent of exposure and thresholding errors, but independent of gap fraction.

Discussion: In photographs, SCBs are recorded as grey pixels which greyness is in between that of sky and canopy pixels. When there are exposure and thresholding errors, grey pixels are those prone to be misclassified in image analysis. Longer (vegetation with taller canopies) and wider (lower image sharpness) SCBs in photographs can both result in a higher amount of grey pixels and ultimately higher HP inaccuracy for a given extent of exposure and threshold errors.

Conclusions: Using lenses with view angle narrower rather than that of fisheye lens can shorten the SCB length in photographs and in turn reduce HP estimation inaccuracy for canopy structure and understorey light environment. Since short SCBs and low levels of exposure and thresholding errors can both result in low HP inaccuracy, to identify the true performance of new exposure and thresholding methods for HP, photographs recording canopies with long SCBs and acquired with fisheye lenses should be used. Because HP inaccuracy in a function of the amount of grey pixels resulting from SCBs, the amount of these pixels in photographs can be used as a universal parameter to quantify canopy properties influential to HP estimation and in turn make cross-study comparisons feasible.

Keywords: Canopy openness, Digital number, Gap fraction, Leaf area index, Mixed sky-canopy pixel

*Correspondence: mikesong@nchu.edu.tw
[1] Department of Soil and Water Conservation, National Chung Hsing University, Taichung 40227, Taiwan
Full list of author information is available at the end of the article

Background

Using hemispherical photography (HP) to estimate canopy parameter (e.g. canopy gap fraction and leaf area index (LAI)) relies on correct exposure and thresholding of canopy hemispherical photographs (CHPs) (Beckschäfer et al. 2013; Jonckheere et al. 2005; Rich 1990). If CHPs are overexposed or analysed with threshold values that are too low, canopy gap fraction will be overestimated and LAI will be underestimated (Zhang et al. 2005). Conversely, the underexposure of CHPs or application of too-high threshold values will result in underestimated gap fraction and overestimated LAI. Achieving correct exposure and thresholding are widely-recognised challenges in HP (Beckschäfer et al. 2013; Jonckheere et al. 2005; Rich 1990). Most previous studies attempted to improve HP accuracy by increasing the preciseness of exposure and thresholding (Rich 1990; Zhang et al. 2005). Identifying mechanisms of how exposure and thresholding errors influence HP accuracy can provide hints for novel approaches to raise HP accuracy.

There are multiple sources of grey pixels in CHPs. In CHPs, the desired sky and canopy pixels should appear white and black respectively. Nevertheless, some sky pixels look grey because of thick cloud cover and low sky illumination created by the direction of sun radiation (Rich 1990). On the other hand, pixels recording bright leaves and light-coloured trunks appear grey. The other main source of grey pixels is the so-called mixed sky-canopy pixels (hereafter mixed pixels) (Leblanc et al. 2005), which record portions of both sky and canopy in single pixels. Using regular grid cells to record digital images, widthless borders between two photographing objects (i.e. sky-canopy borders, SCBs, in the present study) are inevitably recorded as zones in which every single pixel contains images of these two objects (Fisher 1997). In the present study, these zones recording images of SCBs are defined as mixed-pixel zones (MPZs). Mixed pixels also include pixels recording images of linear (e.g. twigs) and small (e.g. small canopy openings) sub-pixel objects (Fisher 1997). Due to the nature of the grid pixel system of digital images, even with optically-perfect lens, the widthless SCBs are captured as MPZs which as wide as at least one pixel. Lens imperfections, image editing and operation errors (e.g. focusing failure, smudged lens, canopies moved by wind or camera shake) can blur SCBs and further increase the width of SCBs images.

Effects of exposure and thresholding errors on HP accuracy are exerted through their interactions with grey pixels in CHPs. Thresholding is a step of image analysis for HP, in which image pixels originally with a continuous greyness gradient are classified into either black or white pixels (binary images) by using a cut-off greyness (e.g. Rich 1990). Due to distinct differences between the greyness of sky and canopy pixels, HP inaccuracy introduced by (manual or automatic) thresholding generally results from the misclassification of grey pixels rather than sky or canopy pixels. Therefore, the extent of HP inaccuracy resulting from thresholding errors should be a function of the grey-pixel fraction. Optimum exposure allows the image capture devices to capture the desired amount of light through manipulating shutter speed, lens aperture and light sensitivity (i.e. ISO value) of the device (Allen and Triantaphillidou 2011). In HP, blooming is an over-exposure effect in which excessive light causes light saturation on the sky pixels and pixels nearby, and in turn canopy gaps appear larger than they should be (Leblanc et al. 2005). In other words, overexposure can result in that all grey pixels located between sky and canopy pixels (i.e. mixed pixels) appear as bright as sky pixels. Consequently, the extent of HP inaccuracy caused by exposure errors should also be a function of the grey-pixel fraction. Even though the misclassification of grey pixels has been widely recognised as a source of HP inaccuracy and various studies have developed new methods to classify them precisely (Hwang et al. 2016; Leblanc et al. 2005; Macfarlane 2011; Macfarlane et al. 2014), few studies quantified their effects.

The effects of exposure and thresholding errors on HP estimation have to be examined together. The digital number (also known as the greyness scale number) of all pixels in CHPs increases with exposure (Zhang et al. 2005). Nevertheless, as long as exposure is not high enough to cause the so-called blooming effect (Leblanc et al. 2005), the effects of higher exposure on HP estimates can be cancelled out by using higher threshold values in CHP analysis (Song et al. 2014). In other words, an estimation inaccuracy cannot be introduced to HP by either exposure or thresholding errors alone, but by the mismatch between exposure and thresholding (hereafter, exposure-thresholding mismatch, ETM).

HP inaccuracy is likely a function of the ETM extent and two canopy properties (gap fraction and SCB length). Macfarlane et al. (2000) and Macfarlane (2011) showed that, in Eucalyptus forests, the estimated LAI changed by 9–13% for each stop deviation from the correct exposure, suggesting that HP inaccuracy is a linear function of the ETM extent. Nevertheless, further studies are needed to show whether this relationship is applicable to other forest types. Macfarlane (2011) had reported a negative linear relationship between the grey-pixel fraction and canopy gap fraction. As mentioned above, pixels recording SCBs appear grey. Therefore, canopy gap fraction and SCB length may influence HP estimation through their effects on the grey-pixel fraction.

To identify whether canopy properties influence HP inaccuracy through their effects on the grey-pixel fraction, a model canopy system was used to minimise the sources of grey pixels and manipulate two canopy properties, SCB length and gap fraction, independently. Model canopies have been used in several HP studies because they are more controllable and manipulable than real canopies (Chianucci 2016; Hwang et al. 2016; Macfarlane et al. 2014; Song et al. 2014). In the present study, model canopies were made with opaque black adhesive vinyl sheet, which diminished the amount of grey pixels caused by bright canopy elements; the greyness variation of sky pixels was controlled by using an artificial light source. The image size of smallest canopy elements and openings of model canopies were larger than the pixel size of the camera we used, which eliminated mixed pixels resulting from sub-pixel objects. Consequently, in our model CHPs, the only source of grey pixels is mixed pixels recording SCBs. This system also allowed us to vary SCB length and gap fraction independently, so that we could distinguish one's effects from the other's.

The present study aimed to improve HP accuracy through quantifying the relationships between canopy properties (SCB length and gap fraction), the grey-pixel fraction, the ETM extent and HP inaccuracy. We asked the following questions: (1) Is the grey-pixel fraction in a CHP a function of canopy gap fraction and SCB length? (2) Is the linear relationship between HP inaccuracy and the ETM extent (reported by Macfarlane (2011)) applicable to all forest types? (3) Can these two canopy properties influence HP accuracy through their effects on the grey-pixel fraction?

Materials and methods

The grid model CHPs and real CHPs used in the present study were those in Song et al. (2014). For photographs showing the design of grid model canopies and image acquisition devices of model CHPs, please refer to Song et al. (2014).

Design of model canopy

Model canopies were constructed from opaque black adhesive vinyl sheet (20 cm by 21 cm) with square openings of different dimensions, stuck to clear acrylic plastic to eliminate distortion during handling. A range of gap fraction and SCB length was simulated through varying the side length (L) of square openings and the separation width (W) of these openings (Fig. 1, Table 1). Gaps in real canopies were simulated by square openings, while black grids were the analogue of canopy elements (Fig. 1b). Designed gap fraction (GF_d) is equal to $L^2/(L+W)^2$. For a desired combination of a given gap fraction and a side length of square opening, W can be acquired with the equation, $W = (L/\sqrt{GF_d}) - L$ (Table 1).

Acquisition of CHPs

Three adapted CHP acquisition procedures were used to take photographs for the two-dimensional model

Table 1 Designed gap fraction (GF_d) and side length of square opening (L) of the 17 model canopies

Designed gap fraction (GF_d)	Side length of square opening (L) (mm)			
	2.50	5.00	10.00	20.00
0.05	8.68, 0.08	17.36, 0.04	34.72, 0.02	–, –
0.10	5.41, 0.16	10.81, 0.08	21.62, 0.04	–, –
0.25	2.50, 0.40	5.00, 0.20	10.00, 0.10	20.00, 0.05
0.50	1.04, 0.80	2.07, 0.40	4.14, 0.20	8.28, 0.10
0.75	–, –	0.77, 0.60	1.55, 0.30	3.09, 0.15

The left and right values in each cell show the separation width between square openings (W; mm) and SCB length per unit area (B; mm mm^{-2}) respectively. $GF_d = L^2/(L+W)^2$. $B = 4 \cdot L/(L+W)^2$

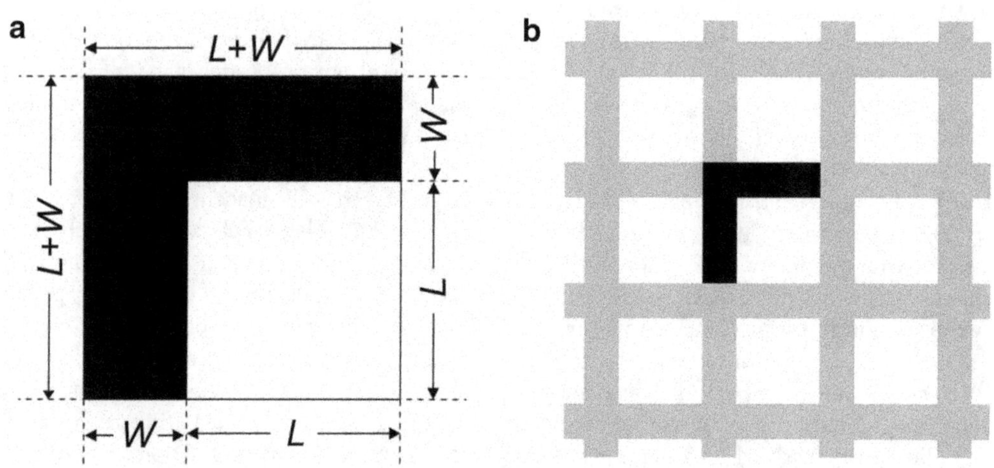

Fig. 1 Design of (**a**) the basic unit for (**b**) a model canopy (gap fraction = 0.5). L is the side length of square openings. W is the separation width between square openings

canopies. First, model canopies backed with a piece of sand-scrubbed acrylic board (simulating the sky) was mounted at an end of a cylindrical pipe (internal diameter of 12.3 cm and length of 41 cm). The internal wall of this pipe was covered with black flannelette to reduce stray radiation. Second, a 28-watt ring fluorescent light (FLC-28, Falcon Eyes, Hong Kong, China) was placed behind the model sky to provide light for the model sky. A spot light meter (Spotmeter F, Konica Minolta, Tokyo, Japan) were used to make sure the illumination was even (illumination variation ≤ 0.2 stop) across the model sky and also measure the reference exposure from the model sky. Third, a Nikon D5000 camera (Tokyo, Japan) attached to Nikon AF-S ED 18–70 mm lens was mounted at the other end of the pipe. During the acquisition of model CHPs, the lens was always zoomed into 70 mm to reduce vignetting associated with wide view-angle lens. Through varying shutter speed but keeping aperture constant, CHPs for the 17 model canopies were taken with nine levels of exposure (−3, −2, −1, 0, +1, +2, +3, +4 and +5 stop) compensated from the reference exposure.

On 12 and 13 of August 2011, real CHPs were acquired in a forest about 500 m east of the Nanjenshan Research Station (22°05′05″N, 120°50′07″E) of Taiwan. *Schefflera octophylla, Psychotria rubra, Castanopsis cuspidata* var. *carlesii* and *Illicium arborescens* were the most dominant tree species in this forest (Chao et al. 2010). CHPs were acquired with a 4.5 mm F2.8 circular fisheye lens (Sigma Corp., Kanagawa, Japan) mounted on the same Nikon D5000 camera which was used for the acquisition of model CHPs. During the acquisition of CHPs, the camera was mounted on a tripod to keep it 1 m above ground and reduce camera shake. To eliminate focusing failure associated with automatic focusing, the focus of the fisheye lens was set to a constant distance, allowing that canopy elements more than 0.2 m away from the lens were included in the depth of field. Because CHPs acquired with the popular under-canopy exposure method (exposure of CHPs measured under canopy) tend to be overexposed (Chen et al. 1991; Zhang et al. 2005), we mainly examined the effects of overexposure on HP. Therefore, real CHPs were acquired with up to +5 stop compensated from the reference exposure. Seven levels of exposure compensation were manipulated (−1, 0, +1, +2, +3, +4 and +5 stop) for every of the 46 photographing locations and totally 322 CHPs were acquired. Before CHP acquisition for every location, the exposure for CHPs was measured with a Nikon D300 camera following the above-canopy exposure method suggested by Zhang et al. (2005). This exposure-measuring camera was placed in an open area not farther than 0.5 km away from the CHP acquisition locations. This camera was attached with a Nikon AF-S ED 18–70 mm lens and mounted on a tripod. The lens was pointed to the sky zenith and zoom out to 18 mm to cover most of the sky area. The desired aperture for exposure compensation of +5 stop was obtained with the following settings: shutter-priority of camera mode, ISO 1600 of light sensitivity, 1/30 s of shutter speed and +5 stop of exposure compensation. Using this aperture coupled with the same setting of light sensitivity and shutter speed on the CHP-acquisition camera could acquire CHPs with exposure compensations of +5 stop. CHPs of exposure compensations of +4, +3, +2, +1, 0 and −1 stop were acquired by keeping aperture constant but using shutter speeds of 1/60, 1/125, 1/250, 1/500, 1/1000 and 1/2000 s respectively. During CHP acquisition, all the following settings remained constant, including image format (Jpeg), image size (4288 × 2848 pixels), image quality (fine), metering mode (matrix), sensitivity to light (ISO 1600), colour mode and white balance (cloudy sky).

Quantifying the grey-pixel fraction and SCB length

The method of Leblanc et al. (2005) was modified to quantify the grey-pixel fraction in our CHPs. Sharp turns of curves in the greyness histograms of CHPs were used to visually identify the low threshold value separating canopy and grey pixels and the high threshold value separating sky and grey pixels. Our method differed from that of Leblanc et al. (2005) in that the scale of the y axis (pixel fraction) of our histograms was linear rather than logarithmic. The grey-pixel fraction was the proportion of pixels with digit numbers between the low and high threshold values. Macfarlane (2011) has proposed an automatic method (two-corner method) using two humps of greyness histograms to identify the low and high threshold values. This method was not used in the present study because many greyness histograms of our CHPs had only one hump due to overexposure or low canopy gap fraction.

When the side length (L) and separation width (W) of square openings are given, the SCB length per unit area (B) is equal to $4{\cdot}L/(L+W)^2$ (Fig. 1, Table 1). Because there is no direct method to assess the SCB length for real canopies, the MPZ length was therefore measured as an approximation of the SCB length. The first step of measuring the MPZ length was to transform real CHPs into correctly-thresholded binary images by following the exposure-corresponding method proposed by Song et al. (2014). Pixels of MPZs in binary images were then identified with the function of "Find edges" of ImageJ version 1.43 (Abramoff et al. 2004). This function uses the Sobel edge detector which lays out a three-by-three convolution kernel vertically and the other horizontally relative to the pixel grid to detect sharp greyness changes for two dimensional images (Ferreira and Rasband 2012). The MPZ length of a CHP is the proportion of pixels in MPZs (hereafter, pixel fraction of MPZ). Analyses showed that the MPZ length obtained through this method was a close approximation of SCB length per unit area (Additional file 1: Figure S1).

Assessing HP inaccuracy associated with the ETM

Because exposure and thresholding of CHPs can have both independent and interacting effects on HP accuracy (Song et al. 2014), these two processes should be examined together. A uniform basis for comparison of CHP exposure and thresholding effects was achieved by using two parameters, exposure manipulation and thresholding manipulation. Exposure manipulation was the extent of exposure deviating from the reference exposure measured under an unobscured overcast sky. Thresholding manipulation was transformed from the grey-scale threshold value using the regression model established by Song et al. (2014), in which the relationship between exposure manipulation (X) and the optimal threshold value (Y) is $Y = 260.542/(1 + e^{(-X+1.422)/1.160})$. This model determines the optimal threshold values for CHPs exposed with specific exposure manipulations. If a CHP is analysed with a threshold value which is the optimal threshold value for CHPs acquired with exposure manipulation of S stop, this CHP is analysed with thresholding manipulation of S stop. For example, the optimal threshold value for CHPs acquired with exposure manipulation of $+3$ stop is 202. When a CHP is analysed with a threshold value of 202, its thresholding manipulation is $+3$ stop. The ETM extent can be quantified by subtracting the thresholding manipulation from the exposure manipulation. When the ETM extent of a CHP is 0 stop, that means this CHP is analysed with its optimal threshold value and it can provide a correct estimate for HP. Positive ETM extents indicate that CHPs are overexposed or analysed with threshold values that are too low, so that gap fraction is overestimated. Negative ETM extents underestimate gap fraction, as a result of underexposing CHPs or analysing CHPs with threshold values that are too high.

Because gap fraction is the HP estimation with the most direct connection between image and estimate (Jonckheere et al. 2005), we used gap fraction to evaluate the influence of grey pixels on HP inaccuracy. Prior to image analysis, the colour model and real CHPs were transformed into 8-bit greyscale images first. For model canopies, the correct gap fraction was obtained by analysing scanned images of model canopies (GF_s) (details in Additional file 1: Text S1). To obtain photographed gap fraction (GF_p) estimated with different thresholding manipulations, the amount of pixels for each digit number was obtained for every model CHP with the function of "Histogram" of ImageJ version 1.43 (Abramoff et al. 2004). Photographed gap fraction for the threshold value i (GF_{pi}) was obtained through dividing the total number of pixels with digit numbers $\geq i$ by the total number of pixels in the circular images of CHPs. Using the relationship between optimal threshold value and threshold manipulation, we could obtain gap fraction under different thresholding manipulations for every model CHP. The estimation inaccuracy of model CHPs (I_m) was obtained by subtracting the correct gap fraction from the photographed gap fraction ($I_m = GF_p - GF_s$).

For real canopies, the correct gap fraction was obtained by processing CHPs with matched exposure manipulation and thresholding manipulation (GF_{match}). The estimation inaccuracy of real CHPs (I_r) was obtained by subtracting the correct gap fraction from the gap fraction obtained with a mismatch combination of exposure and thresholding manipulations ($GF_{mismatch}$) ($I_r = GF_{mismatch} - GF_{match}$). This method for obtaining correct estimates can ensure that the ETM is the only source of estimation inaccuracy. If correct estimates are obtained with other often-used approaches (such as diffuse light transmittance obtained with quantum sensors), additional sources of estimation inaccuracy (e.g. differences between instruments) may obscure the relationship between HP inaccuracy and the ETM.

Data analysis

The F-test was used to evaluate the slope difference of regression lines with zero intercept, including regression lines between the SCB length and HP inaccuracy, and between the MPZ length and HP inaccuracy. It is assumed that there are totally n observations in k groups. The total sum of square (SST) is the sum of square for residuals between observations and the grand regression line. The sum of square within groups (SSW) is the sum of square for residuals between observations in the same groups and the group-specific regression lines. The sum of square between groups (SSB) is obtained by subtracting SSW from SST (Sokal and Rohlf 1995). The degree of freedom within groups (DFW) and between groups (DFB) is equal to $n - k$ and $k - 1$ respectively (Sokal and Rohlf 1995). F values are calculated where.

$$F\text{ value} = \frac{\text{Mean square among groups}}{\text{Mean square within groups}} = \frac{SSB/DFB}{SSW/DFW} \quad (1)$$

Data analyses and significance tests in the present study were conducted with Microsoft Excel 2010 (Redmond, Washington, US).

Results

The relationship between the grey-pixel fraction and canopy gap fraction was hump-shaped (Figs. 2 and 3). The coefficients of determination (r^2) for fitting this relationship with linear models were lower than with hump-shaped models (Additional file 1: Figures S2 and S3). In contrast, the grey-pixel fraction was linearly related to the SCB length and MPZ length, but independent of canopy gap fraction (Figs. 4 and 5).

Panels in the same columns of Fig. 6 show binary images which exposure manipulation is kept constant but thresholding manipulation is varied. Panels in the same rows of Fig. 6 shows binary images which thresholding manipulation is kept constant but exposure manipulation is varied. As long as the ETM extent is kept constant, even though the exposure manipulation and thresholding manipulation paired together vary, the amount and spatial distribution of misclassified pixels are almost identical (e.g. Fig. 6e, j, o, t, m, r, w, ab). This indicates that the amount of misclassified pixels is not determined solely by either exposure manipulation or thresholding manipulation, but the ETM extent (Fig. 6). When the ETM extent is lower than −1 stop, all pixels of model CHPs are classified as canopy pixels (Fig. 6q, u, v, y, z, aa). When the ETM extent ranges from −1 to +3 stop, those misclassified pixels were all located on where sky and canopy elements meet (Fig. 6e, i, j, m, n, o, r, s, t, w, x, ab). That is, misclassified pixels were those pixels recording SCBs. Once the ETM extent reached +5 stop, a great proportion of canopy pixels (pixels of grids) are misclassified as sky pixels (Fig. 6f, k, p). All pixels are classified as sky pixels, when the ETM extent is equal to or greater than +7 stop (Fig. 6g, h, l).

Panels a–c in Figs. 7 and 8 show greyness histograms (the frequency distribution of pixel fraction along the greyness gradient) of model and real CHPs respectively. In Figs. 7a–c, 8a–c, every curve has two humps. Humps at lower digital numbers are contributed by canopy pixels, whereas humps at higher digital numbers are contributed by sky pixels. Pixels in between these two humps are the grey pixels. The grey pixel fraction increases with increasing SCB length (Fig. 7a–c) and MPZ length (Fig. 8a–c). Panels d–f in Figs. 7 and 8 show gap fraction estimated with threshold values from 0 to 255. As long as threshold values within the greyness range of grey pixels (ETM extent from −1 to +3 stop) are used

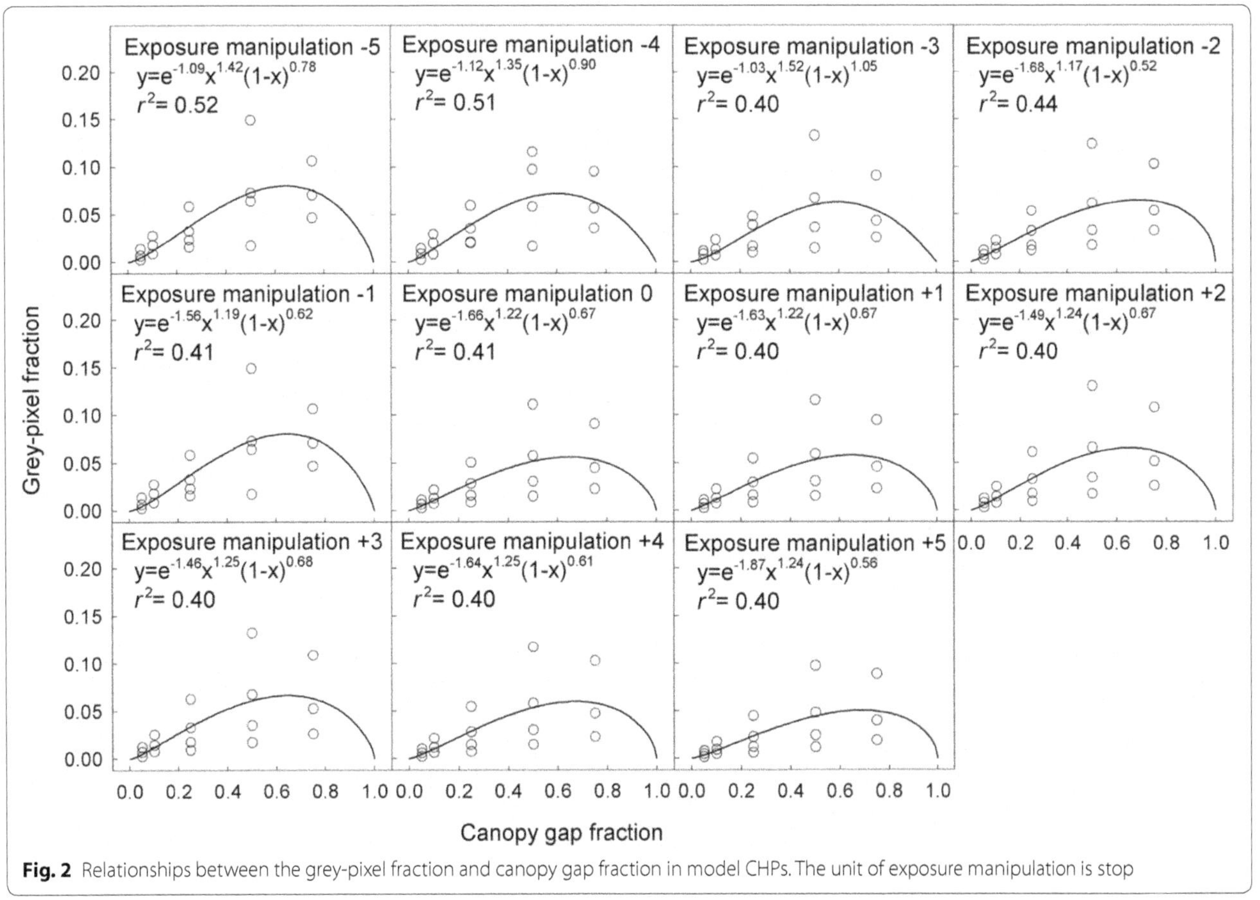

Fig. 2 Relationships between the grey-pixel fraction and canopy gap fraction in model CHPs. The unit of exposure manipulation is stop

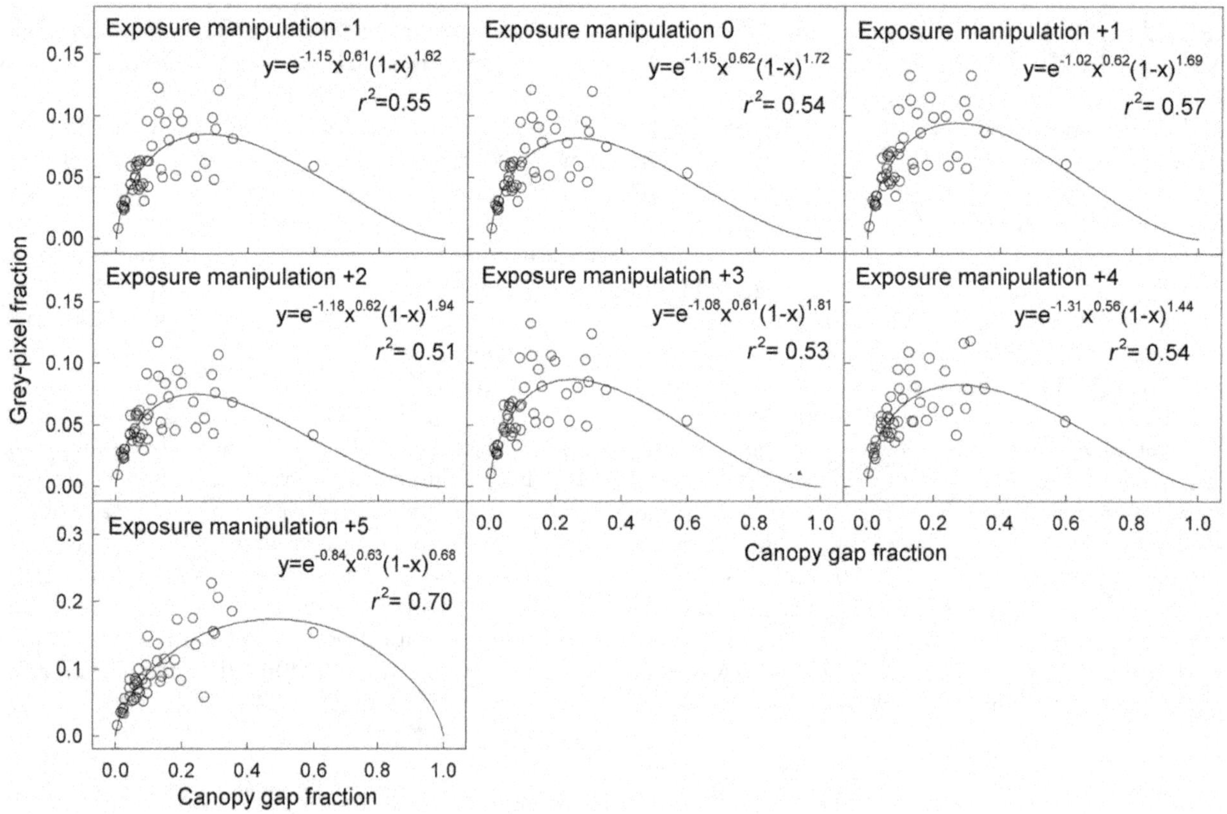

Fig. 3 Relationships between the grey-pixel fraction and canopy gap fraction in real CHPs. The unit of exposure manipulation is stop

for thresholding (i.e. only grey pixels are misclassified), estimated gap fraction increases with ETM extents (Figs. 7d–f and 8d–f).

As long as only grey pixels are misclassified (i.e. the ETM extent is not zero and ranges from -1 to $+3$ stop), HP inaccuracy is positively related to the SCB length (Fig. 9) and MPZ length (Fig. 10). However, the HP inaccuracy is independent of canopy gap fraction for both model and real CHPs (Figs. 9 and 10). Significant tests showed that slopes of most regression lines with the same ETM extents (panels within the same rows in Figs. 9 and 10) differed significant in Fig. 9 (Table 2). This was a surprising result especially for regression lines within the same ETM extents of $+1$ to $+3$ stop in Fig. 9, because their slopes were almost identical. This result was attributed to that our control on the light intensity of the light source was not precise enough during the acquisition of model CHPs. For the detailed explanation, please see Additional file 1: Text S2. In panels within the same columns in Figs. 9 and 10, the slopes of regression lines approximately doubled with each stop increase in the ETM extent, indicating a positive linear relationship between HP inaccuracy and the ETM extent in both model and real CHPs.

Discussion

For ease of understanding, the mechanisms of how HP accuracy for canopy gap fraction is affected by canopy properties (gap fraction and SCB length), exposure and thresholding errors, and other factors are illustrated by a schematic diagram (Fig. 11).

Relationships between canopy properties, the grey-pixel fraction, the ETM and HP inaccuracy

The relationship between canopy gap fraction and grey-pixel fraction is actually mediated by the SCB length (Fig. 11). Although the grey-pixel fraction was related with canopy gap fraction (Figs. 2 and 3), this relationship became negligible if the SCB length (Fig. 4) or MPZ length (Fig. 5) was included. Moreover, in the model canopy system, the SCB length and gap fraction could be varied independently (Table 1, Additional file 1: Figure S4). These results indicated that the relationship between gap fraction and the SCB length in Figs. 2 and 3 is a tendency rather than a cause-and-effect relationship, so is the relationship between gap fraction and the grey-pixel fraction (Fig. 11).

The relationship between canopy gap fraction and the grey-pixel fraction is hump-shaped (Figs. 2 and 3)

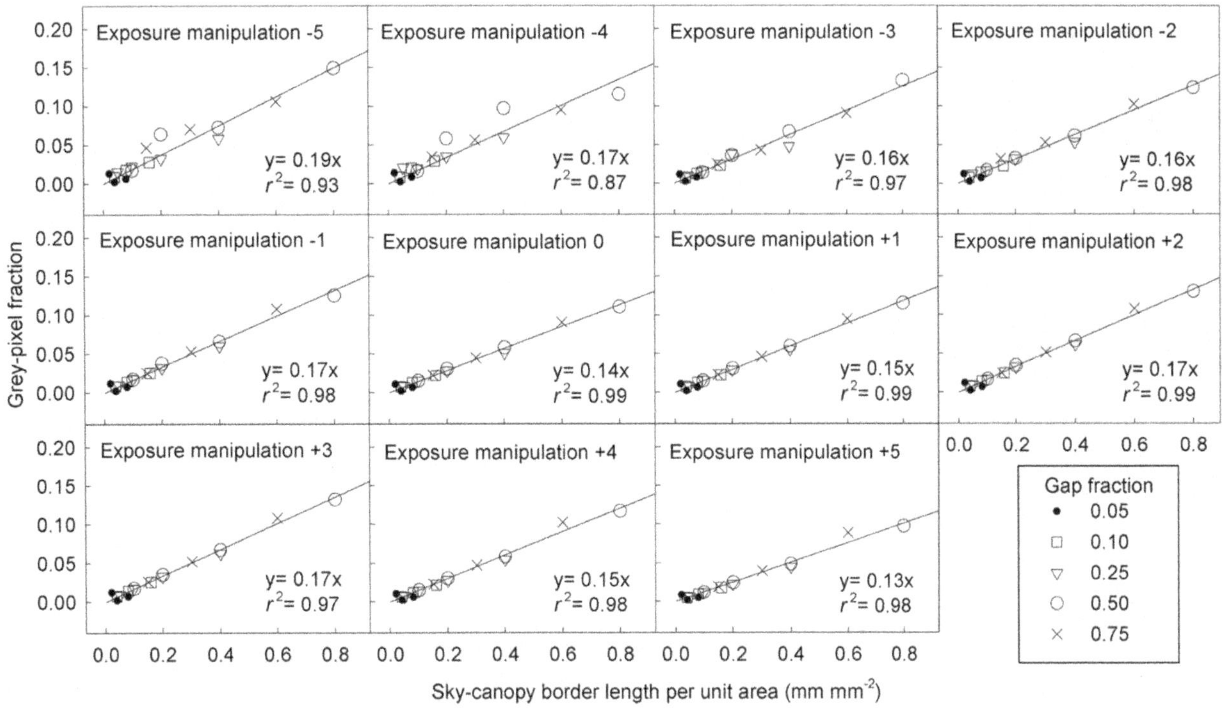

Fig. 4 Relationships between the grey-pixel fraction and the SCB length of model canopies in model CHPs. The unit of the exposure manipulation is stop

rather than linear (as reported by Macfarlane (2011)). When canopy gap fraction is zero or one, because of the absence of either sky or canopy elements, the SCB length has to be zero, and so do the MPZ length in CHPs. Consequently, if CHPs are sorted with gap fraction from zero to slightly higher values, the MPZ lengths of these CHPs increase from zero to greater values. If CHPs are sorted with gap fraction from one to slightly lower values, the MPZ lengths of CHPs also increase from zero to greater values. In summary, the relationship between canopy gap fraction and the grey-pixel fraction has to be hump-shaped. The study of Macfarlane (2011) was conducted in forests with relatively high canopy gap fraction (approximately ranging from 0.15 to 0.6), he therefore observed the negative relationship near the higher end of the gap fraction gradient.

The mixed-pixel fraction is a function of not only the length but also the width of MPZs (Fig. 11). In images, the width of edges between two different objects (i.e. the MPZ width in the present study) decreases with the increase of image sharpness (Allen and Triantaphillidou 2011; Smith 1997) (Fig. 11). As mentioned in the introduction, even with optically-perfect lens, MPZs are at least as wide as one pixel due to the grid pixel system of digital images (Fisher 1997). Because image resolution of the image capturing devices in many up-to-date digital cameras exceeds that of lens (https://www.dxomark.com/Reviews/Looking-for-new-photo-gear-DxOMark-s-Perceptual-Megapixel-can-help-you, 20th April, 2017), blurred SCBs caused by lens imperfections are recorded in details, resulting in that MPZs are even wider. In addition, image sharpness is also influenced by the image-processing algorithms of camera firmware and operation errors during image acquisition (Allen and Triantaphillidou 2011) (Fig. 11). Effects of lens and firmware can be controlled by using the same hard- and soft-ware respectively. In contrast, effects of operations during image acquisition can vary greatly shot by shot. When taking CHPs, HP users should minimise the commonly-encountered operation errors, including focusing failure, smudged lens, canopy movement caused by wind and camera shake (Fig. 11).

In a CHP, the relationships among the overall grey-pixel fraction (F), the SCB length (B), image sharpness (S) and the fraction of grey pixels recording what else other than SCBs (G) can be summarised by Eq. 2. In practice, image sharpness of CHPs can be quantified with the 10–90% edge response distance, which is a parameter measuring the distance required for the edge response rising from 10 to 90% of the greyscale amplitude (Smith 1997).

$$F = B/S + G \tag{2}$$

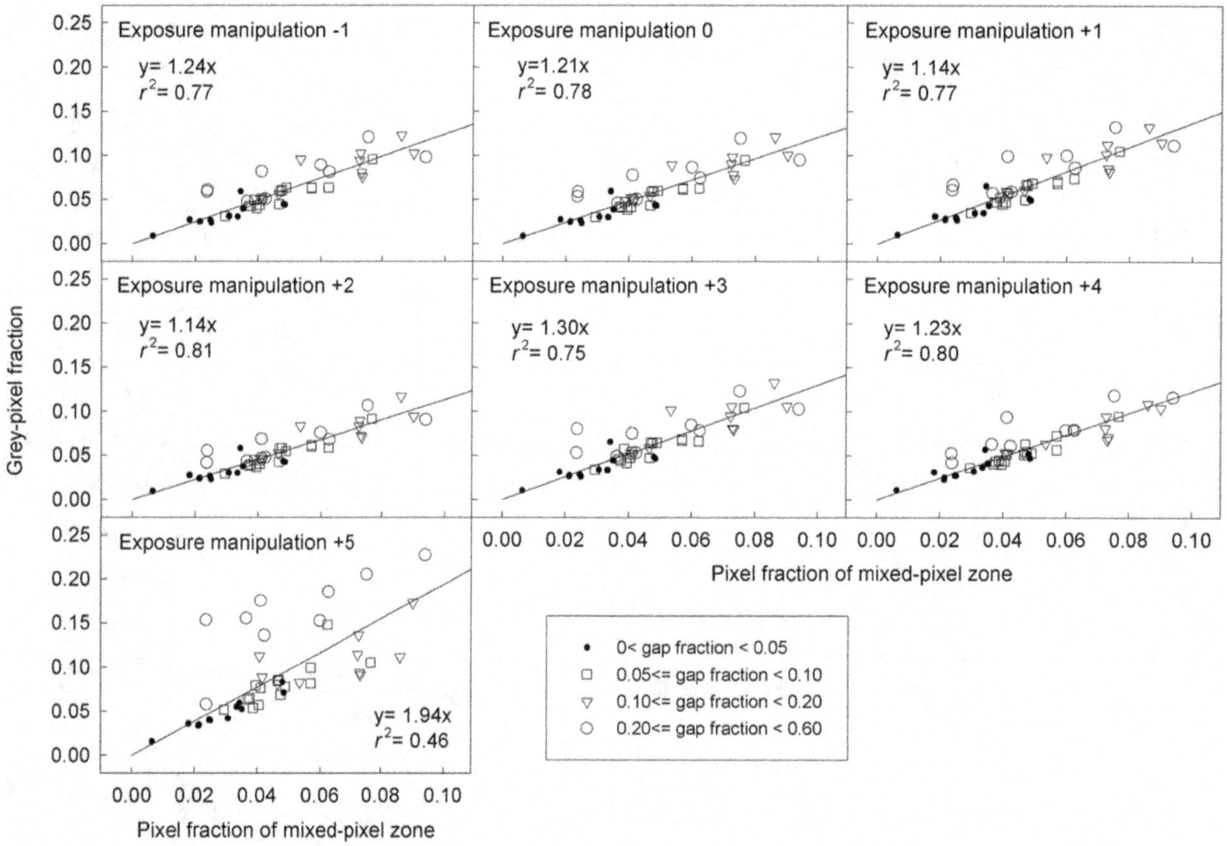

Fig. 5 Relationships between the grey-pixel fraction and the mixed-pixel zone length in real CHPs. The pixel fraction of mixed-pixel zone is an approximation of the SCB length. The unit of the exposure manipulation is stop

Even in real CHPs, grey pixels consist mainly of pixels recording SCBs. Due to the model canopy properties mentioned in the introduction, grey pixels recording what else other than SCBs (G) is eliminated from model CHPs, giving rise to almost perfectly linear relationships between the overall grey-pixel fraction (F) and the SCB length (B) (Fig. 4, Eq. 2). Because bright canopy elements and dark skies are inevitable properties of real CHPs (Rich 1990), G must be higher in real CHPs than in model CHPs. Consequently, coefficients of determination in Fig. 5 are lower than those in Fig. 4. In other words, high G weakens the linear relationship between F and B (Eq. 2). Nevertheless, except exposure manipulation of +5 stop, all coefficients of determination in Fig. 5 were greater than 0.75, indicating that the majority of grey pixels in real CHPs are pixels recording SCBs. In other words, G in real CHPs is low enough so that the linear relationships between F and B are observed in Fig. 5. Exposure manipulation of +5 stop raised the digit number of canopy elements to the level that many canopy pixels were recognised as grey pixels and in turn weakened the linear relationship between F and B (Fig. 5).

(See figure on next page.)

Fig. 6 Cropped binary images of model CHPs with the ETM extents ranging from −7 to +9 stop. (**a**)–(**d**) are the original images of model CHPs acquired with exposure manipulation of −2, 0, +2 and +4 stop. (**e**)–(**ab**) are the binary images obtained with combinations of 4 exposure manipulations (−2, 0, +2, and +4 stop) and 6 thresholding manipulations (−5, −3, −1, +1, +3, and +5 stop). Canopy and grey pixels which are misclassified as sky pixels are shown in red; sky and grey pixels which are misclassified as canopy pixels are blue; sky and grey pixels which are correctly classified as sky are white; canopy and grey pixels which are correctly classified as canopy are black. These CHPs were taken for a model canopy with gap fraction of 0.25 and SCB length per unit area of 0.4 mm mm^{-1} (i.e. side length of square opening of 2.5 mm). The diameter of the original circular images of these model CHPs is 2098 pixels, and the image dimension of cropped binary images shown in this figure is 177 by 177 pixels. Greyness histograms of these CHPs presented in this figure are shown in Fig. 7. The unit of the exposure manipulation, thresholding manipulation and ETM extent is stop

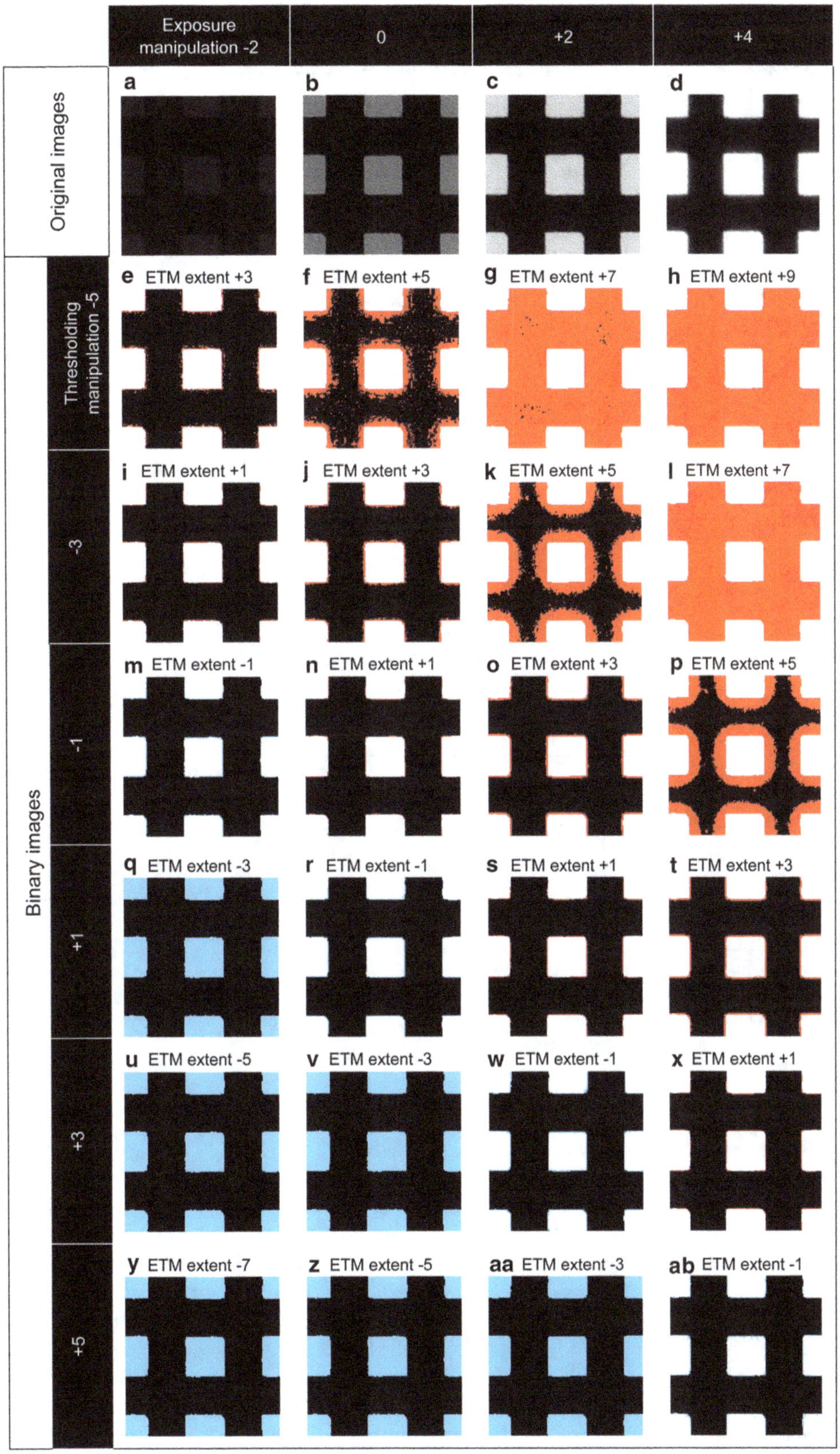

Exposure manipulation -2
0
+2
+4
Original images
Binary images
Thresholding manipulation -5
-3
-1
+1
+3
+5
a
b
c
d
e ETM extent +3
f ETM extent +5
g ETM extent +7
h ETM extent +9
i ETM extent +1
j ETM extent +3
k ETM extent +5
l ETM extent +7
m ETM extent -1
n ETM extent +1
o ETM extent +3
p ETM extent +5
q ETM extent -3
r ETM extent -1
s ETM extent +1
t ETM extent +3
u ETM extent -5
v ETM extent -3
w ETM extent -1
x ETM extent +1
y ETM extent -7
z ETM extent -5
aa ETM extent -3
ab ETM extent -1

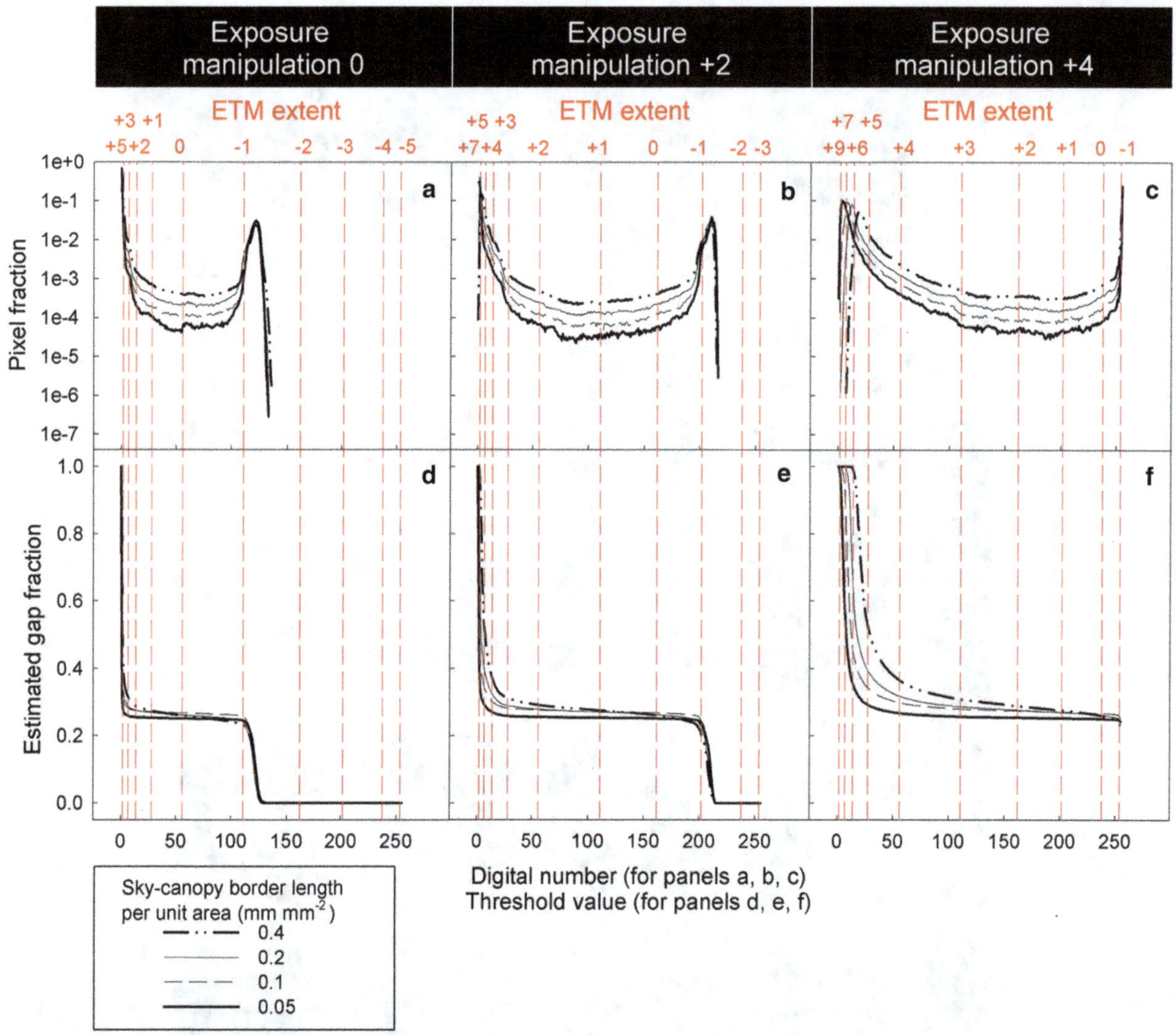

Fig. 7 **a–c** Frequency distribution of pixel fraction along the greyness gradient of model CHPs, and **d–f** estimated gap fraction under different ETM extents. In **a–c**, grey pixels are those in between the humps of canopy and sky pixels. The figure is derived from CHPs of four model canopies with the same designed gap fraction of 0.25 and SCB length per unit area ranging from 0.05 to 0.4 mm mm^{-1}. Note log scale used on y-axis of panels **a–c**. The unit of the exposure manipulation, thresholding manipulation and ETM extent is stop

As long as only grey pixels are misclassified, Eq. 3 bellow can summarise the relationship between HP inaccuracy (I) and its influential factors (M, ETM extent; F, the overall grey-pixel fraction; B, the SCB length; S, image sharpness; G, the fraction of grey pixels resulting from what else other than the SCBs) in Figs. 4, 5, 9 and 10.

$$I = M \cdot F = M \cdot (B/S + G) \tag{3}$$

The linear relationship between the ETM extent and HP inaccuracy is applicable for all forest types. In addition to studies of Macfarlane et al. (2000) and Macfarlane (2011), this relationship has been implicitly reported in Fig. 4b of Hwang (2016), Fig. 4 of Macfarlane et al. (2014), Fig. 3 of Thimonier et al. (2010) and Figs. 4 and 5 of Zhang et al. (2005). Figures of these studies which took CHPs for model canopies or real forests with different species composition all showed that each stop of exposure variation introduces approximately the same magnitude of HP inaccuracy. In the present study, despite the fact that the geometry of model and real canopies differed substantially, linear relationships between the ETM extent and HP inaccuracy were observed in both model and real CHPs (Figs. 9 and 10). Results of the previous

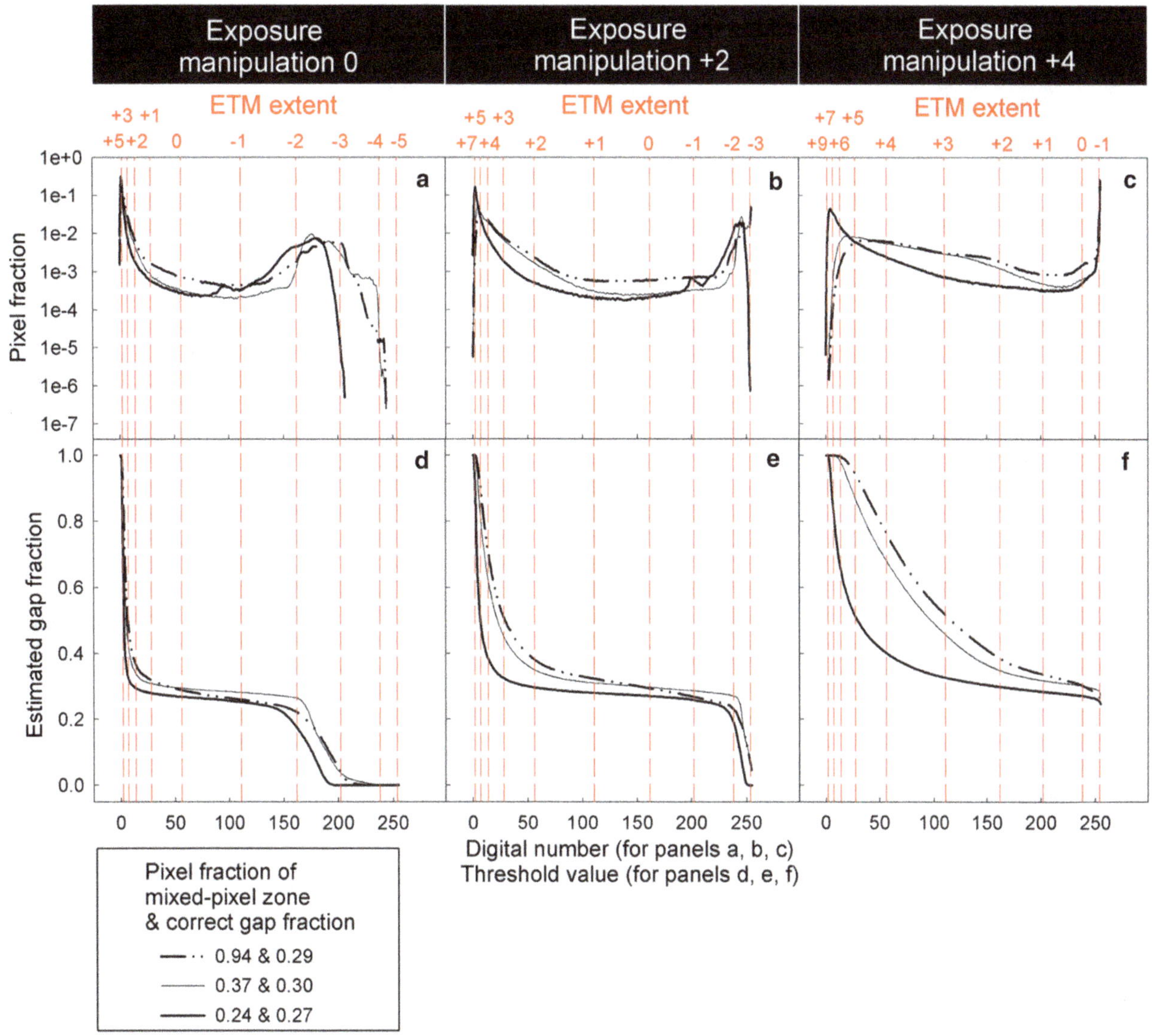

Fig. 8 **a–c** Frequency distribution of pixel fraction along the greyness gradient of real CHPs, and **d–f** estimated gap fraction under different ETM extents. The figure is derived from CHPs of three real canopies with similar gap fraction (0.29, 0.30 and 0.27), with proportion pixel of mixed-pixel zone (i.e. mixed-pixel zone length) of 0.94, 0.37 and 0.24, respectively. Note log scale used on y-axis in panels **a–c**. The unit of the exposure manipulation, thresholding manipulation and ETM extent is stop

and present studies indicated that this linear relationship is applicable to all forest types.

Suggestions for operations and future studies of HP

The lack of precision on exposure and thresholding can lead to higher HP inaccuracy for images with long than short sky-canopy borders (Figs. 9 and 10). There is a need to know what kind of canopies are of long SCBs. Plants with taller canopies tend to have longer SCBs. When doubling the camera-to-canopy distance, the SCB length in captured images increases by a factor of two (Additional file 1: Figure S4), so do the grey-pixel fraction. In other words, the taller the canopies (i.e. the greater camera-to-canopy distance), the higher the grey-pixel fraction of captured CHPs. Therefore, when exposure manipulation was controlled, HP inaccuracy increased with the distance between model canopies and the camera (Fig. 4 of Macfarlane et al. (2014)). Trees with smaller leaf sizes and higher degrees of leaf division may have more openings and in turn longer SCBs in their canopies (Fig. 11). Further studies are needed to identify effects of leaf size and leaf division on the SCB length in acquired images.

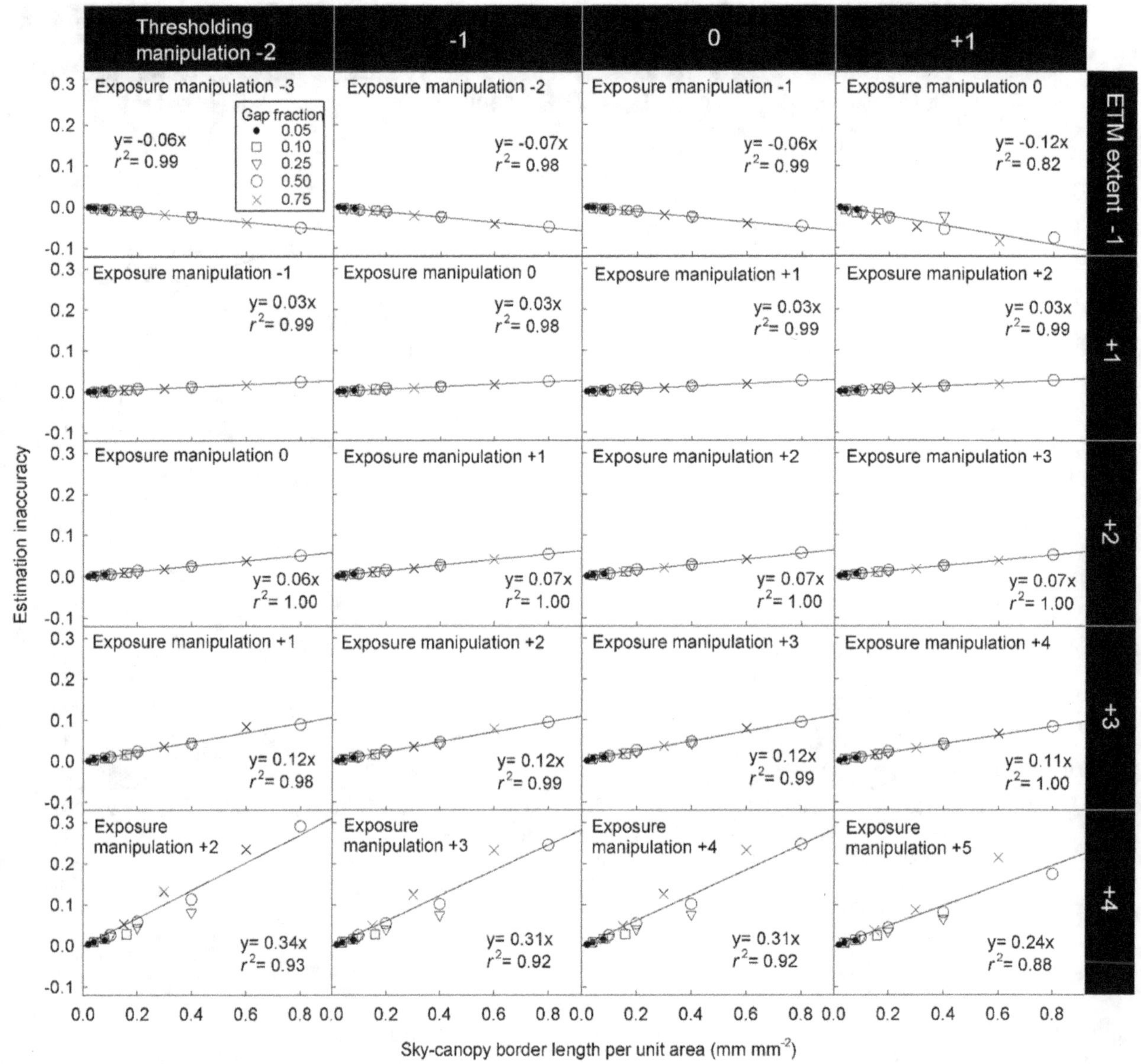

Fig. 9 Effects of the SCB length, canopy gap fraction and the ETM extent on the HP estimation derived from model CHPs of 17 model canopies. The unit of the exposure manipulation, thresholding manipulation and ETM extent is stop

The MPZ length has the potential to be a universal parameter, which enable researchers to quantify the effects of canopy properties which are influential to HP estimation and in turn carry out cross-study comparisons for HP studies. Most HP researchers are aware that HP estimation is likely to be affected by species- and site-specific canopy properties, so they described where CHPs were taken and species composition and structure (e.g. stem density, tree height) of vegetation (e.g. Hwang et al. 2016; Jonckheere et al. 2005; Zhang et al. 2005). Nevertheless, these attributes provide little insight into how and to what extent they may affect HP accuracy. Imperfect exposure and thresholding are among the most frequently encountered errors in HP (e.g. Jonckheere et al. 2005; Rich 1990; Zhang et al. 2005). In the presence of exposure and thresholding errors, the SCB length and MPZ length can ultimately change HP estimates (Figs. 9 and 10) through their effects on the grey-pixel fraction (Figs. 4, 5 and 11). This indicates that the MPZ length, a close approximation for the SCB length, can be used as a parameter to quantify effects of canopy properties on HP estimation. As mentioned previously, the SCB length is affected by at least one canopy property (i.e. canopy height (Additional file 1: Figure S4)). Most important,

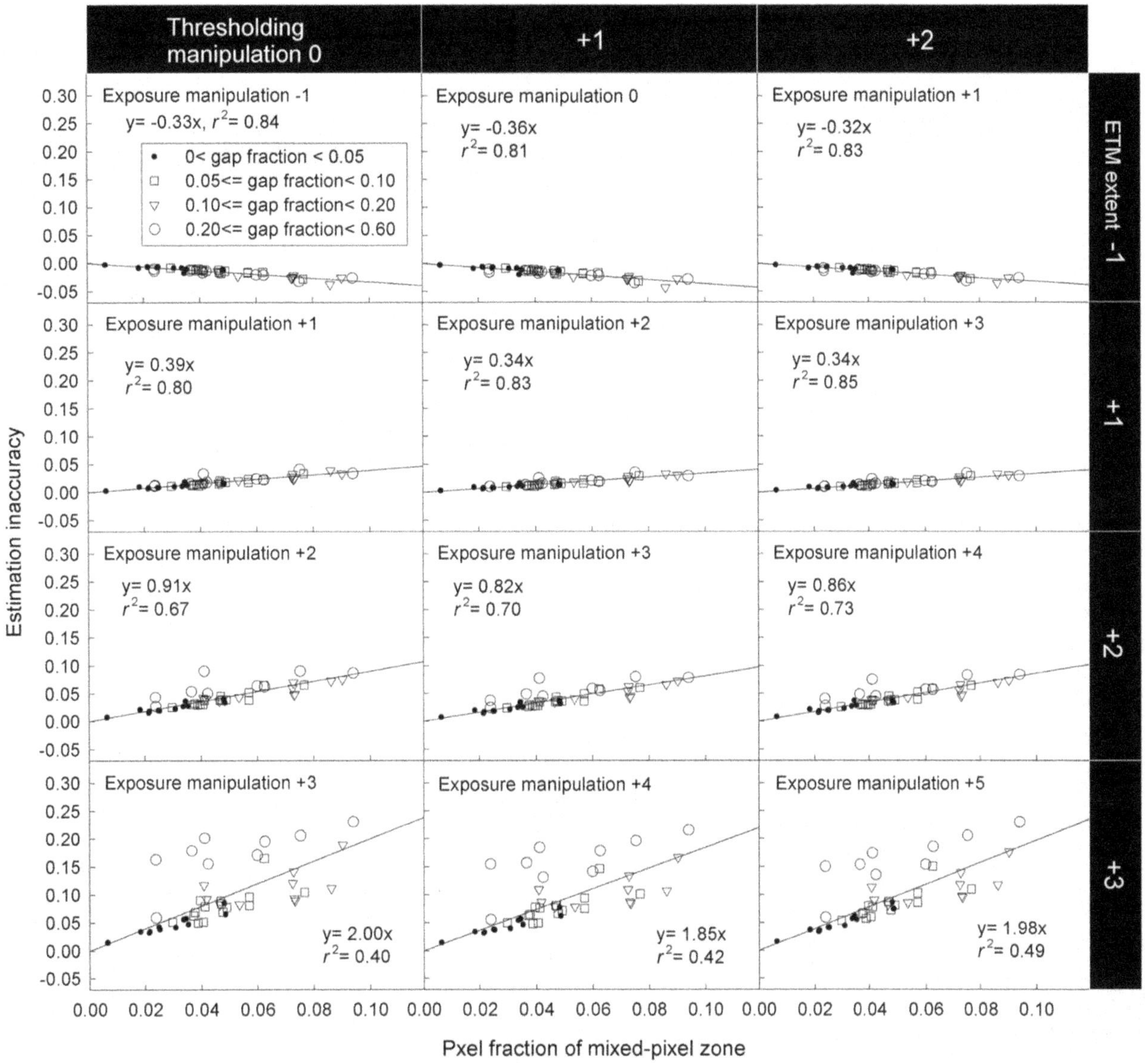

Fig. 10 Effects of the mixed-pixel zone length (i.e. pixel fraction of mixed-pixel zone), canopy gap fraction and the ETM extent on the HP estimation derived from real CHPs taken in 46 locations. The unit of the exposure manipulation, thresholding manipulation and ETM extent is stop

the SCB length can quantify effects of canopy properties which can affect HP estimation. To facilitate cross-study comparisons, the MPZ length of CHPs should be provided in studies which use or develop the technique of HP.

The SCB length in photographs can be reduced through using lens with a narrower view angle. Although SCB length is an unmanipulable canopy property, SCB length captured by photographs can vary with lens' view angle (Additional file 1: Figure S4). Cover photography developed by Macfarlane et al. (2007) uses lens with view angle narrower than that of the fisheye lens to acquire photographs for LAI estimation. When taking photographs for the same canopies, the narrower the lens' view angle, the smaller fraction of canopies is captured, as a result, the shorter SCBs are recorded in photographs (Additional file 1: Figure S4). Because the grey-pixel fraction, a function of the SCB length (Fig. 4), of cover photographs was lower than that of hemispherical photographs (Chianucci 2016; Macfarlane 2011), estimation derived from cover photographs are more accurate than that based on hemispherical photographs (Chianucci 2016).

For photographs with long SCBs, e.g. photographs captured with wide view-angle lenses (fisheye lens), accurate

Table 2 Significant tests for slopes of regression models with the same ETM extents

Type of CHPs	ETM extent	DFW	DFB	*F* value	*P* value
Model CHPs	−1	64	3	34.899	0.007
	+1	64	3	12.444	0.030
	+2	64	3	32.819	0.007
	+3	64	3	9.870	0.041
	+4	64	3	5.244	0.098
	−1	134	2	5.099	0.178
Real CHPs	+1	134	2	9.742	0.097
	+2	134	2	2.087	0.380
	+3	134	2	0.561	0.828

These regression model are shown in Figs. 9 and 10

CHP canopy hemispherical photograph, *DFW* degree of freedom within groups, *DFB* degree of freedom between groups

exposure and thresholding are the key to reduce negative effects of grey pixels on HP estimation. As mentioned in the introduction, it is infeasible to eliminate grey pixels which record SCBs. However, HP inaccuracy caused by the misclassification of grey pixels can be minimised through reducing the ETM extent (Fig. 11; Eq. 3). Although extra efforts needed to establish the relationship between exposure and thresholding in advance, the exposure-corresponding method (Song et al. 2014) is one of the few objective methods which can integrate processes of exposure and thresholding and in turn reduce the ETM extent. Individually carrying out exposure and thresholding with correct methods is another approach to achieve accurate exposure and thresholding. Song et al. (2014) showed that using auto-thresholding methods to analyse CHPs with exposure of one to two stop higher than the reference exposure measured from an unobscured overcast sky (the so-called above-canopy exposure method (Zhang et al. 2005)) can also provide accurate HP estimation. Compared with the exposure-corresponding method, the advantage of this approach is that there is no need to establish the relationship between exposure and thresholding.

Manual thresholding is not recommended for studies which need high HP accuracy (e.g. monitoring temporal dynamics of understorey light environment in the same locations). Our results showed that the misclassified pixels in binary images is hardly detectable even when the ETM extent is up to +3 stop (Fig. 6e, j, o, t). For example, Fig. 6o shows the binary image which was acquired with exposure manipulation of +2 stop (its optimal threshold value is 162 (Song et al. 2014)) and analysed with thresholding manipulation of −1 stop (that is, thresholding value of 28 was used (Song et al. 2014)). Even though the threshold value mismatch is as much as 134 (i.e. the ETM extent is +3), the presence of misclassified grey pixels is barely noticeable if they are not coloured in red (Fig. 6o). On the other hand, the amount of grey pixels can account for as much as 10–15% of the total pixel number in real CHPs (Fig. 5) (Macfarlane 2011). This circumstance coupled with the misclassification of the majority of grey pixels will result in considerable HP inaccuracy. For example, when the ETM extent is +3 stop, for half (23 out of 46) of our real CHPs source locations, the estimated gap fraction is more than twice the correct gap fraction (Additional file 1: Figure S5). In summary, it is very likely that estimates for gap fraction already deviate too much from correct ones before the grey-pixel misclassification reaches to the visually-detectable extent. As mentioned previously, for studies which require high HP accuracy, auto-thresholding methods used with the above-canopy exposure method (Zhang et al. 2005) and the exposure-corresponding method (Song et al. 2014) are recommended.

For studies developing exposure and thresholding methods for HP, photographs with long SCBs (i.e. photograph which are acquired with wide view-angle lenses and record long SCB canopies) should be used. As mentioned above, in terms of LAI estimation, cover photographs outperform CHPs because the former contain fewer grey pixels recording SCBs (Chianucci 2016). Our results also showed that the shorter the SCBs, the lower the HP inaccuracy (Fig. 11; Eq. 3). In other words, low HP inaccuracy is the result of not only the good performance of exposure and thresholding methods, but also short SCBs in photographs used for performance evaluation. To ensure that the true performance of exposure and threshold methods is identified, their performance evaluation should be carried out with photographs with long SCBs rather than cover photographs.

Conclusions

The present study aims to identify how canopy properties (gap fraction and SCB length) and errors of operation (exposure and thresholding) influence an image property (grey-pixel fraction) of CHPs and ultimately affect HP estimation for gap fraction. Grey pixels in

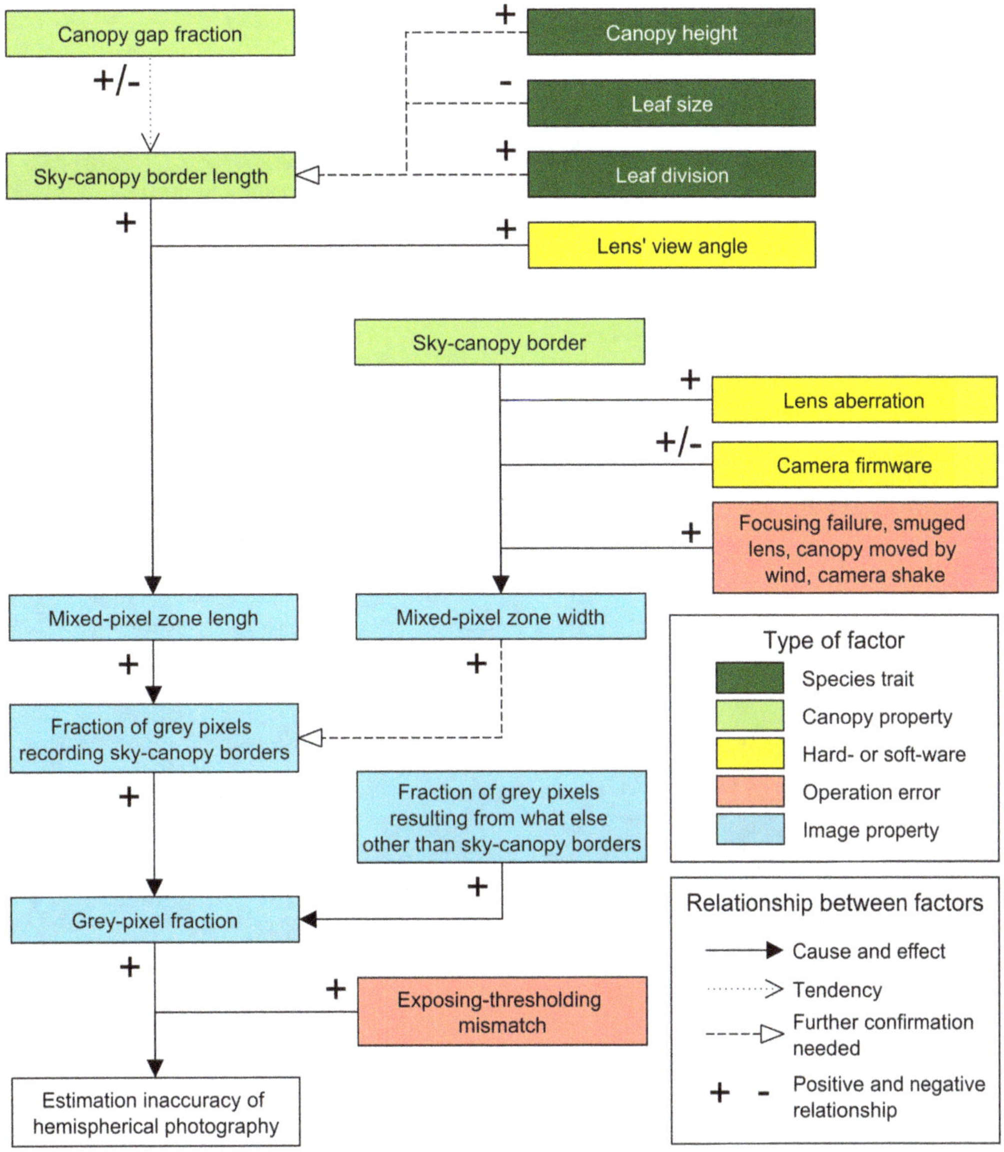

Fig. 11 Relationships between the estimation inaccuracy of HP and its influential factors. Minus and plus signs indicate negative and positive relationships, respectively

CHPs include pixels recording darkened sky, bright canopy elements, sub-pixel canopy elements and SCBs. Our results showed that exposure and thresholding errors cause the misclassification of grey pixels in binary images and in turn result in the HP inaccuracy. Moreover, the grey-pixel fraction of CHPs is a linear function of the SCB length because the majority of grey pixels are those recording SCBs. As a result, HP inaccuracy increases linearly with the SCB length of CHPs once there are exposure and thresholding errors. The relationship between canopy gap fraction and the grey-pixel fraction is, in fact, mediated by the SCB length. Results also showed that the linear relationship between HP inaccuracy and the extent of exposure and thresholding errors is applicable to all forest types. Compared with hemispherical photographs, cover photographs which are acquired with narrow view-angle lenses can provide more accurate estimation for HP, because narrow view-angle lenses capture only a small fraction of canopies and in turn shortens the SCB length in photographs. Due to the fact that short SCBs in photographs and low levels of exposure and thresholding errors can both result in low HP inaccuracy, using photographs with long SCBs rather than cover photographs can ensure that the true performance of new exposure and thresholding methods for HP is identified.

Abbreviations

CHP: canopy hemispherical photograph; ETM: exposure-thresholding mismatch; HP: hemispherical photography; MPZ: mixed-pixel zone; LAI: leaf area index; SCB: sky-canopy border.

Authors' contributions

GZMS designed and conducted the experiments and wrote the manuscript. KJC, DD and DY reviewed the manuscript. All authors read and approved the final manuscript.

Author details

[1] Department of Soil and Water Conservation, National Chung Hsing University, Taichung 40227, Taiwan. [2] International Master Program of Agriculture, National Chung Hsing University, Taichung 40227, Taiwan. [3] Centre for Mined Land Rehabilitation, The University of Queensland, Brisbane, QLD 4072, Australia.

Acknowledgements

We thank Dr. Peter Chesson's suggestions for regression models and statistics. Chien-Hui Liao and Po-Chuen Kung have been very helpful for acquiring CHPs in the field.

Competing interests

The authors declare that they have no competing interests.

Funding

The study was supported by the Ministry of Science and Technology of Taiwan (Grant Nos. NSC102-2313-B-415-001 and MOST104-2311-B-415-002).

References

Abramoff MD, Magalhães PJ, Ram SJ (2004) Image processing with ImageJ. Biophotonics Int 11:36–42

Allen E, Triantaphillidou S (2011) The manual of photography. Focal Press, Oxford

Beckschäfer P, Seidel D, Kleinn C, Xu J (2013) On the exposure of hemispherical photographs in forests. iForest 6:228–237. https://doi.org/10.3832/ifor0957-006

Chao KJ et al (2010) Lowland rainforests in southern Taiwan and Lanyu, at the northern border of Paleotropics and under the influence of monsoon wind. Plant Ecol 210:1–17

Chen JM, Black TA, Adams RS (1991) Evaluation of hemispherical photography for determining plant-area index and geometry of a forest stand. Agric For Meteorol 56:129–143. https://doi.org/10.1016/0168-1923(91)90108-3

Chianucci F (2016) A note on estimating canopy cover from digital cover and hemispherical photography. Silva Fenn 50:1–10

Ferreira T, Rasband W (2012) ImageJ user guide—IJ 1.46

Fisher P (1997) The pixel: a snare and a delusion. Int J Remote Sens 18:679–685

Hwang Y, Ryu Y, Kimm H, Jiang C, Lang M, Macfarlane C, Sonnentag O (2016) Correction for light scattering combined with sub-pixel classification improves estimation of gap fraction from digital cover photography. Agric For Meteorol 222:32–44

Jonckheere I, Nackaerts K, Muys B, Coppin P (2005) Assessment of automatic gap fraction estimation of forests from digital hemispherical photography. Agric For Meteorol 132:96–114. https://doi.org/10.1016/j.agrformet.2005.06.003

Leblanc SG, Chen JM, Fernandes R, Deering DW, Conley A (2005) Methodology comparison for canopy structure parameters extraction from digital hemispherical photography in boreal forests. Agric For Meteorol 129:187–207. https://doi.org/10.1016/j.agrformet.2004.09.006

Macfarlane C (2011) Classification method of mixed pixels does not affect canopy metrics from digital images of forest overstorey. Agric For Meteorol 151:833–840. https://doi.org/10.1016/j.agrformet.2011.01.019

Macfarlane C, Coote M, White DA, Adams MA (2000) Photographic exposure affects indirect estimation of leaf area in plantations of *Eucalyptus globulus* Labill. Agric For Meteorol 100:155–168. https://doi.org/10.1016/S0168-1923(99)00139-2

Macfarlane C, Grigg A, Evangelista C (2007) Estimating forest leaf area using cover and fullframe fisheye photography: thinking inside the circle. Agric For Meteorol 146:1–12

Macfarlane C, Ryu Y, Ogden GN, Sonnentag O (2014) Digital canopy photography: exposed and in the raw. Agric For Meteorol 197:244–253. https://doi.org/10.1016/j.agrformet.2014.05.014

Rich PM (1990) Characterizing plant canopies with hemispherical photographs. Remote Sens Rev 5:13–29. https://doi.org/10.1080/02757259009532119

Smith SW (1997) The scientist and engineer's guide to digital signal processing. California Technical Publishing, San Diego

Sokal RR, Rohlf FJ (1995) Biometry, 3rd edn. W. H. Freeman and Company, New York

Song G-ZM, Doley D, Yates D, Chao K-J, Hsieh C-F (2014) Improving accuracy of canopy hemispherical photography by a constant threshold value derived from an unobscured overcast sky. Can J For Res 44:17–27. https://doi.org/10.1139/cjfr-2013-0082

Thimonier A, Sedivy I, Schleppi P (2010) Estimating leaf area index in different types of mature forest stands in Switzerland: a comparison of methods. Eur J For Res 129:543–562

Zhang YQ, Chen JM, Miller JR (2005) Determining digital hemispherical photograph exposure for leaf area index estimation. Agric For Meteorol 133:166–181. https://doi.org/10.1016/j.agrformet.2005.09.009

13

Identification, sexual dimorphism and aspects of the natural history of *Sapria himalayana* (Rafflesiaceae) on Vietnam's Lang Biang Plateau

Hữu Đăng Trần[1*], Hồng Trường Lưu[1,2], Quốc Đạt Nguyễn[1], Hiếu Cường Nguyễn[1], Parusuraman Athen[3] and K. M. Wong[3]

Abstract

Background: *Sapria* is a distinctive and narrowly host-specific holoparasitic genus belonging to the Rafflesiaceae. *Sapria himalayana*, rare throughout its range from NE India, SW China, Thailand to Vietnam, is a little-understood species first recorded for Vietnam in 1959, and only recollected there over half a century later in February 2017. This has facilitated an assessment of its taxonomic identity and our understanding of its morphology and natural history aspects.

Results: Six populations of *Sapria* at Vietnam's Tuyền Lâm Lake, and another two populations at the Nam Ban Protection Forest and the Cam Ly area were found, in an area of about 20 km in radius. Previously documented size attributes, morphological details and colour patterns allowed clear identification of the Vietnamese taxon as *Sapria himalayana* f. *albovinosa*. A full description of the species for Vietnam is provided. Past authors have distinguished the sexes by column form and structure, colour of the upper disk, details of the inner surface of the perigone tube, and presence of ovarial chambers below the column in the female. We present additional observations that male flowers consistently have more steeply held perigone lobes than females, in which the lobes were more spread out at wider angles in fully open flowers, and that males have a much lower cupule than females. The latter difference, especially, appears to be useful for quick determination of the sex even in the advanced floral bud stage. The host plant was the lianescent *Tetrastigma laoticum* (Vitaceae), but superficially it was not possible to ascertain the clonal relationship of neighbouring host lianas. Male and female flowers were found mixed together in the same cluster from one individual liana. Potential pollinators included Calliphorid and Stratiomyid flies observed visiting open flowers.

Conclusions: Our observations have added to an increased understanding of the morphology of this highly specialized parasitic life form. More than this, we have ascertained its occurrence in Vietnam, with information made available to authorities of the Lâm Đồng Province where our studies were conducted, for the sites to be specially demarcated for conservation and carefully managed tourism use.

Keywords: Da Lat, Lang Biang Plateau, Morphology, Parasitic plants, *Sapria himalayana*, Sexual dimorphism

*Correspondence: tranhuudang@gmail.com
[1] Southern Institute of Ecology, Vietnam Academy of Science and Technology, 1 Mạc Đĩnh Chi Street, District 1, Ho Chi Minh City, Vietnam
Full list of author information is available at the end of the article

Background

Sapria Griff. is one of three genera belonging to the Rafflesiaceae, in which all members are holoparasites on various species of *Tetrastigma* (Vitaceae) lianas. They are all narrowly host-specific and produce either male or female flowers. The most conspicuous part of the saucer-shaped open flower is the perianth complement of fleshy lobes (also known as the perigone lobes) that spread out and recurve, and which are frequently deep red to orange. *Sapria* is well distinguished from the other two genera (Hansen 1973; Meijer 1997). *Rafflesia* R.Br. ex Thomson has 5(−6) imbricate perigone lobes in the flower, *Rhizanthes* Dumort. has 16–18 valvate lobes, but *Sapria* has 10 valvate lobes separated as an outer and an inner series. In *Rhizanthes,* the perianth (or perigone) tube does not develop any membraneous outgrowth ('diaphragm') just below the base of the perigone lobes, which have a smooth inner surface, but in the other two genera the flower develops a broad diaphragm that defines a central aperture and the perigone lobes are conspicuously warty on the inner side. Whereas perigone lobes are rounded in both *Rafflesia* and *Sapria,* those in *Rhizanthes* terminate in an abruptly attenuated bayonet-like apex. Also, small, irregularly knob-, finger-, or tube-shaped fleshy outgrowths develop on the inner perigone tube surface in *Rafflesia,* but these are found on the diaphragm in *Sapria. Sapria* presently consists of three recognized species: *S. himalayana* Griff. (from NE India, SW China, Thailand to Vietnam), *S. poilanei* Gagnep. (Cambodia and Thailand) and *S. ram* Bänziger and Hansen (Bänziger et al. 1997, 2000) (found only in Thailand).

Sapria himalayana was first recorded on the Lang Biang Plateau in Vietnam by Tixier (1959) (Fig. 1). Since then, there have been no other collections made for *S. himalayana* in Vietnam. Subsequent mention of *Sapria himalayana* from Vietnam in the literature (Hansen 1973; Schmid 1974; Pham-hoang 2003) were all based on Tixier's collection. Unfortunately, we could not locate where the specimens were deposited.

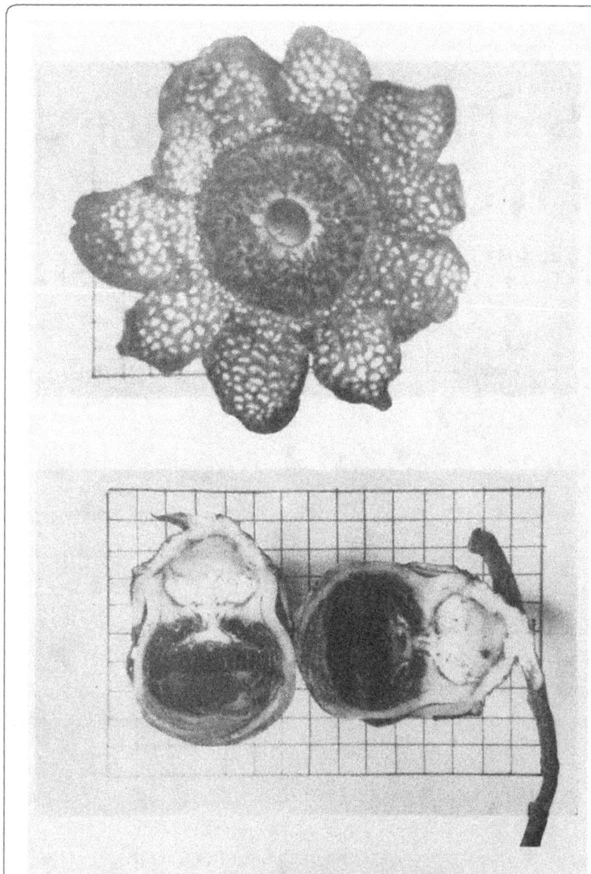

Fig. 1 First ever collection of *Sapria himalayana* Griff. (Tixier s.n.) for Vietnam from the Lang Biang Plateau, as photographed in Tixier (1959), showing female flower in top view (above) and young female flower bud in longitudinal section (below)

In early February 2016, a picture of a beautiful, red, huge, geophytic flower was sent to the first author by a relative from Lang Biang Plateau. Although it was identified as *Sapria himalayana,* no proper collections were made. In February 2017, during a botanical survey in Nam Ban Protection Forest, Lâm Đồng province by the Southern Institute of Ecology, and also during a field trip to Đà Lạt by the first and last authors, several populations of *Sapria himalayana* were again discovered in bloom, and it was then possible to make appropriate botanical specimens.

The recent rediscovery of *Sapria himalayana* populations on Vietnam's Lang Biang Plateau offered the opportunity to verify if it belongs to one of the known forms of Bänziger et al. (2000), whose work has dealt with the morphological variation in some detail with excellent information on the distinction of the two forms that they recognised.

Methods and materials

The surveys resulted in the documentation of six populations of *Sapria* (of which two had open flowers) at Tuyền Lâm Lake, and another two populations with open flowers at the Nam Ban Protection Forest and the Cam Ly area. They all occur in a region of about 20 km in radius.

These populations were examined to determine if they formed a 'cluster' [defined by Bänziger et al. (2000) as individual blooms arising from the same host vine]. Sample open blooms, either fresh or not too badly deteriorated and still with recognisable colour patterns, were taken, examined and various parts were measured according to the protocol used by Bänziger and Hansen (1997) with slight modification (Fig. 2). Notes and photographs of different parts and their colour were taken. A Leica S6D microscope was used to examine a section through the ovarial chamber. Vouchers for both sexes from the Tuyen Lam Lake population (*Tran* et al. *472* for the female, *Tran* et al. *473* for the male) were deposited with the SGN and SING herbaria (herbarium acronyms follow Thiers, continuously updated), and vouchers for the Nam Ban population (*Q.Đ. Nguyễn and H.C. Nguyễn NB-001* for the female, *NB-002* for the male) have been lodged with the SGN herbarium. The identification of the host plant of *Sapria himalayana* was confirmed at the VNM herbarium.

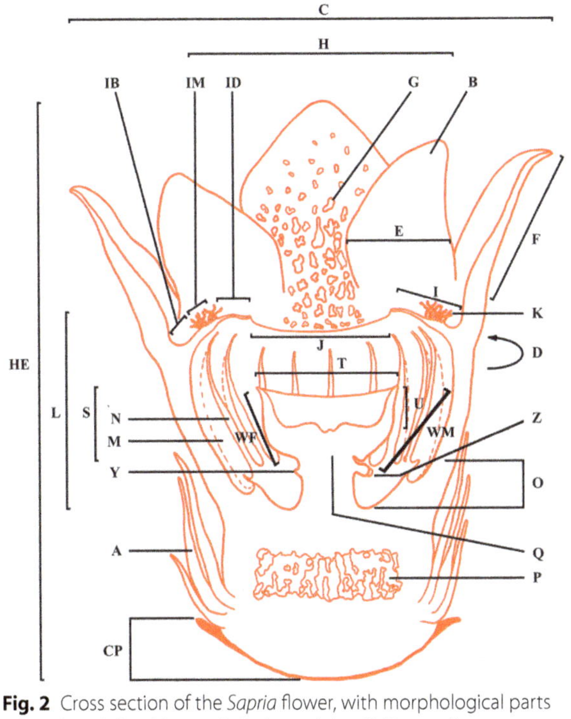

Fig. 2 Cross section of the *Sapria* flower, with morphological parts referred to. A. Basal bracts. B. Perigone lobe. C. Flower diameter. CP. Height of cupule at flower base. D. Circumference of perigone tube at lobe base. E. Perigone lobe width. F. Perigone lobe length. G. Wart. H. Diameter of whole diaphragm. HE. Height of flower. I. Width of diaphragm collar. IB. Basal band. ID. Width of distal band of diaphragm. IM. Width of band of ramenta. J. Width of diaphragm aperture. K. Ramenta. L. Height of perigone tube. M. Ridge on inner surface of perigone tube. N. Flange of ridge on inner surface of perigone tube. O. Fusion area of perigone tube ridges. P. Ovary (in female). Q. Column. S. Height of disk. T. Diameter of crest of disk. U. Depth of disk. WF. Stigmatic fascia (in female). WM. Longitudinal strip of tissue bearing anther (in male). Y. Vestigial anther (in female). Z. Anther (in male) [Based on the original in Bänziger and Hansen (1997), modified by H.Đ. Trần]

Results and discussion

Distribution of *Sapria*

In Vietnam, *Sapria himalayana* is thus far known only on the Lang Biang Plateau, at Tuyền Lâm Lake, Cam Ly and Nam Ban Protection Forest. Outside of Vietnam, the known range of *Sapria himalayana* includes India and Thailand, where the forma *himalayana* (distinguished by yellow warts on blood-red perigone lobes) occurs in India (Assam), Myanmar and Thailand (Elliott 1992; Bendiksby et al. 2010) and the forma *albovinosa* Bänziger and Hansen (distinguished by white warts on wine-red perigone lobes) is documented from Phu Khiew Wildlife Sanctuary, Chaiyaphum Province, Thailand (Bänziger et al. 2000). This distribution is likely to be very incomplete because there are generally very few careful documentations of *Sapria* and their colour expression or variation.

The other two species of *Sapria* known include *S. poilanei*, so far known from the Cardamom mountains area in Cambodia and Chanthaburi in Thailand, and *S. ram* documented from Dawna-Bilauktaung, Phangnga and Suratthani in Thailand (Bänziger and Hansen 1997).

Identification of the Lang Biang *Sapria*

By its size attributes, morphological details (Table 1) and colour patterns (Figs. 3, 4), we were able to identify the Vietnamese taxon as *Sapria himalayana* f. *albovinosa* according to the excellent key and attributes of fresh material provided by Bänziger and Hansen (1997) and Bänziger et al. (2000). Fresh open flowers of *Sapria himalayana* are distinctly larger (previously documented floral span 95–200 mm) than either *Sapria poilanei* (floral span 65–120 mm) or *Sapria ram* (floral span 55–110 mm) (Bänziger and Hansen 1997), and our Vietnamese taxon has a floral span of 110–165 mm, falling in the range for *S. himalayana*. The ramenta apices of the Vietnamese taxon are also bilobed or multilobed (especially in female specimens), or crateriform (i.e., with a shallow recess, especially in males). The aperture of the diaphragm is 14–21 mm, distinctly smaller than the disk of 51–61 mm diameter, as also documented by Bänziger and Hansen for *Sapria himalayana*. The disk is somewhat flat and thins out towards its periphery so that it does not have a distinctive "wall" between upper and lower surfaces, as noted by Bänziger and Hansen (1997) for *S. himalayana*; in contrast, the other two species have a pan- to bowl-shaped (female) or cup-shaped (male) disk with a distinct wall. Other morphological details (Table 1) are also consistent with those of *Sapria himalayana* as described by Bänziger and Hansen (1997). Comparison between these three *Sapria* species is summarized in Table 2.

Table 1 Measurements (in mm) of female and male flowers of *Sapria himalayana* f. *albovinosa* sampled on Lang Biang Plateau for the present study

Flowers	Code	♀1	♀2	♀3	♂1	♂2	♂3
Flower diameter	C	157–160	145–165	130–140	110–116	138–142	117–142
Lobe inclination from the vertical		60°–75°	30°–70°	55°–90°	0°–10°	30°–45°	10°–45°
Height of flower	HE	90	73	71	97	100	107
Circumference of perigone tube at lobe base	D	260	254	250	250	265	255
Height of perigone tube	L	34	32	32	31	37	38
Perigone lobe width/length (outer whorl)	B (E/F)	42–58/48–50	43.5–55/49–53	41–55/38–52	40–50/50–57	40–52/46–53	39–53/45–54
Perigone lobe width/length (inner whorl)	B (E/F)	32–34/43–46	31–37/38–41	32–34/41–46	36–37/44–51	33–38/40–50	31–37/39–44
Diameter of whole diaphragm	H	58–60	54–55	54–61	52–55	53–56	51–55
Width of diaphragm collar	I	21–24	20–21.5	18–21	16–18	18–19	16–17
Width of diaphragm aperture	J	14–16	14–15	19–20	18–20	19.5–20.5	20–21
Width of ramenta band	IM	14–15	12.5–15	11–14.5	9–12	12–13	11–12
Width of distal band	ID	6–7	6–7	5.5–7	5–6	5–6	4–5
Height of column	Q	20	18	16–17	16–17	17–19	Column and disk missing (eaten)
Diameter of column where narrowest		15	14	15	8	8	
Diameter of crest of disk	T	35–39	36–38	37–38	23–24	32–33	
Depth of disk	U	6	5	5–5.5	6–7	7–8	
Height of disk (to just above anthers)		11	10.5	10–11	9–10	8–9	
Width of stigmatic fascia/annular row of anthers	WF/WM	15	14	13.5	12–13	12	
Maximum diameter of stigmatic circle		32	32.5	32.5	–	–	
Maximum height of perigone tube ridges: width of its flange at that point	M:N	5:2	4.8:2.4	5.5:2.5	7	7	7
Height of cupule at flower base	CP	10–13	15–16.6	12–14.5	2–3	3–4.5	4–6
Outer whorl of basal bracts (length)	A	21–35	20–34	21–30	14–17	20–29	20–20
Inner whorl of basal bracts (length)	A	53–59	50–58	43–56	39–62	50–58	44–59

The colour patterns of the Vietnam taxon (Figs. 3, 4, 5 and 7) are those documented for *Sapria himalayana* f. *albovinosa*. The warts are white against the distinctive wine-red colour of the perigone lobes as described for this form.

Description of the Lang Biang *Sapria*

In the following description, the key morphological parts are provided with abbreviations in brackets that correspond to Fig. 2.

Holoparasitic plant on *Tetrastigma laoticum* roots. Flower buds c. 8 cm in diameter near to opening stage. Cupule (CP) 10–16.6 mm high in female, 2–6 mm high in male. Bracts (A) of outer whorl 5, (14–)20–35 mm long; those of inner whorl 5, (39–)43–62 mm long. Diameter (C) × height (HE) of fully open flowers: 130–165 × 71–90 mm in female, 110–142 × 97–107 mm in male. Perigone lobes 10 (inner whorl 5 and outer whorl 5); female lobes when fully open (30–)60° to 90° from the vertical plane through the floral centre; male lobes when fully open 0° to 30(–45)° from the vertical plane through the floral centre; outer lobes ovate, length (F) 38–57 × width (E) 39–58 mm, red with white warts adaxially, with white dots abaxially, glabrous; inner lobes spathulate, 38–51 × 31–38 mm, with white warts adaxially, with white dots abaxially, glabrous. Diaphragm 51–61 mm in diameter (H); 14–21 mm across aperture (J); collar width (I) 16–24 mm; basal band (IB) white with pink radial stripes, turning red then purple-black with age; ramenta band (IM) 9–15 mm wide, reddish black, turning purple-black with age, ramenta (K) mostly bi- or multilobed in females and occasionally in males, mostly crateriform in males and occasionally in females, 5–9 mm long, wine-red at base, turning dark orange to dark red towards their apex, blackening with age; distal band (ID) 4–7 mm wide, dark red turning black with age. Perigone tube circumference (D) around lobe bases 254–265 mm, height (L) 31–38 mm. Central column (Q) dimorphic between sexes: female column 16–20 mm high, 14–15 mm in diameter at narrowest point, disk bowl-shaped, upper surface whitish cream, margin reddish, crest of disk (T) 35–39 mm in diameter, depth of disk (U) 5–6 mm, height of disk (S) to just above vestigial anthers 10–11 mm, stigmatic fascia (WF) 13.5–15 mm wide, sparsely bristly; male column 16–19 mm high, c. 8 mm in diameter at narrowest point, disk bowl-shaped,

Fig. 3 *Sapria himalayana* Griff. f. *albovinosa* Bänziger and Hansen, female flower. **a** Whole flower. **b** Habitat in natural forest at Tuyen Lam. **c** Longitudinal section of flower. **d** Ramenta on upper surface of diaphragm. **e** Cross section of ovary. **f** Lateral view of disk and upper part of column, with vestigial anthers (yellow structures in a ring below the disk) visible. **g** Ovules. Photos: H.Đ. Trần

upper surface wine-red, margin dark red, crest of disk (T) 23–33 mm in diameter, depth of disk (U) 6–8 mm, height of disk to just above anthers 8–10 mm, longitudinal strip of tissue bearing anther (WM) 12–13 mm long, hairy. Anthers (Z) 20, set in a ring, each 2.5 mm wide, filaments 2.0–2.3 mm long. Ovary (P) inferior, 36–38 × 14–20 mm, cavities irregular and bearing many ovules.

Additional features of sexual dimorphism from the Lang Biang observations

Bänziger and Hansen (1997) have summarised the distinction between females and males by differences mainly in the column (stalk and disk) form and structure, the colour of the upper disk surface, the form and dimensions of radial ridges on the inner surface of the perigone tube, and the occurrence of ovarial chambers in an expanded base below the column in the female. The *Sapria himalayana* f. *albovinosa* populations in our study area also conformed in these characteristics.

Additionally, we have observed several more attributes that help in the distinction between the sexes in *Sapria himalayana* f. *albovinosa*. For the Lang Biang populations examined, all males consistently had more steeply held perigone lobes (spreading 0–45° from the vertical) than females, in which the lobes were more spread out at wider angles (spreading 30–90° from the vertical) in fully open flowers (Figs. 3c, 4c, Table 1). Male flowers that were past anthesis and beginning to appear discoloured

Fig. 4 *Sapria himalayana* Griff. f. *albovinosa* Bänziger and Hansen male flower. **a** Whole flower. **b** Open flowers with more steeply held perigone lobes than in the female. **c** Longitudinal section of flower. **d** Ramenta on upper surface of diaphragm. **e** Part of upper disk surface. **f** Lateral view of column and disk, showing anthers (yellow structures) in a ring and longitudinal ridges on the inner surface of the perigone tube. **g** Anthers. Photos: H.Đ. Trần

Table 2 Floral characters and known flowering period compared for three *Sapria* species

Species	*Sapria himalayana*	*Sapria poilanei*	*Sapria ram*
Flower diameter	95–200	65–120	55–110
Ramenta long (mm)	Up to 10	1–3(5)	5–5.5
Lobes colour	Blood-red	Wine-red	Wine-red
Dotted warts colour	Yellow (or white)	White (sometimes pale pink)	White (sometimes pale pink)
Aperture diameter (mm)	18–37	15–25	13–29
Female disk shape	Bowl-shaped	Bowl-shaped	Pan-shaped
Female disk's crest diameter (mm)	35–39	9–18	22–28
Male disk shape	Bowl-shaped	Cup-shaped	Cup-shaped
Male disk's crest diameter (mm)	23–33	12–15.5	14–18
Flowering period	November–February	November–March	November–March

Fig. 5 *Sapria himalayana* Griff. f. *albovinosa* Bänziger and Hansen. **a** Female flower with higher cupule and wider spreading perigone lobes. **b** Male flower with lower cupule and more steeply held perigone lobes. **c** Diaphragm and aperture of female flower with a pale disk. **d** Diaphragm and aperture of male flower, with reddish disk. **e** Post-anthesis female flower. **f** Post-anthesis male flower. Photos: H.Đ. Trần

were also examined, and were found to have similarly steeply held perigone lobes, indicating that this was indeed the full extent of floral opening in males (Fig. 5e, f). We remain uncertain if biomechanical reasons linked to structural variation have caused these differences.

The basal cupules of male and female flowers were also different. Males have a much lower cupule (2–6 mm high) than females (10–16.6 mm high) (see Table 1, Fig. 5). This difference appears to be a significant and easy character for quick determination of the sex even when the flower is in the advanced bud stage, i.e., when the bud begins to reveal the apex of the paler colored developing perigone lobes (and before the other main characters can be assessed), and so might be of potential importance in field investigations that require special preparations for sampling or observation of the different sexes.

Fig. 6 *Tetrastigma laoticum*, the host vine of *Sapria himalayana*. **a** Mature climbing stem with flat structure due to extended secondary thickening of the original stem axis in opposite directions; demonstrated by Parusuraman Athen. **b** Leaves with tendrils arising from petiole bases. **c** Cross section of root near emergence of *S. himalayana* flower. Photos: H.Đ. Trần

Some natural history aspects

The host plant (Fig. 6) was identified as the liana *Tetrastigma laoticum* Gagnep. (Vitaceae). Although the host plant populations at the Lang Biang Plateau sites were dense, we were unsure of their genetic variation and relationships, as lianas often drape onto the forest floor and can vegetatively generate new individuals. More detailed studies of the host vine relationships of *Rafflesia* by Pelser et al. (2016) have demonstrated that the more common species of *Rafflesia* can have multiple host vines, so that host-parasite specificity is not as great as previously supposed. In other words, the potentially clonal nature of neighbouring host lianas was uncertain. It was also not known if the phenological behaviour of the host lianas, e.g., if a state of flowering had any possible effects on their predisposition to infection by and establishment of the parasite. From our observations, male and female flowers can be mixed together in the same cluster from one individual liana. Flowers are red when open, turning black as they senesce. We have not specially studied the pollinators but were able to note that flies belonging to the Calliphoridae and Stratiomyidae families visited open flowers (Fig. 7c, d). The pollination biology of *Sapria* is similar to that of *Rafflesia* in that pollen from the anther is deposited on the visiting flies, which is then carried to other flowers (Heide-Jørgensen 2008).

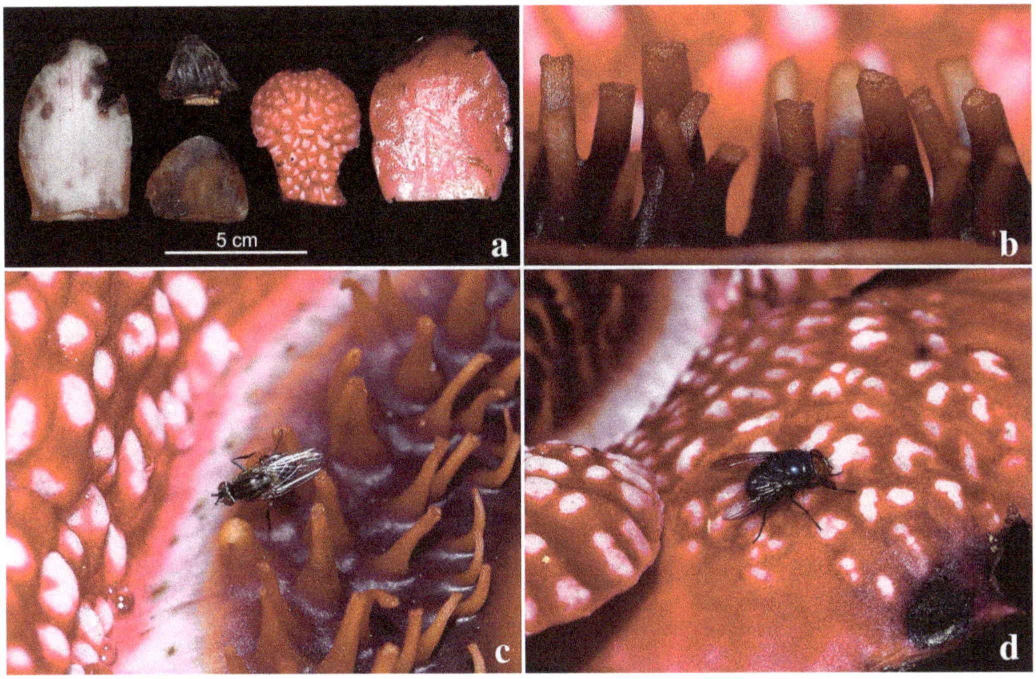

Fig. 7 *Sapria himalayana* Griff. f. *albovinosa* Bänziger and Hansen. **a** Bracts (darker) and perigone lobes (red, patterned) in male flower. **b** Ramenta in male flower. **c** Detail of perigone lobe warts and Stratiomyidae fly. **d** Calliphoridae fly on *Sapria* flower. Photos: H.Đ. Trần

There is currently a dearth of information on seed development, seed dispersal, germination, as well as ecology in terms of infection and establishment for *Sapria* species. Better observations of these aspects will only be possible where a good number of populations are present, allowing comparative work, within secure sites where disturbance is likely to be absent or minimal, such as at the presently known Vietnamese sites. Also, the security and commonness of the parasite on site would enable possibilities for tagging and longer-term observations to be carried out, such as phenology and other ecological attributes, and plans are being formed by our research group for suitable studies on *Sapria himalayana* f. *albovinosa*.

Conservation management

Both known *Sapria* sites on the Lang Biang Plateau are within relatively undisturbed broadleaf evergreen montane forest with *Acer* (Aceraceae), Fagaceae trees and *Tetrastigma* (Vitaceae) lianas commonly occurring. Recent studies of genetic populations of Rafflesiaceae have indicated that loss of a single infected host vine could result in large genetic losses of the parasite (Barkman et al. 2017), thus conservation activities should also focus on the conservation of host vine populations.

Whereas the Nam Ban site is itself a protection forest, the Tuyền Lâm site is within the Lâm Viên Protection Forest where only light tourism activities (e.g., forest trekking and interpretation) are permitted.

Following our investigations, the Southern Institute of Ecology has alerted authorities of the Lâm Đồng Province, under which jurisdiction the sites are administered, for the sites to be specially demarcated as Protected Research Reserves. This special status will no doubt enhance further careful use of designated parts of the Tuyền Lâm site for nature tourism, as well as maintain the integrity of the other strictly protected parts that will, among other things, contribute to the conduct of our longer-term studies.

Authors' contributions

HĐT, QĐN, HCN, PA and KMW made field studies, collected and photographed specimens. HTL and HĐT checked localities and discussed conservation aspects. HĐT, HTL and KMW drafted the manuscript. All authors read and approved the final manuscript.

Author details

[1] Southern Institute of Ecology, Vietnam Academy of Science and Technology, 1 Mạc Đĩnh Chi Street, District 1, Ho Chi Minh City, Vietnam. [2] Graduate University of Science and Technology, Vietnam Academy of Science and Technology, 18 Hoàng Quốc Việt, Cầu Giấy, Hanoi, Vietnam. [3] Singapore Botanic Gardens, National Parks Board, 1 Cluny Road, Singapore 259569, Singapore.

Acknowledgements

We thank Mr. Tô Văn Quang for the identification of flies visiting *Sapria* flowers and Mr. Phạm Hữu Nhân for his information on the species' distribution. Dr. Võ

Thị Phi Giao kindly gave us access to Tixier's paper deposited at Pham Hoang Ho's Bibliothèque in the University of Science, Hồ Chí Minh City. The study was supported through Project ĐTĐL.XH.11/15 funded by Vietnam's Ministry of Science and Technology as well as the National Parks Board, Singapore.

Competing interests

The authors declare that they have no competing interests.

Funding

Vietnam's Ministry of Science and Technology, No: ĐTĐL.XH.11/15.

References

Bänziger H, Hansen B (1997) Unmasking the real identity of *Sapria poilanei* Gagnepain emend., and description of *Sapria ram* sp. n. (Rafflesiaceae). Nat Hist Bull Siam Soc 45:149–170

Bänziger H, Hansen B, Kreetiyutanont K (2000) A new form of the hermit's spittoon, *Sapria himalayana* Griffith f. *albovinosa* Bänziger and Hansen f. nov. (Rafflesiaceae), with notes on its ecology. Nat Hist Bull Siam Soc 48:213–219

Barkman TJ, Klooster MR, Gaddis KD, Franzone B, Calhoun S, Manickam S, Vessabutr S, Sasirat S, Davis CC (2017) Reading between the vines: hosts as islands for extreme holoparasitic plants. Am J Bot 104:1382–1389. https://doi.org/10.3732/ajb.1700117

Bendiksby M, Schumacher T, Gussarova G, Nais J, Mat-Salleh K, Sofiyanti N, Madulid D, Smith SA, Barkman T (2010) Elucidating the evolutionary history of the Southeast Asian, holoparasitic, giant-flowered Rafflesiaceae: pliocene vicariance, morphological convergence and character displacement. Mol Phylogenet Evol 57(2):620–633

Elliott S (1992) Status, ecology and conservation of *Sapria himalayana* Griff. (Rafflesiaceae) in Thailand. J Wildlife Thail 2(1):44–52

Hansen B (1973) Rafflésiacées. In: du Aubréville A (ed) Flore du Cambodge du Laos et du Viêt-nam. Muséum National d'Histoire Naturelle, Paris, pp 60–62

Heide-Jørgensen HS (2008) Parasitic flowering plants. Brill, Leiden

Meijer W (1997) Rafflesiaceae. Flora Malesiana, Series I, 13:1–42

Pelser PB, Nickrent DL, Barcelona JF (2016) Untangling a vine and its parasite: host specificity of Philippine *Rafflesia* (Rafflesiaceae). Taxon 65(4):739–758

Pham-hoang H (2003) An illustrated flora of Vietnam, vol 2. Youth Publication, Hồ Chí Minh City, pp 141–142

Schmid M (1974) Végétation du Viet-Nam - Le massif sud-annamitique et les régions limitrophes. Orstom, Paris, pp 123–126

Thiers B (continuously updated). Index herbariorum: a global directory of public herbaria and associated staff. New York Botanical Garden's Virtual Herbarium. http://sweetgum.nybg.org/ih/. Accessed 12 Apr 2018

Tixier P (1959) *Sur la présence de Sapria himalayana* Griff. *dans la région de Dalat*. Ann Fac Sci Saigon, pp 133–135

14

Taxonomy of *Cyrtochilum*-alliance (Orchidaceae) in the light of molecular and morphological data

Dariusz L. Szlachetko[1], Marta Kolanowska[1,2*], Aleksandra Naczk[3], Marcin Górniak[3], Magdalena Dudek[1], Piotr Rutkowski[1] and Guy Chiron[4]

Abstract

Background: The generic separateness and specific composition of the orchid genus *Cyrtochilum* was discussed for almost two centuries. Over the years several smaller taxa were segregated from this taxon, but their separateness was recently questioned based on molecular studies outcomes. The aim of our study was to revise concepts of morphological-based generic delimitation in *Cyrtochilum*-alliance and to compare it with the results of genetic analysis. We used phylogenetic framework in combination with phenetical analysis to provide proposal of the generic delimitation within *Cyrtochilum*-alliance. Two molecular markers, ITS and *mat*K were used to construct phylogenetic tree. A total of over 5000 herbarium specimens were included in the morphological examination and the phenetical analysis included 29 generative and vegetative characters.

Results: Comparative morphology of the previously recognized genera: *Buesiella, Dasyglossum, Neodryas, Rusbyella, Siederella* and *Trigonochilum* is presented. A new species within the the latter genus is described. Fourteen new combinations are proposed. The key to the identification of the genera of the *Cyrtochilum*-alliance and morphological characteristics of each genus are provided.

Conclusions: A total of six separated genera are recognized within *Cyrtochilum*-alliance. The reasons of the incompatibility between morphological differences observed within studied taxa and phylogenetic tree are argued and the taxonomic implications of such inconsistency, resulting in fragmentation or lumping of taxonomic units, are discussed.

Keywords: *Cyrtochilum*, Monophyly, New combinations, New species, Oncidiinae, Paraphyly, Taxonomy

Background

The genus *Cyrtochilum* was proposed in 1816 by German botanist C.S. Kunth along with descriptions of two new species, *Cyrtochilum flexuosum* Kunth and *Cyrtochilum undulatum* Kunth. Neither was designated as the generitype, which was standard procedure at that time. *C. undulatum* was selected as the type species of the genus by Garay (1974). Since its description, *Cyrtochilum* has been incorporated into the widely circumscribed genera *Oncidium* Sw. or *Odontoglossum* Kunth. by most subsequent taxonomists. The only exception was Kraenzlin (1917), who revitalized the genus a hundred years after its first description.

Cyrtochilum once again became lost for over 80 years till Dalström (2001) reevaluated it and proposed several new nomenclatural combinations. The generitype determines somewhat the generic delimitation. According to this, *Cyrtochilum* should comprise species with flexuose, branching inflorescence, large flowers with broad, unguiculate sepals and petals, and narrow, slender lips covered in the basal part by large, massive, composed callus consisting of keels and digitate segments, and partially connate with a clavate, slender gynostemium, forming a right angle with the lip (Fig. 1).

*Correspondence: martakolanowska@wp.pl
[1] Department of Plant Taxonomy and Nature Conservation, The University of Gdańsk, ul. Wita Stwosza 59, 80-308 Gdańsk, Poland
Full list of author information is available at the end of the article

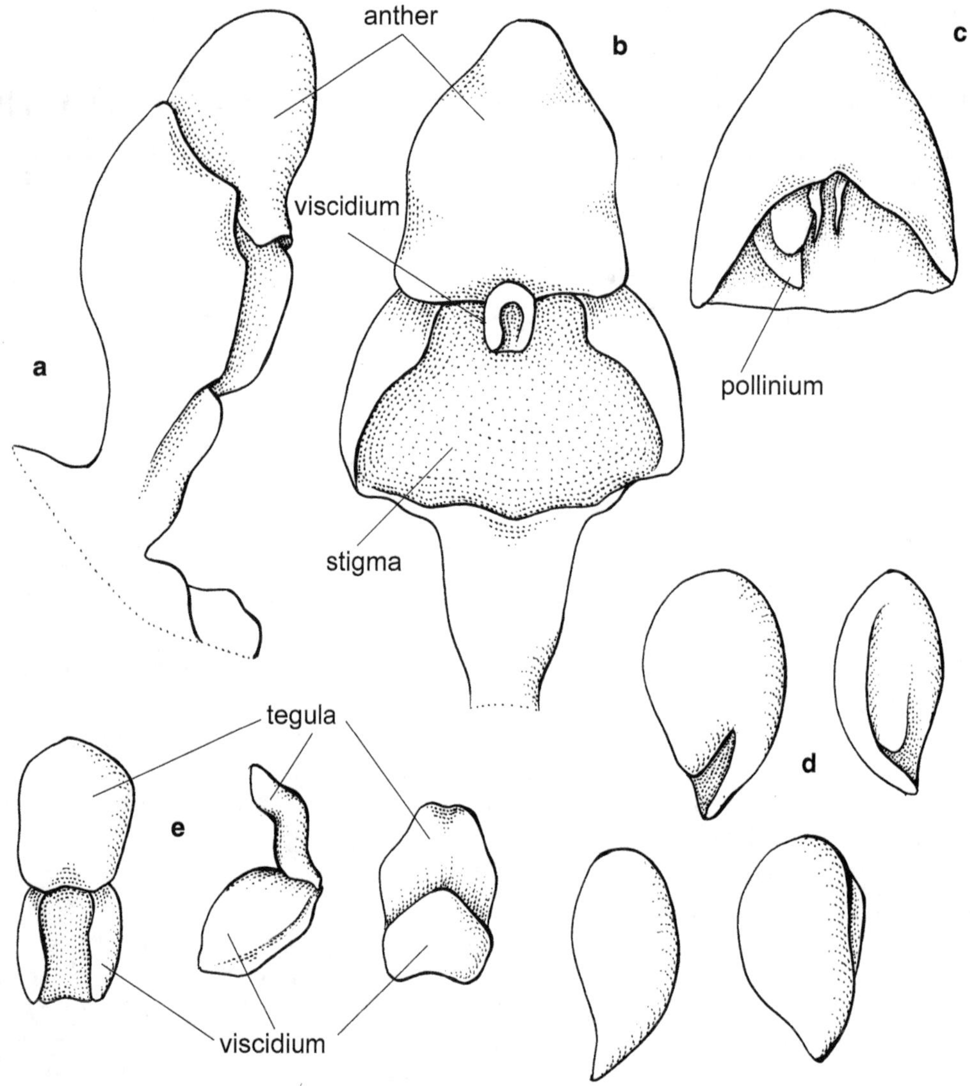

Fig. 1 *Cyrtochilum volubile*. **a** Gynostemium, *side view*. **b** Gynostemium, *bottom view*. **c** Anther. **d** Pollinia, *various views*. **e** Tegula and viscidium, *various views* (Szlachetko & Mytnik-Ejsmont 2009)

On the basis of the sequences of molecular markers Neubig et al. (2012) proposed another circumscription of the genus. The authors included here various species, for example *Odontoglossum myanthum* Lindl. (generitype of *Dasyglossum* Königer & Schildh.), *Cyrtochilum flexuosum* Kunth (generitype of *Trigonochilum* Königer & Schildh.), *Oncidium aureum* Lindl. (generitype of *Siederella* Szlach., Mytnik, Górniak & Romowicz), as well as rspresentatives of *Rusbyella, Buesiella, Neodryas* and *Odontoglossum*. All of them inhabit mainly Ecuadorian Andes with many species also found in Colombian and northern Peruvian mountains. Neubig et al. (2012) created a monophyletic but highly heteromorphic unit, what resulted in the very enigmatic description of the genus (cf. Pridgeon et al. 2009; Dalström 2010).

The aim of presented study was to evaluate and compare morphological differences between taxa of *Cyrtochilum*-complex with the outcomes of molecular studies.

Methods

Morphological study

A total of over 5000 herbarium and liquid preserved specimens of orchids representing *Cyrtochilum s.l.* and related oncidioid genera and deposited in AMES, AMO, B, BM, C, COL, CUVC, F, FLAS, HUA, JAUM, K, MO, NY, P, PMA, UGDA, VALLE and W (Thiers 2015) were examined according to the standard procedures (database of specimens representing *Cyrtochilum s.l.* and *Odontoglossum* is provided in Additional file 1: Appendix S1). Every studied specimen was photographed and the data from the labels were taken. Both vegetative and generative characters of each plant were examined (the shape and size of the pseudobulbs, leaves, inflorescence architecture, shape and size of the floral bracts, flower morphology and gynostemium structure) and compared with existing type material of the most of distinguished species of the subtribe. The nomenclature of morphological characters follows Dressler (1981) and Szlachetko (1995).

Phenetical analysis

Phenetical studies were employed based on 29 characteristics describing the taxonomically important generative and vegetative structures of *Cyrtochilum* species exploited by Neubig et al. (2012). As an outgroup we selected *Odontoglossum epidendroides,* a generitype of the genus *Odontoglossum.* A complete list of these features, as well as selected sets, is given in Additional file 2: Appendix S2. We have used a binary, 0–1, system of coding characteristics, because it is unambiguous and the most often applied in phenetic analyses. The incorporation of each feature for every *Cyrtochilum s.l.* species has resulted in a data matrix containing 1247 characteristics. To create hierarchic phenograms we used the PAST program (Hammer and Harper Ryan 2001). The so-called cluster analysis process is a typical method of analysis used in phenetic research (Stace 1989). We created a distance matrix using the Manhattan measure (Domański and Kęsy 2005; Pandit and Gupta 2011; Madhulatha 2012), which is an average subtraction measured across the dimensions $D = \Sigma i|Xij - Xik|$. We have also used the "middle links rule unweighted pair-group average" (UPGMA) as an amalgamation rule. The resulting phenograms were compared with the results of research conducted by Neubig et al. (2012).

Molecular analyses

Taxon sampling

For the molecular analyses 91 specimens representing genus *Cyrtochilum.* The outgroup includes one species, *Odontoglossum epidendroides*. Sequences of outgroup taxa and for the most representatives of *Cyrtochilum* were downloaded from GenBank (Additional file 3: Appendix S3). DNA sequences of *Cyrtochilum volubile* were obtained in laboratory on the Department of Plant Taxonomy and Nature Conservation University of Gdansk. Sequences for both markers (ITS, *mat*K) were deposited in GenBank. Accession number and information about collector were place in Additional file 3: Appendix S3.

Molecular markers

Nucleotide sequences from one nuclear (ITS) and one plastid (*mat*K) genome region were used in the molecular analyses. The ITS region consisted of the 18S and 26S ribosomal RNA genes, respectively the internal transcribed spacers (ITS1, ITS2) and the intervening gene 5.8S. For the sample of *Cyrtochilum volubile* was amplified part of the ITS region (ITS1 – 5.8S – ITS2) using the primers 101F and 102R (Douzery et al. 1999). For the *mat*K gene, we amplified fragment of approximately 1400 bp using the primers 19F (5′CGTTCTGACCATATTGCACTATG3′) from Molvary et al. (2000) and 1326R (5′TCTAGCACACGAAAGTCGAAGT3′) from Cuénoud et al. (2002).

DNA extraction, amplification and sequencing

DNA was extracted using the Sherlock AX Kit (A&A Biotechnology, Poland) following manufacturer protocol. For the sample homogenization were used precooled in −45 °C lysing Matrix A tube and FastPrep instrument (MP Biomedicals, USA). Pellet of DNA was resuspended in 50 μl of TE buffer.

Amplifications and sequencing were using Eppendorf and Biometra TGradient thermal cyclers. PCR reaction for the both markers (ITS, *mat*K) were performed in a total volume of 25 μl containing 1 μl temple DNA (~10–100 ng), 0.5 μl of 10 μM of each primers, 12.0 μl Start Warm 2X PCR Master Mix (A&A Biotechnology, Poland), water and/or 1.0 μl dimethyl sulfoxide (DMSO) to ITS region/0.5 μl 25 mM MgCl2 only to *mat*K marker. Amplification parameters for nrITS (ITS1 + 5.8S + ITS2) were: 94 °C, 4 min; 30X (94 °C, 45 s; 52 °C, 45 s; 72 °C, 1 min); 72 °C, 7 min. For the part of *mat*K gene were: 95 °C, 3 min; 33X (94 °C, 45 s; 52 °C, 45 s, 72 °C, 2 min 30 s); 72 °C, 7 min. Wizaed SvGel and PCR Clean Up System (Promega, US) was used to clean PCR products following manufacturer protocol. Purified products of PCR reaction were cycle-sequenced using Big Dye Terminator v 3.1 Cycle Sequencing Kit (Applied Biosystems, Icn., ABI, Warrington, Cheshire, UK). Cycle sequencing parameters were: 95 °C, 2 min 40 s; 25X (95 °C, 10 s; 50 °C, 10 s; 60 °C, 4 min). Total volume sequencing reaction of 10 μl containing 1.3 μl of 5X sequencing buffer, 1 μl of Big Dye terminator, 0.4 μl of 10 μM primer (1.6/3.2 pmol), 0.5 μl dimethyl sulfoxide (DMSO), 1 μl of amplified product (30–90 ng/μl) and water. The sequencing reaction products were then purified and sequenced on an ABI 3720 automated capillary DNA sequencer in the Genomed S. A (Warsaw, Poland). DNA sequences chromatograms were inspected/edited in FintchTV and assembled using AutoAssembler (Applied Biosystems, Inc). Sequences for the *Cyrtochilum volubile* were deposited in GenBank (see Additional file 3: Appendix S3).

Data analyses

The consensus sequences, both ITS region and part of *mat*K gene, were done automatically alignment by

Seaview (Galtier et al. 1996) using algorithm MUSCLE (Edgar 2004). Analyses were performed separately on the matrix of each marker separately using PAUP*4.0b10 (Swofford 2002) and MrBayes 3.1.2 (Ronquist and Huelsenbeck 2003).

Maximum parsimony analysis (MP) used a heuristic search strategy with tree-bisection-reconnection (TBR) branch swapping and the MULTREES option in effect, simple addition and ACCTRAN optimization. Gaps were treated as a missing value. All characters were unordered and equally weighted (Fitch 1971). Internal support of clades was evaluated by character bootstrapping (Felsenstein 1985) using 1000 replicates. For bootstrap support levels, we considered bootstrap percentages (BP) of 50–70% as weak, 71–85% as moderate and >85% as strong (Kores et al. 2001). We also performed a Bayesian inference (BA). An evolutionary model for each region (ITS, *mat*K) was calculated with MrModeltest 2.2 (Nylander 2004). For the both data matrix the GTR + I + G model was selected according to the AIC (Akaike Information Criterion). For analyses, two simultaneous runs of four chains each were carried out with the MCMC algorithm, for 10,000,000 generations, sampling one tree for each 100, until the average standard deviation of split ranges was smaller than 0.01. After discarding the initial 25% trees of each chain as the burnin. Majority rule consensus tree was generation for the remaining trees in PAUP to assess topology and clades posterior probabilities (PP). Value of PP in Bayesian analysis are not equivalent to BP, generally are much higher (Erixon et al. 2003).

Results

Morphological analyses

The phenetic similarity of the studied species based on morphological data is presented in Fig. 2. The first group comprises species usually classified to the genus *Dasyglossum* along with *Neodryas/Buesiella.* The species in this complex are characterized by subsimilar tepals, usually free sepals, an entire or 3-lobed lip, united basally with the base of the column, and parellal to it. The upper part of the lip is geniculate and often retrorse. The lip callus is simple, consisting of a pair of fleshy, parallel, adjoining tori, diverging in front, mostly enclosed by the thickened flanks of the gynostemium. The gynostemium is rather short, robust, in the upper half gently upcurved or straight. The generic borderline between *Dasyglossum* and *Neodryas/Buesiella* mostly concerns the character of the lip callus, which is large and variously lobed in the latter.

The next group includes *Cyrtochilum* species, such as "*C. ioplocon*", "*C. ramosissimum*", "*C. revolutum*", "*C. angustatum*" and "*C. pardinum*". All of these species are characterised by rather narrow, acuminate tepals with more or less undulate margins and somewhat twisted apices. Sepals and petals are dissimilar in form. Sepals have long and narrow claw, and petals—relatively short and wide. Lip is sessile, basally parallel to the gynostemium, and then geniculate bent down, the lamina is oblanceolate to oblong obovate in general outline, with acuminate and twisted apex. Lip calli consist of a pair of rather large basal wings with additional digitate or clavate projections below them. Gynostemium is erect, only basally connate with the lip, cylindrical, without any additional projections at the apex or at the base of the stigma. Floral bracts are usually shorter than half of pedicellate ovary. These species are mingled with *Odontoglossum epidendroides* and "*C. macasense*". The former species is the type of the genus. Tepals of *Odontoglossum* are usually subsimilar, either set on prominent claw, or subsessile, but in both situations the claw of sepals and petals are similar. Margins of tepals are smooth, often crispate, and rarely undulate. Lip is basally connate with the gynostemium. In *O. epidendroides* the fusion is prominent and can reach one-fifth of the total lip length. Basal part of the lip is clawed, and lamina is more or less perpendicular to it. The shape of the lamina varies—usually it is oblanceolate to elliptic, often with crispate margins and long acuminate apex. Lip calli form a complicated pattern and consist of numerous digitate or lamellar projections, glabrous or ciliate. The gynostemium is usually somewhat arcuate, and form with the column an acute angle. It is apically adorned by various, filiform, digitate or lamellar projections. Floral bracts are prominently shorter than pedicellate ovary.

"*C. macasense*" is characterised by subsimilar, shortly clawed tepals, and sessile lip, which is prominently 3-lobed. The lip calli is compoused of two pairs of fleshy ridges of various lengths. The shorter pair is bilobed. Gynostemium forms an acute angle with the lip, and is erect, relatively short and massive, without any prominent appendages.

The "*C. midas*" group embraces species with small usually dull-coloured flowers, brownish or greenish-brown, which are usually treated as *Trigonochilum.* Tepals are rather dissimilar, sepals are narrower, with narrow claw, and petals are wider, short-clawed. The lip is triangular-cordate, sessile, diverging from the gynostemium at 70°–90° with a simple, torous, sometimes verrucose or gibbous callus. The lip lamina is centrally convex. The form and position of the gynostemium versus the lip in the species of this group is somewhat similar to *Cyrtochilum s.str.* It is usually elongate, basally much expanded and connate with the lip, slightly sigmoid or upcurved, slender, and the tegula has a prominent roof-like projection on the inner surface above the viscidium. We did not observe this character in any other species of the *Cyrtochilum*-clade. Floral bracts are rudimentary, much

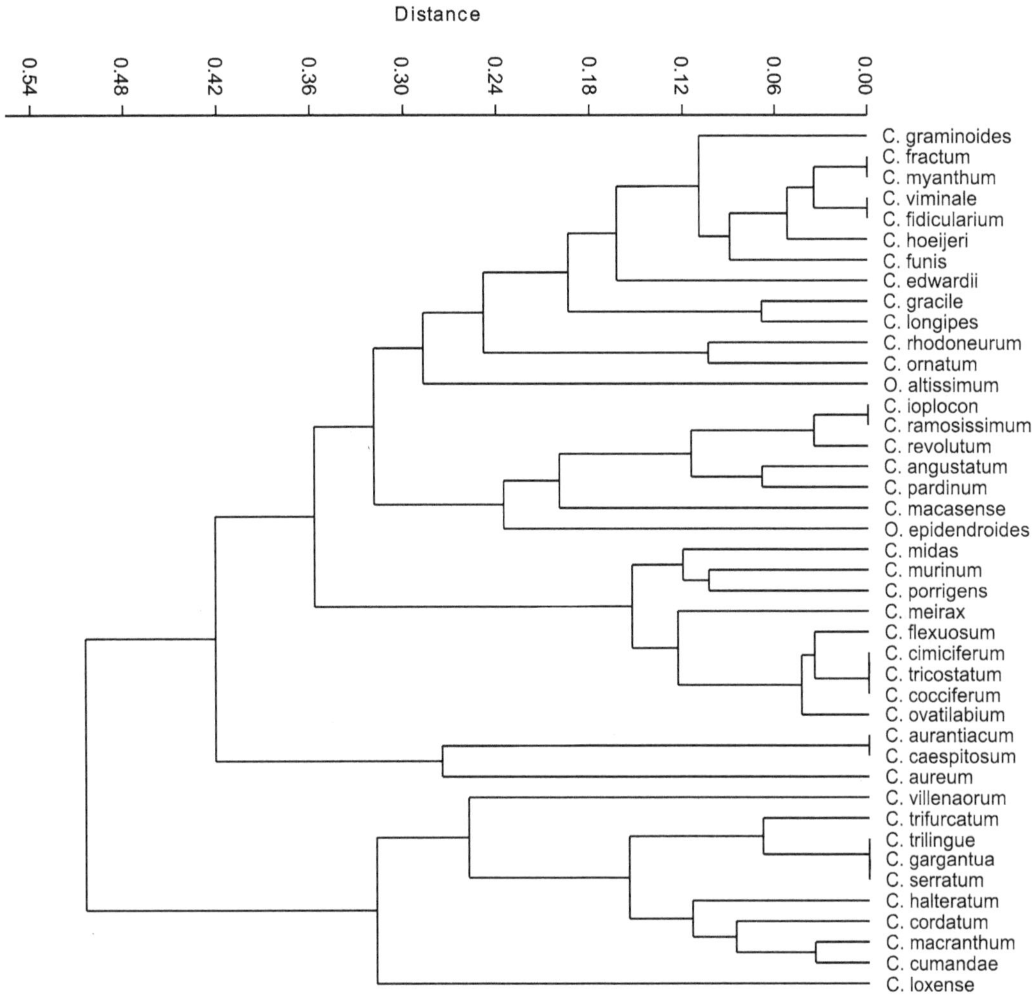

Fig. 2 Phenetic similarity (UPGMA) of *Cyrtochilum s.l.* species based on morphological data

shorter than pedicellate ovary. *Trigonochilum* species are rarely confused with other genera, although the species boundaries are often not clear.

"*C. aurantiacum/caespitosum*" is rather an isolated group, at least as morphology is considered. Both are easily recognisable by the lip structure which has narrow, lower part, more or less canaliculated, with prominent rather simple calli. The apical part of the lip is much expanded forming transversely elliptic lamina. The gynostemium is somewhat similar to that one of *Dasyglossum*, i.e. it is erect, narrowly winged, apically upcurved. What is interesting tegula is narrow, linear 3–4 times longer than viscidium. Both species are included in the genus *Rusbyella*. Interestingly, "*C. aureum*" is linked to this group, although the gynostemium structure of "*C. aureum*" can suggest the affinity of this species to *Cyrtochilum s.str.* The short gynostemium is clavate, somewhat arcuate, with oblong-obovate projections with fringed margins. The gynostemium forms an acute angle with the lip. The lip reminds somewhat "*C. loxense*", i.e. it is clawed, lamina is flat or convex, obscurely 3-lobed or pentagonal in outline, lip calli is missing to prominent, and contain of series of small projections in two rows. Lateral sepals are connate almost to the apex.

The last group contains those species which are included in the genus *Cyrtochilum s.str.* The common character of those species is gynostemium, gently sigmoid, basally prominently connate with the lip, elongated and slender above. The erect part is clavate and forms a right angle with the lip. The column part is slightly thickened just above the base, with two wing-like or digitate projections just below the stigma. Tepals are dissimilar, usually shield-like, obtuse to rounded apically, often undulate. Sepals have long and narrow claw, and petals—short and wide. At the base of the sepals' claw wing-like appendices can be observed in most of the species.

The lip of *Cyrtochilum s.str.* is sessile to shortly clawed, and usually divided into expanded and convex basal part and usually narrow, ligulate, pendent apical part. The lip calli is much complicated and usually consist of massive and variously lobed central part, with various number of additional projections spread all over the basal part. The floral bracts are large, leafy, nearly half as long as pedicellate ovary.

"*C. villenaorum*" is different from the species described above by the subsessile lip which has very large lamina, unequally 3-lobed, the middle lobe is more or less transversely elliptic in outline, with relatively small and simple calli with the middle lobe being somewhat upcurved. The gynostemium is devoid of any projections. Regarding morphology, "*C. volubile*" is very similar to "*C. villenaorum*", but we did not include the former species in our analysis.

Morphologically distinct species in *Cyrtochilum s.str.* is "*C. loxense*". Its tepals are subsimilar, shortly clawed; lateral sepals are connate in the basal fifth or so. Lip is straight, clawed, lamina is very unequally 3-lobed, with both lateral lobes relatively small, and the middle lobe very large, transversely elliptic with truncate apex. The calli is rather obscure and consist of series of irregular small projections near the lip base. The gynostemium is perpendicular to the lip, somewhat arcuate, basally connate with the lip claw, with short digital projections near the stigma.

Molecular analyses

Statistcs for the data matrices (ITS, *mat*K) are separated by "/". The number of analyzed taxa was 80/65 respectively. The aligned length of the matrix was 779/1303 characters of which 88/70 were parsimony informative. The number of the most parsimonious trees were >10.000, tree-length was 219/181, consistency index (CI) = 0.76/0.83 and retention index (RI) = 0.89/0.90. Consensus trees of Bayesian analysis are presented in Figs. 3 and 4.

Topology of MP trees and Bayesian trees are similar. The clades that have low bootstrap support or/and collapse in the strict consensus tree in parsimony analysis often appeared in Bayesian trees with low posterior probabilities too. One of the most parsimonious trees is available from the corresponding author. The combined phylogenetic tree presented by Neubig et al. (2012) is based on the analyses of five DNA regions (ITS, *trnH-psbA*, *5′ycf1*, *3′ycf1*, *matK*).

The first subclade comprises the species of *Cyrtochilum s.str.* (Fig. 5) and "*C. ramosissimum*", and is sister to the next subclade including two species—"*C. angustatum*" (Fig. 6) and "*C. pardinum*". The last three aforementioned species resemble *Odontoglossum* typified by *Odontoglossum epidendroides* Kunth and, in fact, they have usually been assigned to that genus. It is noteworthy that *Odontoglossum epidendroides* is embedded in a separate clade (Fig. 7) and treated by Neubig et al. (2012) as a member of *Oncidium s.l.*. All the *Odontoglossum*-like species of *Cyrtochilum* mentioned above share a series of mutual features with *Odontoglossum*, i.a. gynostemium is slender, erect, forms an acute angle with a narrow lip, and it is fused with it along the midline at the base, creating two basal cavities (Fig. 8). The lip is geniculately bent near the middle exposing multiple calli consisting of narrow, digitate and/or filiform projections. Sepals and petals are narrow and undulate on margins, and sepals are prominently clawed.

In our *matK* tree species constituting this subclade form two groups A and B with posterior propability value 53 and 81, respectively. Group B comprises also *C. volubile* and *C. villenaorum*. The ITS tree does not solve relations between particular groups of species, although some branches are relatively highly or highly supported, e.g. *Cyrtochilum angustatum–C. pardinum* (e) with BS/PP = 62/100. Most other species of *Cyrtochilum s.str.* (a) are grouped together with BS/PP = 55/80.

The subclade "*Cyrtochilum myanthum*" includes species classified in *Dasyglossum* (Fig. 9), the genus established by Königer and Schildhauer (1994) and typified with *Odontoglossum myanthum* Lindl. The key characters of the genus mentioned by the authors are a simple callus, consisting of a pair of fleshy ridges and the lower half of the lip being parallel with the gynostemium, and apically part geniculately bent. Additionally, all species possess a massive, erect gynostemium, prominently winged and lateral sepals being free to the base (Fig. 10). The gynostemium and channeled lip callus form a kind of tube accessible to long-beaked pollinators.

The position of "*C. edwardii*" which is sister to *Dasyglossum* sublcade is unexpected, as it shares characters of the genus *Trigonochilum* rather than *Dasyglossum*, i.e. lip callus consisting of 7 massive projections confined to the central part of lamina, lip being arcuately bent down, and gynostemium and lip form a right angle. The colour of the flower, however, is unique for *Dasyglossum/Trigonochilum* alliance and is deep purple or lilac and lip callus is bright yellow. The gynostemium just below stigma is adorned with a pair of wing-like projections, not found in *Dasyglossum*.

It is interesting to note a position of "*C. flexuosum*". The species is nested in two different places in cladogram; the first one is polytomic with *Dasyglossum* and "*C. edwardii*", and the other one is embedded in *Trigonochilum* subclade. As the species is generitype of *Trigonochilum* we discuss it below.

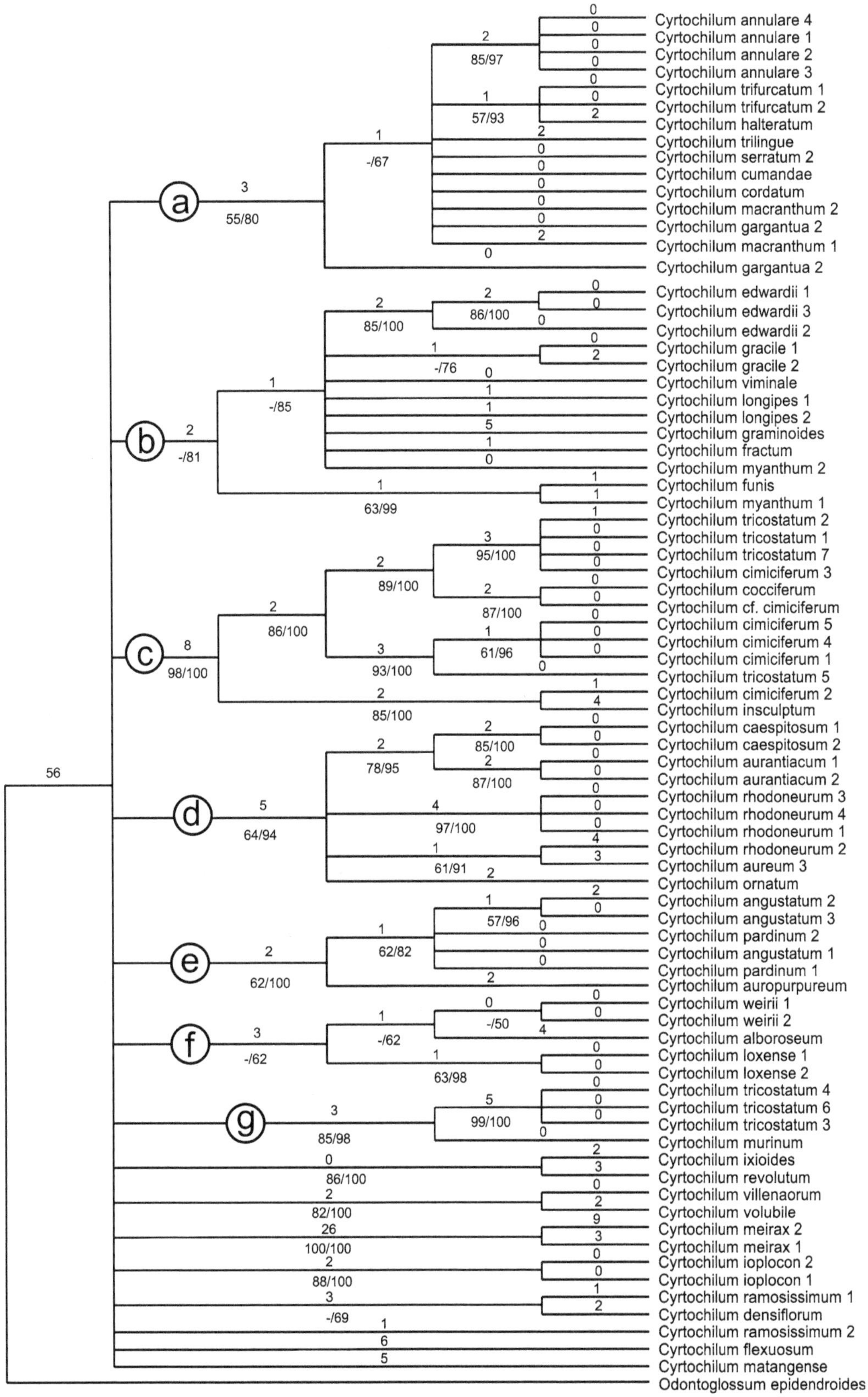

Fig. 3 Bayesian 50% majority-rule tree for genus *Cyrtochilum* from ITS1-5.8S-ITS2 sequences. The *numbers* below the branches are bootstrap percentages (BP) and posterior probability (PP), bootstrap percentages ≥50% are given for supported clades. The branches length is shown above

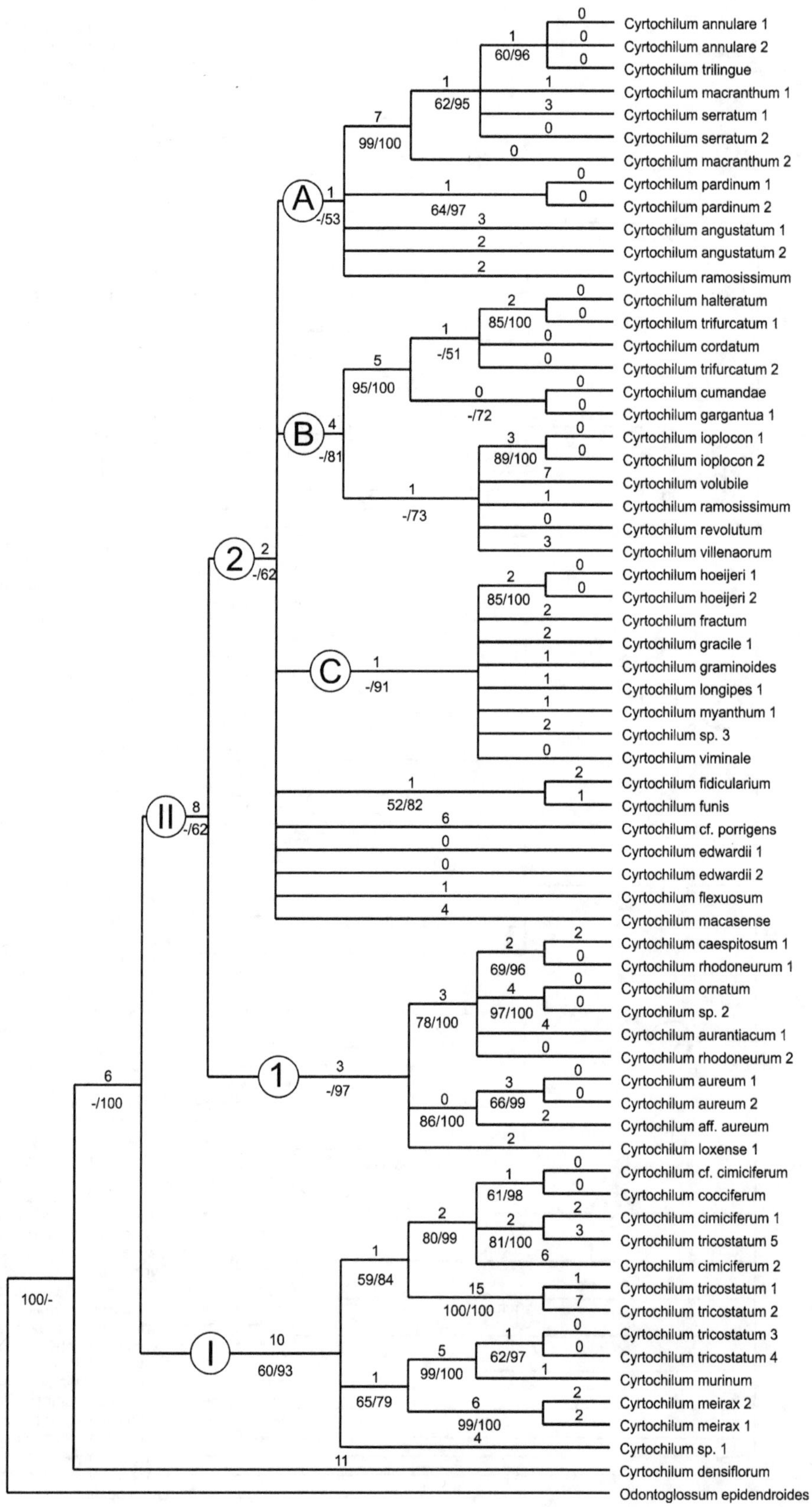

Fig. 4 Majority-rule consensus of 7500 trees obtained in Bayesian analysis of *matK* gene for genus *Cyrtochilum*. *Values* below branches represent bootstrap support (≥50%) from 1000 replicates and posteriori probabilities (≥50%) (BP/PP). The branches length is shown above

Fig. 5 *Cyrtochilum cryptocopis*. Photo: Guido Deburghgraeve

Fig. 6 *Odontoglossum angustatum*. Photo by Guido Deburghgraeve

Fig. 7 *Odontoglossum epidendroides*. Photo by Guido Deburghgraeve

"*C.* cf. *porrigens*" is again polytomic to the subclades mentioned above, and "*C. macasense*" is sister to all aforementioned groups. The first species is similar in all respects to *Trigonochilum* and has more or less triangular-obovate lip with complexed calli, clavate gynostemium basally connate with the lip and then abruptly upcurved in result forming an obtuse angle with it.The general flower architecture of *C. macasense*" reminds somewhat "*C. edwardii*". The gynostemium and the lip form a right angle, lip callus consists of 4 ridges of various length, of which the shorter pair is bilobed. The colour of the flowers is a mixtre of yellow and brown, likes in *Trigonochilum*. The unique character of this species is prominently 3-lobed lip with much elongate middle lobe.

The *matK* tree does not solve relation between species of this subclade—some of them—e.g. *Cyrtochilum myanthum, C. viminale, C. gracile*, etc.—are grouped together (C) and highly supported (PP = 91). The others are polytomic, e.g. *C. edwardii, C. macasense* or *C. flexuosum*. All those species form a mutual subclade (b) in the ITS analysis (PP = 81).

The subclade "*Cyrtochilum flexuosum*" embraces species assigned to the genus *Trigonochilum* (Fig. 11). The genus was described in 1994 by Königer and Schildhauer to encompass Oncidiinae species characterized by a subtriangular lip diverging from the gynostemium at 70°–90° and a short, stout, clavate gynostemium (Fig. 12) with distinct swellings below the stigma. The lip callus is a large mass of variously, but shallowly lobed tissue occupying the central part of the lamina. The authors designated *T. flexuosum* (Kunth) Königer & Schildh. as a generitype and presented a list of 22 species transferred to the newly established taxon from *Cyrtochilum* Kunth, *Odontoglossum* Kunth and *Oncidium* Sw. In the following years, Königer (1996, 1999, 2000) described some new species of *Trigonochilum* and other species were reassigned to the genus or described by Senghas (2001, 2003). The latter author, however, synonymized all the species of *Dasyglossum* Königer under *Trigonochilum*. With some additional transfers made by Königer (2008, 2010) and a description of the new species, the genus currently includes about 60 species with a distribution from Peru and Bolivia to Colombia and Venezuela. The border between them is very often very difficult to define. Dualistic position of "*C. flexuosum*" on the Neubig et al. (2012) phylogenetic tree is probably caused by misidentification of one of the samples.

The species constituting this subclade are on the mutual branch (I) in *matK* tree and has 60/93 BS/PP. This branch is sister to all other *Cyrtochilum*-alliances. The ITS tree analysis gives somewhat different pattern of relation between aforementioned species—this subclade is divided into two groups c and g, with high bootstrap support and posterior propability—98/100 and 85/98, respectively. Relations between these groups are not solved.

The last subclade of the *Cyrtochilum*-group is composed of a mixture of species included in various genera, whose common features are more or less connate lateral

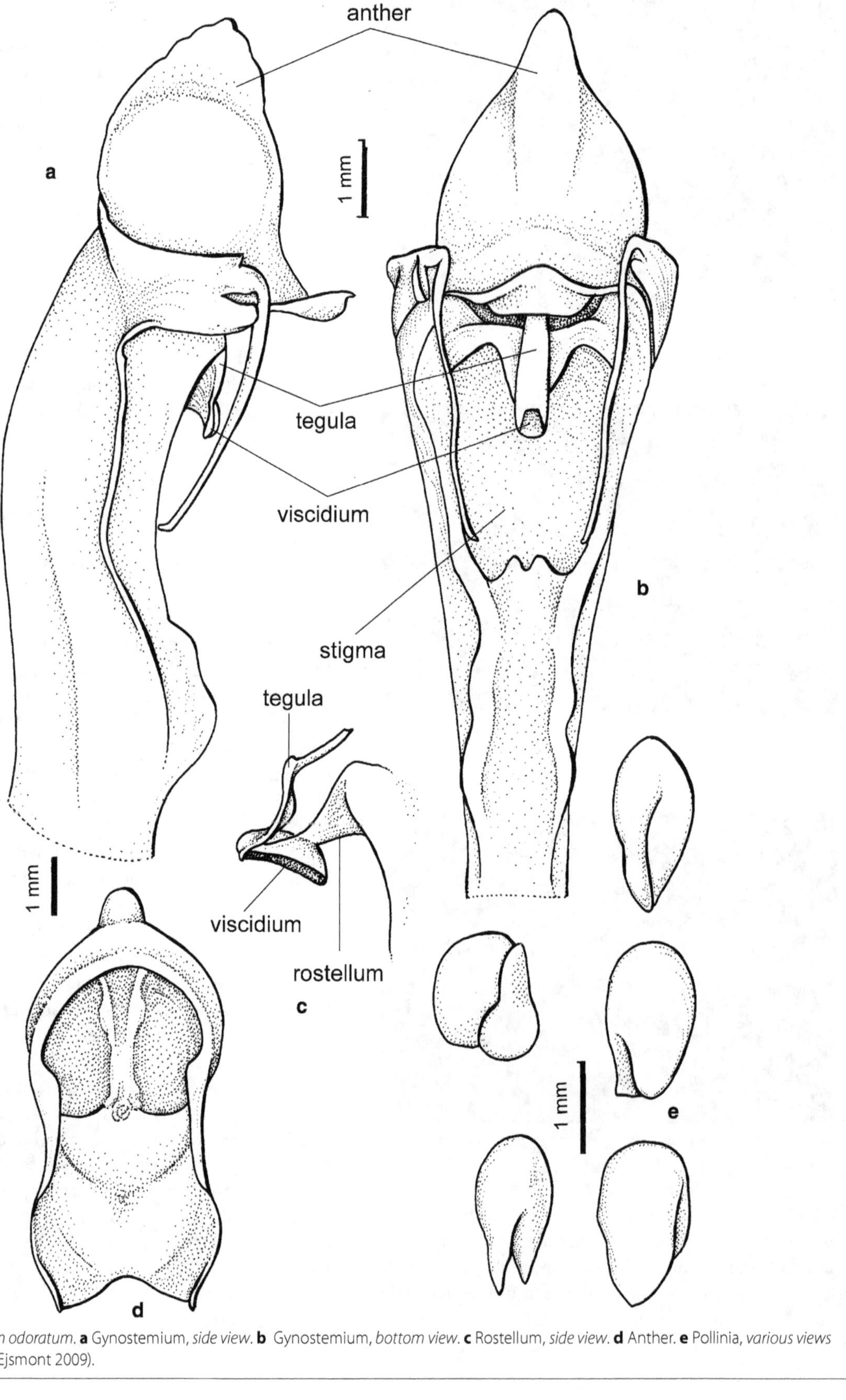

Fig. 8 *Odontoglossum odoratum*. **a** Gynostemium, *side view*. **b** Gynostemium, *bottom view*. **c** Rostellum, *side view*. **d** Anther. **e** Pollinia, *various views* (Szlachetko & Mytnik-Ejsmont 2009).

Fig. 9 *Dasyglossum myanthum*. Photo: Guido Deburghgraeve

sepals, stout gynostemium, usually parallel to the lower part of the lip, and bent in the geniculate manner above base. "*C. aurantiacum*" and "*C. caespitosum*" are easily distinguishable from all other *Cyrtochilum* species by their lip structure, i.e. a narrow, canaliculated claw occupied by an oblong callus, expanded apically in transversely elliptic lamina. The gynostemium is straight and apically reflexed. These species have been classified in the genus *Rusbyella* (Fig. 13). "*C. rhodoneurum*" differs from the aforementioned species in its oblong-ligulate lip with a prominent central callus. It has been assigned to the genus *Neodryas* (Figs. 14, 15). "*C. ornatum*", usually included in the genus *Buesiella*, are distinguished from the above species by their digitate projections near

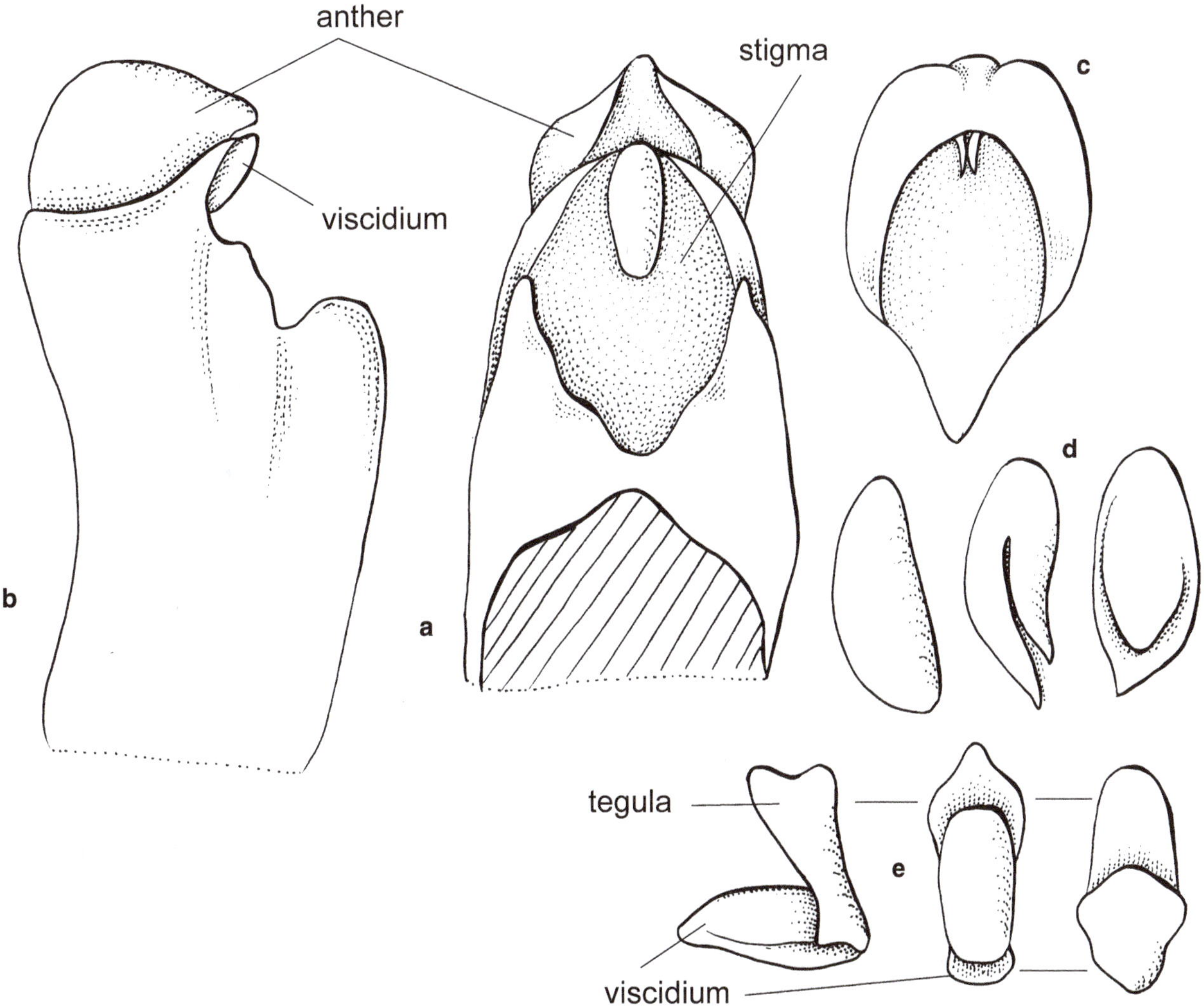

Fig. 10 Dasyglossum myanthum. **a** Gynostemium, *bottom view*. **b** Gynostemium, *side view*. **c** Anther. **d** Pollinia, *various views*. **e** Tegula and viscidium, *various views* (Szlachetko & Mytnik-Ejsmont 2009)

Fig. 11 *Trigonochilum meirax*. Photo: Guido Deburghgraeve

the receptive surface and a hastate lip. "*C. aureum*" was the only species of the genus *Siederella* characterized by a narrowly clawed lip with greatly expanded lamina (Fig. 16), a rather obscure central callus and digitate projections near the stigma (Fig. 17). The gynostemium forms an angle of ca 30° with the lip (Fig. 18). In both analysed trees based on ITS (d) and *matK* (1) aforementioned species are grouped together with high PP value—94 and 97, respectively. In this case bootstrap suppor is low (64 and <50).

The last species in the group is "*C. loxense*" (Figs. 19, 20), which in habit, type of inflorescence and clawed tepals is reminiscent of *Cyrtochilum s.str.* Even though its gynostemium is perpendicular to the lip, the labellum is unique in the genus—it is short-clawed, 3-lobed with the middle lobe being the largest, transversely elliptic and concave. The lip callus is relatively small and confined to the basal part of the lip. In *matK* tree *C. loxense* is attached to *C. caespitosum*-alliance (1), and in the ITS tree this species is connected with *C. alboroseum* and *C. weirii* (f). In the first case value of posterior propabilty is high (97) and in the second—only 62. The *matK* shows that *C. loxense* is only distantly related with *C. villeanorum* and *C. volubile*, with which it is very similar morphologically. The relations between these species are not solved in out ITS analysis.

Discussion

Until recently, it appeared that DNA fragment sequencing would enable the reconstruction of the phylogeny of organisms with a high degree of accuracy. Almost all data obtained from any sources other than genetic material began to be discarded. Numerous articles presenting a completely new approach to the taxonomy of plants and other organisms were published (e.g. Chase et al. 2000; Asmussen et al. 2006; Friesen et al. 2006; Lefébure et al. 2006). In many cases, the new classifications overturned those proposed earlier. Interestingly, one can note a disagreement between molecular based systems and morphological ones. Usually, priority was given to those based on the results of DNA fragment analyses, even though relatively often it was difficult or even impossible to interpret the topology of the tree in terms of its morphology. Yet, no systems based on limited datasets reflect the evolution of the whole organisms; rather, they focus just on the evolutionary modifications of the data in question. Using phylogenetic data to study speciation requires that potential limitations be kept in mind. The approach assumes that we have an accurate and complete understanding of the evolutionary relationships within a clade. Solid phylogenetic methods and markers are needed to reconstruct the phylogeny, which is often difficult, especially among recently diverged taxa.

The utility of nuclear gene sequences in intraspecific phylogenetic analyses appears to be limited by increased coalescence time as compared to chloroplast genes. In addition, the potential for reticulate evolution among nuclear alleles due to recombination is likely to further limit their utility for phylogenetic studies (Bermingham and Moritz 1998). When using organellar genes in combination with nuclear genes, several factors contribute towards an increase in the genetic structure encountered within plant species. For phylogenetic purposes, it would be desirable to consider multiple gene trees based on chloroplast and nuclear genomes, because independently derived gene trees may not be congruent (Schaal et al. 1998). However, Doyle (1997) notes that when the history of the organellar genome is different from that of the nuclear genome (e.g. in lineage sorting or introgression) every comparison sequence in these genomes will give a false phylogenetic pattern for those taxa, and this can confound phylogenetic reconstruction. Plant molecular phylogenetic studies at species levels are generally limited by the availability of sequences with levels of resolution suitable for the construction of well-supported trees (Doyle et al. 1996).

Defining *Cyrtochilum s.l.* Neubig et al. (2012) stated that "vegetatively *Cyrtochilum* are distinguished by dull pseudobulbs that are round or ovoid in cross section with two to four apical leaves and two to six leaf-bearing sheaths and relatively thick roots, in contrast *Oncidium* spp. have glossy, ancipitous (two-edged) pseudobulbs and thin roots". Unfortunately, characters mentioned by Neubig et al. (2012) do not warrant proper identification of *Cyrtochilum*, since the features selected by the authors as disciminative can be found also in other Oncidiinae, for example in *Brassia s.l.*

A problem has emerged as to how to explain the similarity between molecular marker sequences in morphologically different species, such as *Cyrtochilum s.str.* and "*Cyrtochilum ramosissimum*" or "*C. angustatum*", which

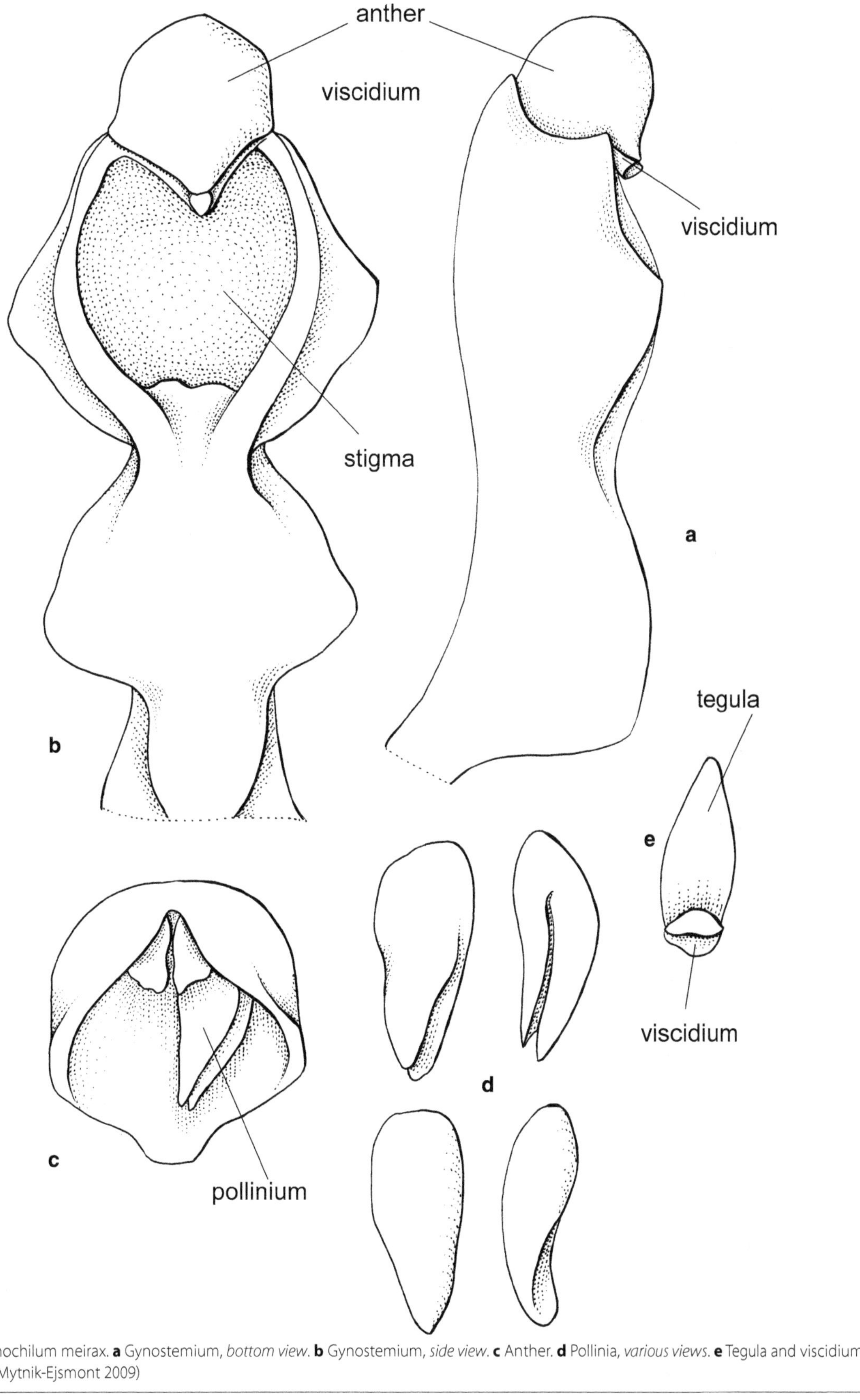

Fig. 12 Trigonochilum meirax. **a** Gynostemium, *bottom view*. **b** Gynostemium, *side view*. **c** Anther. **d** Pollinia, *various views*. **e** Tegula and viscidium (Szlachetko & Mytnik-Ejsmont 2009)

Fig. 13 *Rusbyella aurantiaca*. Photo: Guido Deburghgraeve

Fig. 14 *Neodryas rhodoneura*. Photo: Eric Hunt

Fig. 15 *Neodryas schildhaueri*. Photo: Guido Deburghgraeve

together form a common phylogenetic branch. Neubig et al. (2012) stated that great variability in the flower architecture in Oncidiinae probably reflect a shift in pollinators. On the other hand, morphological similarity between phylogenetically distantly related taxa can be explained by homoplasy. It cannot be excluded, however, that the explanation is much more complicated.

There are at least some phenomena which can usher generate a disturbance to the topology of the phylogenetic tree. Ancestral hybridization, polyploidization and hybrid speciation are significant evolutionary forces in the Orchidaceae. Numerous examples of hybrids are noted in this group of plants. Interspecific hybrids occur in Orchidaceae, but they are typically sporadic and local (e.g. Cozzolino and Aceto 1994; Cozzolino et al. 1998). On the other hand, some putative orchid hybrids are more widespread and stabilized (e.g. Hedrén 1996, 2001; Arft and Ranker 1998; Bullini et al. 2001). Most polyploid species have formed recurrently from genetically-distinct diploid progenitors, representing a potentially great gene pool for the derivative polyploid. Relatively recent hybrid-derived species disclose some degree of morphological intermediacy between putative parents or a similarity to one of the parents. Furthermore, such deviation from intermediacy may be expected in a stablilized hybrid that has been under various selective pressures (Goldman et al. 2004).

A genomic investigation has demonstrated that polyploidization and hybridization are highly effective evolutionary mechanisms for introducing new plant species, promoting their persistence, and ultimately increasing the diversity of plant species (Cook et al. 1998; Ramsey and Schemske 1998; Soltis and Soltis 1999; Otto and Witton 2000; Wendel 2000; Hewitt 2001). While hybridization can be a threat to species integrity, it can also be a source of new variation and a source of new species, especially through polyploidy (Grant 1981).

The stability of the polyploid genome depends on nonrandom genetic changes, including chromosome and genome gains and losses of loci. This genomic reorganization seems to proceed quickly (Rieseberg et al. 1996; Rieseberg 1997; Buerkle and Rieseberg 2008), for example, after 10–60 generations in the case of *Helianthus anomalus* (Ungerer et al. 1998).

Hybrid speciation appears to be facilitated by several additional factors, for example, availability of a suitable ecological niche or development of appropriate fitness (Rieseberg 1997; Mallet 2007). To be evolutionarily successful, even fertile and stable hybrids must be reproductively isolated from the parental species either by chromosomal sterility factors, or evolution of reproductive barriers, or divergence into a new ecological niche (Grant 1981; Rieseberg 1997, 2001; Wu 2001; Paun et al. 2009).

In the case of species of hybrid origin, we expect the conflict of the topology between nuclear and plastid genes. Below we explain the mechanisms leading to

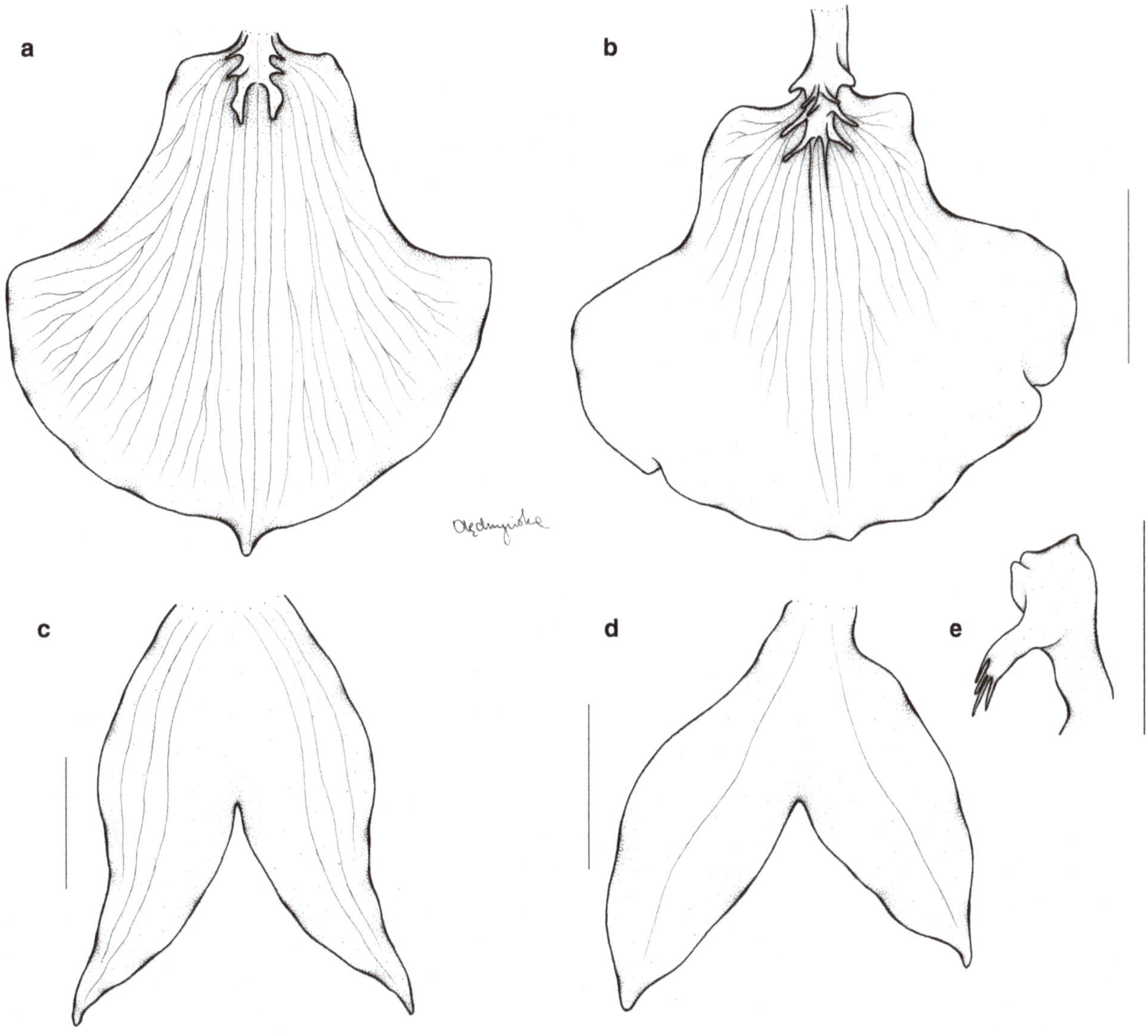

Fig. 16 *Siederella aurea*. **a**, **b** Lip, **c**, **d** lateral sepals, **e** gynostemium. Drawn by N. Olędrzyńska. *Scale bar* 5 mm

Fig. 17 *Siederella aurea*. Photo: Guido Deburghgraeve

these conflicts. Consider a situation of conflict between molecular and morphological data. There are two possibilities of such cases. The first concerns the situation where two related taxa differ morphologically due to the divergent evolution resulting as adaptation to different habitats and/or pollinators. The second case explains this phenomenon by referring to the convergence that results from adaptation to a common pollinators. But there is a third solution to the conflict—reticulate evolution. To make its detection should be compared to a tree topologies based on plastid and nuclear sequences. The phylogenetic analyses conducted by Neubig et al. (2012) used a nuclear marker: ITS1-5.8S-ITS2, which is part of a family of genes coding for ribosomal DNA (rDNA).

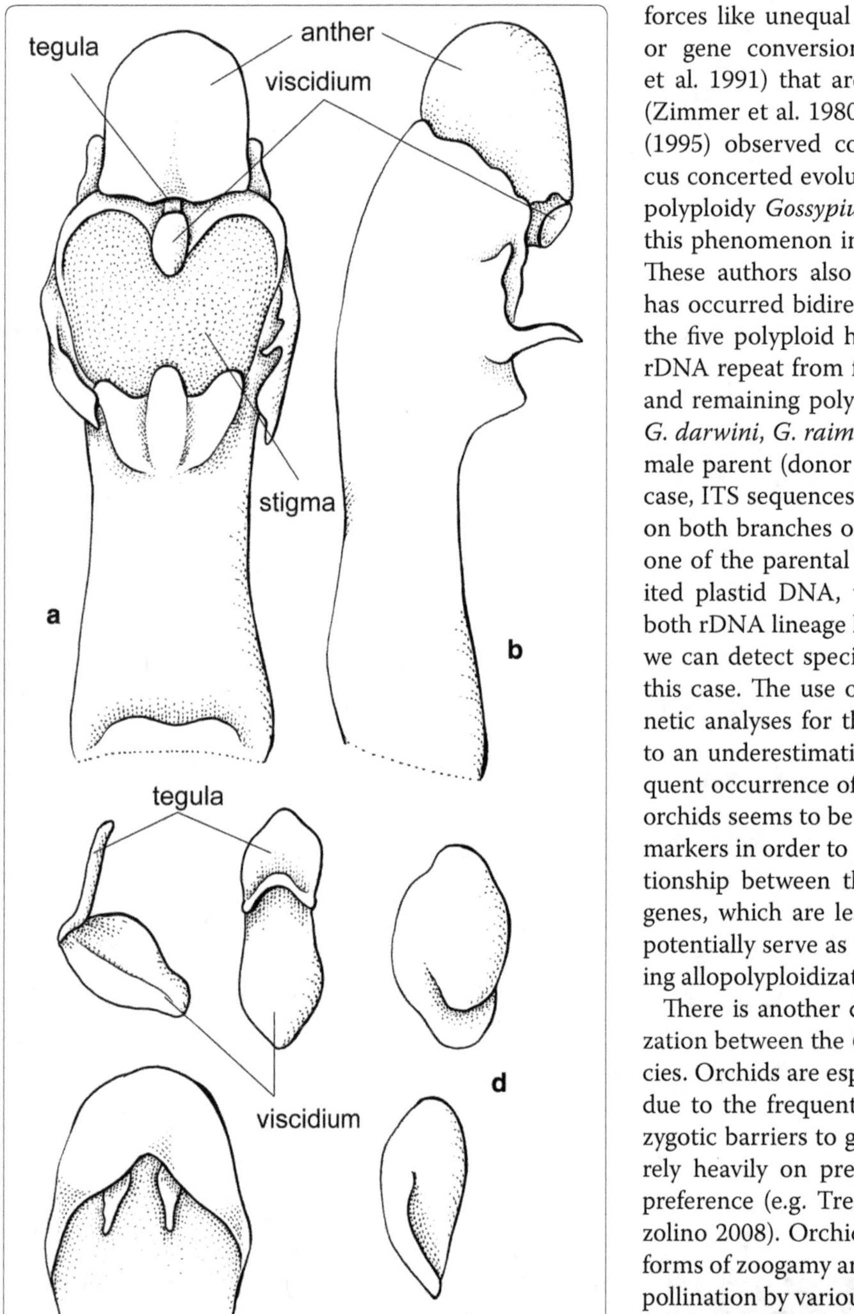

Fig. 18 Siederella aurea. **a** Gynostemium, *bottom view*. **b** Gynostemium, *side view*. **c** Anther. **d** Pollinia, *various views* (Szlachetko & Mytnik-Ejsmont 2009)

In higher plant rDNA is organized into arrays at one or more chromosomal locations (Rogers and Bendich 1987; Hillis and Dixon 1991). Each array contains hundreds to thousands of identical to near-identical repeats. The repeats having become homogenized by evolutionary forces like unequal crossing-over (Seperack et al. 1988) or gene conversion (Enea and Corredor 1991; Hillis et al. 1991) that are referred to as concerted evolution (Zimmer et al. 1980; Arnheim et al. 1980). Wendel et al. (1995) observed complete or nearly complete interlocus concerted evolution of the ITS region in diploid and polyploidy *Gossypium* species. Moreover, they observed this phenomenon in other components of rDNA repeat. These authors also observed, that concerted evolution has occurred bidirectionally in analysed species. One of the five polyploid hybrids (*Gossypium mustelinum*) had rDNA repeat from female parent (receiver of the pollen) and remaining polyploids (*G. tomentosum, G. hirsutum, G. darwini, G. raimondii*) has become homogenized to a male parent (donor of the pollen) rDNA repeat. In such case, ITS sequences from the allopolyploid species occur on both branches of phylogenetic tree, each close to the one of the parental species. In case of maternally inherited plastid DNA, which occurs in most angiosperms, both rDNA lineage has the same plastid DNA lineage. So we can detect species of hybrid origin only with one of this case. The use of nuclear ITS sequences in phylogenetic analyses for the species of hybrid origin may lead to an underestimation of the phylogeny. Due to the frequent occurrence of hybridization in plants, especially in orchids seems to be a reasonable use of the other nuclear markers in order to properly assess the phylogenetic relationship between the analysed taxa. Low-copy nuclear genes, which are less liable to concerted evolution, can potentially serve as a very useful marker for reconstructing allopolyploidization (Small et al. 1998).

There is another question about the potential hybridization between the *Cyrtochilum* and *Odontoglossum* species. Orchids are especially prone to hybridization, partly due to the frequent weakness or even absence of post-zygotic barriers to gene exchange. Instead, many orchids rely heavily on pre-zygotic barriers, notably pollinator preference (e.g. Tremblay et al. 2005; Schiestl and Cozzolino 2008). Orchids, in general, are adapted to various forms of zoogamy and their flowers are accommodated to pollination by various factors, which is one of the reasons for the high degree of variability in the flower, androecium and gynoecium structures. Unfortunately, there are only a few reports concerning this phenomenon in *Cyrtochilum* and *Odontoglossum*. Van der Pijl and Dodson (1966) noted *Bombus hortulanum* and *Centris* bees pollinating *Cyrtochilum macranthum*. Van der Cingel (2001) assumed that the bright-flowered, high-elevation "*C. retusum*" might be hummingbird-pollinated. The flowers of *Odontoglossum s.str.* have also been observed to be pollinated by *Bombus hortulanum* (van der Pijl and Dodson 1966) trying to find reward in the cavity formed by the basal part of the lip and gynostemium. The presence

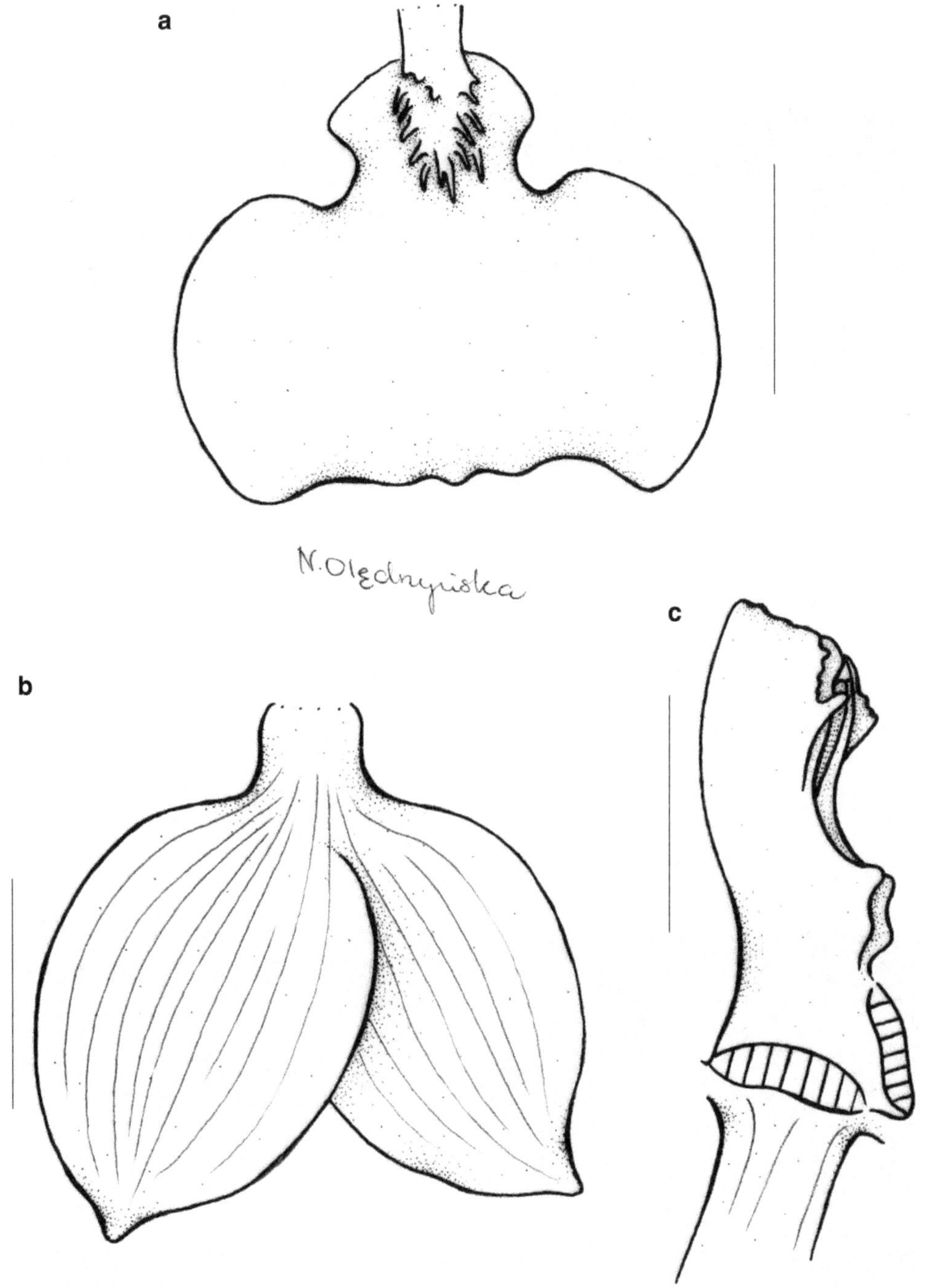

Fig. 19 *Oncidium loxense*. **a** Lip, **b** lateral sepals, **c** lip. Drawn by N. Olędrzyńska. *Scale bar* **a**, **b** = 10 mm, **c** = 5 mm

of a hairy lip in some *Odontoglossum* species might suggest pollination conducted by certain callus-collecting insects. In other words, there is a distinct possibility that the species of both *Cyrtochilum* and *Odontoglossum* coul share the same or similar pollination agents, which enable gene flow between plants representing different genera. We have assumed that *Odontoglossum*-like *Cyrtochilum*, i.a. "*C. angustatum*", "*C. pardinum*", "*C. ramosissimum*", could be of hybrid origin or, at least, demonstrate a stronger influence of genetic materials from *Odontoglossum*. It is noteworthy that in 2004 Shaw registered an artificial hybrid between *Odontoglossum* and *Cyrtochilum* named × *Cyrtoglossum* (Fig. 21). The hybrid is characterized by clawed tepals (with petals

Fig. 20 *Oncidium loxense.* Photo: Eric Hunt

claws short and wide), and slightly sigmoid gynostemium with winged projections on both sides of stigma. Interestingly, the form of the lip, i.e. oblong obtriangular, with exposed callus reminds *Cyrtochilum macranthos*.

The evolution of sister species is not always combined with parallel evolutionary shifts in pollination syndromes (Cozzolino and Widmer 2005). As a consequence, these closely related species, usually growing in sympatry and having overlapping flowering periods and non-specific pollination are exposed to ample opportunities for interspecific hybridization (van der Cingel 2001; Jersáková et al. 2006).

It is worthy to mention that the paper published by Stegemann et al. (2012) concerning horizontal gene transfer (HGT), which sheds new light on an incongruity between molecular datasets of various origin. The authors discovered that chloroplast genomes can be readily transferred between relatively closely related species by natural grafting, thus also providing a possible explanation for why chloroplast sequences frequently provide trees that disagree with canonical phylogeny and/or trees constructed with nuclear markers.

It is also possible that portions of genomes can be transferred between even very distantly related lineages via, for example, viral vectors (Won and Renner 2003; Bergthorsson et al. 2004). This process was noticed as a possible cause of erroneous tree topologies by Tyteca and Klein (2008). On the other hand, Tsai et al. (2010) obtained heterogeneous plastid DNA within individuals of two *Phalaenopsis* species—*P. lowii* and *P. gibbosa*. The short DNA fragments in both the *trn*L intron and the *atp*B-*rbc*L intergenic spacer also had indels based on the sequence alignment. One explanation for the different copies of plastid DNA within an individual is the fact that the short form of the plastid DNA might be maintained in the nuclear or mitochondrial genome through horizontal gene transfer (Ellis 1982; Cheung and Scott 1989; Ayliffe and Timmis 1992; Ayliffe et al. 1998). Horizontal transfer can occur in large fragments coming from the organellar genome into the nuclear genome (Yuan et al. 2002; Huang et al. 2005).

Taking into consideration the aforementioned reports, both HGT as well as hybridization might be responsible

Fig. 21 *Cyrtoglossum* flowers. Photos provided by Ecuagenera

for the topology of phylogenetic trees and the puzzling position of some species. Reconstruction of evolutionary history in genera strongly affected by reticulation and polyploidization is definitely not an easy task. Importantly, *Cyrtochilum* and *Odontoglossum* species inhabit similar geographical regions and plant communities, i.e. Andean humid montane and premontane forests, often growing together, which facilitates gene/genome transfer. The Andes are a well-known centre of biodiversity, where the process of speciation can take place freely as a result of various geographic conditions (cf. Bates et al. 2008; Richter et al. 2009), and ecological factors. Of course, it could be one of many other explanations for the observed topology of the *Cyrtochilum* phylogenetic tree.

Intraspecific gene evolution cannot always be represented by a bifurcating tree model. Linder and Rieseberg (2004) and Vriesendorp and Bakker (2005) have pointed to the fact that the evolutionary history of many plant groups does not follow divergent evolutionary patterns, and hardly can be unravelled in a tree-building procedure. Rather, it is like a network, which displays a number of reticulate evolutionary events. The family Orchidaceae is not an exception, and poliploidy and hybridization are events, which often result in a reticulated pattern of evolution.

The incompatibility of gene trees does not necessarily constitute evidence for reticulate evolution, as gene phylogenies may conflict with other processes: the gene trees may not be historically accurate due to model misspecification or inappropriate methodology, or due to sampling effects (i.e. insufficient sites to compensate for site saturation or short interior edges); alternatively, the gene phylogenies may be historically correct but differ from the species tree due to the population-genetic effect known as lineage sorting (Maddison 1997; Sang and Zhong 2000; Rosenberg 2002).

To most taxonomists, classification depends on characteristics and we have assumed that to cladists the features are no longer important. A classification should be able to recognize distinctive characteristics which have evolved in a group and if we cannot do that it is in consequence impossible to reflect evolution (Brummitt 2006). Brummitt's point of view is shared by many other authors (see 150 scientists who signed a letter by Nordal and Stedje 2005), who state that the traditional classification is the optimal tool for cataloguing biodiversity and requires the recognition of paraphyletic taxa.

The dilemma with which every taxonomist has to struggle is fragmentation or lumping of taxonomic units. Both of them can generate various problems. The effect of the fragmentation of taxa is the creation of numerous smaller, but morphologically well-defined genera, whereas integration produces fewer taxa, well-established genetically, but poorly circumscribed morphologically. The borders between taxonomic units are rather a matter of taxonomic philosophy than scientific objectivity. In our opinion, reasonable fragmentation of taxa can be accepted as long as it leads to separation of well-defined entities. We deal with the latter situation in the case of *Cyrtochilum s.l.* We hypothesise that groups of species forming the particular subclades presented above make well-matched genera which evolve in various directions in response to pollinator pressure, which is manifested, e.g. in various spans between lip and gynostemium and the type of the lip calli.

Interestingly, molecular taxonomists attempt to appoint their taxa using morphological characteristics. Unfortunately, they often create ill-defined units and *Cyrtochilum* sensu *latissimo* is an example. Until more data become available concerning the influence of horizontal gene transfer, as well as hybridization and polyploidy on speciation in the *Cyrtochilum* alliance, thus enabling the solution of the problem of incongruity between molecular and morphological datasets, we suggest maintaining a narrower generic concept.

As stated at the beginning of this chapter, Neubig et al. (2012) definition of *Cyrtochilum* does not warrant the proper identification of the genus representatives and can lead to the confusion. We propose to recognize at the generic level smaller, monophyletic and morphologically well-defined taxa, what in our opinion assure stability in taxonomy of this interesting oncidoid group.

Taxonomic treatment

Key to the genera of *Cyrtochilum*-complex

1. Gynostemium and the lower part of the lip form more or less a right angle ... 2
1*. Gynostemium parallel with the lower part of the lip ... 4
2. Lip unguiculate, lamina more or less transversely elliptic, callus rather obscure, lateral sepals basally connate ... *Siederella*
2*. Lip sessile to subsessile, lamina cordate, sagittate to hastate, callus prominent, lateral sepals free to the base or almost to the base ... 3
3. Lip much smaller than tepals, callus very large, complexed, composed of horns and various digitate projections ... *Cyrtochilum*
3*. Lip as large as tepals, triangular-cordate in outline, callus large, composed of large mass of tissue divided into 4 or more lobes ... *Trigonochilum*
4. Lip callus simple, consisting of 2 fleshy, parallel, adjoining torus, diverging in front ... *Dasyglossum*
4*. Lip callus not as above ... 5

5. Gynostemium with digitate projections on each sides of the stigma, lip sessile, callus prominent ... *Neodryas*
5*. Gynostemium with or without very obscure projections, lip long clawed, callus obscure ... *Rusbyella*
Cyrtochilum Kunth
Nov. Gen. Sp. 1: 279. 1816; Generitype: *Cyrtochilum undulatum* Kunth.

Epiphytic or terrestrial plants. Pseudobulbs ovoid, usually round in cross-section, distributed on elongate creeping rhizome, sometimes caespitose, but clusters of pseudobulbs usually distantly remote along the rhizome, with 2–6 foliaceous bracts. Leaves 2–4 per pseudobulb, conduplicate, articulate. Inflorescence flexuose, usually very long, branched, branches with few to many flowers. Flowers resupinate, showy, white, yellow, pink, brown or purple. Floral bracts large, leafy. Tepals free, prominently unguiculate, similar in size and shape or petals much wider. Lip triangular to ovate rarely hastate to panduriform, apex reflexed, callus complex, often digitate, tuberculate or horned. Gynostemium gently sigmoid to erect, elongated, slender, clavate, usually forms right angles with lip. Anther subventral, incumbent, operculate, ellipsoid-ovoid. Pollinia 2, oblong ellipsoid, hard, unequally and deeply cleft, empty inside. Stigma large, elliptic, deeply concave. Rostellum short. Viscidium single, relatively large, elliptic, very thick, concave in the centre of the outer surface. Tegula single, very small, transversely elliptic-obtriangular, obscurely bilobulate at the apex, thin, lamellate. Rostellum remnant with oblique shallowly concave plate at the apex, canaliculate on the upper surface (Fig. 1). Capsule triangular.

Taxonomic notes—Species of the genus *Cyrtochilum*, as treated here, are easily distinguishable from all other genera of the clade by having elongated, flexuose inflorescence, large, showy flowers with clawed, large sepals and petals, a relatively small lip widest at the base, attenuated towards apex, with a composed callus occupying a large portion of the lip lamina, and gynostemium gently sigmoid to erect, slender, usually perpendicular to the lip base, or even deflexed. We assume that a group of *C. angustatum*-group can be of hybrid origin.

Dasyglossum Königer & Schildh.
Arcula 1: 5 1994; Generitype: *Dasyglossum myanthum* (Lindl.) Königer & Schildh. [≡*Odontoglossum myanthum* Lindl.].

Epiphytic plants. Pseudobulbs approximate, ovoid or elliptic-oblong, compressed, enveloped at the base by papery or foliaceous sheaths. 1–3 leaves, coriaceous or fleshy. Inflorescence usually elongated, erect or arching, few to many flowers, racemose or paniculate. Flowers small. Floral bracts rudimentary. Sepals free, occasionally basally connate, subequal, spreading. Petals usually subequal to the dorsal sepal. Lip entire or 3-lobed, at the base united with the base of the column, lower half of the lip parallel to the column; callus simple, often consisting of 2 fleshy, parallel, adjoining tori, diverging in front, mostly enclosed by the thickened flanks of the gynostemium. Gynostemium rather short, in the upper half gently upcurved or straight, rather robust. Column part ca. 2.5 times longer than anther, fused with the lip in the basal half, firmly winged, wings entire on margins, surrounding lip callus. Anther subdorsal to apical, operculate, ellipsoid, obscurely 2-chambered. Pollinia 2, oblong ellipsoid-ovoid, hard, unequally and deeply cleft. Stigma oblong to transversely elliptic, slightly concave. Rostellum suberect to pendent, rather short, ovate, rounded at the apex. Viscidium single, oblong ellipsoid, very thick, fleshy. Tegula single, as long as viscidium, oblong, thin, lamellate, flat. Rostellum remnant bilobulate at the apex (Fig. 10).

Taxonomic notes—All representatives of the genus are easily separable from other taxa of the *Cyrtochilum* alliance by the presence of a short, massive gynostemium, winged on the ventral surface, and parallel with the lower part of the lip.

The following new combinations are validated below:

Dasyglossum colobium (Dalström) Szlach., Kolan. & Chiron, *comb. nov.*
Basionym: *Cyrtochilum colobium* Dalström *in* Dodson & Luer, Fl. Ecuador, Orchidaceae 87: 51. 2010. Type: Ecuador. Sucumbíos, cloud forest W of La Bonita. 29 Mar 1992. *Dalström & Höijer 1687* (Holotype: SEL).
Dasyglossum ferrugineum (Dalström & D. Trujillo) Szlach., Kolan. Chiron, *comb. nov.*
Basionym: *Cyrtochilum ferrugineum* Dalström & D. Trujillo *in* Dodson & Luer, Fl. Ecuador, Orchidaceae 87: 74. 2010. Type: Ecuador. *Sine loc. hort. Beckendorf sub Dalström 2377* (Holotype: SEL).
Dasyglossum fidicularium (Dalström) Szlach., Kolan. & Chiron, *comb. nov.*
Basionym: *Odontoglossum fidicularium* Dalström, Lindleyana 14(3): 168. 1999. Type: Ecuador. Zamora-Chinchipe. Along road between Yanga and Valladolid. 21 Feb 1982. *Luer* et al. *7134* (Holotype: SEL, Isotype: K).
Dasyglossum hoeijeri (Dalström) Szlach., Kolan. & Chiron, *comb. nov.*
Basionym: *Odontoglossum hoeijeri* Dalström, Lindleyana 14(3): 171. 1999. Type: Ecuador. Loja. North of Loja, along road to Saraguro. 8 Feb 1993. *Dalström* et al. *1871* (Holotype: SEL, Isotype: K).

Dasyglossum sphinx (Dalström & G.Calat.) Szlach., Kolan. & Chiron, *comb. nov.*
Basionym *Cyrtochilum sphinx* Dalström & G.Calat., *in* Dodson & Luer, Fl. Ecuador, Orchidaceae 87: 170. 2010. Type: Peru. Cajamarca. Prov. San Ignacio, San José de Lourdes. *Calatayud 746* (Holotype: CUZ).
Dasyglossum verrucosum (Dalström) Szlach., Kolan. & Chiron, *comb. nov.*
Basionym: *Cyrtochilum verrucosum* Dalström *in* Fl. Ecuador 87: 183. 2010. Type: Ecuador. Morona-Santiago. *Ecuagenera sub Whitten 3219* (Holotype: QCA).
Neodryas Rchb.f.
Bot. Zeit. 10: 835. 1852; Generitype: *Neodryas rhodoneura* Rchb.f.
Buesiella C.Schweinf., Bot. Mus. Leafl., Harvard Univ. 15: 153. 1952; Generitype: *Buesiella pusilla* C.Schweinf.

Plants caespitose. Pseudobulbs cylindrical-ovoid, enclothed basally with 1–3 leafy sheaths, unifoliate. Inflorescence more or less branching, branches with few to many flowers. Flowers medium-sized, campanulate, somewhat laterally compressed. Sepals narrower than petals, both sessile. Lateral sepals variously connate. Lip sessile, entire, oblong-ovate to ligulate-subcordate, geniculate near the middle, callus large, variously lobed. Gynostemium almost erect, stout, parallel to the lip. Column part ca. twice longer than anther, widened in the middle, wings spread, oblong-elliptic, with digitate projections on each side of the stigma. Pollinia 2, obliquely obovoid-ellipsoid, slightly dorsiventrally compressed, hard, unequally and shallowly cleft at the apex. Apical clinandrium forms a narrow collar surrounding the anther base. Stigma rather large, transversely elliptic, deeply concave. Rostellum suberect, short, ligulate. Viscidium single, small, elliptic-ovate, thin. Tegula single, oblong-elliptic, thin, lamellate. Rostellum remnant bilobulate at the middle, slightly concave between acute lobules, canaliculate on the dorsal surface.

Taxonomic notes—We did not identify any morphologically crucial differences between *Neodryas* and *Buesiella*. Therefore, we have combined them together which is also supported by the results of the analyses of the DNA markers. The genus is characterised by the presence of a gynostemium parallel with a sessile lip furnished with a large callus.

Neodryas fredericae (Dalström) Szlach., Kolan. & Chiron, *comb. nov.*
Basionym *Cyrtochilum fredericae* Dalström, *in* Dodson & Luer, Fl. Ecuador, Orchidaceae 87: 86. 2010. Type: Ecuador. Loja. Cajanuma. *Ecuagenera sub Dalström 2478* (Holotype: SEL).
Rusbyella Rolfe *ex* Rusby
Mem. Torrey Bot. Club 6: 122. 1896; Generitype: *Rusbyella caespitosa* Rolfe *ex* Rusby.

Plants caespitose. Pseudobulbs oblong to cylindrical-ovoid, slightly laterally compressed, enclothed basally with 1–2 leafy bracts, unifoliate. Inflorescence erect, simple or branching. Flowers medium-sized. Sepals and petals subsimilar, narrow. Lateral sepals variously connate. Lip clawed, claw narrow, channelled, furnished at the apex with quadrilobed callus, lamina transversely elliptic to reniform, abruptly reflexed. Gynostemium elongated, slightly swollen and bent back at the apex, parallel with the lower part of the lip. Column part ca. 1.5 times longer than anther, glabrous, with two wing-like projections on both sides of the stigma. Anther subdorsal, incumbent, operculate, dorsiventrally compressed, obovoid-ellipsoid. Connective narrow, thin, forming prominent, apical, roof-like projection in front. Pollinia 2, obliquely obovoid, shallowly cleft at the apex, hard. Apical clinandrium forms a narrow collar-like structure surrounding the anther base. Stigma rather small, oblong elliptic, concave. Rostellum suberect, ligulate, rounded at the apex. Viscidium very small, single, elliptic, thin. Tegula single, linear, thin, lamellate. Rostellum remnant with oblique, apical plate on the inner surface surrounded by obscure fovea.

Taxonomic notes—*Rusbyella* is easily separable from other genera of the *Cyrtochilum* alliance by having a lip with a long, channelled claw, and parallel gynostemium with an apical part and anther bent back.

Siederella Szlach., Mytnik, Górniak & Romowicz
Biodiv. Res. Cons. 1–2: 5. 2006; Generitype: *Siederella aurea* (Lindl.) Szlach., Mytnik, Górniak & Romowicz [≡*Oncidium aureum* Lindl.]

Plants caespitose. Pseudobulbs ovoid, somewhat laterally compressed, enclothed basally in some leafy sheaths, usually unifoliate. Inflorescence erect, loosely several-flowered. Flowers rather large, showy; sepals and petals subsimilar, sessile. Lateral sepals connate almost to the apex. Lip shortly unguiculate, callus variously developed—almost missing or two parallel ridges, lamina obovate to pandurate. Gynostemium slightly arched, elongated, slender, forming a 30° angle with the lip. Column part ca. 3 times longer than anther, basally joined with the lip, with two projections near the stigma, more or less finger-like. Anther subventral, incumbent, operculate, ellipsoid-obovoid, obscurely 2-chambered. Connective narrow, indistinctly apically elongate. Pollinia 2, obliquely ellipsoid, hard, unequally and deeply cleft, empty inside. Caudiculae sticky, amorphous. Apical clinandrium forms a narrow collar-like structure around the anther base. Stigma large, elliptic, deeply concave.

Rostellum shortly conical-digitate in the middle, ligulate, blunt. Viscidium single, rather large, oblong elliptic, very thick. Tegula single, small, linear, thin, and lamellate (Fig. 18).

Taxonomic notes—This genus differs from *Rusbyella* in terms of gynostemium and lip morphology. The lip claw possesses two, parallel elevated calli. The gynostemium forms an acute angle with the lip, it is terete below the stigma, and with small, finger-like projections near the stigma.

Incertae sedis

Oncidium loxense Lindl. is similar to *Cyrtochilum* Kunth with which it shares a similar habit, creeping rhizome, long, flexuose, branching inflorescence, large flowers, clawed sepals and petals and gynostemium forming a right angle with the lip. Differs from most species of the genus by having connate lateral sepals, a large, prominently unguiculate lip, with an obscure callus, transversely elliptic or obreniform, concave lip lamina. The lip is somewhat similar to *C. volubile* (Fig. 22) and *C. villenaorum*, but despite these species it is prominently clawed and callus is obscure. Temporarily, we propose to maintain this species in *Siederella*.

Siederella loxense (Lindl.) Szlach., Kolan. & Chiron, *comb. nov.*
Basionym: *Oncidium loxense* Lindl., Paxt. Fl. Gard. 2: 128. 1851. Type (Dalström 2010): Ecuador. *Hartweg s.n.* (Lectotype: K-L, Isolectotype: W).
Trigonochilum Königer & Schildh.
Arcula 1: 13. 1994; Generitype: *Trigonochilum flexuosum* (H.B.K.) Königer & Schildh. [≡*Cyrtochilum flexuosum* H.B.K.].

Epiphytic plants. Pseudobulbs approximate, ovoid or elliptic-oblong, compressed, enveloped at the base by papery or foliaceous sheaths. 1–3 leaves, coriaceous or fleshy. Inflorescence from the base of pseudobulb, usually very long, twining, frequently with short, zigzag branches. Flowers relatively small. Floral bracts small. Sepals free, subequal, spreading. Petals usually subequal to the dorsal sepal. Lip triangular-cordate in outline, diverging from the gynostemium at 70°–90° with a simple, torous, sometimes verrucose or gibbous callus. Gynostemium elongate, slightly sigmoid to erect, rather slender. Column part 2–4 times longer than anther, with deltoid tabula infrastigmatica, obscurely winged near the stigma, glabrous, wings triangular, entire on margins. Anther subventral, incumbent, operculate, ellipsoid-ovoid, obscurely 2-chambered. Pollinia 2, oblong obovoid, slightly dorsiventrally flattened, hard, unequally and deeply cleft. Stigma elliptic, deeply concave. Rostellum short, conical-digitate in the middle, blunt. Viscidium single, very small, elliptic, rather thick. Tegula single, oblong ovate, thin, lamellate, with roof-like projection above viscidium. Rostellum remnant bilobulate at the middle (Fig. 12).

Taxonomic notes—Considering the flower structure, the genus appears to be similar to *Cyrtochilum*. In both taxa, the gynostemium is more or less perpendicular to the lip base, but unlike *Cyrtochilum*, in *Trigonochilum* the petals and sepals are sessile and the lip is wider than the tepals, lip lamina is triangular-cordate in outline in major part occupied by a massive callus.

Our course of study on the North Andean orchids revealed the existence of an undescribed *Trigonochilum* species from the Colombian Department of Putumayo, as well as the necessity of including into the genus Peruvian, Ecuadorian and Colombian species of *Cyrtochilum*.

Trigonochilum corniculatum (Dalström) Szlach., Kolan. & Chiron, *comb. nov.*
Basionym: *Cyrtochilum corniculatum* Dalström *in* Dalström & Perez, Lankesteriana 12(3): 147. 2012. Type: Colombia. Antioquia. Yarumal, Km 87 along road Medellín-Yarumal, Llanos de Cuiba [Cuiva]. 12 Sep 1984. *Dodson* et al. *15264* (Holotype: RPSC, Isotype: MO).
Trigonochilum midas (Dalström) Szlach., Kolan. & Chiron, *comb. nov.*
Basionym: *Cyrtochilum midas* Dalström, *in* Dodson & Luer, Fl. Ecuador, Orchidaceae 87: 136. 2010. Type: Ecuador. Morona-Santiago. Macas-Guamote. Jan 1989. *Hirtz* et al. *4061* (Holotype: RPSC).
Note: Orchid collection of RPSC was transferred to MO.
Trigonochilum russellianum (Dalström & Ruíz-Pérez) Szlach., Kolan. & Chiron, *comb. nov.*
Basionym: *Cyrtochilum russellianum* Dalström & Ruíz-Pérez, Lankesteriana 12(3): 149. 2012. Type: Peru. Ayacucho. La Mar, Aina, Calicanto. *Peruflora sub Dalström 3415* (Holotype: USM).
Trigonochilum sharoniae (Dalström) Szlach., Kolan. & Chiron, *comb. nov.*
Basionym: *Cyrtochilum sharoniae* Dalström, Selbyana 28(2): 106. 2007. Type: Peru. *Sine loc. Peruflora sub Daltröm 2638* (Holotype: SEL).
Trigonochilum tanii (Dalström) Szlach., Kolan. & Chiron, *comb. nov.*
Basionym: *Cyrtochilum tanii* Dalström *in* Dodson & Luer, Fl. Ecuador, Orchidaceae 87: 172. 2010. Type: Ecuador. Manabí, 11 km east of Manta, summit of Cerro Monte Cristi. *Cult. SEL sub Tan 1361* (Holotype: SEL).
Trigonochilum tricornis (Dalström & Ruíz-Pérez) Szlach., Kolan. & Chiron, *comb. nov.*

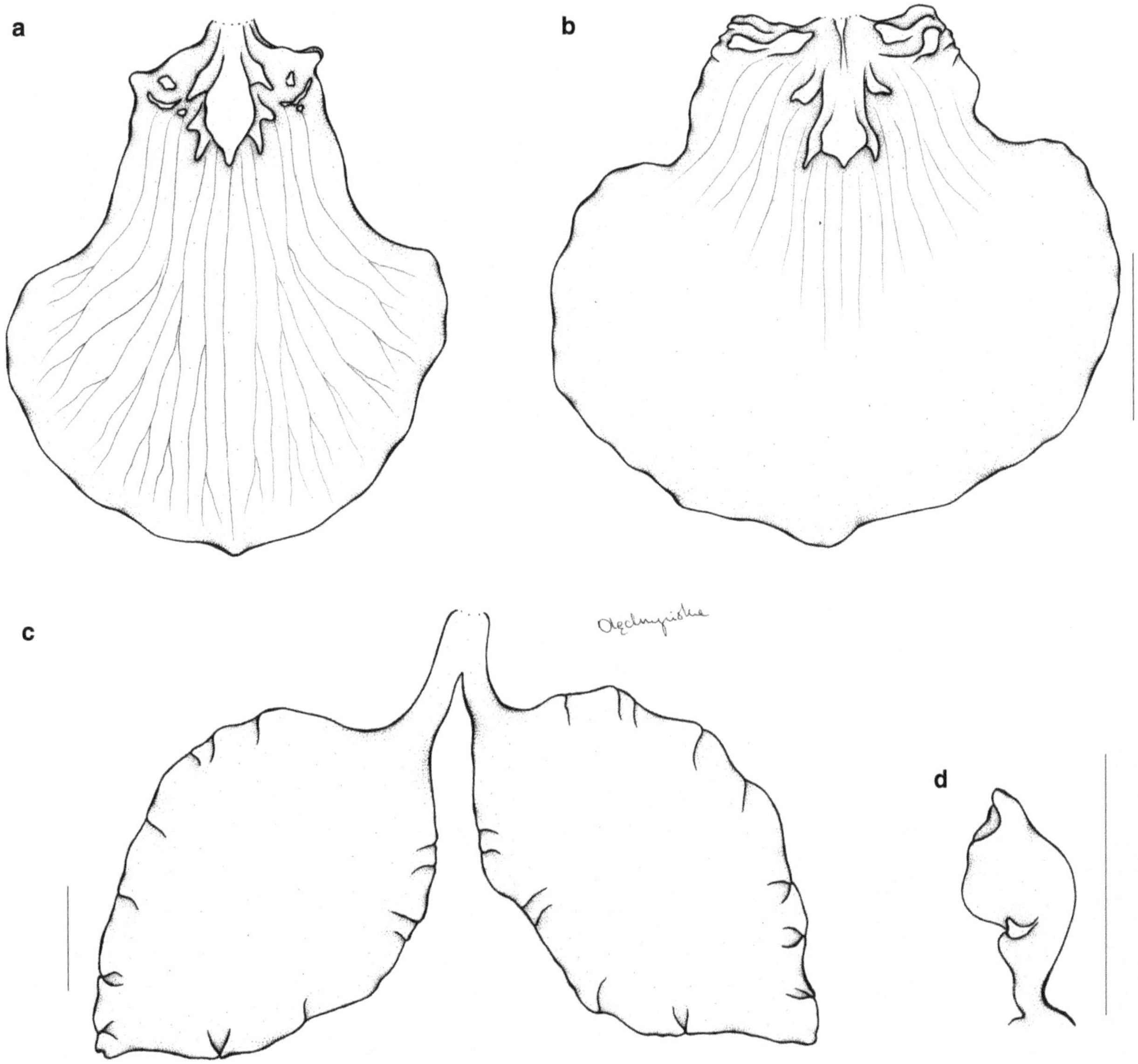

Fig. 22 *Cyrtochilum volubile*. **a**, **b** Lip, **c** lateral sepals, **d** gynostemium. Drawn by N. Olędrzyńska. *Scale bar* 5 mm

Basionym: *Cyrtochilum tricornis* Dalström & Ruíz-Pérez, Lankesteriana 12(3): 151. 2012. Type: Peru. Cusco, Quillabamba, Rio Chullapi Reserva. *Valenzuela* et al. *sub* *Dalström* et al. *2699* (Holotype: CUZ).

Trigonochilum koenigerii Szlach., Kolan. & Chiron, *sp. nov.* (Figs. 23, 24).

The species is similar to *T. cimiciferum* and *T. midas*, but easily separable from both by the flower color, which is brown or brownish, and complicated lip callus. Additionally, it differs from *T. cimiciferum* by oblong elliptic-obovate, acute petals and large subquadrate stigma, from *T. midas* by acute lip.

Type: COLOMBIA. Dept. Putumayo. Valle de Sibundoy. Vereda San Pablo Bajo en Km 3 lado izquierdo de la carretera. Cult. *R. Medina 325* (Holotype: MEDEL!).

Pseudobulbs about 7 × 4 cm, oblong-cylindrical, apically attenuate, almost rounded in cross section, unifoliate, enclothed basally by 4–6, large, leafy bracts. Leaves up to 38 × 3.5 cm, linear-oblanceolate to linear-oblong, acute, attenuate towards the base. Inflorescence basal, very long, with short, fractiflex, laxly few-flowered branches. Floral bracts 6–8 mm long, ovate-elliptic, obtuse. Pedicellate ovary about 28 mm long, almost straight. Flowers brown or brownish with

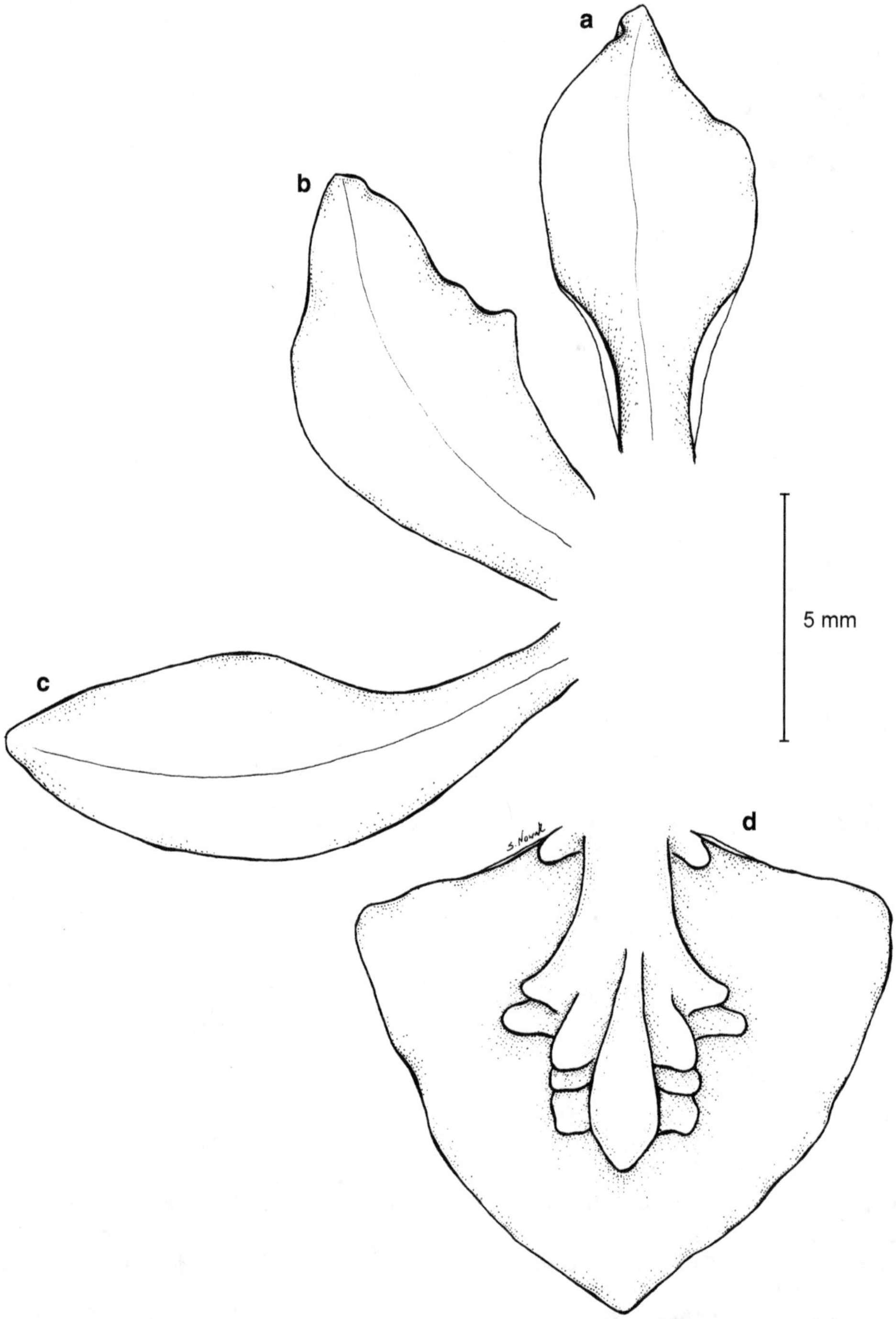

Fig. 23 *Trigonochilum koenigerii*—dissected perianth. **a** Dorsal sepal, **b** petal, **c** lateral sepal, **d** lip. Drawn by S. Nowak from the holotype

yellow apices of the petals and yellow lip callus. Sepals free, subequal, spreading. Dorsal sepal about 10 × 5 mm, concave, unguiculate, elliptic-oblanceolate, subacute. Lateral sepals 13 × 6 mm, obliquely spathulate, elliptic-oblanceolate, subobtuse, distinctly clawed. Petals 11 × 0.8 mm, oblong elliptic-obovate, acute, sessile. Lip sessile, 9.7 × 11 mm, entire, convex, widely triangular to ovate-triangular, acute; disc with a fleshy, complicated callus occupying the basal 2/3 and two subglobose thickenings at the base. Gynostemium elongated, slightly sigmoid, slender, forming a right angle with the lip.

Etymology: Dedicated to Willibald Königer, German orchidologist who contributed to our knowledge and understanding of Oncidiinae.

Fig. 24 *Trigonochilum koenigerii*. **a** Flower (*side view*), **b** gynostemium, **c** flower (*front view*), **d** fragment of the inflorescence. Photos: R. Medina

Table 1 Comparative morphology of *Trigonochilum cimiciferum*, *T. koenigerii*, and *T. midas*

Character	*T. cimiciferum*	*T. koenigerii*	*T. midas*
Pseudobulbs	Rounded-ovoid, 5–8 × 1.5–2.5 mm	Oblong-cylindrical, 7 × 4 cm	Ovoid, 4–10 × 2–3 cm
Leaves	Narrowly ovate, 20–40 × 1.5–3 cm	Linear-oblanceolate to linear-oblong, 38 × 3.5 cm wide	Elongate-obovate, acuminate, up to 65 × 2 cm
Pedicellate ovary	10–30 mm	About 28 mm	Up to 20 mm
Flower	Brownish-yellow with brown spots, with yellow lip callus	Brown with yellow lip callus	Almost completely white with whitish to yellow lip callus
Sepals	Spathulate to rotundate, rounded-acute, acute or acuminate	Unguiculate to spathulate, elliptic-oblanceolate, subacute or subobtuse	Spathulate, ovate, acuminate
Petals	Subsessile, obovate-elliptic to narrowly obovate or oblanceolate, acute	Subsessile, oblong elliptic-obovate, acute	Subsessile, ovate, acuminate
Lip	Ovate, ovate-triangular to elliptic-obovate; disc with a fleshy callus in the basal half	Widely triangular to ovate-triangular; disc with a fleshy, complicated callus in the basal 2/3 and two subglobose thickenings at the base	Cordate to truncate; callus a central, longitudinal, fleshy structure, extending from the base to 2/3 of the disc length, terminating in digitate knobs or denticles

Distribution and habitat: So far, the new species is known exclusively from the Colombian department of Putumayo. It was found growing in disturbed humid montane forest at about 2400 m a.s.l. In cultivation, it flowered in November.

Taxonomic notes: The new species is similar to *T. cimiciferum* (Rchb.f. *ex* Lindl.) Königer from which it differs on the basis of the brown flowers with a yellow lip callus (vs brownish-yellow flowers with brown spots), a complicated lip callus and oblong elliptic-obovate, acute petals. In petal shape, the new species resembles Ecuadorian *T. midas* from which it is easily separable not only in terms of the flower color (almost completely white with a yellow lip callus in *T. midas*) and acute lip (vs lip apiculate), but also the significantly longer ovaries (Table 1).

Incertae sedis

The position of "*C. edwardii*" and "*C.* cf. *porrigens*" in the phylogenetic tree is in conflict with results of morphological study, what we discuss above. Partial influence of *Dasyglossum* genetic material can not be excluded in both cases. Because morphological congruence of aforementioned species to other *Trigonochilum* representatives we propose to maintain them temporarily in this genus.

Authors' contributions

DLS: designing research, morphological data collection, data analysis, data interpretation, writing manuscript. MK: morphological data collection, data analysis, data interpretation, writing manuscript. AN: molecular analysis, writing manuscript, data interpretation. MG: molecular analysis, writing manuscript. MD: molecular analysis. PR: phenetical analysis, writing manuscript. GC: morphological data collection, data interpretation. All authors read and approved the final manuscript.

Author details

[1] Department of Plant Taxonomy and Nature Conservation, The University of Gdańsk, ul. Wita Stwosza 59, 80-308 Gdańsk, Poland. [2] Department of Biodiversity Research, Global Change Research Institute AS CR, Bělidla 4a., 603 00 Brno, Czech Republic. [3] Department of Molecular Evolution, The University of Gdańsk, Wita Stwosza 59, 80-308 Gdańsk, Poland. [4] Herbiers, Université de Lyon I, 69622 Villeurbanne Cedex, France.

Acknowledgements

The lead author wishes to express his gratitude to the owner and staff of Ecuagenera (Ecuador), Peruflora (Peru) and Orquivalle (Colombia) for their hospitality during fieldwork and access to their rich orchid collections. We are grateful to Dr. Henrik Pedersen and Dr. Finn N. Rasmussen (Copenhagen University) for their comments and suggestions. We are also grateful to Ecuagenera, G. Deburghgraeve, E. Hunt and R. Medina T. for providing photos of the *Cyrtochilum*-complex species. We wish to thank Sławomir Nowak for preparing the illustration of new *Trigonochilum* species and to Mario Camilo Barrera G. and Ramiro Medina T. for their cooperation during the studies and providing information about habitat of the new species.

Competing interests

The authors declare that they have no competing interests.

Funding

The project has been supported by the Polish Ministry of Science and Higher Education (research Grant No. 8124/B/PO1/2011/40), European Commission's Research Infrastructure Action via the SYNTHESYS Project (GB-TAF-2445) and the Foundation for Polish Science (Fundacja na rzecz Nauki Polskiej, FNP).

References

Arft AM, Ranker TA (1998) Allopolyploid origin and population genetics of the rare orchid *Spiranthes diluvialis*. Am J Bot 85:110–122

Arnheim N, Krystal M, Schmickel R, Wilson G, Ryder O, Zimmer E (1980) Molecular evidence for genetic exchanges among ribosomal genes on nonhomologous chromosomes in man and apes. Proc Natl Acad Sci USA 77:7323–7327

Asmussen CB, Dransfield J, Deickmann V, Barfod AS, Pintaud J-C, Baker WJ (2006) A new subfamily classification of the palm family (Arecaceae): evidence from Plastid DNA. Bot J Linn Soc 151:15–38

Ayliffe MA, Timmis JN (1992) Tobaco nuclear DNA contains long tracts of homology to chloroplast DNA. Theor Appl Genet 85:229–238

Ayliffe MA, Scott NS, Timmis JN (1998) Analysis of plastid DNA-like sequences within the nuclear genome of higher plants. Mol Biol Evol 15:738–745

Bates JM, Cadena CD, Tello JG, Brumfield RT (2008) Diversification in the neotropics: phylogenetic patterns and historical processes. Ornitol Neotrop 19:427–432

Bergthorsson U, Richardson AO, Young GJ, Goertzen LR, Palmer JD (2004) Massive horizontal transfer of mitochondrial genes from diverse plant donors to the basal angiosperm *Amborella*. Proc Natl Acad Sci USA 101:17747–17752

Bermingham E, Moritz C (1998) Comparative phylogeography: concepts and applications. Mol Ecol 7:367–369

Brummitt RK (2006) Am I a bony fish? Taxon 55:268–269

Buerkle CA, Rieseberg LH (2008) The rate of genome stabilization in homoploid hybrid species. Evolution 62:266–275

Bullini L, Cianchi R, Arduino P, de Bonis L, Mosco MC, Verrardi A, Porretta D, Corrias B, Rossi W (2001) Molecular evidence for allopolyploid speciation and a single origin of the western Mediterranean orchid *Dactylorhiza insularis* (Orchidaceae). Biol J Linn Soc 72:193–201

Chase MW, Soltis DE, Soltis PS, Rudall PJ, Fay MF, Hahn WJ, Sullivan S, Joseph J, Givnish T, Sytsma KJ, Pires JC (2000) Higher-level systematics of the monocotyledons: an assessment of current knowledge and a new classification. In: Wilson KL, Morrison DA (eds) Systematics and evolution of monocots. CSIRO, Collingwood

Cheung WY, Scott NS (1989) A contiguous sequence in spinach nuclear DNA is homologous to three separated sequences in chloroplast DNA. Theor Appl Genet 77:625–633

Cook LM, Soltis PS, Brunsfeld SJ, Soltis DS (1998) Multiple independent formations of *Tragopogon* tetraploids (Asteraceae): evidence from RAPD markers. Mol Ecol 7:1293–1302

Cozzolino S, Aceto S (1994) Morphological and molecular characterization of *Orchiaceras bergonii* (Nanteuil) E. G. Cam. G Bot Ital 128:861–867

Cozzolino S, Widmer A (2005) The evolutionary basis of the reproductive isolation in Mediterranean orchids. Taxon 54:977–985

Cozzolino S, Aceto S, Caputo P, Menale B (1998) Characterization of *Orchis × dietrichiana* Bogenh., a natural orchid hybrid. Plant Biosyst 132:71–76

Cuénoud P, Savolainen V, Chatrou LW, Powell M, Grayer RJ, Chase MW (2002) Molecular phylogenetics of Caryophyllales based on nuclear 18S rDNA and plastid *rbc*L, *atp*B, and *mat*K DNA sequences. Am J Bot 89:132–144

Dalström S (2001) New species and combinations in the Oncidiinae (Orchidaceae) and a synopsis of the Cochlioda clade (Oncidiinae). Selbyana 22:135–145

Dalström S (2010) *Cyrtochilum* Kunth. In: Dodson CH, Luer CA (eds) Flora of Ecuador 225(3): Orchidaceae; genera *Cyrtochiloides–Epibator*. Department of Plant and Environmental Sciences, University of Gothenburg, Gothenburg

Domański Z, Kęsy J (2005) Distribuition of Manhattan distance in square and triangular lattices. Sci Res Inst Math Comput Sci 4:34–37

Douzery EJP, Pridgeon AM, Kores P, Linder HP, Kurzweil H, Chase MW (1999) Molecular phylogenetics of Diseae (Orchidaceae): a contribution from

nuclear ribosomal ITS sequences. Am J Bot 86:887–899
Doyle JJ (1997) Within trees: genes and species, molecules and morphology. Syst Biol 46:537–553
Doyle JJ, Kanazin V, Schoemaker RC (1996) Phylogenetic utility of histone H3 intron sequences in the perennial relatives on the soybean (*Glycine*: Leguminosae). Mol Phylogenet Evol 6:438–447
Dressler RL (1981) The Orchids–natural history and classification. Harvard University Press, Cambridge
Edgar RC (2004) MUSCLE: multiple sequence alignment with high accuracy and high throughput. Nucleic Acids Res 32:1792–1797
Ellis J (1982) Promiscuous DNA–chloroplast genes inside plant mitochondria. Nature 299:678–679
Enea V, Corredor V (1991) The evolution of plasmodial stage-specific rRNA genes is dominated by gene conversion. J Mol Evol 32:183–186
Erixon P, Svennblad B, Britton T, Oxelman B (2003) Reliability of Bayesian posterior probabilities and bootstrap frequencies in phylogenetics. Syst Biol 52:665–673
Felsenstein J (1985) Confidence limits on phylogenies: an approach using the bootstrap. Evolution 39:783–791
Fitch WM (1971) Towards defining the course of evolution: minimum change for a specific tree topology. Syst Zool 20:406–416
Friesen N, Fritsch RM, Blattner FR (2006) Phylogeny and new intrageneric classification of *Allium* L. (Alliaceae) based on nuclear ribosomal DNA ITS sequences. Aliso 22:372–395
Galtier N, Gouy M, Gautier C (1996) SEAVIEW and PHYLO_WIN: two graphic tools for sequence alignment and molecular phylogeny. Comput Appl Biosci 12:543–548
Garay LA (1974) On the systematics of the monopodial orchids II. Bot Mus Leafl Harv Univ 23:369–376
Goldman DH, Jansen RK, van den Berg C, Leitch IJ, Fay MF, Chase MW (2004) Molecular and cytological examination of *Calopogon* (Orchidaceae, Epidendroideae): circumscription, phylogeny, polyploidy, and possible hybrid speciation. Am J Bot 91:707–723
Grant V (1981) Plant speciation. Columbia University Press, New York
Hammer Ø, Harper DAT, Ryan PD (2001) PAST: Paleontological Statistics Software Package for Education and Data Analysis. Palaeontol Electronica 4:1-9
Hedrén M (1996) Genetic differentiation, polyploidization and hybridization in northern European *Dactylorhiza* (Orchidaceae): evidence from allozyme markers. Plant Syst Evol 201:31–55
Hedrén M (2001) Systematics of the *Dactylorhiza euxina/incarnata/maculata* polyploidy complex (Orchidaceae) in Turkey: evidence from allozyme data. Plant Syst Evol 229:23–44
Hewitt GM (2001) Speciation, hybrid zones and phylogeography—or seeing genes in space and time. Mol Ecol 10:537–549
Hillis DM, Dixon MT (1991) Ribosomal DNA: molecular evolution and phylogenetic inference. Q Rev Biol 66:411–453
Hillis DM, Moritz C, Porter CA, Baker RJ (1991) Evidence for biased gene conversion in concerted evolution of ribosomal DNA. Science 251:308–310
Huang CY, Grunheit N, Ahmadinejad N, Timmis JN, Martin W (2005) Mutational decay and age of chloroplast and mitochondrial genomes transferred recently to angiosperm nuclear chromosomes. Plant Physiol 138:1712–1733
Jersáková J, Johnson SD, Kindlmann P (2006) Mechanisms and evolution of deceptive pollination in orchids. Biol Rev 81:219–235
Königer W (1996) *Portillia*—eine neue Gattung in der Subtribus Pleurothallidinae, neue Arten der Gattungen *Portilia*, *Odontoglossum*, *Oncidium*, *Sigmatostalix*, *Trichocentrum* und *Trigonochilum*. Arcula 6:154–185
Königer W (1999) Neue Arten der Gattungen *Dracula*, *Masdevallia*, *Oncidium*, *Sigmatostalix* und *Trigonochilum*. Arcula 8:218–249
Königer W (2000) Neue Arten der Gattungen *Masdevallia*, *Oncidium*, *Sigmatostalix* and *Trigonochilum*. Arcula 10:266–280
Königer W (2008) Neue Arten der Gattungen *Oncidium* und *Scelochilus*. Arcula 16:376–391
Königer W (2010) Neue Arten der Gattungen *Oncidium* und *Trigonochilum*. Arcula 19:422–437
Königer W, Schildhauer H (1994) *Dasyglossum* and *Trigonochilum*—two new genera in subtribus Oncidiinae. Arcula 1:1–28
Kores PJ, Molvray M, Weston PH, Hopper SD, Brown AP, Cameron KM, Chase MW (2001) A phylogenetic analysis of Diurideae (Orchidaceae) based on plastid DNA sequence data. Am J Bot 88:1903–1914
Kraenzlin F (1917) Cyrtochilum. Notizbl Bot Gart Berl Dahl 7:81–101
Lefébure T, Douady CJ, Gouy M, Gibert J (2006) Relationship between morphological taxonomy and molecular divergence within Crustacea: proposal of a molecular threshold to help species delimitation. Mol Phylogenet Evol 40:435–447
Linder CR, Rieseberg LH (2004) Reconstructing patterns of reticulate evolution in plants. Am J Bot 91:1700–1708
Maddison W (1997) Gene tress in species trees. Syst Biol 46:523–536
Madhulatha TS (2012) An overview on clustering methods. IOSR J Eng 2(4):719–725
Mallet J (2007) Hybrid speciation. Nature 446:279–283
Molvary MP, Kores PJ, Chase MW (2000) Polyphyly of mycoheterotrophic orchids and functional influences on floral and molecular characters. In: Wilson KL, Morrison DA (eds) Monocots: systematics and evolution. CSIRO, Collingwood
Neubig KM, Whitten WM, Williams NH, Blanco MA, Endara L, Burleigh JG, Silvera K, Cushman JC, Chase MW (2012) Generic recircumscriptions of Oncidiinae (Orchidaceae: Cymbidieae) based on maximum likelihood analysis of combined DNA datasets. Bot J Linn Soc 168:117–146
Nordal I, Stedje B (2005) Paraphyletic taxa should be accepted. Taxon 54:5–6
Nylander JAA (2004) MrModeltest v2. Program distributed by the author. Evolutionary Biology Centre, Uppsala University, Uppsala
Otto SP, Witton J (2000) Polyploid incidence and evolution. Annu Rev Genet 34:401–437
Pandit S, Gupta S (2011) A comparative study on distance measuring approaches for clustering. IJORCS 2:29–31
Paun O, Forest F, Fay MF, Chase MW (2009) Hybrid speciation in angiosperms: parental divergence drives peloidy. New Phytol 182:507–518
Pridgeon AM, Chase MW, Cribb PJ, Rasmussen FN (2009) Genera Orchidacearum, vol 5: Epidendroideae (part two). Oxford University Press, Oxford
Ramsey J, Schemske DW (1998) Pathways, mechanisms, and rates of polyploidy formation in flowering plants. Annu Rev Ecol Syst 29:467–501
Richter M, Diertl K-H, Emck P, Peters T, Beck E (2009) Reasons for an outstanding plant diversity in the tropical Andes of Southern Ecuador. Landsc Online 12:1–35
Rieseberg LH (1997) Hybrid origins of plant species. Annu Rev Ecol Syst 28:359–389
Rieseberg LH (2001) Chromosomal rearrangements and speciation. Trends Ecol Evol 16:351–358
Rieseberg LH, Sinervo B, Linder CR, Ungerer M, Arias DM (1996) Role of gene interactions in hybrid speciation: evidence from ancient and experimental hybrids. Science 272:741–745
Rogers SO, Bendich AJ (1987) Ribosomal RNA genes in plants: variability in copy number and in the intergenic spacer. Plant Mol Biol 9:509–520
Ronquist F, Huelsenbeck JF (2003) MrBayes 3: Bayesian phylogenetic inference under mixed models. Bioinformatics 19:1572–1574
Rosenberg NA (2002) The probability of topological concordance of gene trees and species trees. Theor Popul Biol 61:225–247
Sang T, Zhong Y (2000) Testing hybridization hypotheses based on incongruent gene trees. Syst Biol 49:422–424
Schaal BA, Hayworth DA, Olsen KM, Rauscher JT, Smith WA (1998) Phylogeographic studies in plants: problems and prospects. Mol Ecol 7:465–474
Schiestl F, Cozzolino S (2008) Evolution of sexual mimicry in the orchid subtribe Orchidinae: the role of preadaptations in the attraction of male bees as polinators. BMC Evol Biol 8:27
Senghas K (2001) Nachtraege 19. Tribus Oncidieae. Schlechter, Die Orchideen, Ed.3, Vol.1c, Lieferung 44/45:2784–2810
Senghas K (2003) Aus dem Beiratsressort Wissenschaft, Neues zur Gattung *Odontoglossum*. Journal fuer den Orchideenfreund 10:277–286
Seperack P, Slatkin M, Arnheim N (1988) Linkage disequilibrium in human ribosomal genes: implications for multigene family evolution. Genetics 119:943–949
Small RL, Ryburn JA, Cronn RC, Seelanan T, Wendel JF (1998) The tortoise and the hare: choosing between noncoding plastome and nuclear *Adh* sequences for phylogeny reconstruction in a recently diverged plant group. Am J Bot 85:1301–1315
Soltis DE, Soltis PS (1999) Polyploidy: recurrent formation and genome evolution. Trends Ecol Evol 14:348–352

Stace CA (1989) Plant taxonomy and biosystematics. 2nd edn. Edward Arnold, Cambridge
Stegemann S, Keuthe M, Greiner S, Bock R (2012) Horizontal transfer of chloroplast genomes between plant species. Proc Natl Acad Sci USA 109:2434–2438
Swofford DL (2002) PAUP*: phylogenetic analysis using parsimony (*and Other Methods). Version 4. Sinauer Associates, Sunderland
Szlachetko DL (1995) Systema Orchidalium. W. Szafer Institute of Botany, Polish Academy of Sciences, Kraków
Szlachetko DL, Mytnik-Ejsmont J (2009) Gynostemia Orchidalium Vol. 4. Orchidaceae - Vandoideae. Acta Bot Fenn 180:1–313
Tremblay RL, Ackerman JD, Zimmerman JK, Calvo RN (2005) Variaton in sexual reproduction in orchids and its evolutionary consequences: a spasmodic journey to deversification. Biol J Linn Soc 84:1–54
Tsai CC, Chiang YC, Huang SC, Chen CH, Chou CH (2010) Molecular phylogeny of *Phalaenopsis* Blume (Orchidaceae) on the basis of plastid and nuclear DNA. Plant Syst Evol 288:77–98
Tyteca D, Klein E (2008) Genes, morphology and biology—the systematics of *Orchidinae* revisted. J Eur Orchid 40:501–544
Ungerer MC, Baird SJE, Pan J, Rieseberg LH (1998) Rapid hybrid speciation in wild sunflowers. Proc Natl Acad Sci USA 95:11757–11762
van der Cingel NA (2001) An atlas of orchid pollination: America, Africa, Asia and Australia. A. A. Balkema, Rotterdam
van der Pijl L, Dodson CH (1966) Orchid flowers: their pollination and evolution. University of Miami Press, Coral Gables
Vriesendorp B, Bakker FT (2005) Reconstructing patterns of reticulate evolution in angiosperms: what can we do? Taxon 54:593–604
Wendel JF (2000) Genome evolution in polyploids. Plant Mol Biol 42:225–249
Wendel JF, Schnabel A, Seelanan T (1995) Bidirectional interlocus concerted evolution following allopolyploid speciation in cotton (*Gossypium*). Proc Natl Acad Sci USA 92:280–284
Won H, Renner S (2003) Horizontal gene transfer from flowering plants to *Gnetum*. Proc Natl Acad Sci USA 100:10824–10829
Wu CI (2001) The genic view of the process of speciation. J Evol Biol 14:851–865
Yuan Q, Hill J, Hsiao J, Moffat K, Ouyang S, Cheng Z, Jiang J, Buell CR (2002) Genome sequencing of a 239-kb region of rice chromosome 10L reveals a high frequency of gene duplication and a large chloroplast DNA insertion. Mol Genet Genomics 267:713–720
Zimmer EA, Martin SL, Beverley SM, Kan YW, Wilson AC (1980) Rapid duplication and loss of genes coding for the alpha chains of hemoglobin. Proc Natl Acad Sci USA 77:2158–2162

15

Phellinus noxius: molecular diversity among isolates from Taiwan and its phylogenetic relationship with other species of *Phellinus* based on sequences of the ITS region

Jyh-Nong Tsai[1], Pao-Jen Ann[1], Ruey-Fen Liou[2], Wen-Hsui Hsieh[3] and Wen-Hsiung Ko[3*]

Abstract

Background: Analysis of phylogenetic relationship of 91 isolates of *Phellinus noxius* obtained from 46 plant species in Taiwan did not show distinct grouping based on ITS sequences.

Results: However, the ITS nucleotides showed 20 different kinds of variations including single nucleotide polymorphisms, deletion and insertion in ITS1 and ITS2, but none in 5.8 S. The Taiwanese isolates of *P. noxius* were dividable into long (type L), median (type M) and short (type S) groups based on ITS sequence length. Two isolates with identical ITS sequence belonged to types L. Type M with 72 isolates was further divided into 33 subtypes, while types S with 17 isolates was further divided into two subtypes.

Conclusion: Phylogenetic analysis of ITS sequences among *Phellinus* species showed that isolates of *P. noxius* were in the same clade distinctly separated from other *Phellinus* species.

Keywords: Deletion, Insertion, ITS sequence, Nucleotide variation, Sequence length, Single nucleotide polymorphism

Background

Brown root rot caused by *Phellinus noxius* (Corner) G. H. Cunn. is widespread among tropical countries in Southeast Asia, Africa, Oceania, Central America and the Caribbean (Pegler and Waterston 1968). In China, it has been reported from the tropical Hainan Island (Tai 1979). In Japan, it was found on the subtropical island of Okinawa (Abe et al. 1995). The pathogen attacks more than 120 species of fruit and ornamental trees in both topical and subtropical districts in Taiwan (Ann et al. 1999; Chang and Yang 1998). Among the approximately 200 plant species listed as hosts of *P. noxius* in the world, about half of them were reported for the first time from Taiwan (Ann et al. 2002). Even though the fungus lacks air-borne spores for efficient dissemination, it is very widespread and occurs on so many kinds of hosts at very different geographic locations on the island of Taiwan (Ann et al. 2002). It is, therefore, conceivable that *P. noxius* may be an ancient residence of the island where diverse isolates of this fungus may have existed. There are very few morphological characters in *P. noxius* available for testing this hypothesis because the fungus rarely produces basidiocarps on diseased trees in the fields (Ann et al. 1999; Chang 1995, 1996).

In this study, molecular variation in the ITS (ITS1, 5.8S and ITS2) region among isolates of *P. noxius* from Taiwan was investigated and compared with the ITS sequences reported from other countries available in the GenBank. We also investigated the ITS phylogenetic relationship of

*Correspondence: kowh@nchu.edu.tw
[3] Department of Plant Pathology, National Chung Hsing University, Taichung, Taiwan
Full list of author information is available at the end of the article

P. noxius with other species of *Phellinus*. Details of the study are reported herein.

Methods

Isolation and storage of the pathogen

Main roots of trees showing quick or slow decline symptoms (Ann et al. 2002) were exposed and examined. Those showing typical brown discoloration were cut and brought back to the laboratory. Small pieces (5 × 2 × 1 mm) of tissue were obtained from the advancing margins of the diseased roots, surface-sterilized with 0.5% NaClO for 1 min, plated on potato dextrose agar (PDA) supplemented with 100 ppm streptomycin sulfate and 10 ppm benomyl for inhibition of growth of bacteria and other fungi, and incubated at room temperature (24–30 °C). Fungal mycelia growing from diseased tissue were transferred to 2% water agar. Single-hyphal tips obtained from the fungus growing on water agar were cultured on PDA and stored in sterile distilled water in test tubes at room temperature (Boesewinkle 1976; Ko 2003). From each diseased tree only one isolate was saved for the study. The cultures were identified as *P. noxius* based on the production of brown colonies with irregular dark brown zone lines on PDA and formation of arthrospores and trichocysts (Ann and Ko 1992).

DNA extraction, amplification and sequencing

Each isolate of *P. noxius* was grown on cellophane placed on PDA (Ko et al. 2011). After incubation at 25 °C for 10 days, mycelia were harvested, lyophilized and stored at −20 °C until use. About 20 mg lyophilized mycelia were ground in liquid nitrogen and used for extraction of DNA using the genomic DNA extraction kit (GenMark Technology Co., Taichung, Taiwan). The ITS (ITS1-5.8S-ITS2) region was amplified with primer pair of ITS4 and ITS5 (White et al. 1990). The 25 μl reaction mixture consisting of 0.2 μg template DNA, 0.2 μM each primer, 200 μM each dNTP, 2 μl 2X polymerase chain reaction (PCR) buffer and 1.0 U ZyM Taq DNA polymerase (Zymeset, Taiwan) was subjected to thermal cycling in a Perkin-Elmer Thermal Cycler 9700 (Perkin-Elmer Applied Biosystem, USA). Cycling conditions for amplification were an initial denaturation at 94 °C for 3 min, followed by 35 cycles at 94 °C for 45 s, 50 °C for 45 s, 72 °C for 45 s, and a final elongation at 72 °C for 7 min. The PCR products were electrophoresed on a 1.5% agarose gel. Direct sequencing of the PCR products was performed by the Seeing Bioscience Company (Taipei, Taiwan), using ITS4, ITS5 (White et al. 1990), PN-5.8S-1 (5′-GCA GCG AAA TGC GAT AAG TA-3′), or PN-5.8S-2 (5′-CAT GAC ACT CAA ACA GGC AT-3′) as the primer. The sequences of ITS region obtained from the sequencing process were assembled, trimmed and edited using the Vector NT1 software v. 10.0 (InforMax Inc., USA). The sequence of ITS tail was determined using the ITS 2 annotation tool (Keller et al. 2009). The polymorphic portions were marked by IUPAC ambiguity codes. The ITS sequences of 36 isolates of *P. noxius*, representing all ITS types found in Taiwan, were submitted to NCBI (National Center for Biotechnology Information; http://www.ncbi.nlm.mih.gov).

Phylogenetic analysis

The ITS sequences of 91 isolates of *P. noxius* from Taiwan were analyzed in order to understand the phylogenetic relationship among these isolates. Multiple alignments and minor adjustments of the sequences of these isolates were performed using clustal X 1.81 (Thompson et al. 1977) followed by BioEditor software. Sequence alignment was deposited at TreeBase (http://purl.org/phylo/treebase/phylows/study/TB2:S16384). Phylogenetic relationships were analyzed using the Philip 3.67 software (Phylogeny Inference Package, Version 3.67) and the neighbor joining program with 1000 bootstrap replicates. The program of Treeview was used to view phylogenetic trees.

In order to study the phylogenetic relationship between isolates of *P. noxius* from Taiwan and other countries and other *Phellinus* species, the ITS sequences of all *Phellinus* species in the GenBank were retrieved. A total of 58 isolates belonging to 39 species of *Phellinus* was obtained and used for phylogenetic analysis (Table 1). The ITS types L, M and S divided based on ITS length were used as local strains for analysis. The method described above was used for phylogenetic relationship analysis.

Results

Phylogenetic relationship among Taiwanese isolates of P. noxius

A total of 91 isolates of *P. noxius* was obtained from 46 species of plants distributed in different geographic locations in Taiwan from 1991 to 2009 (Table 2). Analysis of the phylogenetic relationship of these Taiwanese isolates did not show distinct grouping based on ITS sequences. The bootstrap values on the branches were very low and were all below 50% (data not shown) with accession number JN836346-JQ003229 (Tables 1, 2).

Nucleotide variation in ITS region among Taiwanese isolates of *P. noxius*

The examination of ITS nucleotide variation revealed the existence of 20 different kinds of variants, designated

Table 1 List of taxa of *Phellinus noxius* and related species from GenBank used for phylogenetic analysis

Species	Geographic origin[a]	Strain no.[b]	ITS sequence length (bp)[c]	Accession no.
1. *Phellinus alni*		TW322	610	AY340041
2. *P. badius*		CBS 449.76	663	AY558609
3. *P. baumii*		MPNU 7006	711	AF200231
4. *P. bicuspidatus*		KCTC 6651	621	AY558610
5. *P. calcitratus*			584	JF894115
6. *P. chrysoloma*			644	AF055370
7. *P. cinereus*		05-37	620	AM931248
8. *P. conchatus*		CBS 167.29	708	AY558614
9. *P. ferrugineovelutinus*		CBS 218.48	542	AY558618
10. *P. gilvus*		ATCC26729	613	AF250932
11. *P. hartigii*		CBS 162.30	692	AY558621
12. *P. h ippophaeicola*		CBS 252.50	705	AY558622
13. *P. igniarius*		CFMR 5698	609	AY558623
P. igniarius		KCTC6228	598	AF056192
14. *P. igniarius var. trivialis*		CBS 512.63	596	AY558624
15. *P. johnsonianus*		ATCC60051	702	AF250931
16. *P. laevigatus*		CFMR 5640	588	AY558626
17. *P. linteus*		MPNU 7002	670	AF200228
18. *P. lundellii*		CBS 540.72	605	AY558630
19. *P. merrillii*		PM950703-1	707	EU035310
20. *P. nigricans*		CBS 213.48	611	AY558631
P. nigricans		H6002112	621	GQ383726
21. *P. noxius*	Taiwan	PN72.1	613-L	JQ003239
P. noxius	Taiwan	PNP1.2	609-M	JN836341
P. noxius	Taiwan	PN29.1	609-M	JN836344
P. noxius	Taiwan	PNA4.1	609-M	JN836346
P. noxius	Taiwan	PN5.2	608-M	JQ003233
P. noxius	Taiwan	PNP4.2	607-M	JQ029276
P. noxius	Taiwan	PN22.1	601-S	EF065630
P. noxius		CBS170.32	601-S	EF065631
P. noxius,	Japan	Tf566	601-S	JQ003238
P. noxius	Malaysia	FRIM638	610-M	HQ400698
P. noxius	Malaysia	FRIM618	602-S	HQ400699
P. noxius	Malaysia	FRIM613	602-S	HQ400700
P. noxius	Malaysia	FRIM551	603-S	HQ400702
P. noxius	Malaysia	FRIM154	601-S	HQ400703
P. noxius	Malaysia	FRIM147	599-S	HQ400704
P. noxius	India	–	608-M	AB639022
22. *P. occidentalis*		CBS 196.55	706	AY558634
23. *P. pachyphloeus = Inonotus pachphloeus*		CBS 193.37	571	AY558635
24. *P. pini*		ATCC12240	635	AF250930
25. *P. pini* var. *cancriformans*		IMSNU 32031	636	AF200242
26. *P. pomaceus*		25	599	FR686572
27. *P. populicola*		CBS 638.75	599	AY558638
28. *P. punctatus*		CBS 386.66	649	AY558640
29. *P. repandus*		CBS 616.89	658	AF534076
30. *P. rhabarbarinus*		CBS 282.77	714	AY558642
31. *P. ribis f. ulicis*		CBS 579.50	653	AY558644
32. *P. rimosus*		MDJCBS86	608	DQ103885

Table 1 continued

Species	Geographic origin[a]	Strain no.[b]	ITS sequence length (bp)[c]	Accession no.
33. *P. robustus*		KCTC 6657	679	AY558645
34. *P. senex*		CBS 442.76	578	AY558647
35. *P. spiculosus*		KTCC 6658	641	AY558648
36. *P. tremulae*		CBS 123.40	595	AY558650
37. *P. tropicalis*		CBS 617.89	636	AF534077
38. *P. tuberculosus*		CBS 171.32	600	AY558652
39. *P. weirii*		CNU 6017	620	AF251438

[a] The country where *P. noxius* was isolated

[b] *CBS* Centraalbureau voor Schimmelcultures, *NPMU* National Programme Management Unit, *KCTC* Korean Collection for Type Cultures, *ATCC* American Type Culture Collection, *CFMR* Colegiul Fizicienilor Medicali din România, *FRIM* Forest Research Institute Malaysia, *IMSNU* Institute of Microbiology, Seoul National University, *CNU* Collection of Newcastle University

[c] L: ITS type L; M: ITS type M; S: ITS type S

Table 2 List of hosts, locations, ITS information, GenBank accession no. of strains of *Phellinus noxius* from Taiwan used in the study

Scientific name (common name)	Isolate	Location	Year of isolation	GenBank accession no.	Sequence ITS1/5.8S/ITS2 (bp)	ITS type
1. *Annona squamosa* (custard apple)	PNA4.1	Taitung County	1996	JN836346	609	M6
2. *Araucaria cunninghamii* (hook pine)	PN29.1	Taichung City	2004	JN836344	609	M4
A. cunninghamii	PN30.1	Taichung City	2004		609	M4
3. *Averrhoa carambola* (carambola)	PNS1.1	Tainan City	1992		607	M32
4. *Bauhinia* × hybrid (butterfly tree)	PN40.2	Changhua County	2005	JQ003235	606	M33
5. *Bauhinia variegata* (orchid tree)	PN7.1	Nantou County	1996		609	M6
B. variegata	PN35.1	Hualian County	2005	JN836349	609	M9
B. variegata	PN35.2	Hualian County	2005		609	M9
6. *Calocedrus formosana* (Taiwan incense cedar)	PN70.2	Taichung City	2009	JQ003232	608	M23
7. *Casuarina equisetifolia* (ironwood tree)	PN22.1	Nantou County	1998	EF065630.1	601	S1
8. *Cinnamomum kotoensis* (botel tobago cinnamon tree)	PN74.2	Taitung County	2009		608	M24
C. kotoensis	PN74.1	Taitung County	2009	JQ029271	608	M27
9. *C. osmophloeum* (Taiwan cinnamon)	PN50.1	Nantou County	2006		601	S1
C. osmophloeum	PN51.1	Nantou County	2006		608	M24
10. *Cinnamomun camphora* (camphor)	PN32.1	Taichung City	2005		608	M24
C. camphora	PN32.2	Taichung City	2005		608	M24
C. camphora	PN94001.1	Nantou County	2005		607	M32
C. camphora	94001.2	Nantou County	2005		607	M32
11. *Citrus limon* (lemon)	PNC1.1	Tainan City	2003		609	M1
C. limon	PNC1.2	Tainan City	2003		609	M1
C. limon	PNC4.1	Chiayi County	2006		607	M32

Table 2 continued

Scientific name (common name)	Isolate	Location	Year of isolation	GenBank accession no.	Sequence ITS1/ 5.8S/ITS2 (bp)	ITS type
12. *Delonix regia* (flame tree)	PN37.1	Hualian County	2005		601	S1
D. regia	PN37.2	Hualian County	2005	JQ029275	607	M31
D. regia	PN42.1	Hualian County	2005		601	S1
13. *Dimocarpus. longan* (longan)	PNLn5.1	Tainan City	1992	JQ003236	601	S1
D. longan	PNLn9.2	Changhua County	1998	JQ003226	608	M17
D. longan	PNLn10.1	Tainan City	2003		609	M6
D. longan	PNLn10.2	Tainan City	2003	JQ003222	609	M13
D. longan	PNLn14.2	Changhua County	2006		601	S1
14. *Diospyros kaki* (persimmon)	PNPe1.1	Chiayi County	1991		609	M1
15. *Duranta repens* (creeping sky flower)	PN3.1	Nantou County	1996	JQ003231	608	M22
16. *Eriobotrya japonica* (loquat)	PNLo3.1	Taitung County	1997		601	S1
E. japonica	PNLo5.1	Taitung County	2009		609	M1
17. *Eucalyptus citriodora* (lemon gum eucalyptus)	PN6.1	Nantou County	1996		608	M22
18. *Ficus microcarpa* (small-leafed banyan)	PN21.1	Miaoli County	2003		608	M24
F. microcarpa	PN21.2	Miaoli County	2003		608	M24
F. microcarpa	PN12.1	Taichung City	1996	JQ029274	607	M30
F. microcarpa	PN26	Nantou County	2003		608	M24
F. microcarpa	PN28.2	Taichung City	2004		608	M19
F. microcarpa	PN49.2	Taichung City	2005	JQ003227	608	M18
F. microcarpa	PN57.1	Taichung City	2005		609	M7
F. microcarpa	PN75.1	Taichung City	2009		607	M30
F. microcarpa	PN76.1	Taichung City	2009		609	M4
19. *Ficus pumila* var. *awkeotsang* (jellyfig)	PN10.1	Chiayi County	1991	JQ029272	608	M28
20. *F. religiosa* (botree fig)	PN90.1	Taichung	2009	JN836342	609	M2
21. *Juniperus chinensis* var. kaizuka (dragon juniper).	PN65.1	Nantou County	2007		609	M4
22. *Kigelia pinnata* (sausage tree)	PN14.1	Nantou County	1998	JN836348	609	M8
23. *Koelreuteria henryi* (flame gold-rain tree)	PN94.1	Taichung City	2009	JQ003237	601	S2
K. henryi	PN33.2	Hualian County	2005	JQ003223	609	M14
K. henryi	PN41.1	Hualian County	2005		609	M14
K. henryi	PN41.2	Hualian County	2005		609	M14
24. *Liquidambar formosana* (maple)	R9218	New Taipei city	1992		608	M15
25. *Litchi chinensis* (litchi)	PNL2.1	Chiayi County	1992	JN836347	609	M7
L. chinensis	PNL2.2	Chiayi County	1992		601	S1
L. chinensis	PNL5.1	Kaohsiung City	2003	JQ029273	607	M29
L. chinensis	PNL5.2	Kaohsiung City	2003		607	M29
26. *Mangifera indica* (mango)	PNM4.1	Changhua County	2009		609	M6
27. *Melaleuca bracteata* 'Revolution Gold' (white cloud tree)	PN73.2	Taichung City	2009		609	M6

Table 2 continued

Scientific name (common name)	Isolate	Location	Year of isolation	GenBank accession no.	Sequence ITS1/ 5.8S/ITS2 (bp)	ITS type
M. bracteata 'Revolution Gold'	PN73.1	Taichung City	2009		609	M6
28. *Murraya paniculata* (orange jasmine)	PN5.1	Nantou County	1996		608	M24
M. paniculata	PN5.2	Nantou County	1996	JQ003233	608	M24
M. paniculata	PN25.1	New Taipei city	2004		601	S1
M. paniculata	PN25.2	New Taipei city	2004		601	S1
29. *Oncidium* Gower Ramsey	PN44	Yungling County	2005		601	S1
30. *Osmanthus fragrans* (osmanthus)	PN140.1	Changhua County	2009	JQ003221	609	M12
31. *Psidium guajava* (guava)	PN98007	Kaohsiung City	2009	JQ029270	608	M26
32. *Podocarpus macrophyllus* (long-leaved podocarpus)	PN98.3	Taichung City	2009		601	S1
33. *Prunus armeniaca* (apricot)	PN72.1	Taichung City	2009	JQ003239	613	L1
P. armeniaca	PN72.2	Taichung City	2009		613	L1
34. *Prunus campanulata* (Taiwan cherry)	PN71.1	Taichung City	2009	JN386350	609	M10
P. campanulata	PN71.2	Taichung City	2009		609	M10
35. *Prunus mume* (plum)	PNP1.2	Kaohsiung City	1991	JN836341	609	M1
P. mume	PNP2.1	Nantou County	1996	JN836345	609	M5
36. *Prunus persica* (peach)	PNP5.1	Nantou County	1999	JQ003228	608	M19
P. persica	PNP10.1	Changhua County	2005	JQ003225	608	M16
37. *Pterocarpus indicus* (rose wood)	PN104.1	Taichung City	2009	JQ003220	609	M11
38. *Pyrus pyrifolia* (pear)	PNP4.1	Miaoli County	2003		601	S1
P. pyrifolia	PNP4.2	Taichung City	1998	JQ029276	607	M32
P. pyrifolia	PNP9.1	Nantou County	2004	JN836343	609	M3
39. *Schinus terebinthifolius* (Brazilian peppertree)	PN48.1	Taichung City	2005		601	S1
S. terebinthifolinus	PN48.2	Taichung City	2005		601	S1
40. *Spathodea campanulata* (African tulip tree)	PN147	Changhua County	2009		601	S1
41. *Sterculia nobilis* (ping-pong)	PN17.1	Nantou County	1999	JQ003224	608	M15
S. nobilis	PN84.1	Taichung City	2009		607	M32
S. nobilis	PN124.1	Taichung City	2009		609	M6
42. *Syzygium samarangense* (wax apple)	PNW1.1	Chiayi County	1991	JQ003234	608	M25
43. *Terminalia catappa* (Indian almond)	PN2.1	Chiayi County	1996		609	M4
44. *T. catappa*	PN63.1	Changhua County	2007		607	M32
45. *Toona sinensis*, *Cedrela sinensis* (Chinese cedar)	PN64.1	Taichung City	2007	JQ003230	608	M21
46. *Vitis vinifera* (grape)	PNG1.1	Nantou County	1999		609	M6
Zizyphus mauritiana (Indian jujube)	PNZ2.1	Kaohsiung City	2001	JQ003229	608	M20

as V1 to V20 in ITS1 and ITS2 but not 5.8S in the 91 Taiwanese isolates of *P. noxius* obtained in this study (Table 3) . The variation included insertion, deletion and single nucleotide polymorphism. Some isolates showed single nucleotide polymorphism among chromosomes in the same isolate.

Grouping based on ITS sequence length

The examination of ITS nucleotide variation also revealed the possible division into three distinct groups based on sequence length among the 91 isolates of *P. noxius* from Taiwan (Table 4). Isolates with long sequence of 613 bp were termed type L. Only two isolates belonged to this type. Isolates with median sequence length of 606–609 bp were termed type M. The majority of the Taiwanese isolates with a total of 72 isolates belonged to this type. Type M was further divided into 33 subtypes based on single nucleotide polymorphisms, single nucleotide deletion (V12), double nucleotide deletion (V19) and single nucleotide insertion (V20) (Tables 3, 4) . Isolates with short sequence of 601 bp were termed type S. Type S was further divided into two subtypes as a result of a single nucleotide polymorphism at position 114. Seventeen isolates belonged to this type.

Isolates of *P. noxius* from GenBank fitted or nearly fitted the M or S ITS types (in Taiwan; Table 1). Isolate CBS170.32 of unknown origin belonged to type S, so was the Japanese isolate. The isolate from India belonged to type M. Among the six isolates from Malaysia, isolate FRIM154 fitted the type S and isolate FRIM 638 nearly fitted the type M with 1 bp more than the Taiwanese type M. Isolates FRIM 618, FRIM 613 and FRIM 551 nearly fitted type S with 1–2 bp more than the Taiwanese type S, while isolates FRIM 147 was also close to type S with 2 bp less than the Taiwanese type S. However no isolates from other countries were founded to fit the Taiwanese type L in this study.

Relation between ITS types and hosts and locations from where *P. noxius* was found in Taiwan

Type L was detected only in Taichung City (Fig. 1). Type M was found in three cities and seven counties, while type S was found in two cities and eight counties. *P. noxius* was not found in Yilan County, Taoyuan County, Hsinchu Tounty and Pingtung County during this study.

Subtype S1 was found on 12 plant species located in three cities and seven counties, while subtype S2 was found only on flame gold-rain tree in Taichung City (Table 2). Other isolates found on flame gold-rain tree in

Table 3 Nucleotide variation in ITS detected among isolates of *Phellinus noxius* in Taiwan

Kind of variant	Sequence position	Nucleotide variation		
		Single nucleotide polymorphism	Deletion	Insertion
ITS1				
V1/V1*	19	G, A/R		
V2	31	T, C		
V3/V3*	32	G, C/S		
V4	114	G, C		
V5	116–117			GGAGAG
V6	117–118	TG, AT		
V7	125–126	TC, AT		
V8/V8*	129	T, A/C		
V9	135–142		ATTTATTC	
V10	152	A, G		
V11	168	C, T		
V12	193		A	
V13	197	T, C		
ITS2				
V14/V14*	420	T, C/Y		
V15/V15*	442	G, A/R		
V16/V16*	469	A, G/R		
V17/V17*	546	A, G/R		
V18/V18*	593	C, G/S		
V19	594–595		AC	
V20	600–601			C

*The variants with an asterisk symbol represent isolates with single nucleotide polymorphisms among chromosomes in the same isolate

Table 4 ITS types and subtypes among *Phellinus noxius* isolates from Taiwan

ITS type and subtype	ITS sequence length (bp)	Nucleotide variation	Representative isolate (total no.)
Type L			
L1	613	V1, V3, V5, V15, V16, V19	PN72.1 (2)
Type M			
M1	609		PNP1.2 (5)
M2	609	V16	PN90.1 (1)
M3	609	V17	PNP9.1 (1)
M4	609	V18	PN29.1 (5)
M5	609	V1, V3	PNP2.1 (1)
M6	609	V14, V15	PNA4.1 (8)
M7	609	V15, V16	PNL2.1 (2)
M8	609	V3, V14, V15	PN14.1 (1)
M9	609	V1, V3, V16, V17	PN35.1 (2)
M10	609	V1, V3, V11, V15, V17	PN71.1 (2)
M11	609	V15*, V18*	PN104.1 (1)
M12	609	V1, V15*, V16	PN140.1 (1)
M13	609	V14*, V15, V17*	PNLn10.2 (1)
M14	609	V1*, V3*, V15*, V16*	PN33.2 (3)
M15	608	V3, V12	PN17.1 (2)
M16	608	V3, V12, V17	PNP10.1 (1)
M17	608	V10, V12, V15, V16	PNLn9.2 (1)
M18	608	V1, V3, V11, V12	PN49.2 (1)
M19	608	V1, V3, V8, V11, V12, V16	PNP5.1 (2)
M20	608	V1, V3, V6, V11, V12, V15	PNZ2.1 (1)
M21	608	V1, V2, V3, V11, V12, V15, V16	PN64.1 (1)
M22	608	V1, V2, V3, V8, V11, V12, V13, V16	PN3.1 (2)
M23	608	V1, V2, V3, V6, V7, V11, V12, V15, V16	PN70.2 (1)
M24	608	V15, V16, V19, V20	PN5.2 (9)
M25	608	V1, V3, V15, V16, V19, V20	PNW1.1 (1)
M26	608	V3, V12, V15*, V16*	PN98007 (1)
M27	608	V15, V16*, V19, V20	PN74.1 (1)
M28	608	V1, V3*, V15, V19, V20	PN10.1 (1)
M29	607	V19	PNL5.1 (2)
M30	607	V16, V19	PN12.1 (2)
M31	607	V15, V16, V19	PN37.2 (1)
M32	607	V1, V3, V15, V16, V19	PNP4.2 (7)
M33	606	V1, V2, V3, V6, V7, V11, V12, V15, V16, V19	PN40.2 (1)
Type S			
S1	601	V9	PNLn5.1 (16)
S2	601	V4, V9	PN94.1 (1)

*The variants with an asterisk symbol represent isolates with single nucleotide polymorphisms among chromosomes in the same isolate

Hualian County belonged to subtype M14. This study also revealed that isolates of *P. noxius* obtained from the same plant species in the same location may belong to different subtypes. In Taichung City, isolates of *P. noxius* found on small-leafed banyan consisted of subtypes M4, M7, M18, M19 and M30. Similarly, isolates obtained from longan in Tainan City contained subtypes S1, M6 and M13. It was also found that isolates obtained from the same host in different locations may belong to the same subtype. For examples, subtype S1 on longan was found in Tainan City and Changhua County, while subtype M24 on small-leafed banyan was found in Miaoli County and Nantou County. Isolates obtained from different hosts in different locations may also belong to the same subtype. For examples, subtype M1 was found on lemon in Tainan City and on persimmon in Chiayi County, while subtype M6 was found

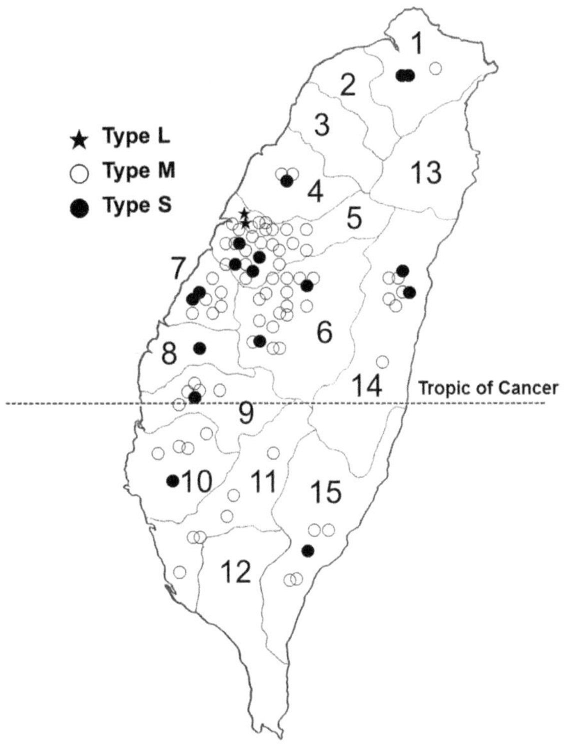

Fig. 1 Distribution of different ITS types of *Phellinus noxius* from Taiwan. Number referred to *1* New Taipei City, *2* Taoyuan County, *3* Hsinchu County, *4* Miaoli County, *5* Taichung City, *6* Nantou County, *7* Changhua County, *8* Yunlin County, *9* Chiayi County, *10* Tainan City, *11* Kaohsiung City, *12* Pingtung County, *13* Yilan County, *14* Hualian County, *15* Taitung County

on custard apple in Taitung County and on orchid tree in Nantou County.

Phylogenetic analysis based on ITS sequences among *Phellinus* species

The ITS sequences of 58 isolates belonging to 39 species of *Phellinus* retrieved from GenBank and seven *P. noxius* isolates representing type L, type M and type S of ITS sequences from Taiwan were used in the analysis of the phylogenetic relationship among *Phellinus* species. The result showed that all the isolates of *P. noxius* including isolates from Taiwan and other countries were in the same clade with 100% bootstrap support (Fig. 2). The sequence similarity between *P. noxius* and other *Phellinus* species was less than 85%. The species most closely related to *P. noxius* was *P. pachphloeus* with 83% similarity, whereas the most distant species was *P. badius* with only 67% similarity.

Discussion

Results from this study showed that the isolates of *P. noxius* from Taiwan can be divided into type L, type M and type S based on ITS sequence length. From 1991 to 2009, 2 type L isolates, 82 type M isolates and 17 type S isolates were found on 46 plant species in Taiwan (Table 2). To our best knowledge, this is the first report of division of isolate from the same fungal species into different groups based on ITS length. *P. noxius* was reported from Taiwan as early as 1928 (Sawada 1928). It is conceivable that type M and type S may have existed in Taiwan for a very long period of time and that type M may have evolved in Taiwan earlier and became the predominant type. Only two isolates of type L was obtained from apricot at Taichung City. It is possible that type L may be a recent mutation from subtype M25 through an 6 bp insertion at position 116–117 (V5), and deletion at position 600–601 (V20) (Tables 3, 4). However, the possibility that it may be due to host specificity of type L has not been ruled out.

The results also suggested the possibility that type S may originate from type M through an 8 bp deletion at the position between 135 and 142 (V9) (Tables 3, 4). After the deletion, the ITS sequences seem to become stable because there were only two subtypes among 17 isolates of type S obtained in this study. Moreover, the difference between subtype S1 and subtype S2 was the occurrence of a single nucleotide polymorphism at sequence position 114 (V4) in the latter.

Phellius noxius is one of the plant pathogens with a very wide host range. Among the more than 200 plant species representing 59 families listed as hosts of *P. noxius* in the world, about half of them were reported for the first time from Taiwan (Ann et al. 2002). This is compatible with the discovery of great nucleotide variation in ITS region among isolates of *P. noxius* found in Taiwan in this study. The variation included 15 kinds of single nucleotide polymorphisms, three kinds of deletions and two kinds of insertions (Table 3).

Analysis of the ITS sequences of the Taiwanese isolates of *P. noxius* revealed that the 5.8 S region was identical in all isolates, while significant sequence variation was observed in ITS regions. This is in agreement with those reported with powdery mildews (Hirata and Takamatsu 1996) and *Fusarium* species (Naqvi et al. 2013). Our studies showed that the ITS1 was more variable than ITS 2 (Table 3). The former contained 10 single nucleotide polymorphisms, one 8 bp deletion, one 1 bp deletion and one 6 bp insertion, while the latter consisted of only five single nucleotide polymorphisms, one 2 bp deletion and one 1 bp insertion.

Phylogenetic analysis of ITS sequences among *Phellinus* species showed that isolates of *P. noxius* were in the same clade distinctly separated from other *Phellinus* species (Fig. 2). Phylogenetic relationship among *Phellinus* species based on ITS sequences has been reported previously (Shin 2001; Wagner and Fischer 2002; Jeong et al. 2005; Decock et al. 2006). However, none of them

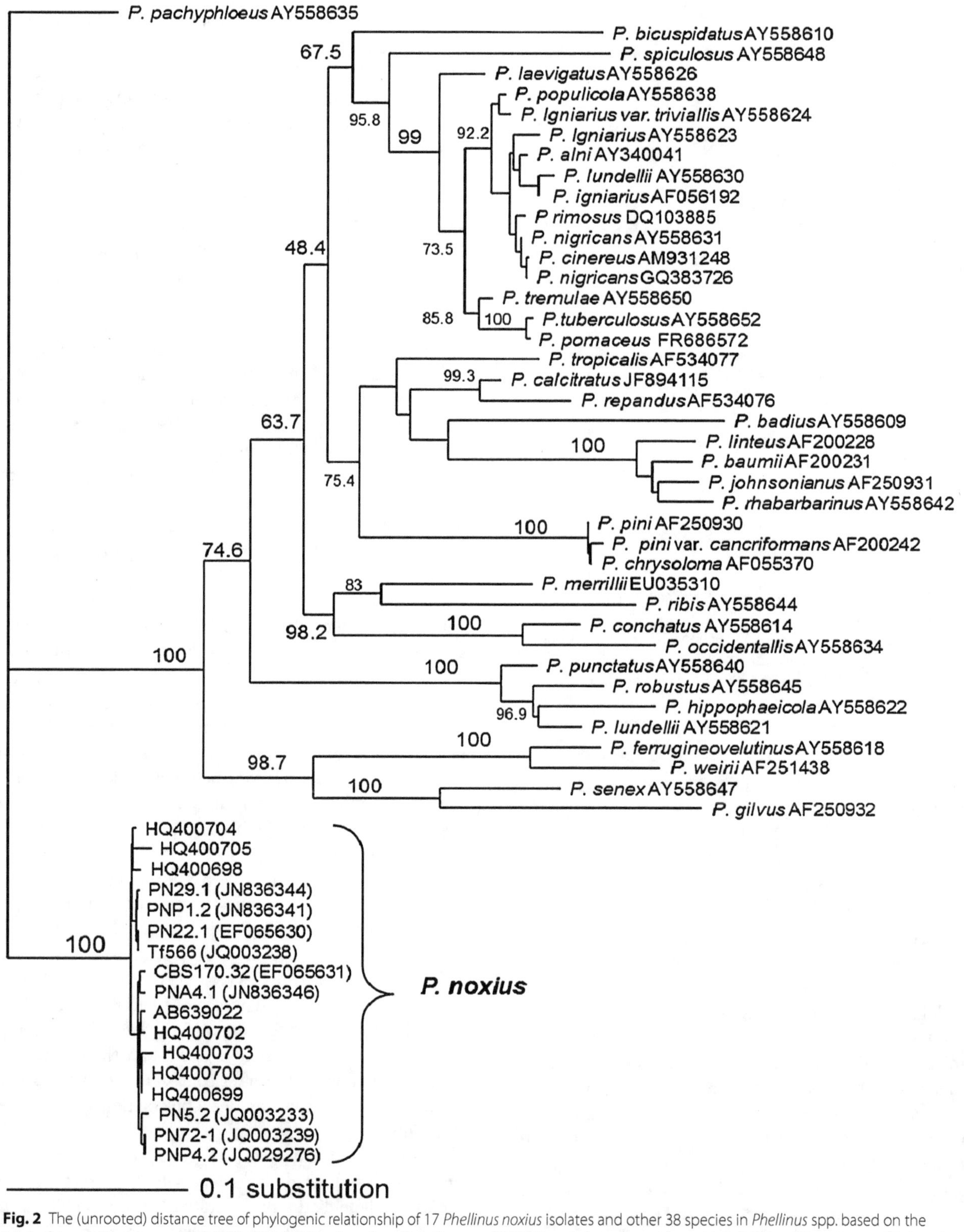

Fig. 2 The (unrooted) distance tree of phylogenic relationship of 17 *Phellinus noxius* isolates and other 38 species in *Phellinus* spp. based on the internal transcribed spacer (ITS1-5.8Sr DNA-ITS2) region of nuclear ribosomal DNA. Branch lengths and boot strap (1000 replicates) were displayed in the distance tree by using neighbor-joining and bootstrap methods of Philip 3.67 software. The distance bar is corresponded to 10 substitutions per 100 nucleotide sites

has included *P. noxius* in their studies. *P. noxius* has been transferred to *Phellinidium noxium* (Corner) Bondartseva & S. Herrera in 1992 (Bondartseva et al. 1992). However, *Phellinidium noxium* was distinctly separated phylogenetically from other *Phellinidium* species (Dai 2010), indicating that more study is needed in the future.

During this study, *P. noxius* was not found in the counties of Yilan, Taoyuan, Hsinchu and Pingtung (Fig. 1). This does not mean that the fungus was not present in those areas because detection of *P. noxius* in those counties had been reported previously (Ann et al. 2002).

Conclusion

The 91 isolates of *Phellinus noxius* obtained from 46 plant species in Taiwan showed 20 different kinds of variation including single nucleotide polymorphisms, deletion, insertion in ITS1 and ITS2, but none in 5.8S. The Taiwanese isolates of *P. noxius* were dividable into long (type L), median (type M) and short (type S) groups based on ITS sequence length. Phylogenetic analysis of ITS sequence among *Phellinus* species showed the isolate of *P. noxius* were in the same clade distinctly separated from other *Phellinus* species.

Authors' contributions

WHK designed the experiments and wrote the manuscript; PJA, RFL and WHH conceived the experiments; JNT performed the experiments. All authors read and approved the final manuscript.

Author details

[1] Division of Plant Pathology, Taiwan Agricultural Research Institute, Wufeng, Taichung, Taiwan. [2] Department of Plant Pathology and Microbiology, National Taiwan University, Taipei, Taiwan. [3] Department of Plant Pathology, National Chung Hsing University, Taichung, Taiwan.

Acknowledgements

The study was supported by the Grants from Council of Agriculture (101N-10.2.1-N-C7(3)) and Ministry of Science and Technology (NSC 101-2321-B-005-004) of Taiwan.

Competing interests

The authors declare that they have no competing interests.

References

Abe Y, Kobayashi T, Onuki M, Hattori T, Tsurumachi M (1995) Brown root rot of trees caused by *Phellinus noxius* in windbreaks on Ishigaki Island, Japan-incidence of disease, pathogen and artificial inoculation. Ann Phytopathol Soc Jpn 61:425–433

Ann PJ, Ko WH (1992) Decline of longan trees: association with brown root rot caused by *Phellinus noxius*. Plant Pathol Bull 1:19–25

Ann PJ, Lee HL, Tsai JN (1999) Survey of brown root disease of fruit and ornamental trees caused by *Phellinus noxius* in Taiwan. Plant Pathol Bull 8:51–60

Ann PJ, Chang TT, Ko WH (2002) *Phellinus noxius* brown root rot of fruit and ornamental trees in Taiwan. Plant Dis 86:820–826

Boesewinkle HJ (1976) Storage of fungal cultures in water. Trans Br Mycol Soc 66:183–185

Bondartseva MA, Herrera SD, Sandoval FC (1992) Taxonomical problems of the Cuban Hymenochaetaceous fungi. Mikologiia i Fitopatologiia 26:1–13

Chang TT (1995) Decline of nine tree species associated with brown root rot caused by *Phellinus noxius* in Taiwan. Plant Dis 79:962–965

Chang TT (1996) Survival of *Phellinus noxius* in soil and in the roots of dead host plants. Phytopathology 86:272–276

Chang TT, Yang WW (1998) *Phellinus noxius* in Taiwan: distribution, host plants and the pH and texture of the rhizosphere soils of infected host. Mycol Res 102:1085–1088

Dai YC, Zhou LW, Cui BK, Chen YQ, Decock C (2010) Current advances in *Phellinus* sensu lato: medicinal species, functions, metabolites and mechanisms. Appl Microbial Biotech 87:1587–1593

Decock C, Figueroa SH, Robledo G, Castillo G (2006) *Phellinus caribaeo-quercicolus* sp. nov. parasitic on *Quercus cubana*: taxonomy and preliminary phylogenetic relationships. Mycologia 298:265–274

Hirata T, Takamatsu S (1996) Nucleotide sequence diversity of rDNA internal transcribed spacers extracted from conidia and cleistothecia of several powdery mildew fungi. Mycoscience 39:283–288

Jeong WJ, Lim YW, Lee JS, Jung HS (2005) Phylogeny of *Phellinus* and related genera inferred from combined data of ITS and mitochondrial SSU rDNA sequences. J Microbiol Biotechnol 15:1028–1038

Keller A, Schleicher T, Schultz J, Müller T, Dandekar T, Wolf M (2009) 5.8S-28S rRNA interaction and HMM-based ITS2 annotation. Gene 430:50–57

Ko WH (2003) Long-term storage and survival structure of three species of *Phytophthora* in water. J Gen Plant Pathol 69:186–188

Ko WH, Yang CH, Lin MJ, Chen CY, Tsou YJ (2011) *Humicola phialophorides* sp. nov. from soil with potential for biological control of plant diseases. Bot Stud 52:197–202

Naqvi SKB, Ahmed S, Rauf CA, Naqvi SMS (2013) Application and sequencing of interval transcribed regions 1 & 2, and 5.8S rDNA from local isolates of *Fusarium* species. Pak J Bot 45:301–307

Pegler DN, Waterston JM (1968) *Phellinus noxius* No. 195. In: Descriptions of pathogenic fungi and bacteria. Commonwealth Mycological Institute, Kew

Sawada K (1928) Camphor tree decline. Descr Catal Formos Fungi 4:86–91

Shin KS (2001) Identification of some *Phellinus* spp. Mycobiology 29:190–195

Tai FL (1979) Sylloge fungorum sinicorum. Science Press, Peking

Thompson JD, Gibson TJ, Plewniak F, Jeanmougin F, Higgins DG (1997) The clustal X windows interface: flexible strategies for multiple sequence alignment aided by quality analysis tools. Nucleic Acids Res 25:4876-4882. doi:10.1093/nar/25.24.4876

Wagner T, Fischer M (2002) Proceeding towards a natural classification of the worldwide taxa *Phellinus* s. l. and *Inonotus* s. l., and phylogenetic relationships of allied genera. Mycologia 94:998–1016

White TJ, Bruns T, Lee S, Taylor J (1990) Amplification and direct sequencing of ribosomal RNA genes for applications. In: Innis MA, Gelfand DH, Snirsky JJ, White TJ (eds) PCR protocols: a guide to methods and applications. Academic Press, San Diego, pp 315–322

Chilling susceptibility in mungbean varieties is associated with their differentially expressed genes

Li-Ru Chen[1*], Chia-Yun Ko[2], William R. Folk[3] and Tsai-Yun Lin[4*]

Abstract

Background: Mungbean (*Vigna radiata* L. Wilczek) is an economically important legume of high nutritional value, however, its cultivation is limited by susceptibility to chilling. Varieties NM94 and VC1973A, with differential susceptibility to stress, serve as good materials for uncovering how they differ in chilling tolerance. This study aimed to identify the ultrastructural, physiological and molecular changes to provide new insights on the differential susceptibility to chilling between varieties VC1973A and NM94.

Results: Chilling stress caused a greater reduction in relative growth rate, a more significant decrease in maximum photochemical efficiency of PSII and DPPH scavenging activity and more-pronounced ultrastructural changes in VC1973A than in NM94 seedlings. Comparative analyses of transcriptional profiles in NM94 and VC1973A revealed that the higher expression of chilling regulated genes (*COR*s) in NM94. The transcript levels of lipid transfer protein (*LTP*), dehydrin (*DHN*) and plant defensin (*PDF*) in NM94 seedlings after 72 h at 4 °C was higher than that in its parental lines VC1973A, 6601 and VC2768A.

Conclusions: Our results suggested that *LTP*, *DHN* and *PDF* may mediate chilling tolerance in NM94 seedlings.

Keywords: Chilling regulated genes (*CORS*), Chilling tolerance, DHN, LTP, PDF, *Vigna radiata*

Background

Mungbean (*Vigna radiata* L. Wilczek) seeds and its sprouts contain high levels of proteins rich in essential amino acids and various phytochemicals with beneficial activities, such as antioxidant, antimicrobial, anti-inflammatory, antidiabetic, antihypertensive, lipid metabolism accommodation and antitumor effects (Tang et al. 2014). According to Food and Agriculture Organization of the United Nations, worldwide production of mungbean increased fourfold to 21.4 million tons between 1990 and 2014, with the majority produced in Asia (85.4%).

Being sensitive to cold, the geographical and seasonal distribution is limited by low temperatures. Cold temperatures in the early growing season greatly affect its cultivation. Exposure of mungbean seedlings to 4 °C for 2 days resulted in irreversible cellular electrolyte leakage, but cold acclimation at 10 °C reduced the damage (Chang et al. 2001).

Chilling/cold stress limits plant growth and causes significant crop loss. Stress responses are influenced by duration of exposure (Ercoli et al. 2004), species (Carvallo et al. 2011) and the stage of plant development (Ohnishi et al. 2010). Plants native to temperate regions exhibit varying degrees of cold tolerance and can acquire freezing tolerance after cold acclimation, while many tropical and subtropical plants are sensitive to chilling at 0–15 °C (Miura and Furumoto 2013). To cope with low temperatures, plants often upregulate the expression of some cold related genes (*CORs*). At the molecular level, a C-repeat (*CRT*) or dehydration responsive element (*DRE*) is commonly found in the promoter regions of *COR*, and is bound by CRT binding factor (CBF)/DRE

*Correspondence: clr2@faculty.pccu.edu.tw; tylin@life.nthu.edu.tw
[1] Department of Horticulture and Biotechnology, Chinese Culture University, Taipei 11114, Taiwan
[4] Department of Life Science & Institute of Bioinformatics and Structural Biology, National Tsing Hua University, Hsinchu 30013, Taiwan
Full list of author information is available at the end of the article

binding (DREB) protein (Carvallo et al. 2011). Expression of *VrDREB2A* in mungbean seedlings was markedly induced by drought, high-salt stress and ABA treatment, but only slightly by cold stress (Chen et al. 2016). Although the CBF pathway has been one of the dominant signal mechanisms mediating cold acclimation and is widely conserved in higher plants (Miura and Furumoto 2013), it may function differently in mungbean plants.

Different mungbean cultivars vary in their physio-biochemical responses to UV-B (Choudhary and Agrawal 2014), heat stress (Kaur et al. 2015) and bruchid damage (Hong et al. 2015). Genome size of different mungbean varieties ranged from 494 to 554 Mb and at least 52,739 genes have been annotated (Kang et al. 2014; Liu et al. 2016). Transcriptomic comparison between bruchid-resistant and -susceptible mungbean lines identified nucleotide variations caused by differential expressed genes and sequence-changed-protein genes of mungbean and transposon elements, besides bruchid-resistant (*Br*) genes, as putative modifier factors for bruchid resistance (Liu et al. 2016). Nevertheless, little is known about the molecular and physiological mechanisms underlying the intrinsic susceptibility of mungbean to chilling stress. We previously isolated 1198 mungbean expressed sequence tags (ESTs) informative to early seedling development and chilling response, and showed that variety NM94 maintained better membrane integrity than VC1973A under chilling/cold stress (Chen et al. 2008).

Lipid transfer proteins (LTPs) regulate diverse lipid-mediated cellular processes and accelerate transport of lipid monomers between membranes in vitro (Lev 2010). Salt and dehydration stress resulted in increased mRNA levels of mungbean VrLTPs (Liu and Lin 2003). A maize dehydrin (DHN) was found to bind to small lipid vesicles containing acidic phospholipids (Koag et al. 2003) which may stabilize lipid membranes. A broad spectrum of DHN proteins accumulate in response to chilling in *Arabidopsis*, cauliflower and yellow lupin (Rurek 2010). DHNs associate with large unilamellar vesicles emulating the lipid compositions of plasma and organelle membranes in *Thellungiella salsuginea* (Rahman et al. 2010). The mungbean VrDhn1, a Y_2K-type DHN, can interact with DNA (Lin et al. 2012). Genes encoding plant defensins (PDFs) are up-regulated in *Oxytropis* (*Fabaceae*) species adapted to the arctic (Archambault and Strömvik 2011), and during cold acclimation in winter wheat (Gaudet et al. 2003) and *Arabidopsis* (Oono et al. 2006). Our objective was to examine how ultrastructural and physiological changes and differential gene expression of *COR*s varied between NM94 and VC1973A, and testing the hypothesis that VrLTP, VrDhn1 and PDF play important roles under chilling stress. Temporal gene expression patterns of seedlings during chilling stress were compared in order to decipher *COR*s underlying chilling tolerance.

Methods

Plant materials and growth conditions

Seedlings of mungbean varieties (Additional file 1: Figure S1; Shanmugasundaram et al. 2009) were prepared and the membrane damage was determined with electrolyte leakage as described in our previous report (Chen et al. 2008). At 3 days after imbibition (DAI), VC1973A and NM94 seedlings were randomly separated into two groups for cDNA microarray analysis: one group was chilled (4 °C for 72 h) then allowed to recover at 25 °C for 72 h; the other group was maintained at 25 °C as a control. Plants were harvested at 1, 2, 4, 24, 48, 72 h after chilling, and after 3 days recovery following 72 h chilling.

Measurement of plant relative growth rate (RGR) and chlorophyll fluorescence

RGR was calculated as $(\ln Wt_n - \ln Wt_0)/t$. Wt_n is the fresh weight of seedlings at the indicated time, Wt_0 is the initial fresh weight of 3-DAI seedlings, and t indicates the time difference between t_n and t_0. Each experiment contained five replications. Mungbean seedlings at 4 DAI were exposed to 4 °C and Fv/Fm ratios were measured at indicated times with a portable Pulse Amplitude Modulated (PAM) fluorometer (PAM 2100, Walz, Germany) after 30 min of dark adaptation. The maximum quantum efficiency of photosystem II, Fv/Fm, was calculated as (Fm − Fo)/Fm (Kitajima and Butler 1975). The injury index of Fv/Fm was defined as $100 \times [1 - (Fv/Fm_t/Fv/Fm_0)]$ and electrolyte leakage as $100 \times [(EC_t - EC_0)/(1 - EC_0)]$, in which t equals each time point and 0 equals the initial time before chilling (Eriksson et al. 2011). The correlation coefficient between the injury indices of Fv/Fm and electrolyte leakage was analyzed. The experiment included 15 replicates.

Detection of radical scavenging activity and reactive oxygen species (ROS)

Radical scavenging activity of mungbean seedlings was measured using a modified method of Brand-Williams et al. (1995). Each mungbean seedling (~0.3 g fresh weight) was extracted with 1 ml of methanol and 50 μl of the extract was mixed with 150 μl of *1,1-diphenyl-2-picrylhydrazyl* (DPPH) solution for measuring the absorbance at 517 nm with a spectrophotometer (Powerwave XS2, BioTek, Bad Friedrichshall, Germany). The same volume of methanol was used as a blank. The experiment was replicated five times. DPPH scavenging activity (%) was defined as $(Aa - Ab)/Aa \times 100\%$, where *A*a is the absorption of the blank; *A*b is the absorption of extract solution.

Hydrogen peroxide was detected in situ by staining with 3,3′-diaminobenzidine (DAB, Sigma-Aldrich) according to the method of Daudi et al. (2012). Mungbean seedlings at 3 DAI were treated at 4 °C and sampled at the indicated times. Each experiment contained six individual plants as biological replicates, and the experiment was duplicated.

Electron microscopy

Leaves of 3-DAI seedlings subjected to 4 or 25 °C for 24 h were examined with transmission electron microscopy (TEM). Fixed specimens were dehydrated in a graded acetone series and embedded in Spurr's resin according to the method of Lin et al. (2012). Sample blocks were polymerized at 70 °C for 24 h and ultrathin sections (100 nm) were mounted on 50-mesh copper grids. The ultrathin sections were stained with saturated uranyl acetate in 50% methanol for 20 min and then with 0.2% (w/v) aqueous lead citrate for 4 min. Samples were examined with a Philips CM100 BioTwin TEM (FEI Company Electron Optics, Eindhoven, Netherlands). Both number and area distribution of the vacuoles were calculated using Nikon NIS-element software.

RNA isolation and cDNA microarray analysis

Total RNA was extracted as described in our previous study (Chen et al. 2008) and quantified at 260 nm with a NanoDrop 2000 spectrophotometer (NanoDrop Technologies, Wilmington, DE, USA). A cDNA microarray containing 735 mungbean early developmental and chilling stress-responsive uniESTs (Chen et al. 2008) was used to probe the mungbean RNAs, with λDNA (TX803; Takara, Kyoto, Japan) as external and *VrActin* (accession no. AM910789) as internal controls. DNA fragments from a cDNA library of VC1973A seedlings were amplified with T7 primer (5′-TAATACGACT CACTATAGGG-3′) and reverse primer (5′-TCACACAG GAAACAGCTATGAC-3′), and those from subtractive cDNA libraries were amplified with nested primers 1 (5′-TCGAGCGGCCGCCCGGGCAGGT-3′) and 2R (5′-AGCGTGGTCGCGGCCGAGGT-3′). Purification of PCR products, microarray preparation, hybridization and data analysis were performed according to Chen et al. (2007). RNA samples of 50 μg from each time point were labeled with cyanine 3 (Cy3; control) and cyanine 5 (Cy5; 4 °C-treated plants) dyes using a 3DNA™ Array 50 Kit (Genisphere, Montvale, NJ) with 500 pg polyA$^+$ λRNA (TX802; Takara, Kyoto, Japan) as an external control. The fluorescence intensity of each clone was divided by its corresponding control, and then normalized to the expression of the *VrActin* gene. Fold change for each gene was calculated by dividing the average intensity of 4 °C-treated samples by the average intensity of control samples at the corresponding time point.

Relative quantification in real-time PCR (qRT-PCR)

qRT-PCR was performed as described in Chen et al. (2007) using primers (Additional file 2: Table S1) designed with Primer Express 2.0 Software (Applied Biosystems, Foster City, CA). The expression ratio between a target gene and the endogenous control gene *VrActin* was determined based on the $2^{-\Delta\Delta Ct}$ method (Livak and Schmittgen 2001).

Statistical analysis

Data of RGR, Fv/Fm and DPPH scavenging activity were subjected to two-way analysis of variance (ANOVA) using CoStat statistical software (Cohort Berkeley, Monterey, CA). Significant differences among mungbean varieties under chilling stress were determined using Student–Newman–Keuls test.

Results

Chilling more profoundly decreased RGR, Fv/Fm and free radical scavenging capacity in VC1973A than in NM94

When grown at 25 °C, NM94 and VC1973A did not differ in RGR, and both reached a growth plateau before 6 DAI. NM94 showed more positive RGR values than VC1973A at 9–10 DAI ($P < 0.05$). Seedling growth was arrested by chilling after exposure to 4 °C for 1 day, as indicated by a negative RGR during 5–7 DAI (Fig. 1a). NM94 seedlings resumed growth after recovery from chilling, displaying positive RGR values during 7–10 DAI, while the RGR of VC1973A remained negative ($P < 0.01$).

Exposure to 4 °C for 72 h reduced Fv/Fm to 0.35 in VC1973A and 0.45 in NM94 ($P < 0.001$) from 0.75 to 0.76, respectively (Fig. 1b), showing a more severe decrease in the maximum photochemical efficiency of PSII in VC1973A. The injury index calculated from Fv/Fm values positively correlated ($r^2 = 0.98$) with the electrolyte leakage in our previous study (Chen et al. 2008). Visible DAB polymerization appeared in leaves after 24–72 h chilling, indicating a stronger H_2O_2-dependent DAB reaction in leaves of VC1973A than NM94 under stress (Fig. 1c). Accumulation of H_2O_2 in roots may not be affected by chilling, as H_2O_2 was generated in mungbean roots without chilling. At 25 °C, VC1973A and NM94 showed a high DPPH scavenging activity of 81 and 79%, respectively. No significant difference was detected after exposure to 4 °C for 48 h (Fig. 1d). DPPH scavenging activity decreased after 72 h-chilling, but was higher in NM94 (73%) than VC1973A (50%) (Fig. 1d, $P < 0.001$), indicating a better maintenance of redox homeostasis in NM94.

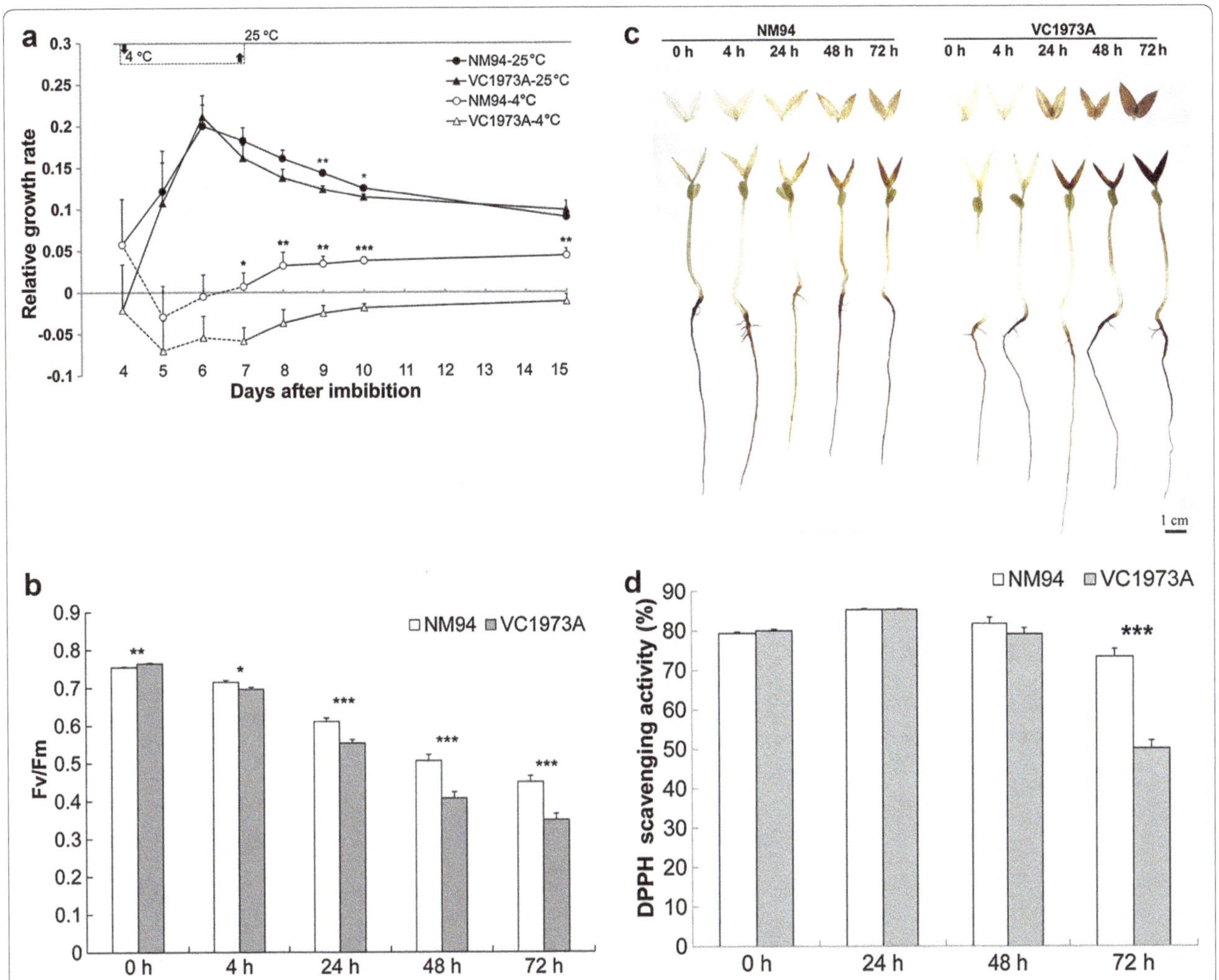

Fig. 1 Effect of chilling/cold stress on RGR, Fv/Fm, H_2O_2 generation and DPPH scavenging capacity in mungbean. Leaves of 4-DAI seedling were used to measure RGR (**a**), Fv/Fm (**b**) and whole seedlings of 3-DAI were used for in situ H_2O_2 detection (**c**) and measurement of DPPH scavenging capacity (**d**). Treatment time in **a** is indicated by a *dash line* with starting (*downward*) and ending (*upward*) *arrows*. Data were analyzed with two-way ANOVA and Student–Newman–Keuls test. *Symbols* indicate significant difference between NM94 and VC1973A; $*P < 0.05$, $**P < 0.01$, $***P < 0.001$

Chilling damaged the photosynthetic apparatus in mungbean seedlings

VC1973A (Fig. 2a) and NM94 (Fig. 2c) seedlings at 3-DAI contained intact chloroplasts and well-developed granal stacks at 25 °C. Exposure to 4 °C for 1 day resulted in fragmentation of the large central vacuole in both VC1973A (Fig. 2b) and NM94 (Fig. 2d). Percentage of the cell space occupied by vacuoles in palisade mesophylls of VC1973A and NM94 at 4 °C decreased to 20.6 and 26.8%, respectively (Additional file 3: Table S2). Yet vacuoles collapsed in palisade mesophylls of VC1973A (28.6%) more than in NM94 (9.7%). Chilling also caused grana unstacking and disintegration of thylakoid membranes in spongy mesophyll of VC1973A (Fig. 2e) and NM94 (Fig. 2g). In addition, chilling triggered formation of small vesicles in chloroplast (Fig. 2f, h) and accumulation of lipid droplets (Fig. 2i, k) in both VC1973A and NM94. Importantly, chilling resulted in fractured plasma membranes in VC1973A mesophylls (21.7%, n = 60) (Fig. 2j) but not in NM94 mesophylls (Fig. 2l). This further supports our previous findings that NM94 seedlings restored plasma membrane integrity better than VC1973A (Chen et al. 2008).

NM94 maintained a better membrane integrity than its parental lines under chilling

The variety NM94 has been bred by the Asian Vegetable Research and Development Center (AVRDC) through a reciprocal cross between VC1973A and a local variety 6601, and seeds of the F1 progeny were irradiated with

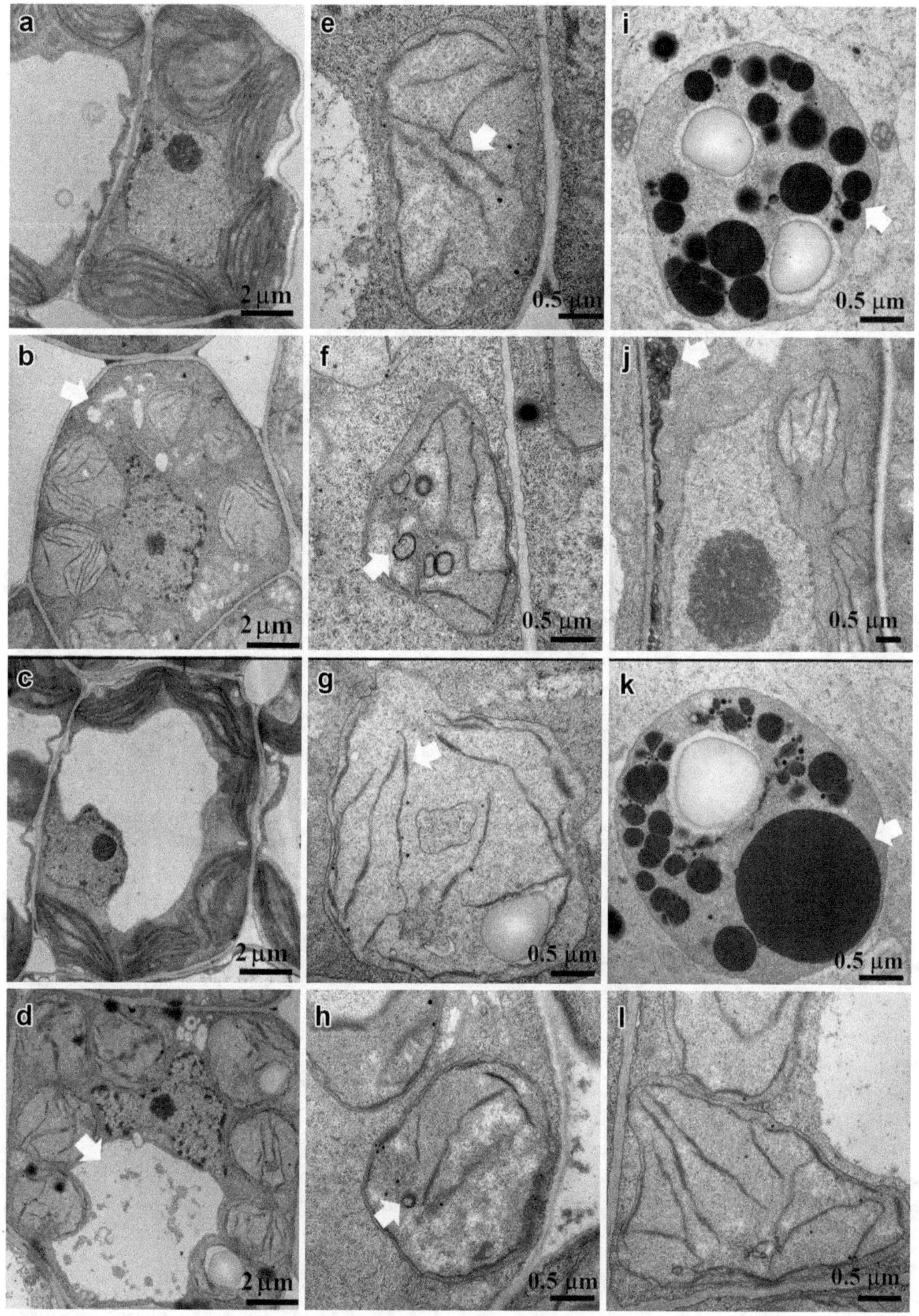

Fig. 2 Ultrastructural changes in mesophylls and chloroplasts under chilling. 3-DAI mungbean seedlings grown at 25 °C were exposed to 25 °C (**a**, **c**) or 4 °C (**b**, **d**–**l**) for 1 day to examine ultrastructural changes (shown by *white arrows*) in VC1973A (*upper panel*) and NM94 (*lower panel*) with TEM. **a**–**d** palisade cells; **e**–**l** chloroplasts

10 k-rad gamma rays. Up to F12, NM36 was selected for resistance to mungbean yellow mosaic virus (MYMV), and crossed with large-seeded line VC2768A to develop the MYMV-resistant and high-yield variety NM94 (Additional file 1: Figure S1). The 6601, VC2768A, VC1973A and NM94 seedling at 5-DAG were exposed to 5, 10, 15 and 20 °C for 48 h and the electrolyte leakage was examined. Leakage of electrolytes in mungbean seedlings was significantly affected by both variety and temperature based on a two-way ANOVA (Table 1). An exposure to 5 °C caused less electrolyte leakage in NM94 than that in its parental lines 6601, VC2768A, VC1973A, indicating that NM94 can maintain a better membrane integrity than its parental lines under chilling (Fig. 3a).

Differentially expressed *CORs* between NM94 and VC1973A seedlings

Of the 735 uniESTs examined using microarray analysis, 91 in NM94 and 40 in VC1973A were found to be *COR*s with a fold change ≥3 (up-regulated) or ≤0.3 (down-regulated) at one or more time points. Early up- (EU) or early down-regulated (ED) *COR*s were detected at 1–4 h, late up (LU) or late down (LD) *COR*s were detected at 24–72 h, and CU indicated a set of *COR*s up-regulated during 1–72 h. High correlations between qRT-PCR and microarray expression values were validated (Additional file 4: Table S3).

A venn diagram was used to compare similarities and differences between NM94 and VC1973A *COR*s gene sets, including 31 overlapping, 60 NM94 and 9 VC1973A *COR*s (Additional file 5: Table S4). NM94 *COR*s involved in photosynthesis, cellular redox homeostasis, disease resistance and membrane stabilization are listed in Table 2.

COR expression after recovery

Both varieties resumed growth (Fig. 1) and reduced electrolyte leakage after recovery at 25 °C for 3 days (Chen et al. 2008). *COR*s with more than a threefold

Table 1 Leakage of electrolytes in mungbean seedlings affected by variety (V) and temperature (T) based on a two-way ANOVA

Source of variance	Degree of freedom	Mean square	F value	*P* value
Variety (V)	3	123.86	2.89	0.041
Temperature (T)	3	435.10	267.10	<0.00001
V × T	9	152.19	3.55	0.0011
Error	71	42.81	–	–

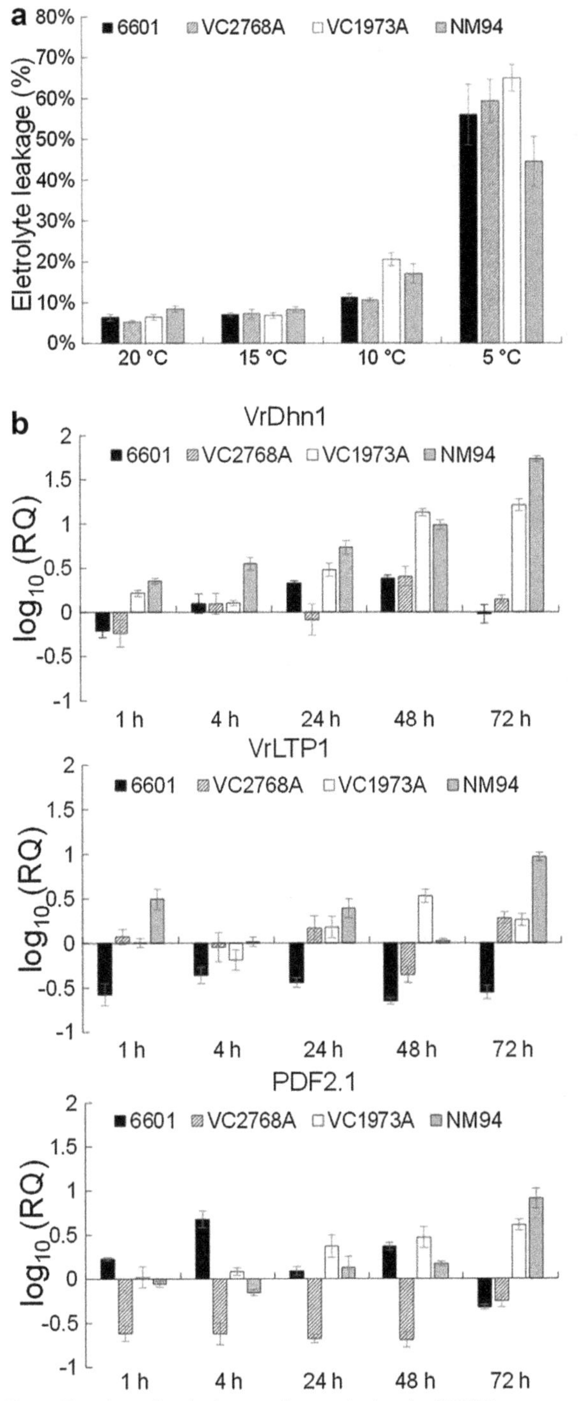

Fig. 3 The electrolyte leakage and transcript levels of VrLTP1, VrDhn1 and PDF2.1 in mungbean seedlings under chilling. The 6601, VC2768A, VC1973A and NM94 seedling at 5-DAG were exposed to 5, 10, 15 and 20 °C for 48 h and the electrolyte leakage (**a**) were measured as the conductivity of electrolyte from leaves. *Error bar* represents standard error (n = 6). qRT-PCR (**b**) was performed to validate the relative quantification (log_{10} scale) at five time points for genes encoding VrLTP1, VrDhn1 and PDF2.1. *Error bar* represents SE

Table 2 Dynamic expression patterns of NM94 *COR*s under chilling stress

ID	DP[b]	NM94[a]						DP[b]	VC1973A[a]						Description
		1 h	4 h	24 h	48 h	72 h	R3d		1 h	4 h	24 h	48 h	72 h	R3d	
Photosynthesis															
MBC214	EU	2.2	*4.9*	0.9	0.5	0.5	1.4	–	1.0	1.7	0.8	0.7	1.0	1.6	Chlorophyll a/b binding protein
Contig043	EU	1.5	*4.0*	1.0	0.5	0.5	1.4	–	0.8	2.2	0.6	0.4	0.6	1.7	Chlorophyll a/b binding protein
Cellular redox homeostasis and detoxification															
VG3C101	CU	1.6	*3.5*	*5.0*	1.1	1.4	*3.1*	–	0.9	0.6	0.5	0.3	1.2	*4.2*	PDI
NG1C087	LU	1.2	1.9	2.1	2.5	*5.1*	2.4	–	1.1	0.9	1.4	1.0	2.6	*7.1*	Thioredoxin
NG3H004	CU	1.2	*11.9*	*9.1*	1.0	1.6	*7.2*	–	0.8	1.4	0.8	0.2	1.4	1.8	Glyoxalase II
Disease resistance															
Contig074	CU	*4.7*	*34.7*	*31.3*	1.4	1.7	*58.1*	–	0.4	0.9	1.0	0.0	2.7	*8.8*	PDF2.1
VG3H087	CU	2.0	*8.9*	*12.9*	0.9	1.4	*17.3*	–	1.1	0.7	0.2	0.1	2.4	*34.9*	PDF2.3
NG1C003	EU	1.5	*4.5*	2.4	0.2	0.5	*4.6*	–	0.3	0.8	0.2	0.1	0.5	*3.9*	PDF precursor
Membrane stabilization															
Contig009	CU	*4.4*	*35.4*	*34.1*	2.4	1.9	*45.9*	–	0.4	0.8	1.0	0.1	2.7	*6.9*	VrLTP1
Contig061	CU	1.1	*22.9*	*18.5*	1.1	1.8	*7.9*	–	0.5	0.7	0.6	0.1	1.0	1.2	VrDhn1

[a] Fold change of each *COR* is the ratio of average intensity of chilling-treated to average intensity of the corresponding control. The italic indicates fold change ≥3 with a *t* test *P* value <0.05

[b] 'EU' contains genes up-regulated by chilling only in 1–4 h; 'LU' contains genes up-regulated only in 24–72 h; 'CU' indicates genes up-regulated in 1–72 h. DP, dynamic pattern. Signs of minus indicates non-*COR*s

change after recovery (R3d) were selected as the recovery responsive *COR*s (*RCOR*s), including 36 *RCOR*s in NM94 and 8 in VC1973A (Additional file 5: Table S4). NM94 and VC1973A shared 5 *RCOR*s with higher transcript levels in NM94 than VC1973A at R3d, including those encoding cysteine proteinases, proline-rich protein and peroxisomal-coenzyme A synthetase. Among the 31 NM94 *RCOR*s, the most significantly up-regulated (12- to 58-fold at R3d) genes encode ACT domain-containing protein, 8S β-globulin, VrLTP1 and PDF2.1. Transcripts of three *RCOR*s increased 3- to 11-fold at R3d in VC1973A but not NM94, one encodes ADH and two had no known homolog. Although ACT domain-containing protein, PDF2.1, PDF2.3, VrLTP1 and nuclease domain-containing protein transcripts were not significantly induced in VC1973A under cold, those transcripts (6- to 35-fold) were detected at elevated levels after recovery (Additional file 5: Table S4).

Discussion

Low temperature at 10 °C completely suppresses chloroplast development of etiolated leaves in mungbean (cv. 2937) seedlings and severely inhibits the expression of 7 genes encoding chlorophyll a/b-binding proteins (Yang et al. 2005). In our study, NM94 seedlings maintained a more positive RGR than VC1973A at 25 °C after 6-DAI, showing a genotypic variation during seedling development. Chilling brought more damage to the photosynthetic activity (Fig. 1b) and apparatus of VC1973A (Fig. 2j) seedlings than NM94 (Fig. 2l), indicating that NM94 maintains not only better membrane integrity under chilling stress, but also maximum photochemical efficiency of PSII. Genes encoding chlorophyll a/b-binding proteins were upregulated at 4 h in NM94 under chilling (Table 1) which may contribute to better efficiency of photochemical energy conversion, similar to the finding in the desert evergreen shrub, *Ammopiptanthus mongolicus* (Wu et al. 2014).

Application of H_2O_2 or cold acclimation increased chilling tolerance of mungbean seedlings through ABA-independent glutathione accumulation (Yu et al. 2003) and influx of extracellular Ca^{2+} (Hung et al. 2007). NM94 seedlings under chilling stress showed less accumulation of H_2O_2 and higher DPPH scavenging activity than VC1973A. NM94 maintained a better balance between ROS generation and scavenging, providing greater antioxidant protection. Upon exposure to chilling, higher levels of protein disulfide isomerase (PDI) and thioredoxin transcripts accumulated in NM94 (Table 1). Early activation of the PDI gene and later activation of thioredoxin gene suggest that these enzymes cooperate to combat the oxidative stress caused by chilling. PDI reduces dehydroascorbate to L-ascorbic acid and acts against oxidative stress in chloroplasts to maintain membrane integrity and photosynthetic capacity (Huang et al. 2005). Chloroplast thioredoxins regulate photosynthetic enzymes via light-dependent reactions, playing an important role in redox regulation (Arnér and Holmgren

2000) and photosynthetic CO_2 assimilation (Meyer et al. 2009).

A significant increase of glyoxalase II mRNA in NM94 (12-fold at 4 h, Table 1) suggests its function in detoxification of methylglyoxal (MG) which was produced under chilling stress (Hoque et al. 2012). Similarly, a rice glyoxalase II functions in salinity adaptation by maintaining better photosynthetic efficiency and anti-oxidation (Ghosh et al. 2014). Overexpression of glyoxalase pathway genes curbs the stress-induced MG level, regulates glutathione homeostasis, and helps plants survive under various abiotic stresses (Mustafiz et al. 2010).

In our experiments after 72 h at 4 °C, the *VrDhn1* mRNA level in NM94 was higher than that in VC1973A, 6601 and VC2768A (Fig. 3b; Table 1). Formation of small vesicles may increase the surface area of the chloroplast inner membrane to sustain adequate metabolite transport across the chloroplast under chilling (Kratsch and Wise 2000). We suggest that VrDhn1 may bind and stabilize small vesicles in chloroplast envelopes caused by chilling, in order to protect organelle membranes in NM94.

Transcript levels of mungbean *LTP*s were increased by salt, dehydration and exogenous ABA (Liu and Lin 2003). Interestingly, chilling resulted in rapid and marked accumulation of *VrLTP1* transcripts in NM94 but less in its parental lines VC1973A, 6601 and VC2768A (Fig. 3b; Table 1). The greater level of LTP may facilitate lipid shuttling between membranes (Douliez et al. 2000) and contribute to better membrane integrity in NM94 at 4 °C with no fractured plasma membrane in mesophyll.

NM94 has greater resistance to MYMV and cercospora leaf spot disease than VC1973A (Shanmugasundaram et al. 2009). PDFs are small, cysteine-rich peptides with antifungal, antibacterial and insect gut α-amylase inhibitory activities (Vriens et al. 2014). The transcript levels of *PDFs* are induced after 72 h at 4 °C in a greater amount in NM94 than in VC1973A, 6601 and VC2768A (Fig. 3b; Table 1). We speculate that PDFs may contribute to chilling tolerance in addition to plant defense against phytopathogenic microorganisms.

NM94 seedlings after chilling for 72 h restored growth better at 25 °C than VC1973A. The transcriptional activation of *VrDhn* and the higher levels of PDF2.1 and VrLTP1 transcripts in NM94 seedlings at R3d (Table 1) may be also involved in the recovery of growth subsequent to chilling exposure.

Conclusions

Mungbean varieties NM94 and VC1973A possess differential susceptibility to chilling and resistance to pathogens. Chilling reduced RGR and damaged the photosynthetic apparatus in both varieties, however NM94 seedlings can sustain the maximum photochemical efficiency of PSII and DPPH scavenging activity better than VC1973A under chilling. *CORs* differentially expressed between NM94 and VC1973A upon chilling/cold stress likely play important roles in determining susceptibility to chilling. We suggested that *CORs* involved in photosynthesis, cellular redox homeostasis, disease resistance and membrane stabilization could mediate the susceptibility to chilling in NM94 seedlings. Moreover, the significant transcriptional activation of *LTP*, *DHN* and *PDFs* in NM94 seedlings under chilling may cause their greater tolerance to chilling/cold.

Abbreviations

ANOVA: analysis of variance; *COR*: chilling/cold regulated gene; CRT: C-repeat; DAB: 3,3-diaminobenzidine; DAI: days after imbibition; DHN: dehydrin; DRE: dehydration responsive element; DPPH: 1,1-diphenyl-2-picrylhydrazyl; DREB: CRT binding factor (CBF)/DRE binding; ED: down-regulated regulated; EU: early up-regulated; EST: expressed sequence tag; Fv/Fm: the maximum quantum efficiency of photosystem II; LD: late down-regulated; LTP: lipid transfer protein; LU: late up-regulated; MG: methylglyoxal; PDF: plant defensin; PDI: protein disulfide isomerase; qRT-PCR: real-time PCR; RGR: relative growth rate; TEM: transmission electron microscopy.

Authors' contributions

LRC designed and performed the experiments, and wrote the manuscript. CYK contributed with data collection and analysis. WRF provides advice and guidance. TYL supervised the research design and wrote the paper. All authors read and approved the final manuscript.

Author details

[1] Department of Horticulture and Biotechnology, Chinese Culture University, Taipei 11114, Taiwan. [2] Institute of Plant and Microbial Biology, Academia Sinica, Taipei 11529, Taiwan. [3] Department of Biochemistry, University of Missouri, Columbia, MO 65211, USA. [4] Department of Life Science & Institute of Bioinformatics and Structural Biology, National Tsing Hua University, Hsinchu 30013, Taiwan.

Acknowledgements

This work was supported by the Ministry of Science and Technology [NSC101-2313-B-007-001-MY3] and the Council of Agriculture [105AS-9.6.1-FD-Z1] in Taiwan, ROC. We thank World Vegetable Center for providing seeds of NM94 and VC1973A, Michel Delseny (University of Perpignan) and Cindy Buschena (University of Minnesota) for critical reading and discussion of the manuscript, W.N. Jane in Institute of Plant and Microbial Biology, Academia Sinica for technical support of TEM and K. C. Yeh in Agricultural Biotechnology Research Center, Academia Sinica for the support of printing DNA probes on glass slides with Cartesian SynQUOD PixSys4500.

Competing interests

The authors declare that they have no competing interests.

References

Archambault A, Strömvik MV (2011) PR-10, defensin and cold dehydrin genes are among those over expressed in *Oxytropis* (*Fabaceae*) species adapted to the arctic. Funct Integr Genomics 11:497–505

Arnér ES, Holmgren A (2000) Physiological functions of thioredoxin and thioredoxin reductase. Eur J Biochem 267:6102–6109

Brand-Williams W, Cuvelier ME, Berset C (1995) Use of free radical method to evaluate antioxidant activity. LWT Food Sci Technol 28:25–30

Carvallo MA, Pino MT, Jeknić Z, Zou C, Doherty CJ, Shiu SH, Chen TH, Thomashow MF (2011) A comparison of the low temperature transcriptomes and CBF regulons of three plant species that differ in freezing tolerance: *Solanum commersonii*, *Solanum tuberosum*, and *Arabidopsis thaliana*. J Exp Bot 62:3807–3819

Chang MY, Chen SL, Lee CF, Chen YM (2001) Cold-acclimation and root temperature protection from chilling injury in chilling-sensitive mungbean (*Vigna radiata* L.) seedlings. Bot Bull Acad Sin 42:53–60

Chen LR, Chen YJ, Lee CY, Lin TY (2007) MeJA-induced transcriptional changes in adventitious roots of *Bupleurum kaoi*. Plant Sci 173:12–24

Chen LR, Markhart AH, Shanmugasundaram S, Lin TY (2008) Early developmental and stress responsive ESTs from mungbean, *Vigna radiata* (L.) Wilczek, seedlings. Plant Cell Rep 27:535–552

Chen H, Liu L, Wang L, Wang S, Cheng X (2016) VrDREB2A, a DREB-binding transcription factor from *Vigna radiata*, increased drought and high-salt tolerance in transgenic *Arabidopsis thaliana*. J Plant Res 129:263–273

Choudhary KK, Agrawal SB (2014) Cultivar specificity of tropical mung bean (*Vigna radiata* L.) to elevated ultraviolet-B: Changes in antioxidative defense system, nitrogen metabolism and accumulation of jasmonic and salicylic acids. Environ Exp Bot 99:122–132

Daudi A, Cheng Z, O'Brien JA, Mammarell N, Khan S, Ausubel FM, Bolwell GP (2012) The apoplastic oxidative burst peroxidase in *Arabidopsis* is a major component of pattern-triggered immunity. Plant Cell 24:275–287

Douliez JP, Michon T, Elmorjani K, Marion D (2000) Structure, biological and technological functions of lipid transfer proteins and indolines, the major lipid binding proteins from cereal kernels. J Cereal Sci 32:1–20

Ercoli L, Mariotti M, Masoni A, Arduini I (2004) Growth responses of sorghum plants to chilling temperature and duration of exposure. Eur J Agron 21:93–103

Eriksson SK, Kutzer M, Procek J, Gröbner G, Harryson P (2011) Tunable membrane binding of the intrinsically disordered dehydrin Lti30, a cold-induced plant stress protein. Plant Cell 23:2391–2404

Gaudet DA, Laroche A, Frick M, Huel R, Puchalski B (2003) Cold induced expression of plant defensin and lipid transfer protein transcripts in winter wheat. Physiol Plant 117:195–205

Ghosh A, Pareek A, Sopory SK, Singla-Pareek SL (2014) A glutathione responsive rice glyoxalase II, OsGLYII-2, functions in salinity adaptation by maintaining better photosynthesis efficiency and anti-oxidant pool. Plant J 80:93–105

Hong MG, Kim KH, Ku JH, Jeong JK, Seo MJ, Park CH, Kim YH, Kim HS, Kim YK, Baek SH, Kim DY, Park SK, Kim SL, Moon JK (2015) Inheritance and quantitative trait loci analysis of resistance genes to bruchid and bean bug in Mungbean (*Vigna radiata* L. Wilczek). Plant Breed Biotechnol 3:39–46

Hoque TS, Uraji M, Ye W, Hossain MA, Nakamura Y, Murata Y (2012) Methylglyoxal-induced stomatal closure accompanied by peroxidase-mediated ROS production in *Arabidopsis*. J Plant Physiol 169:979–986

Huang DJ, Chen HJ, Lin YH (2005) Isolation and expression of protein disulfide isomerase cDNA sweet potato (*Ipomoea batatas* [L.] Lam 'Tainong 57') storage roots. Plant Sci 169:776–784

Hung SH, Wang CC, Ivanov SV, Alexieva V, Yu CW (2007) Repetition of hydrogen peroxide treatment induces a chilling tolerance comparable to cold acclimation in mung bean. J Am Soc Hortic Sci 132:770–776

Kang YJ, Kim SK, Kim MY, Lestari P, Kim KH, Ha BK, Jun TH, Hwang WJ, Lee T, Lee J, Shim S, Yoon MY, Jang YE, Han KS, Taeprayoon P, Yoon N, Somta P, Tanya P, Kim KS, Gwag JG, Moon JK, Lee YH, Park BS, Bombarely A, Doyle JJ, Jackson SA, Schafleitner R, Srinives P, Varshney RK, Lee SH (2014) Genome sequence of mungbean and insights into evolution within *Vigna* species. Nat Commun 5:5443

Kaur R, Bains TS, Bindumadhava H, Nayyar H (2015) Responses of mungbean (*Vigna radiata* L.) genotypes to heat stress: effects on reproductive biology, leaf function and yield traits. Sci Hortic 197:527–541

Kitajima M, Butler WL (1975) Quenching of chlorophyll fluorescence and primary photochemistry in chloroplasts by dibromothymoquinone. Biochim Biophys Acta 376:105–115

Koag MC, Fenton RD, Wilkens S, Close TJ (2003) The binding of maize DHN1 to lipid vesicles. Gain of structure and lipid specificity. Plant Physiol 131:309–316

Kratsch HA, Wise RR (2000) The ultrastructure of chilling stress. Plant Cell Environ 23:337–350

Lev S (2010) Non-vesicular lipid transport by lipid-transfer proteins and beyond. Nat Rev Mol Cell Biol 11:739–750

Lin CH, Peng PH, Ko CY, Markhart AH, Lin TY (2012) Characterization of a novel Y_2K-type dehydrin VrDhn1 from *Vigna radiata*. Plant Cell Physiol 53:930–942

Liu KH, Lin TY (2003) Cloning and characterization of two novel lipid transfer protein I genes in *Vigna radiata*. DNA Seq 14:420–426

Liu MS, Kuo TCY, Ko CY, Wu DC, Li KY, Lin WJ, Lin CP, Wang YW, Schafleitner R, Lo HF, Chen CY, Chen LFO (2016) Genomic and transcriptomic comparison of nucleotide variations for insights into bruchid resistance of mungbean (*Vigna radiata* [L.] R. Wilczek). BMC Plant Biol 16:46

Livak KJ, Schmittgen TD (2001) Analysis of relative gene expression data using real-time quantitative PCR and the $2^{-\Delta\Delta CT}$ method. Methods 25:402–408

Meyer Y, Buchanan BB, Vignols F, Reichheld JP (2009) Thioredoxins and glutaredoxins: unifying elements in redox biology. Annu Rev Genet 43:335–367

Miura K, Furumoto T (2013) Cold signaling and cold response in plants. Int J Mol Sci 14:5312–5337

Mustafiz A, Sahoo KK, Singla-Pareek SL, Sopory SK (2010) Metabolic engineering of glyoxalase pathway for enhancing stress tolerance in plants. Methods Mol Biol 639:95–118

Ohnishi S, Miyoshi T, Shirai S (2010) Low temperature stress at different flower development stages affects pollen development, pollination, and pod set in soybean. Environ Exp Bot 69:56–62

Oono Y, Seki M, Satou M, Iida K, Akiyama K, Sakurai T, Fujita M, Yamaguchi-Shinozaki K, Shinozaki K (2006) Monitoring expression profiles of *Arabidopsis* genes during cold acclimation and deacclimation using DNA microarrays. Funct Integr Genomics 6:212–234

Rahman LN, Chen L, Nazim S, Bamm VV, Yaish MW, Moffatt BA, Dutcher JR, Harauz G (2010) Interactions of intrinsically disordered *Thellungiella salsuginea* DHNs TsDHN-1 and TsDHN-2 with membranes—synergistic effects of lipid composition and temperature on secondary structure. Biochem Cell Biol 88:791–807

Rurek M (2010) Diverse accumulation of several DHN-like proteins in cauliflower (*Brassica oleracea* var. *botrytis*), *Arabidopsis thaliana* and yellow lupin (*Lupinus luteus*) mitochondria under cold and heat stress. BMC Plant Biol 10:181

Shanmugasundaram S, Keatinge JDH, Hughes J (2009) The mungbean transformation: diversifying crops, defeating malnutrition. IFPRI Discuss Pap 922:1–43

Tang D, Dong Y, Ren H, Li L, He C (2014) A review of phytochemistry, metabolite changes, and medicinal uses of the common food mung bean and its sprouts (*Vigna radiata*). Chem Cent J 8:4

Vriens K, Cammue BPA, Thevissen K (2014) Antifungal plant defensins: mechanisms of action and production. Molecules 19:12280–12303

Wu Y, Wei W, Pang X, Wang X, Zhang H, Dong B, Xing Y, Li X, Wang M (2014) Comparative transcriptome profiling of a desert evergreen shrub, *Ammopiptanthus mongolicus*, in response to drought and cold stresses. BMC Genomics 15:671

Yang MT, Chen SL, Lin CY, Chen YM (2005) Chilling stress suppresses chloroplast development and nuclear gene expression in leaves of mung bean seedlings. Planta 221:374–385

Yu CW, Murphy TM, Lin CH (2003) Hydrogen peroxide-induced chilling tolerance in mung beans mediated through ABA-independent glutathione accumulation. Funct Plant Biol 30:955–963

17

Morphological, physiological, cytological and phytochemical studies in diploid and colchicine-induced tetraploid plants of *Fagopyrum tataricum* (L.) Gaertn

Lin-Jiao Wang[1,3], Mao-Yin Sheng[1,2,3*], Pei-Cai Wen[1,4] and Jia-Ying Du[3,4]

Abstract

Background: Tartary buckwheat are very popular as an important functional food material and its cultivation is very widespread in our whole world, but there obviously lack works in the researches of genetic breeding for agricultural and medicinal utilization. The aim of this study is to obtain good germplasm resources for agricultural and medicinal use of tartary buckwheat (*Fagopyrum tataricum*) by inducing the tetraploid plants.

Results: Four cultivars of *F. tataricum*, that is, Qianwei 2#, Jinku 2#, Chuanqiao 1#, and Liuqiao 1# were selected to experiment. The tips of seedlings with two true leaves were treated by 0.25% (w/v) colchicine solution for 48, 72, and 96 h, respectively. The chromosome number of treated plants was determined by metaphase chromosome counting of root tip cells and PMCs (pollen mother cells) meiosis observation. Tetraploid induction successfully occurred in all three treatments with an efficiency ranging from 12.13 to 54.55%. The chromosome number of the diploid plants was $2n = 2x = 16$, and that of the induced tetraploid plants was $2n = 4x = 32$. The typical morphological and physiological qualities were compared between the control diploid and corresponding induced tetraploid plants. Results showed that the induced tetraploid plants had obviously larger leaves, flowers, and seeds. Moreover, the content of seed protein and flavonoid were also increased in the tetraploid plants. The pollen diameter and capsule size of diploid plants were significantly smaller than those of tetraploid plants.

Conclusion: *Fagopyrum tataricum* can be effectively induced into tetraploids by colchicines. The tetraploid induction can produce valuable germplasm resources for breeding and is a practicable breeding way in *F. tataricum*.

Keywords: *Fagopyrum tataricum*, Colchicine, Tetraploid, Chromosome counting, Morphological characteristics

Background

Buckwheat as an important, highly nutritional nonpoaceous crop has great potential as source of food, forage, and medicine (Lin 1994; Holasova et al. 2002; Tang and Wang 2010; Pan and Chen 2010). Buckwheat originates from China and is widely distributed in Asia and Europe (Hou et al. 2016). The tartary buckwheat (*Fagopyrum tataricum* (L.) Gaertn), a diploid species ($2n = 2x = 16$), belongs to the big-achene group of genus *Fagopyrum* (Chen et al. 2004; Sheng et al. 2011). Tartary buckwheat grain, as an important functional food material, contains proteins with high biological value and balanced amino acid composition; relatively high crude fiber and vitamins B1, B2, and B6; and more rutin than common buckwheat (*F. esculentum* Moench) (Pan and Chen 2010; Wang et al. 2015). The products of tartary buckwheat are very popular and its cultivation is widespread in whole world (Lin 1994). Although lots of cultivation and utilization have been done in tartary buckwheat, there obviously lack studies in the genetic breeding of tartary buckwheat for

*Correspondence: shmoy@163.com
[3] National Engineering Research Center for Karst Rocky Desertification Control, Guizhou Normal University, Baoshanbei Road 180, Guiyang 550001, P. R. China
Full list of author information is available at the end of the article

agricultural and medicinal utilization (Ohnishi and Ohta 1987; Chen et al. 2007).

The polyploidy induction is a significant method for the production of new germplasm resources applicable for plant genetic breeding (Tang et al. 2010). Polyploidy induction has been used as a breeding method in many plants, especially, in crop, horticulture and medical botany. Induced polyploidy can enhance qualities for agriculture, medicine and horticulture utilization (Tang et al. 2010) and improve resistibility to environmental stresses and diseases (Ahloowalia 1967). Besides, polyploidy induction can lead to content change of secondary metabolites (Majdi et al. 2010). Colchicine has been found to have a significant effect on polyploid induction and is widely used for inducing polyploidy in plants (Tang et al. 2010) because it can effectively arrests mitosis at the anaphase stage (Kunitake et al. 1998; Kermani et al. 2003). Until now, numerous researchers reported that germplasm resources have been innovated using polyploidization techniques in many crop or medicinal plants (Ahloowalia 1967; Kermani et al. 2003; Tang et al. 2010; Ye et al. 2010; Kanoktip et al. 2014) including *F. esculentum* (Zhu and Gao 1988; Lian and Chen 2013). Limited works regarding tartary buckwheat, polyploidy induction works have been reported (Hou et al. 2002; Zhou et al. 2012). In the present study, the tetraploid *F. tataricum* plants were successfully induced by colchicine. The chromosome number of the induced plants was confirmed by cytological identification, and the morphological features of the induced tetraploid plants were recorded. The germplasm characteristics and breeding utilization of such tetraploid materials were discussed. The aim of this present study was to produce novel and valuable tetraploid germplasm resources for the utilization of agriculture and medicine in *F. tataricum*.

Methods

Plant material

Mature seeds of the following four *F. tataricum* cultivars ($2n = 2x = 16$): Qianwei 2#, Jinku 2#, Chuanqiao 1#, and Liuqiao 1#, were collected from Guizhou, Shanxi, Sichuan, and Yunnan province of China, respectively. Before experiment in this study, they were stored in a freezer at temperature of 0–2 °C. The field testing was conducted during 2012–2013 year in the experimental field site of Guizhou Normal University, Guiyang, China. The laboratory test was conducted in the State Key Laboratory Incubation Base for Karst Mountain Ecology Environment of Guizhou Province, Guiyang, China.

Tetraploid induction

Before germination, seeds, firstly, were disinfected by 0.3% K_2MnO_4 for 20 min, then washed with distilled water for three times and dried at a loft drier with 25 °C for about 24 h and finally sown into flower pots. The pots were covered with transparent plastic film and incubated under greenhouse conditions. Colchicine was applied to the apical meristem of seedlings in the stage of two true leaves. Three treatment durations (i.e. 48, 72 and 96 h) were studied to determine the optimum colchicine exposure times at concentration of 0.25% (w/v) for tetraploid induction. Absorbent cotton balls sucked with a solution of colchicines (0.25%, w/v) were wrapped around the apical tips of seedlings when two true leaves of the emerged seedlings appeared. The experiments were conducted in a greenhouse and continued for the three treatment durations (i.e. 48, 72 and 96 h); the colchicines solution was dropped to the absorbent cotton balls twice a day, at 8:30 and 17:30 o'clock. After the treatment durations, the absorbent cotton balls were removed from the seedling tips, and then the seedling tips were cleaned by distilled water and nursed further in the greenhouse. 50 seedlings were experimented for each treatment. After 2 weeks of the durations of colchicines treatment, death rate on young plants was recorded.

Chromosome counting and meiosis observation

The ploidy levels were determined by chromosome counting of root tip cells, and confirmed by PMCs (pollen mother cells) meiosis observation. Before induction of polyploidy, the chromosome number was determined in root tips of germinated seeds. For the induced plants, seeds were obtained by self-pollination of each treated plant and the chromosome counting was also conducted in root tips. When root tips grew to 1–2 cm long, root rips were cut and pretreated by the method of soaking in a saturated solution of α-bromine naphthalin at room temperature for 5 h. After pretreated, root tips were washed by distilled water for three times and then immersed in Carnoy's liquid (ethanol:acetic acid = 3:1) for 12–20 h. After three rinses by distilled water, the root tips were soaked in 1 N HCl at 60 °C for 8 min and then moved to distilled water. After that, the root tips were stained by improved Carbolfuchsin, and squashed for chromosome observation. For meiotic investigation, PMCs in young flowers of about one week age were fixed in Carnoy II solution and stained in 2% (w/v) aceto-carmine solution. Images of mitotic chromosomes in the root tip and meiosis were photo by a DP70 digital camera in a BX51 microscope (Olympus, Japan).

Morphological and physiological characteristics determinations

The morphological and physiological characteristics were compared between the control diploid plants and corresponding induced tetraploid plants. Parameters of plant height, plant breadth, leaf length, leaf width, inflorescence length and flower size were directly measured by a ruler. A digital caliper was used to measure the characteristics of leaf thickness and diameter of basal stem. Pollen grains were collected from flowers at anthesis, fixed in Carnoy II solution overnight, and then stained by 1% (v/v) I_2-KI solution to detect pollen vitality. The pollen diameter (average of 30 pollens) was measured by a light microscope (magnification of 40×). Seed length was measured by a digital caliper and weight of 1000 seeds was determined by a balance scale.

Determination of seed protein and flavonoid content

The seed protein and flavonoid content was determined and compared between the control diploid plants and corresponding induced tetraploid plants. Mature seeds were harvested, dried to constant weight, and then ground to a fine powder for the determination of seed protein and flavonoid content. An automatic micro-Kjeldahl procedure was used to determine seed protein content using a conversion factor of 6.25 times the N content. Seed flavonoid content of each sample was determined following colorimetric method (Chang et al. 2002). Briefly, 1 mL solution of each plant extracts (at 0.4% w/v) in methanol were separately mixed with 4 mL of methanol, 2 mL of 10% aluminium chloride, and 3 mL of 1 M potassium acetate, and left at room temperature for 30 min. The absorbance of the reaction mixture was measured at 420 nm with a double beam Perkin Elmer UV/Visible spectrophotometer (USA). The calibration curve was prepared by preparing rutin solution at concentrations 1.25–20 μg ml^{-1} in methanol.

Statistical analysis

Comparisons of quantitative data between two variable groups were made using Student's unpaired *t* test (two-sample *t* test), and morphological variation rates among >2 groups were analysed using ANOVA and multiply comparison among levels of each factor and among combinations were analysed using Duncan's test (Sheng et al. 2011). All statistical analyses were performed by SPSS 16.0 software (Du 1999).

Results

Tetraploid induction

The effects of different durations (48, 72, and 96 h) with 0.25% (w/v) colchicines solution on death rate and induction efficiency of tetraploid induction in four chosen cultivars of *F. tataricum*, were examined and recorded in Table 1. Results showed that the solution of colchicines was detrimental to growth and survival of seedlings. The growth of untreated plants (control) was faster than the plants treated with colchicines solution. The death rate (%) and tetraploid induction efficiency (%) were obvious differences among three different durations of colchicines treatment (Table 1). All the four cultivars can be induced tetraploids successfully, but the induction efficiency were obvious different. Among the twelve treatment combinations, the tetraploid induction efficiency ranged from 12.13 to 54.55%. Multi-comparison analysis indicated that the induction efficiency of the treatments of Qianwei 2# and Chuanqiao 1# with the treatment duration of 72 h are the highest (Table 1). The mortality rates of treated seedling ranged from 2 to 10%, and along with the treatment duration increase, the mortality rates become higher.

Cytological differences

Metaphase chromosome observation of root tip cells showed that the control plants are diploid with 8 pairs of chromosomes ($2n = 2x = 16$) (Fig. 1a–d). As expected, there are 16 chromosome pairs in the putative tetraploid ($2n = 4x = 32$) (Fig. 1e–h). Observations of the meiosis in PMCs confirmed the chromosome number of the putative tetraploid plants. At meiotic MI stage, chromosomes of diploid plants formed 8 bivalents ($2x = 8II$, Fig. 2c, d), while the putative tetraploid plants mostly formed 8 quadrivalents ($4x = 8IV$, Fig. 2a), occasionally 4 quadrivalents and 8 bivalents ($4x = 4IV + 8II$, Fig. 2b) were observed. At meiotic telophase II, some abnormal phenomenons, such as micronucleus (Fig. 3a–d), triads (Fig. 3d), pentads (Fig. 3e), and hexads (Fig. 3f) were observed in both of the control diploid and induced tetraploid plants, but the occurrence of abnormal phenomenons in the control diploid plants are obviously less than in those induced tetraploid plants (Table 2).

Morphological and physiological differences

The comparison of vegetative characteristics and reproductive traits between the confirmed tetraploid plants and corresponding diploid plants of the four studied cultivars were presented in Tables 3 and 4, respectively. Results of two-sample *t* test statistical analysis indicated significant ($p = 0.01$) differences between the control diploid and induced tetraploid lines of *F. tataricum* with regard to the length, width and thickness of the leaves and the leaf index (Fig. 4). Compared with the corresponding diploid plants, the length, width and thickness of the leaves in all of the induced tetraploid plants had increased. The averages of increasing rate were 13.98, 37.40 and 139.33%, respectively. There were no significant differences ($p = 0.05$) in plant height, crown size or diameter of the basal stem between the control diploid and corresponding tetraploid lines in each of the four studied cultivars. However, regarding to transverse diameter of the flower bud, flower diameter, length of petal, and width of petal, significant

Table 1 Effects of colchicines (0.25%) treated duration on death rate, and tetraploid induction efficiency in treated cultivars of *Fagopyrum tataricum*

Cultivar	Duration of treatment (h)	Number of seedlings	Death rate (%)	Tetraploid induction efficiency (%)
Qianwei 2#	48	50	2.00	29.15 D
	72	50	6.00	53.23 A
	96	50	10.00	41.45 BC
Jinku 2#	48	50	2.00	12.13 E
	72	50	4.00	31.24 D
	96	50	6.00	19.35 E
Chuanqiao 1#	48	50	4.00	50.34 B
	72	50	6.00	54.55 A
	96	50	8.00	48.56 B
Liuqiao 1#	48	50	4.00	18.34 E
	72	50	8.00	30.45 D
	96	50	10.00	20.04 E

Death rate was recorded after 2 weeks of colchicines treatment

Different capital letters indicate significance of the difference between two mean values at the $p = 0.01$ level, as tested by Duncan's test

differences ($p = 0.01$) were observed (Fig. 5). The diameters of pollen grains in the induced tetraploid plants were significantly larger ($p = 0.01$) than those in each corresponding diploid cultivars (Fig. 5), with an increasing rate of 28.13% in "Qianwei 2#", 45.81% in "Jinku 3#", 28.35% in "Chuanqiao 1#", and 42.67% in "Liuqiao 1#" cultivar, respectively. Contrarily, the pollen vitality in the induced tetraploid plants were significantly lower ($p = 0.05$) than those in the corresponding diploid cultivars (Fig. 6). The seed size and thousand-seed weight in the induced tetraploid lines were obviously greater ($p = 0.05$) than those in the corresponding diploid cultivars (Fig. 7). Regarding to parameters of the average length of inflorescence, inflorescence number per individual, and floret number per inflorescence, no significant differences were observed between the control diploids and corresponding induced tetraploid plants.

In the process of plant cultivation, the resistance of the diploid and induced teteraploid lines to pathogens and insects were observed by the method of Mohammad et al. (2013). The main diseases and pests of *F. tataricum* in the area of Guizhou, China are powdery mildew, aphids and scale insects. Results showed that the resistance of the induced tetraploid lines to these pathogens and insects is stronger than those of the corresponding diploid plants. The average infection rate of the induced teraploid lines (7.93%, ranging from 3.1 to 10.5%) is significance less ($p = 0.05$) than those of the corresponding diploid plants (10.08%, ranging from 5.1 to 12.5%).

Differences of seed protein and flavonoid content

The seed protein and flavonoid content of induced tetraploid and corresponding diploid lines of each of the four cultivars were determined and compared by the method of two-sample t test (Table 5). Results showed there are significant ($p = 0.01$) differences of seed protein and flavonoid content between the control diploid and corresponding tetraploid lines of *F. tataricum*. The seed protein content in the confirmed tetraploid plants were significantly higher ($p = 0.01$) than the corresponding diploid plants. The average increasing rates were: 12.75% in "Qianwei 2#", 39.57% in "Jinku 2#", 14.60% in "Chuanqiao 1#", and 27.59% in "Liuqiao 1#" cultivar, respectively. Also, the seed flavonoid content in the tetraploid lines was significantly higher ($p = 0.01$) than the corresponding diploid lines, with an average increasing rates: 79.92% in "Qianwei 2#", 28.86% in "Jinku 2#", 10.10% in "Chuanqiao 1#", and 51.15% in "Liuqiao 1#" cultivar, respectively.

Discussion

Tetraploid induction and identification in *F. tataricum*

Polyploidization is a very important and valuable method for plant breeding to improve various important morphological and physiological characteristics and to generate valuable germplasm resources, and it has been successfully applied in many plant species (Tang et al. 2010; Dhooghe et al. 2011). Colchicine, as a chromosome doubling agent, has been widely applied to induce plant polyploidy because it disturbs the mitosis at the anaphase stage to result in chromosome doubling (Nigel et al. 2007; Pour et al. 2012). The tetraploid plants of common buckwheat were successfully induced with colchicines treatment by Zhu and Gao (1988) and Zhu et al. (1992); however, such research is very lacking in tartary buckwheat. In the present study, treatments with 0.25% (w/v) colchicines solution for three different

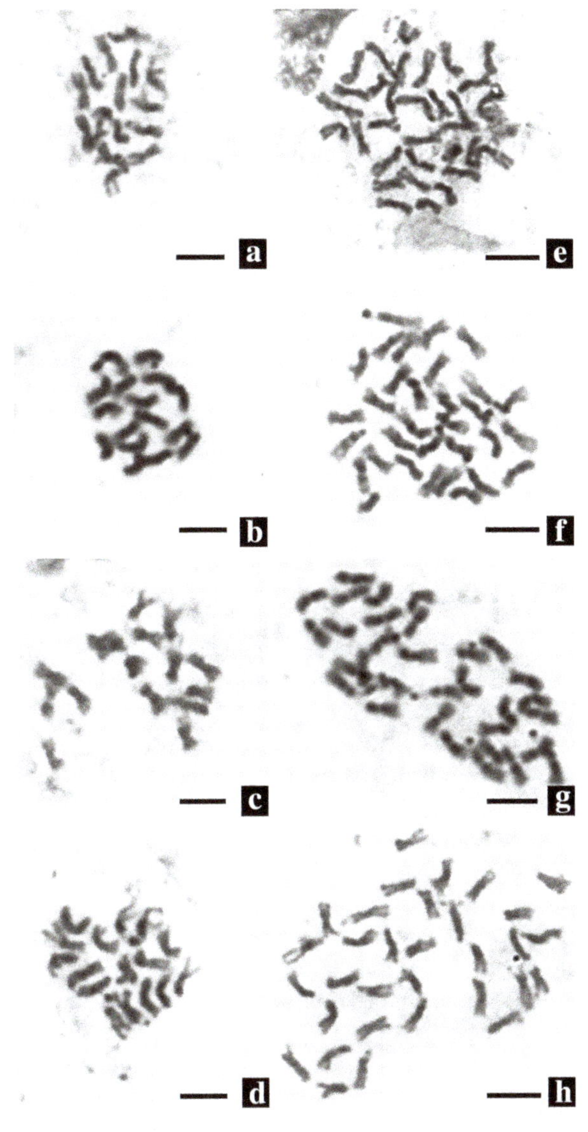

Fig. 1 Chromosomes of root tip cell of the diploid *F. tataricum* cultivars (**a**–**d**) and their corresponding induced tetraploids (**e**–**h**) (**a**/**e** Qianwei 2#; **b**/**f** Jinku 2#; **c**/**g** Chuanqiao 1#; **d**/**h** Liuqiao 1#). *Bars* 5 μm

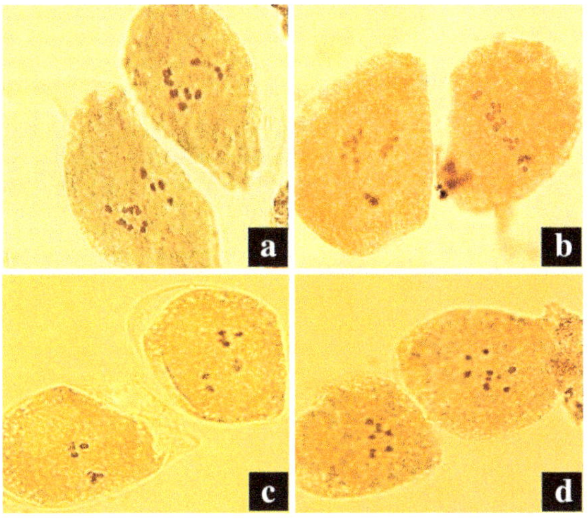

Fig. 2 Chromosome configuration of meiotic MI stage of the diploid and induced tetraploid *F. tataricum* (**a** $4x = 8IV$, **b** $4x = 4IV + 8II$, **c** and **d** $2x = 8II$)

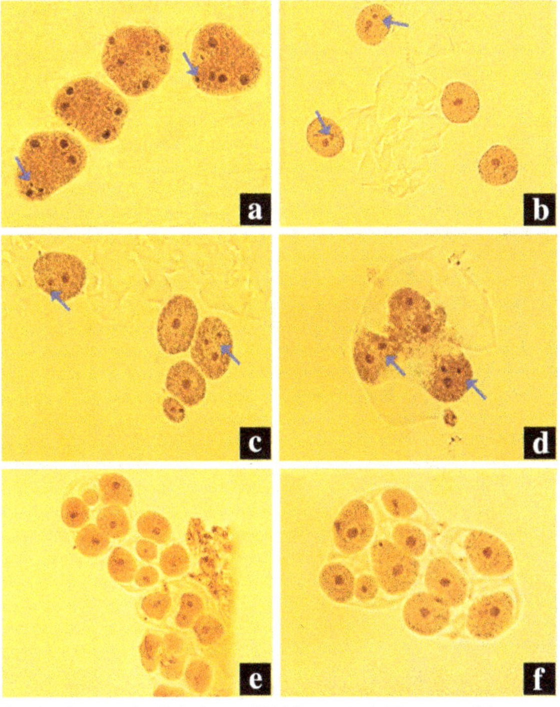

Fig. 3 Abnormal cell division of PMC at meiotic TII stage of the induced tetraploid *F. tataricum* plants (**a** Tetrads with micronucleus; **b** tetrad with multiple nucleoli; **c** Pentad with multiple nucleoli; **d** triad with multiple nucleoli; **e** pentads; **f** hexads). *Arrows* show multiple nucleoli

durations of 48, 72, and 96 h were used to induce tetraploid plants in *F. tataricum*. From results, we conclude that the treated duration would obviously affect the efficiency of tetraploid induction and death rate of seedlings. Numerous previous studies indicated that the duration of colchicines treatment was positive correlation to the seedlings death rate in many plant species (Sikdar and Jolly 1994; Mohammad et al. 2013). Our study also showed the similar results in *F. tataricum*, indicating such effect may be common in plant. However, the duration of colchicines treatments which may cause higher death rate also resulted in higher efficiency to induce tetraploids (Chakraborti et al. 1998; Mohammad et al. 2013). Therefore, the suitable duration of the colchicines solution treatment is very important to obtain successful

Table 2 Comparisons of meiotic TII stage between the induced tetraploids and their corresponding diploid *F. tataricum* cultivars

Cultivar	Tetraploid			Diploid			t^a	t^b
	Sum of cell	Abnormal cell frequency (%)	Micronucleus to cell ratio	Sum of cell	Abnormal cell frequency (%)	Micronucleus to cell ratio		
Qianwei 2#	132	30.30	0.5000	112	5.36	0.0536	4.9645**	7.6190**
Jinku 2#	107	25.23	0.4486	107	4.67	0.0467	4.2171**	6.8096**
Chuanqiao 1#	140	18.57	0.3857	159	4.40	0.0440	3.9015**	7.3160**
Liuqiao 1#	194	28.87	0.4897	124	4.03	0.0403	5.4857**	8.4178**
Average		25.05	0.4494		4.45	0.0445		

t^a Indicate the t value of abnormal cell frequency (%)

t^b Indicate the t value of micronucleus to cell ratio

** Indicate significance of the difference at the $p = 0.01$ level

tetraploid induction. In the present study, the death rate of plants treated by colchicines ranging 2.0–10.0% was recorded. Ahloowalia (1967) reported that seedling mortality rate was 74.6% in *Lolium perenne* plants treated with 0.2% colchicines for 3 h, while Liu et al. (2007) reported that seedling mortality rate was 100% in *Platanus acerifolia* plants treated with 0.5% colchicines for 24 h. Due to a low mortality rate observed in the present study, we concluded that *F. tataricum* plants may have a high level of resistance to the colchicines toxicity. It can be also deduce that colchicines should be effective for producing tetraploids in other species of *Fagopyrum* genus because *F. tataricum* was effectively induced into tetraploids by colchicines.

Until now, several approaches have been used for the polyploidy identification in plants (Tandon and Bali 1957; Pei 1985; Li and Zhang 1991; Yang et al. 2006). Numerous early studies confirmed that the size and number of stomata and the number of chloroplasts within the guard cells of the induced tetraploid plant were significantly different from the corresponding diploid plants (Speckmann et al. 1965; Tan and Dunn 1973; Cohen and Yao 1996; Beck et al. 2003). In the event of chromosome doubling, morphological characteristics of plants can change significantly. Thus, morphological characteristics were useful markers in the identification of induced tetraploids. First, a large population of treated seedlings was pre-screened for putative tetraploid lines by morphological markers in the present study. Subsequently, the real tetraploid lines were confirmed by chromosome counting of root tip cells and meiosis in PMCs. This technological process is considered fast and effective for the identification of induced tetraploid plants. In the present study, the length, width, and thickness of leaves of the tetraploid plants induced from the four different *F. tataricum* cultivars significantly increased in comparison to the corresponding control diploid plants. Parameters of the flower diameter, seed size and weight of 1000 seeds also significantly increased in the induction tetraploid plants, in comparison to the control diploid plants. The chromosome number of *F. tataricum*, $2n = 2x = 16$, was reported (Sheng et al. 2013). In the present study, as expected, the chromosome

Table 3 Comparisons of vegetative characteristics between the induced tetraploids and their corresponding diploid *F. tataricum* cultivars

Vegetative characteristic (cm)	Qianwei 2#		Jinku 2#		Chuanqiao 1#		Liuqiao 1#	
	4x	2x	4x	2x	4x	2x	4x	2x
Plant height	66.30 ± 20.20a	55.03 ± 5.35a	62.21 ± 15.34a	56.00 ± 10.23a	59.80 ± 21.02a	61.67 ± 2.34a	75.00 ± 23.12a	69.90 ± 11.20a
Plant breadth	40.23 ± 12.34a	45.89 ± 9.21a	39.36 ± 12.89a	51.24 ± 18.21a	45.34 ± 18.23a	43.89 ± 8.90a	53.34 ± 20.01a	50.34 ± 15.78a
Diameter of basal stem	1.03 ± 0.38a	0.98 ± 0.31a	1.11 ± 0.38a	1.02 ± 0.30a	0.99 ± 0.20a	0.96 ± 0.13a	1.20 ± 0.14a	1.21 ± 0.33a
Leaf length	6.12 ± 0.12A	5.28 ± 0.28B	6.39 ± 0.22A	5.63 ± 0.26B	6.56 ± 0.17A	5.86 ± 0.15B	7.23 ± 0.32A	6.31 ± 0.18B
Leaf width	4.02 ± 0.28A	3.34 ± 0.17B	4.21 ± 0.21A	3.25 ± 0.13B	3.99 ± 0.15A	3.05 ± 0.11B	5.32 ± 0.21A	3.15 ± 0.15B
Leaf thickness	0.95 ± 0.05A	0.38 ± 0.13B	0.83 ± 0.04A	0.35 ± 0.11A	0.88 ± 0.14A	0.41 ± 0.10B	0.92 ± 0.09A	0.36 ± 0.16B
Leaf index (length/width)	1.43 ± 0.22a	1.21 ± 0.25b	1.48 ± 0.31a	1.11 ± 0.32b	1.32 ± 0.18a	1.39 ± 0.20a	1.42 ± 0.23a	1.35 ± 0.19a

Different small-case/capital letters indicate the significance of difference between two mean values at the $p = 0.05/0.01$ level, respectively, as tested by two-sample t test, the same below

Table 4 Comparisons of reproductive characteristics between the induced tetraploids and their corresponding diploid *F. tataricum* cultivars

Reproductive characteristic	Qianwei 2#		Jinku 2#		Chuanqiao 1#		Liuqiao 1#	
	4x	2x	4x	2x	4x	2x	4x	2x
Vertical length of flower bud (mm)	4.12 ± 0.16A	3.23 ± 0.22B	4.01 ± 0.24a	3.36 ± 0.25b	3.99 ± 0.20a	3.37 ± 0.15b	4.19 ± 0.23A	3.30 ± 0.11B
Transverse diameter of flower bud (mm)	2.35 ± 0.18A	1.88 ± 0.21B	2.22 ± 0.12A	1.79 ± 0.20B	2.31 ± 0.15a	2.01 ± 0.11b	2.68 ± 0.17A	1.90 ± 0.23B
Flower diameter (mm)	5.15 ± 0.26A	3.68 ± 0.20B	6.35 ± 0.32A	3.24 ± 0.19B	5.45 ± 0.25A	3.56 ± 0.18B	6.41 ± 0.20A	3.29 ± 0.12B
Length of petal (mm)	2.69 ± 0.11A	1.63 ± 0.12B	3.01 ± 0.18A	1.51 ± 0.18B	2.23 ± 0.13A	1.25 ± 0.20B	2.95 ± 0.09A	1.17 ± 0.18B
Width of petal (mm)	0.90 ± 0.03A	0.52 ± 0.06B	0.97 ± 0.09A	0.49 ± 0.02B	0.88 ± 0.07A	0.40 ± 0.05B	0.95 ± 0.03A	0.38 ± 0.08B
Pollen grain diameter (μm)	5.01 ± 0.34A	3.91 ± 0.28B	5.57 ± 0.34A	3.82 ± 0.23B	4.98 ± 0.31A	3.88 ± 0.28B	5.55 ± 0.27A	3.89 ± 0.31B
Pollen-vigorous (%)	20.35 ± 1.38B	49.36 ± 3.21A	18.89 ± 1.34B	42.66 ± 2.89A	19.67 ± 1.89B	45.89 ± 2.92A	17.89 ± 1.25B	40.96 ± 3.01A
No. of inflorescences per plant	42 ± 6a	39 ± 8a	37 ± 7a	35 ± 8a	43 ± 7a	39 ± 5a	38 ± 9a	33 ± 5a
Average length of inflorescence (cm)	8.91 ± 4.25a	10.35 ± 4.36a	7.89 ± 5.02a	10.78 ± 5.21a	7.34 ± 4.56a	9.67 ± 3.89a	6.99 ± 4.23a	10.36 ± 5.01a
No. of flowers per inflorescence	34 ± 21a	67 ± 31a	42 ± 20a	73 ± 26a	39 ± 25a	71 ± 30a	49 ± 22a	77 ± 31a
Thousand-seed weight (g)	24.41 ± 3.67A	19.38 ± 2.25B	22.66 ± 1.89A	19.11 ± 1.12B	31.59 ± 1.35A	18.12 ± 0.98B	26.84 ± 2.01A	17.57 ± 1.03B
Seed size (mm)	8.29 ± 0.84A	5.86 ± 0.73B	8.10 ± 0.35A	6.01 ± 0.34B	7.20 ± 0.17A	5.63 ± 0.22B	7.17 ± 0.24A	5.42 ± 0.11B

Different small-case/capital letters indicate the significance of difference between two mean values at the $p = 0.05/0.01$ level, respectively, as tested by two-sample *t* test, the same below

Fig. 4 Leaves of the induced tetraploids and their corresponding diploid *F. tataricum* cultivars (**a** Qianwei 2#; **b** Jinku 2#; **c** Chuanqiao 1#; **d** Liuqiao 1#)

Fig. 5 Flowers of the induced tetraploids and their corresponding diploid *F. tataricum* cultivars (**a** Qianwei 2#; **b** Jinku 2#; **c** Chuanqiao 1#; **d** Liuqiao 1#)

number of control diploid plants was $2n = 2x = 16$ and the induced tetraploid plants $2n = 4x = 32$ (Fig. 1). The autotetraploids, pre-screened by morphological markers, were confirmed by chromosome counting and PMCs meiosis observation. Results showed that applying the morphological characteristics as preliminary screening markers for the identification of induced tetraploid lines was feasible.

Germplasm characteristics and utilization of tetraploid in *F. tataricum*

In the present study, all the four *F. tataricum* cultivars were successfully induced into tetraploid plants by colchicine. In comparison to the control diploid plants, several morphological characteristics were significantly increased in the induced tetraploid plants, such as leaf size, chlorophyll content, flower size and pollen diameter. Particularly, the important agronomic parameters of seeds, including seed size and thousand-seed weight, of the induced tetraploid plants were significantly greater than the control diploid plants. In addition, the seed protein and flavonoid content also significantly increases in the induced tetraploid plants. The improvement of morphology, physiology, and phytochemistry in the induced tetraploid lines obtain from the four *F. tataricum* cultivars suggest that tetraploids are useful materials for generating innovative germplasm resources in the genetics and breeding of genus *Fagopyrum*. Generally, polyploids have higher genetic adaptability and resistance to environmental stresses than diploids. Li et al. (2007) reported that induced tetraploids significantly increased their cold tolerance than corresponding diploid

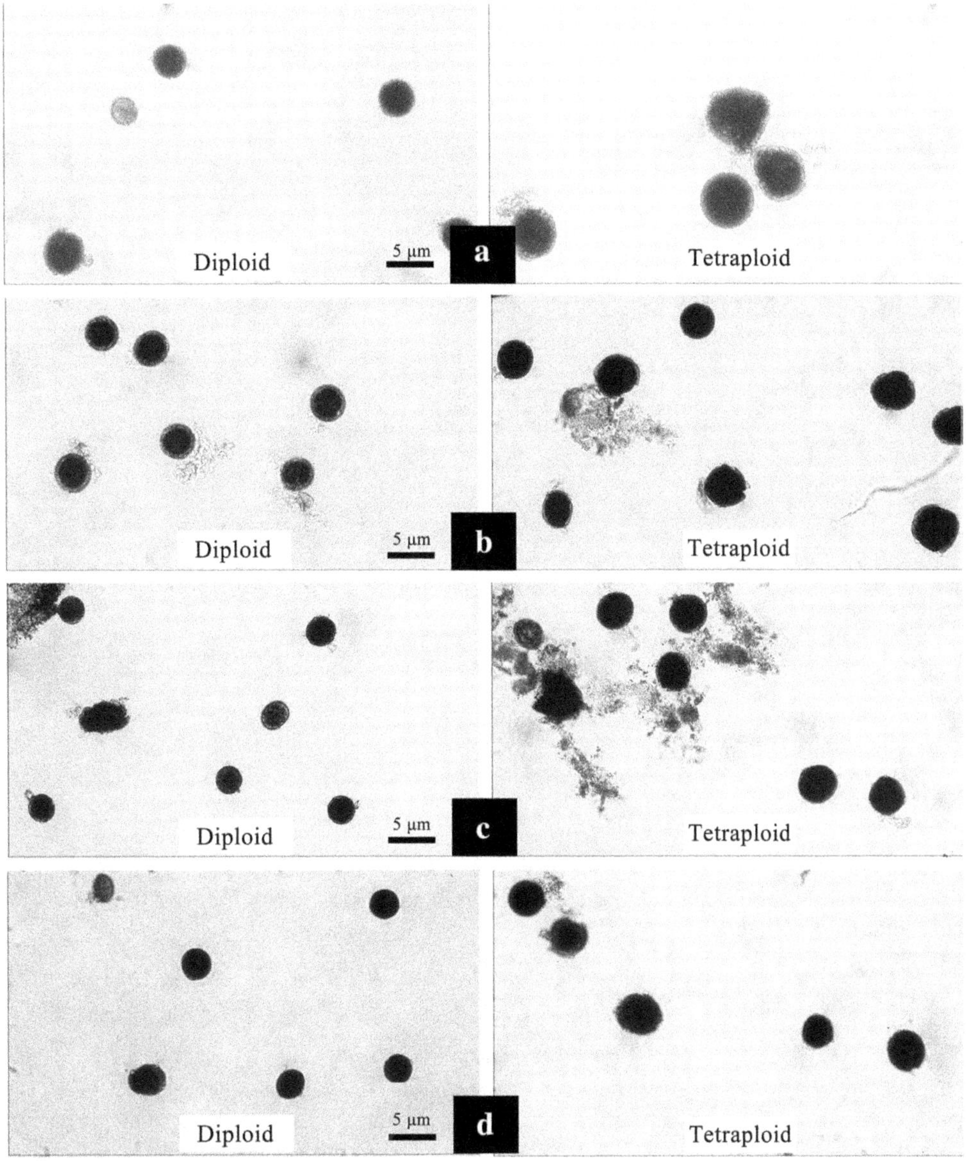

Fig. 6 Pollen grains of the induced tetraploids and their corresponding diploid *F. tataricum* cultivars (**a** Qianwei 2#; **b** Jinku 2#; **c** Chuanqiao 1#; **d** Liuqiao 1#)

plants in the studied ornamental plants. While in the group of agricultural and medical plants, the induced tetraploids significantly increased biomass or effective compounds content. The present study also shows those induced tetraploid plants of *F. tataricum* display stronger resistance to environmental stress than corresponding diploids.

To further evaluate the potential value of induced tetraploids in the *F. tataricum* breeding, the meiosis in PMCs of the induced tetraploids and corresponding control diploids were analysed. The abnormities of meiosis in PMCs mainly happened at the MI and TII stag. The abnormal chromosome behaviours observed at MI and TII stages may imply the occurrence of disturbances in the process of meiosis at PMCs (Chen et al. 2004). In the present study, at MI and TII stage at the meiosis, a few of abnormal chromosomal behaviors and cell divisions, such as micronucleus, triads, pentads, and hexads, were observed in both of the control diploid and induced tetraploid plants, but the frequency of abnormities of the PMCs meiosis in induced tetraploid plants are significantly more than the control diploid plants. This result explains the low seed-setting percentage of the induced tetraploid lines. At MI stage of the

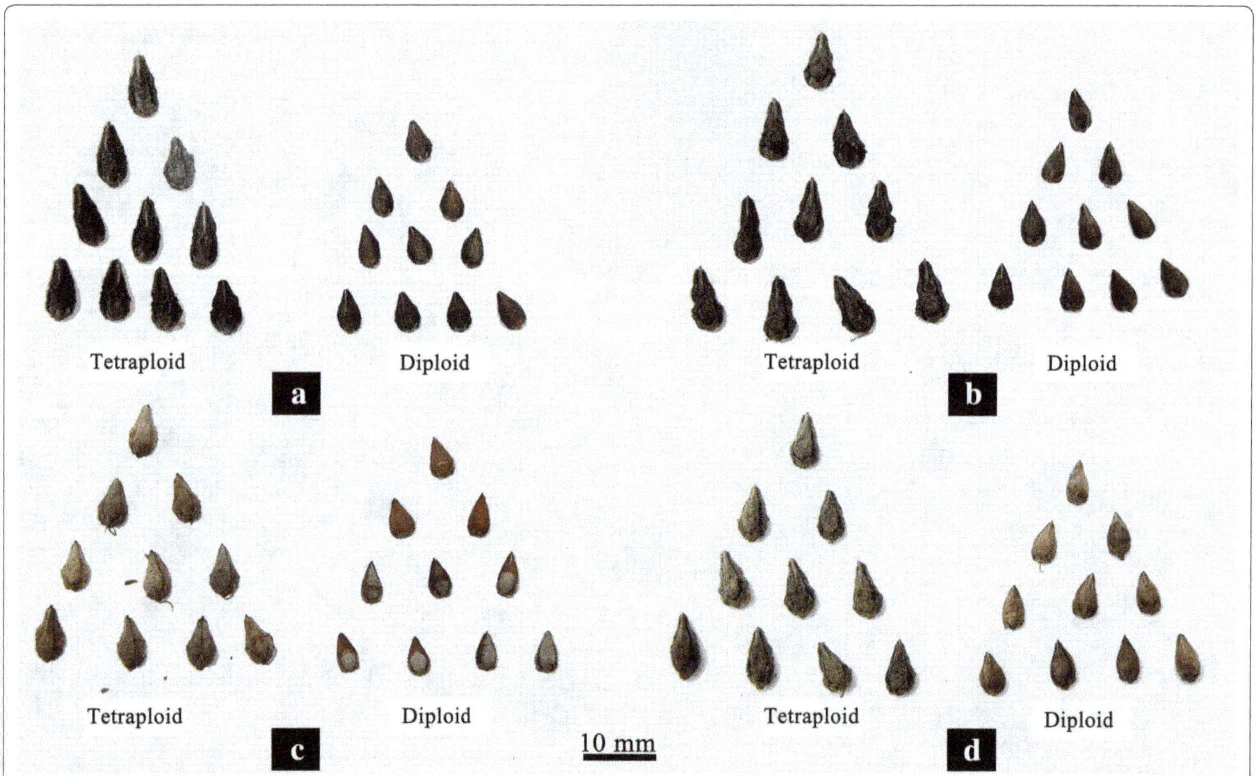

Fig. 7 Seeds of the induced tetraploids and their corresponding diploid *F. tataricum* cultivars (**a** Qianwei 2#; **b** Jinku 2#; **c** Chuanqiao 1#; **d** Liuqiao 1#)

Table 5 Comparisons of seed protein content and seed flavonoid content between induced tetraploids and their corresponding diploid *F. tataricum* cultivars

Content	Qianwei 2#		Jinku 2#		Chuanqiao 1#		Liuqiao 1#	
	4*x*	2*x*	4*x*	2*x*	4*x*	2*x*	4*x*	2*x*
Seed protein (%)	16.4 ± 0.32A	14.9 ± 0.45B	19.4 ± 1.56A	13.9 ± 1.14B	15.7 ± 1.14A	13.7 ± 1.03B	18.5 ± 1.21A	14.5 ± 0.96 B
Seed flavonoid (%)	4.57 ± 0.25 A	2.54 ± 0.18B	3.84 ± 0.38A	2.98 ± 0.27B	3.23 ± 0.22A	2.91 ± 0.16B	3.96 ± 0.31A	2.62 ± 0.11B

meiosis of tetraploid plants, various chromosome configurations including bivalents, univalent, trivalent, and quadrivalent of chromosome synapsis were observed, showing that the meiotic chromosome synapsis of tetraploid plants is very complicated. This can be attributed to the chromosome doubling by colchicines treatment to form a autotetraploid which have four sets of each chromosome in the genome. The problems in chromosome synapsis among four homologous chromosomes of autotetraploid cause the unbalance of chromosome segregation and result in high genetic instability in pollen grains.

Conclusions

Fagopyrum tataricum can be effectively induced into tetraploids by colchicines. In comparison to the control diploid plants, various agronomic traits may be significantly increased in the induced tetraploid plants, including seed size, thousand-seed weight, leaf size, chlorophyll content, flower size and pollen diameter. The content of seed protein and flavonoid may also be increased in the tetraploid plants. Moreover, the genetic instability in tetraploid plants also increased. The induced tetraploid materials can be used to generate innovative germplasm for the *Fagopyrum* species breeding.

Abbreviations

PMCs: pollen mother cells; ANOVA: analysis of variance.

Authors' contributions

LJW carried out the experimental work; MYS provided funding, designed the experiments, and wrote the manuscript; PCW participated in the experimental work and data analysis; JYD collected the plant materials, participated in the experimental work and helped to draft the manuscript. All authors read and approved the final manuscript.

Author details

[1] Karst Research Institute, Guizhou Normal University, Baoshanbei Road 180, Guiyang 550001, P. R. China. [2] School of Life Science, Nanjing University, Hankou Road 32, Nanjing 210093, P. R. China. [3] National Engineering Research Center for Karst Rocky Desertification Control, Guizhou Normal University, Baoshanbei Road 180, Guiyang 550001, P. R. China. [4] State Key Laboratory Incubation Base for Karst Mountain Ecology Environment of Guizhou Province, Baoshaobei Road 180, Guiyang 550001, P. R. China.

Acknowledgements

The authors are grateful to the National Natural Science Fund of China (31660136), the Key Project of Guizhou Science and Technology Fund (Qiankehejichu [2016]1414), the Guizhou R&D Program for Social Development (Qiankehe SY Zi [2014] No. 3036; Qiankeheshe G Zi [2012] No. 4008), the Support Plan for Science and Technology Top-notch Talents of Guizhou Higher Education Institutions (Qianjiaohe KY Zi [2016] No. 064) and the Guizhou Science and Technology Cooperation Program (Qiankehe LH Zi [2015] No. 7779) for their financial support.

Competing interests

The authors declare that they have no competing interests.

References

Ahloowalia BS (1967) Colchicine-induced polyploids in ryegrass. Euphytica 16:49–60

Beck SL, Dunlop RW, Fossey A (2003) Stomatal length and frequency as a measure of ploidy level in black wattle, *Acacia mearnsii* (de Wild). Bot J Lin Soc 41:177–181

Chakraborti SP, Vijayan K, Roy BN, Qadri SMH (1998) In vitro induction of tetraploidy in mulberry (*Morus alba* L.). Plant Cell Rep 17:799–803

Chang C, Yang M, Wen H, Chern J (2002) Estimation of total flavonoid content in propolis by two complementary colorimetric methods. J Food Drug Anal 10:178–182

Chen QF, Hsam SLK, Zeller FJ (2004) A study of cytology, isozyme and interspecific hybridization on the big-achene group of buckwheat species (*Fagopyrum*, Polygonaceae). Crop Sci 44:1511–1518

Chen QF, Sai LK, Fredrich JZ (2007) Cytogenetics studies on diploid and autotetraploid common buckwheat and their autotriploid and trisomics. Crop Sci 47:2340–2345

Cohen D, Yao JL (1996) In vitro chromosome doubling of nine *Zantedeschia* cultivars. Plant Cell Tiss Org 47:43–49

Dhooghe E, Van LK, Eeckhaut T, Leus L, Van HJ (2011) Mitotic chromosome doubling of plant tissues in vitro. Plant Cell Tiss Org 104:359–373

Du RQ (1999) Biostatistics. Higher Education Press and Springer—Verlag Press, Beijing, pp 33–65

Holasova M, Fiedlerova V, Smrcinova H, Orsak M, Lachman J, Vavreinova S (2002) Buckwheat—the source of antioxidant activity in functional foods. Food Res Int 35:207–211

Hou X, Wang ZR, Zhou MD, Zhao ZC (2002) Genetic diversity and differentiation of *Fagopyrum tataricum* and its related species in China. Acta Genet Sin 29:723–734

Hou SY, Sun ZX, Linghu B, Xu DM, Wu B, Zhang B, Wang XC, Han YH, Zhang LJ, Qiao ZJ, Li HY (2016) Genetic diversity of buckwheat cultivars (*Fagopyrum tartaricum* Gaertn.) assessed with SSR markers developed from genome survey sequences. Plant Mol Biol Rep 34:233–241

Kanoktip P, Ratchada S, Ikuo N, Masahiro M, Kanyaratt S (2014) Tetraploid induction of *Mitracarpus hirtus* L. by colchicines and its characterization including antibacterial activity. Plant Cell Tiss Org 117:381–391

Kermani MJ, Sarasan V, Roberts AV, Yokoya K, Wentworth J, Sieber VK (2003) Oryzalin-induced chromosome doubling in *Rosa* and its effect on plant morphology and pollen viability. Theor Appl Genet 107:1195–1200

Kunitake H, Nakashima T, Mori K, Tanaka M (1998) Somaclonal and chromosomal effects of genotype, ploidy and culture duration in *Asparagus officinalis* L. Euphytica 102:309–316

Li MX, Zhang XF (1991) Plant chromosome technology. Northeast Forestry University Publisher, Harbin

Li W, Li DN, Hu H, Chen XY (2007) Polyploid induction of *Lespedeza formosa* by colchicine treatment. Forest Stud China 9:283–286

Lian LK, Chen QF (2013) A comparative study of seed protein content and seed flavonoid content between diploid and tetraploid tartary buckwheat. Seed 32:1–5

Lin RF (1994) Buckwheat in China. Agricultural Publisher, Peking

Liu DF, Li ZN, Bao MZ (2007) Colchicine-induced chromosome doubling in *Platanus acerifolia* and its effect on plant morphology. Euphytica 157:145–154

Majdi M, Karimzadeh G, Malboobi MA, Omidbaigi R, Mirzaghaderi G (2010) Induction of tetraploidy to feverfew (*Tanacetum parthenium* Schulz-Bip.): morphological, physiological, cytological, and phytochemical changes. HortScience 45:16–21

Mohammad A, Ahmad M, Hassanali NB (2013) Morphological, physiological, cytological and phytochemical studies in diploid and colchicines-induced tetraploid plants of *Echinacea purpurea* (L.). Acta Physiol Plant 35:2075–2083

Nigel ARU, Jennie H, Therese M (2007) Generation and characterization of colchicine-induced autotetraploid *Lavandula angustifolia*. Euphytica 156:257–266

Ohnishi O, Ohta T (1987) Construction of a linkage map in common buckwheat, *Fagopyrum esculentum* Moench. Jap J Genet 62:397–414

Pan SJ, Chen QF (2010) Genetic mapping of common buckwheat using DNA, protein and morphological markers. Hereditas 147:27–33

Pei XS (1985) Polyploidy induction and breeding. Shanghai Science and Technology Press, Shanghai

Pour MP, Moieni A, Ebrahimi A, Javidfar F (2012) Doubled haploid plants following colchicine treatment of microspore-derived embryos of oilseed rape (*Brassica napus* L.). Plant Cell Tiss Org 108:251–256

Sheng MY, Wang LJ, Xiong KN (2011) Isozyme diversity and genetic structure of buckwheat populations. Biochem Syst Ecol 39:458–470

Sheng MY, Wang LJ, Tian XJ (2013) Identification of the full set of *Fagopyrum esculentum* trisomics by heterochromatin banding analysis and rDNA physical mapping. Sci Hortic 155:15–23

Sikdar AK, Jolly MS (1994) Induced polyploidy in mulberry (*Morus* spp.): induction of tetraploids. Sericologia 34:105–116

Speckmann GJ, Post JJ, Dijkstra H (1965) The length of stomata as an indicator for polyploidy in rye grasses. Euphytica 14:225–230

Tan GY, Dunn GM (1973) Relationship of stomatal length and frequency and pollen grain diameter to ploidy level in *Bromus inermis* Leyss. Crop Sci 13:332–334

Tandon SL, Bali PN (1957) Morphological and cytological studies of the diploid and the colchicine-induced tetraploid in *Linaria vulgaris*. Genetica 29:101–109

Tang CH, Wang XY (2010) Physicochemical and structural characterization of globulin and albumin from common buckwheat (*Fagopyrum esculentum* Moench) seeds. Food Chem 121:119–126

Tang ZQ, Chen DL, Song ZJ, He YC, Cai DT (2010) In vitro induction and identification of tetraploid plants of *Paulownia tomentosa*. Plant Cell Tiss Org 102:213–220

Wang H, Chen RF, Takashi L, Shen RF, Ma JF (2015) Physiological characterization of aluminum tolerance and accumulation in tartary and wild buckwheat. New Phytol 205:273–279

Yang XM, Cao ZY, An LZ, Wang YM, Fang XW (2006) In vitro tetraploid induction via colchicine treatment from diploid somatic embryos in grapevine (*Vitis vinifera* L.). Euphytica 152:217–224

Ye MY, Tong J, Shi XP, Yuan W, Li GR (2010) Morphological and cytological studies of diploid and colchicines-induced tetraploid lines of crape myrtle (*Lagerstroemia indica* L.). Sci Hortic 124:95–101

Zhou ML, Bai DQ, Tang Y, Zhu XM, Shao JR (2012) Genetic diversity of four new species related to southwestern Sichuan buckwheat as revealed by karyotype, ISSR and allozyme characterization. Plant Syst Evol 298:751–759

Zhu BC, Gao LR (1988) Study on autotetraploid common buckwheat I. comparison of morphology and cytology between autotetraploid and diploid common buckwheat. Hereditas 10:6–8

Zhu BC, Tian XH, Gao LR (1992) A study of cytogenetics on autotetraploid common buckwheat. Hereditas 14:1–4

18

Embryology of two mycoheterotrophic orchid species, *Gastrodia elata* and *Gastrodia nantoensis*: ovule and embryo development

Yuan-Yuan Li[1], Xiao-Mei Chen[1], Shun-Xing Guo[1*] and Yung-I Lee[2,3*]

Abstract

Background: *Gastrodia elata*, a famous herbal medicine, has been received great attention on its treatments of headache, vertigo and epilepsy. *Gastrodia nantoensis* is a newly described species from central Taiwan with potential medicinal value. *Gastrodia* species are fully mycoheterotrophic orchids, and the courses of their seed development are more rapid as compared to the chlorophyllous orchids. A better understanding of their reproductive biology would provide insights into the propagation and conservation of the mycoheterotrophic orchid species.

Results: Based on the histological and histochemical investigations, we observed some notable features in ovule and embryo development. First, only the archesporial cell and/or megasporocyte are present within their ovaries at the time of anthesis. Second, their suspensors consist of a single cell and their mature embryos consist of a gradient of small to large cells. Nile red staining of a globular embryo reveals the presence of cuticular material in the surface wall of embryo proper and the lateral walls of suspensor cell, indicating that the basal wall of suspensor cell is the major route for nutrient supply from maternal tissues to embryo proper. Third, their seed coats are derived from a single integument, and lignin but not cuticular material is present in the outer most layer of seed coat and persists through seed maturation.

Conclusions: The faster seed development of *Gastrodia* species is due to the speedy courses of ovule and embryo development. In the mature seeds, the presence of a differentiated apical zone in embryo proper suggests the easy-to-germinate character. This study provides basic knowledge for further molecular studies on embryo development and symbiotic germination of *Gastrodia* species.

Keywords: Ovule, Embryo, Mycoheterotrophic orchids, *Gastrodia*, Suspensor

Background

The tiny orchid seed has a rudimentary embryo and lacks endosperm; therefore, the germination of orchid seed requires the mycorrhizal association, which supplies nutrients for the germinating seed until the seedling differentiates green leaves and becomes autotrophic (Rasmussen 1995). Although a large amount of orchids are chlorophyllous, some orchids remain achlorophyllous and reply on their mycorrhizal partners for nutrient supplies throughout the entire life cycle, and they are known as mycoheterotrophic plants (Leake 1994; Merckx 2013; Lee et al. 2015). For many fully mycoheterotrophic orchids, their growth patterns are unique. They stay underground year-round, but after sprouting, the phases of blooming and then fruit setting complete within a few weeks. The course of seed development is more rapid than chlorophyllous orchids (Afzelius 1954; Arekal and Karanth 1981).

The genus *Gastrodia* comprises more than 50 species, distributing mainly in Asia and Oceania regions (Hsu 2008). All *Gastrodia* species are fully

*Correspondence: sxguo1986@163.com; leeyungi@hotmail.com
[1] Institute of Medicinal Plant Development, Chinese Academy of Medical Sciences & Peking Union Medical College, Beijing 100193, People's Republic of China
[2] Biology Department, National Museum of Natural Science, No 1, Kuan-Chien Rd, Taichung 40453, Taiwan
Full list of author information is available at the end of the article

mycoheterotrophic, and they are considered as the largest mycoheterotrophic genus in Orchidaceae. *Gastrodia elata*, a representative *Gastrodia* species, is an important traditional Chinese medicine (known as Tian Ma) for treatment of headache, vertigo and epilepsy (Xu and Guo 2000). Besides *G. elata*, there are more than 20 *Gastrodia* species native in Taiwan (Hsu and Kuo 2010). Some of them may have medicinal values and are under the threats of over-collection for the medicinal market. *Gastrodia. nantoensis* is a newly described endemic species with only a few known populations in Taiwan. As compared to the growth of stalk of *G. elata*, the growth of stalk of *G. nantoensis* is quite different. *G. nantoensis* has a very short stalk at the time of anthesis, and the further elongation of stalk is triggered by pollination. Although the structure of *G. elata* embryology has been documented in previous studies by the line drawing or paraffin sections (Kusano 1915; Abe 1976; Liang 1981, 1984), information concerning the development of ovule and embryo of *G. nantoensis* is lacking. For the commercial production and the conservation works of *Gastrodia* species, information concerning its reproductive biology is essential. Basic knowledge of seed development will assist in the improvement of artificial propagation, as demonstrated in our previous studies of *Cypripedium formosanum* (Lee et al. 2005) and *Cypripedium macranthos* (Zhang et al. 2013).

The objectives of this study are to document the key anatomical events in the embryogenesis of two *Gastrodia* species during the courses of megasporogenesis, megagametogenesis, fertilization and seed maturity. In this study, we used the resin-embedded sections, providing high resolution, light microscopic interpretations of various developmental stages. The observations would shed light on the embryogenesis of *Gastrodia* species and add to the current literature.

Methods

Plant materials

Plants of *G. elata* were cultivated in the greenhouse at Institute of Medicinal Plant Development, Chinese Academy of Medical Sciences & Peking Union Medical College, Beijing, China, while plants in a natural population of *G. nantoensis* in the bamboo forest located at Nantou County, Taiwan were selected for this study. Anthesis of *G. elata* occurred from April to May, and anthesis of *G. nantoensis* occurred from September to October. To ensure a good fruit set and seed quantity, the flowers were hand-pollinated. Developing ovaries and fruits were harvested at regular intervals after pollination. Approximately 30 developing ovaries and fruits of each species were gathered for this study.

Light microscopy and histochemical observations

Development ovaries and capsules were fixed with 2 % paraformaldehyde and 2.5 % glutaraldehyde in 0.1 M phosphate buffer, pH 6.8, at 4 °C overnight. After fixation, the samples were dehydrated using an ethanol series, and embedded in Technovit 7100 (Kulzer & Co., Germany) as described by Yeung and Chan (2015). Serial, 3 µm-thick sections were cut with glass knives using a Reichert-Jung 2040 Autocut rotary microtome. These sections were stained with Periodic acid–Schiff's reaction for total insoluble carbohydrates, and counterstained with either 0.05 % (w/v) toluidine blue O (TBO) in benzoate buffer for general histology or 1 % (w/v) amido black 10B in 7 % acetic acid for protein (Yeung 1984). The presence of cuticular material was detected using Nile red as detailed in Lee et al. (2006). The sections were stained with 1 µg ml-1 of Nile red (Sigma Chemical Co., St. Louis, Mo.) for 5 min, briefly washed in distilled water for 1 min, and mounted in a solution containing 0.1 % n-propyl gallate (Sigma Chemical Co.), an antifading compound. The fluorescence signal was examined using an epifluorescence microscope (Axioskop 2, Carl Zeiss AG) equipped with the Zeiss filter set 15 (546/12 nm excitation filter and 590 emission barrier filter). These sections were viewed and the images were captured digitally using a CCD camera attached to the light microscope.

Results and discussion

The major post-pollination developmental events that occurred in the developing fruits of two *Gastrodia* species are summarized in Table 1. Both of them have similar patterns in ovule and embryo developments. The seed maturation of *G. elata* and *G. nantoensis* only took 16 and 20 days after pollination (DAP), respectively. In several mycoheterotrophic orchid species, the above-ground part remained for a few weeks, and then disappeared (Arekal and Karanth 1981; Lin 1977). Accordingly, the

Table 1 Major microscopic structural events in the developing fruits of *G. elata* and *G. nantoensis*

Developmental stage	Days after pollination	
	G. elata	*G. nantoensis*
Archesporial cell and megasporocyte	0	0
Megasporogenesis and megagametogenesis	2	2–4
Mature embryo sac	4	6
Fertilization and zygote	6	8
Proembryo	8	12
Globular embryo	10	14–16
Late globular embryo	12–14	18
Mature seed	16	20

seed development of mycoheterotrophic orchid species was usually faster than most orchid species. As compared to the growth of stalk of *G. elata*, the growth of stalk of *G. nantoensis* was quite different. The stalk of *G. elata* had elongated before anthesis, while the stalk of *G. nantoensis* remained short at the time of anthesis (Figs. 1, 2). It was noticeable that the stalks of *G. nantoensis* started to elongate rapidly by 6 DAP. At maturity, the seeds were released from the split capsule and the stalk could reach to more than 40 cm in length (Fig. 3). In *G. nantoensis*, the rapid elongation of stalk triggered by pollination was helpful for the wind dispersal of seeds to long distance.

Fig. 2 The flowers and developing fruits of *G. nantoensis*. **a** The stalks (*arrow*) are short and the fruits are small (*arrowhead*) at the time of anthesis. *Scale bar* 1 cm. **b** At 6 DAP, the stalks start to elongate rapidly (*arrow*) and the fruits have enlarged (*arrowhead*). *Scale bar* 1 cm

Ovule development

In Orchidaceae, one of the unique features of ovule development is that mature ovules are usually not present at the time of pollination (Wirth and Withner 1959). The ovule development triggered by a successful pollination usually takes several weeks to complete (Arditti 1992, Yeung and Law 1997). In *G. elata* and *G. nantoensis*, at the time of anthesis, the placenta tissues were highly branched and the tip of a nucellar filament had differentiated into an archesporial cell (Figs. 4a, 5a). Prior to pollination, the archesporial cell had enlarged and differentiated into the megasporocyte (Figs. 4b, 5b). As Vij and Sharma (1986) pointed out, at the time of anthesis, three major types of ovule development could be observed in orchids. These types include (1) the ovule development has not been initiated prior to pollination, such as in *Cymbidium* (Swamy 1942), *Dendrobium* (Swamy 1943), *Epidendrum* (Yeung and Law 1989), *Phalaenopsis* (Duncan and Curtis 1942a; Zhang and O'Neill 1993); (2) the ovules contain well differentiated archesporial cells or megasporocytes, such as in *Cypripedium* (Duncan and Curtis 1942b; Law and Yeung 1993) and *Paphiopedilum* (Duncan and Curtis 1942b; Lee and Yeung 2012); (3) the mature ovules are ready for fertilization, such as in *Epipogium aphyllum* (Afzelius 1954), *Epipactis papillosa* (Sato 1974) and *G. elata* (Kusano 1915). In this study, it is clear that only the archesporial cell and/or megasporocyte could be observed within the ovaries of *G. elata* and *G. nantoensis* at the time of anthesis. The developmental stage of ovule in *Gastrodia* species at the time of anthesis is similar to those in slipper orchids, e.g. *Cypripedium* and *Paphiopedilum* (Law and Yeung 1993; Lee and Yeung 2012). However, the seed maturity in several *Cypripedium* and *Paphiopedilum* species usually takes 6–12 months. The faster seed development in

Fig. 1 The flower and developing fruits of *G. elata*. **a** The stalk has elongated at the time of anthesis. *Scale bar* 1 cm. **b** The developing fruit at 6 DAP. *Scale bar* 1 cm

Fig. 3 The elongating stalks and developing fruits of various stages of *G. nantoensis*. After 20 DAP, the seeds (*double arrowhead*) have matured and released from the split fruit. *Scale bar* 1 cm

Gastrodia species is not attributed to the presence of mature embryo sac, but the speedy courses of ovule and embryo development.

The changes in cell dimensions during embryo sac development of *G. elata* and *G. nantoensis* were summarized in Table 2. After pollination, the megasporocyte soon underwent the first meiotic division (Figs. 4c, 5c). The first meiotic division produced an unequal dyad: a smaller micropylar dyad and a larger chalazal dyad (Fig. 4d). The functional dyad at the chalazal end enlarged and performed the second meiotic division, while the micropylar dyad had become highly compressed (Fig. 5d). The second meiotic division resulted in the formation of two megaspores with unequal size: a smaller micropylar megaspore and a larger chalazal megaspore (Figs. 4e, 5e). The chalazal megaspore continued to enlarge and served as the functional megaspore, while the micropylar megaspore had compressed and become a thin dark-staining layer (Figs. 4f, 5f). The functional megaspore divided further and formed a two-nucleate

Fig. 4 Light micrographs of megasporogenesis and megagametogenesis of *G. elata*. **a** Light micrograph showing the placental ridges with branches differentiated into nucellar filaments at the time of anthesis. The formation of archesporial cell (*arrows*) occurs at the terminus of the nucellar filament. *Scale bar* 50 µm. **b** The archesporial cell enlarges and differentiates into the megasporocyte. *Scale bar* 10 µm. **c** The megasporocyte undergoes the first meiotic division (*arrow*). *Scale bar* 10 µm. **d** After the first meiotic division, the functional dyad is located at the chalazal end of the ovule, and the non-functional dyad at the micropylar end degenerates gradually. *Scale bar* 10 µm. **e** After the second meiotic division, the smaller non-functional megaspore at the micropylar end degenerates soon. At the same time, the initiation of integument tissue has become visible (*arrow*). *Scale bar* 10 µm. **f** Light micrograph showing a functional megaspore with a few vacuoles and the non-functional megaspores have degenerated completely (*arrow*). *Scale bar* 10 µm. **g** The functional megaspore divides once resulting in the formation of a two-nucleate embryo sac, and the two-nucleate embryo sac expands further by vacuolation. *Scale bar* 10 µm. **h** At 4 DAP, a mature embryo sac showing one antipodal nucleus (*arrow*) and the egg apparatus, including the egg cell *E* and two synergids *S*. The integument tissue has not enclosed the embryo sac. *Scale bar* 10 µm

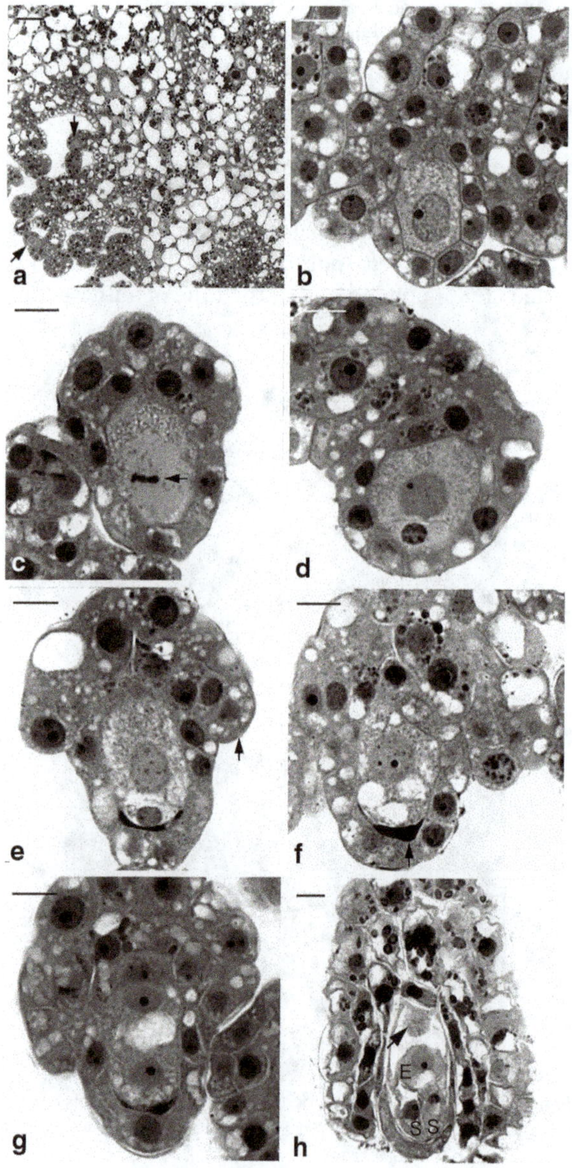

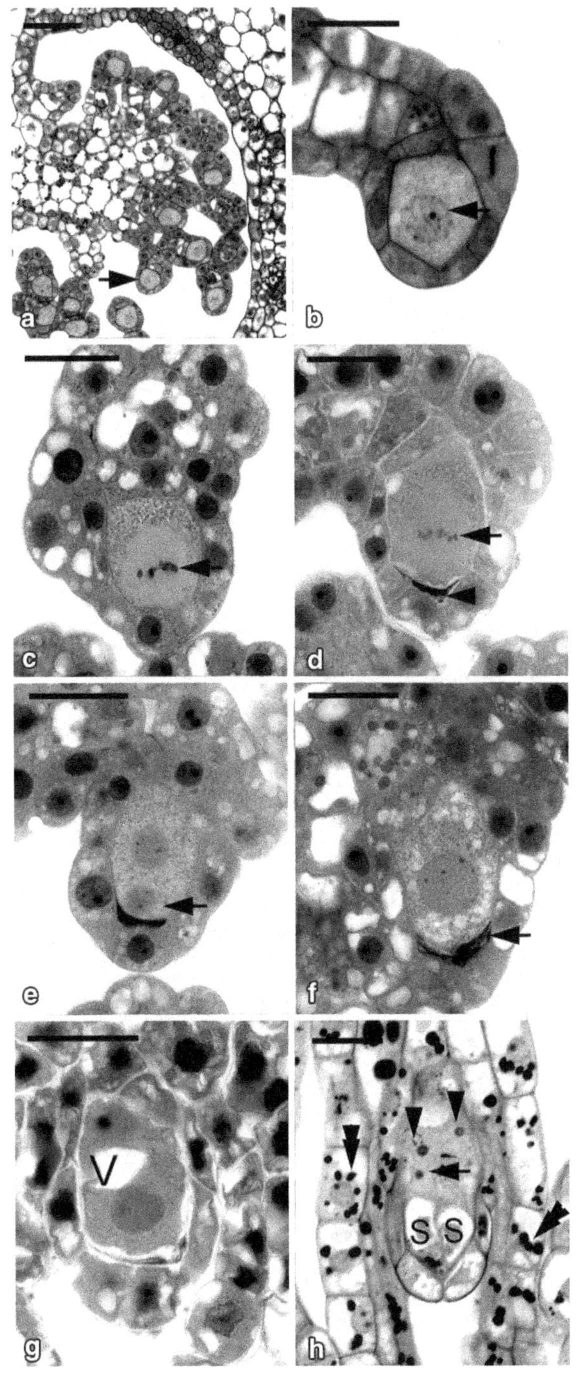

Fig. 5 Light micrographs of megasporogenesis and megagametogenesis of *G. nantoensis*. **a** Light micrograph showing the placental ridges with several branches differentiated into nucellar filaments at the time of anthesis. The archesporial cell (*arrow*) can be observed at the terminus of the nucellar filament. *Scale bar* 50 µm. **b** The archesporial cell has differentiated into the megasporocyte. Before the first meiotic division, the chromatin (*arrow*) is going to condense. *Scale bar* 20 µm. **c** Soon, the megasporocyte undergoes the first meiotic division (*arrow*). *Scale bar* 20 µm. **d** After the first meiotic division, the functional dyad at the chalazal end of the ovule successively undergoes the second meiotic division (*arrow*), and the non-functional dyad at the micropylar end has degenerated (*arrowhead*). *Scale bar* 20 µm. **e** The second meiotic division results in the formation two megaspores of unequal size. The smaller non-functional megaspore (*arrow*) at the micropylar end will degenerate. *Scale bar* 20 µm. **f** Light micrograph showing a functional megaspore with several small vacuoles and the non-functional megaspores have degenerated completely (*arrow*). *Scale bar* 20 µm. **g** Light micrograph showing the two-nucleate embryo sac with a prominent vacuole (*V*) located at the center of cell. *Scale bar* 20 µm. **h** At 6 DAP, a longitudinal section through a mature embryo sac showing the egg apparatus, including the egg cell (*arrow*) and two synergids (*S*) at the micropylar end and two antipodal nuclei at the chalazal end (*arrowheads*). At this stage, the integument has completely enclosed the embryo sac. Several starch grains (*double arrowheads*) start to accumulate in the integument tissue. *Scale bar* 20 µm

embryo sac. The embryo sac had enlarged in size through the process of vacuolation at the center of the cell (Figs. 4g, 5g). In the mature embryo sacs of *G. elata* and *G. nantoensis*, four- to five-nucleate embryo sacs could be observed. Within the embryo sac, the egg apparatus containing one egg cell and two synergids was located at the micropylar end (Figs. 4h, 5h). In the central cell, one polar nucleus existed. At the chalazal pole, the antipodal nuclei were usually absent or occasionally one antipodal nucleus existed (Fig. 5h). The reduction of in the number of nuclei within the embryo sac has been reported as the 'striking phenomenon' (Harling 1950). This phenomenon has been observed in the ovule development of several orchids (Abe 1976; Arekal and Karanth 1981; Law and Yeung 1993). The reduction in the number of nuclei within the embryo sac may be due to the anomalies during the cell division process, such as the failure of mitosis, the fusion of mitotic spindles or the degeneration of existing chalazal nuclei (Yeung and Law 1997).

Embryo development

Most of the ovules had been fertilized at 6 DAP in *G. elata* and at 8 DAP in *G. nantoensis* respectively, and the embryo development commenced (Table 1). The zygote was highly polarized with the nucleus located toward the chalazal end (Figs. 6a, 7a). In both species, their endosperms failed to develop. Within the endosperm cavity, the endosperm nuclei and the chalazal nuclei usually fused to a complex structure and it was eventually

Table 2 Changes in cell dimensions during embryo sac development of *G. elata* and *G. nantoensis*

Developmental stage	Taxon	X-axis[a]	Y-axis[a]
Megasporocyte	*G. elata*	21.7 ± 0.7	29.4 ± 1.3
	G. nantoensis	20.4 ± 0.6	29.1 ± 1.5
Functional dyad	*G. elata*	20.8 ± 1.0	26.6 ± 1.7
	G. nantoensis	20.5 ± 0.8	28.7 ± 1.1
2-nucleate embryo sac	*G. elata*	25.1 ± 1.7	35.7 ± 2.5
	G. nantoensis	26.4 ± 1.1	36.1 ± 2.2
4-nucleate embryo sac	*G. elata*	28.5 ± 1.2	42.8 ± 1.8
	G. nantoensis	27.3 ± 1.7	44.6 ± 2.6
Mature embryo sac	*G. elata*	31.0 ± 1.1	53.1 ± 2.4
	G. nantoensis	31.7 ± 1.5	54.6 ± 1.9

[a] X, Y standard deviation in μm

absorbed by the developing embryo. The first division of the zygote was unequal, giving rise to two daughter cells: a smaller terminal cell and a larger basal cell (Figs. 6b, 7b). The additional transverse cell division resulted in the formation of a three-celled embryo (Figs. 6c, 7c). The basal cell toward the micropylar end would become the suspensor, and the upper cells gave rise to the embryo proper. The suspensor of both species consisted of a single cell that was vacuolated with starch grains accumulated (Figs. 6e, 7e). Throughout the embryo development, no further division or enlargement was observed in the suspensor. As the embryo approached maturity, the suspensor became highly compressed and degenerated (Figs. 6h, 7h).

In *G. elata*, an anticlinal divisions in the terminal cell of the three-celled embryo resulted in the formation of a four-celled embryo (Fig. 6d), and in *G. nantoensis*, the occurrences of an anticlinal division in the terminal cell and transverse division in the subterminal cell of the three-celled embryo gave rise to a five-celled embryo (Fig. 7d). Additional divisions in the inner cell tiers of embryo proper resulted in the formation of an early globular embryo (Figs. 6e, 7e). Further anticlinal and periclinal cell divisions in the inner cell tiers of embryo proper gave rise to a globular embryo (Figs. 6f, 7f). Meanwhile, a large amount of starch grains accumulated in the developing embryo proper. As the embryo approached maturity (Figs. 6g, 7g), cell division had ceased within the embryo proper, and the accumulation of large starch grains was prominent. The composition and structure of these large starch grains remained to be investigated. At maturity, the embryo of *G. elata* took on an ovoid shape and was only seven cells long and four cells wide (Fig. 6h), while in *G. nantoensis*, the embryo had an ellipsoidal shape and was eight cells long and four cells wide (Fig. 7h). Like the mature seeds of many orchids, in *G. elata* and *G. nantoensis*, most starch grains had disappeared; instead, protein bodies and lipid bodies became the major storage products in their embryos (Yam et al. 2002). It is noteworthy that the mature embryos of *G. elata* and *G. nantoensis* consisted of a gradient of small to large cells (Figs. 6h, 7h). The marked gradient of cell size in the embryo proper can be observed in the easy-to-germinate species, such as *Phalaenopsis amabilis* var. *formosa* (Lee et al. 2008). On the contrary, in the difficult-to-germinate species, such as *Calypso bulbosa* (Yeung and Law 1992) and *C. formosanum* (Lee et al. 2005), no marked gradient of cell size within the embryo proper is observed. The presence of marked gradient of cell size within the globular embryo in orchids may speed the germination process. In the asymbiotic cultures, mature seeds of *G. elata* and *G. nantoensis* could readily germinate and reached the average germination percentage over 80 %.

Nile red staining indicated that the cuticular material was deposited in the surface wall of embryo proper and the lateral walls of suspensor cell at the globular stage (Fig. 8a). As the seed matured, the accumulation of cuticular material was observed on the surface wall of embryo proper (Fig. 8b). Since the embryo of *Gastrodia* is housed within a thin layer of seed coat without the endosperm development, the deposition of cuticular materials in the surface wall of embryo proper could protect the globular embryo from early desiccation. The suspensor is a short-lived embryonic structure, and it is responsible for nutrient uptake for the developing embryo (Yeung and Meink 1993; Kawashima and Goldberg 2010). In orchids, the diverse of their suspensors provides distinguishing characteristics for a classification scheme for orchid embryo development (Swamy 1949). *G. elata* and *G. nantoensis* had a simple single-celled suspensor which belong to the type I group (Figs. 6f, 7f). It is remarkable that only the basal wall connecting to the nucellus reacts negatively to Nile red stain, while the lateral walls of suspensor react positively to Nile red stain (Fig. 8a), indicating that there is no cuticular material accumulated in the basal wall. The suspensor connects the embryo proper to the maternal tissues, and there are no direct symplastic connections between these two compartments. Therefore, in *Gastrodia*, the suspensor serves as the major conduit to supply the nutrients for the developing embryo proper as previously reported by Lee et al. (2006) and Lee and Yeung (2010).

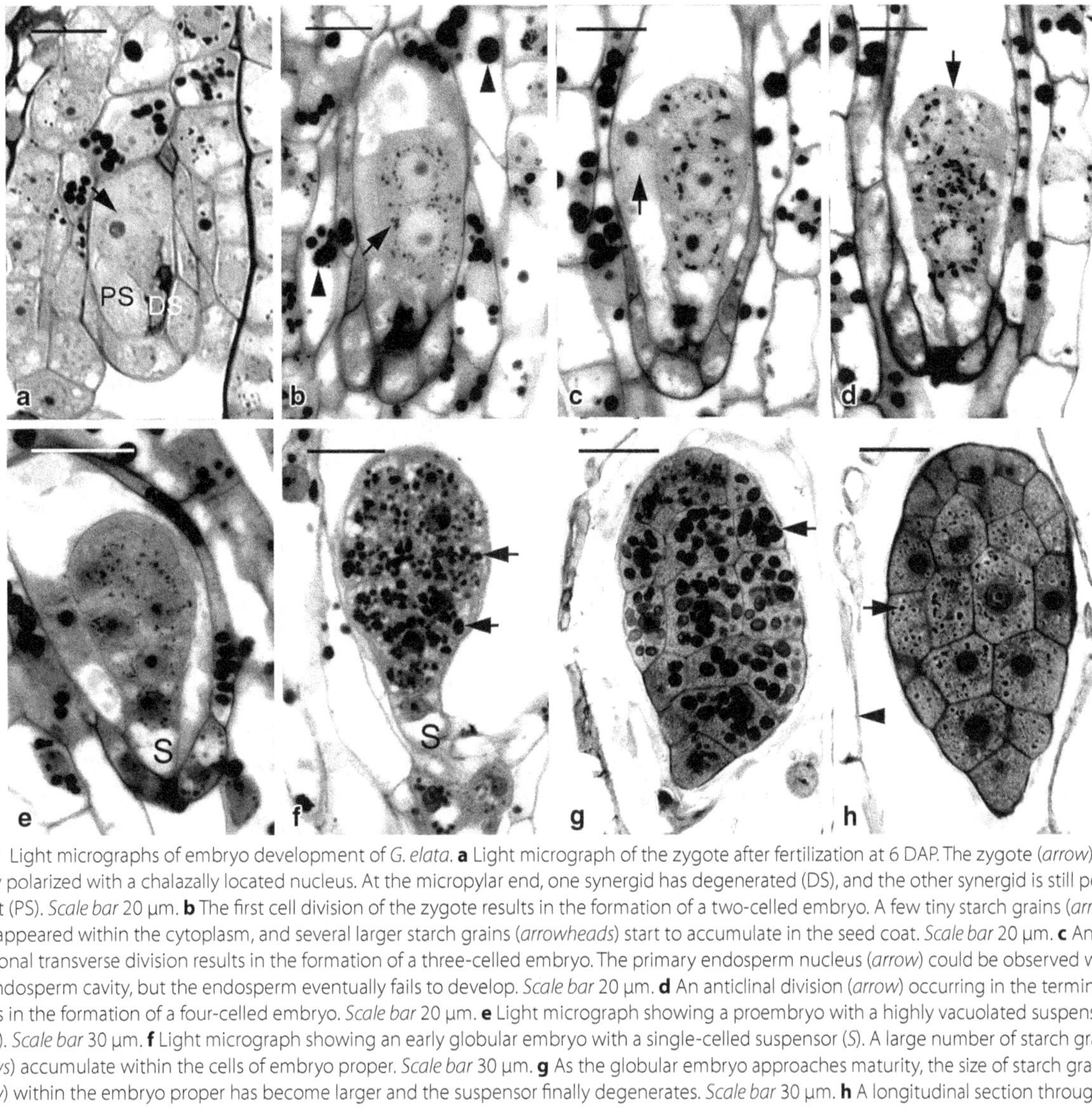

Fig. 6 Light micrographs of embryo development of *G. elata*. **a** Light micrograph of the zygote after fertilization at 6 DAP. The zygote (*arrow*) is highly polarized with a chalazally located nucleus. At the micropylar end, one synergid has degenerated (DS), and the other synergid is still persistent (PS). *Scale bar* 20 μm. **b** The first cell division of the zygote results in the formation of a two-celled embryo. A few tiny starch grains (*arrow*) have appeared within the cytoplasm, and several larger starch grains (*arrowheads*) start to accumulate in the seed coat. *Scale bar* 20 μm. **c** An additional transverse division results in the formation of a three-celled embryo. The primary endosperm nucleus (*arrow*) could be observed within the endosperm cavity, but the endosperm eventually fails to develop. *Scale bar* 20 μm. **d** An anticlinal division (*arrow*) occurring in the terminal cell results in the formation of a four-celled embryo. *Scale bar* 20 μm. **e** Light micrograph showing a proembryo with a highly vacuolated suspensor cell (*S*). *Scale bar* 30 μm. **f** Light micrograph showing an early globular embryo with a single-celled suspensor (*S*). A large number of starch grains (*arrows*) accumulate within the cells of embryo proper. *Scale bar* 30 μm. **g** As the globular embryo approaches maturity, the size of starch grains (*arrow*) within the embryo proper has become larger and the suspensor finally degenerates. *Scale bar* 30 μm. **h** A longitudinal section through a mature seed at 16 DAP. At this stage, the starch grains have disappeared, and a number of small protein bodies (*arrow*) can be seen within the embryo proper. Although the lipid cannot be preserved in the historesin, the spaces between the protein bodies could be the storage lipid bodies. At maturity, the embryo is enveloped by a shriveled seed coat (*arrowhead*). *Scale bar* 30 μm

Seed coat development

In most orchids, their ovules are covered by the outer and inner integuments (known as bitegmic) (Law and Yeung 1993; Mayer et al. 2011), while in *G. elata* and *G. nantoensis*, their seed coats developed from a single integument (known as unitegmic), which were composed of two-celled layers (Figs. 6c, 7c). Species with simple ovule structures, e.g. the unitegmic ovule and the lack of a distinct micropyle are common features in some mycoheterotrophic orchids (Tohda 1967; Abe 1976). It has been proposed that the unitegmic ovules might be more advanced than the bitegmic ovules (Abe 1972). After fertilization, the cells of seed coat elongated rapidly and enclosed the embryo sac entirely. The cells of seed-coat were highly vacuolated and contained several starch grains (Figs. 6a, 7a). At maturity, the seed coat became dehydrated and compressed into a thin layer that enveloped the embryo (Figs. 6h, 7h). In the developing seed

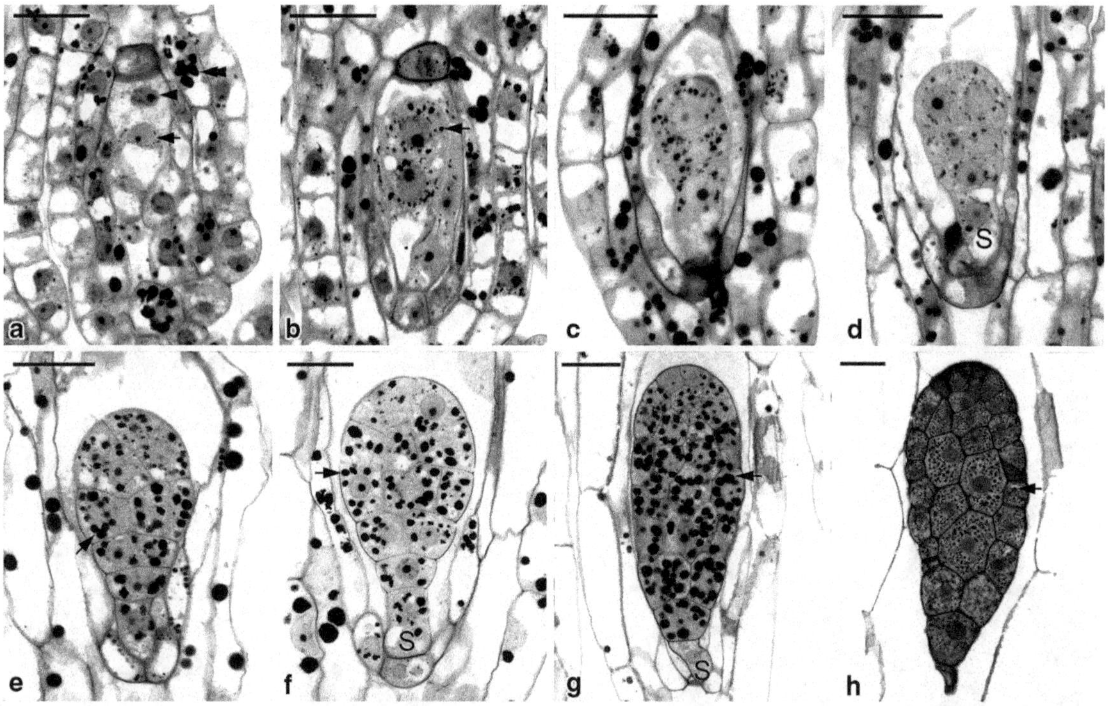

Fig. 7 Light micrographs of embryo development of *G. nantoensis*. **a** Light micrograph of the zygote after fertilization at 8 DAP. The zygote (*arrow*) is highly polarized with a chalazally located nucleus and a prominent vacuole occupying the micropylar end. The polar-chalazal complex (*arrowhead*) includes the chalazal nuclei and the polar nuclei, but the endosperm eventually fails to develop. A number of starch grains (*double arrowhead*) could be seen in the seed coat. *Scale bar* 30 μm. **b** The first cell division of the zygote is unequal and results in the formation of a smaller terminal cell and a larger basal cell. Several tiny starch grains (*arrow*) have appeared within the cytoplasm. *Scale bar* 30 μm. **c** Light micrograph showing a three-celled embryo. The basal cell of the three-celled embryo will differentiate into the suspensor. *Scale bar* 30 μm. **d** Light micrograph showing a proembryo with a highly vacuolated suspensor cell (*S*). *Scale bar* 30 μm. **e** Light micrograph showing an early globular embryo. The size of starch grains (*arrow*) within the embryo proper becomes larger. *Scale bar* 30 μm. **f** A longitudinal section through a developing globular embryo. Large starch grains (*arrow*) are abundant and tend to congregate around the nucleus. The suspensor cell (*S*). *Scale bar* 30 μm. **g** As the seed approaches maturity, the cell division within embryo proper has ceased. At this stage, starch grains (*arrow*) are prominent within the embryo cells. The suspensor cell (*S*) has reduced its size and begins to degenerate. *Scale bar* 30 μm. **h** A longitudinal section through a mature seed at 20 DAP. The embryo has the smaller cells near the chalazal end and the larger cells in the micropylar end. At this stage, the starch grains have disappeared and numerous tiny protein bodies (*arrow*) have accumulated within the embryo proper. In this preparation, the lipid cannot be preserved, however the spaces between the protein bodies could be the storage lipid bodies. At maturity, the suspensor has degenerated and the embryo proper is enveloped by a shriveled seed coat. *Scale bar* 30 μm

of *G. nantoensis*, the outermost wall of seed coat gave a weak fluorescence from Nile red staining, suggesting the presence of little cuticular material (Fig. 8a, b). However, the weak fluorescence in the seed coat can be easily quenched by pre-staining of sections with TBO, indicating that a distinct cuticle may be absent (see Holloway 1982; Yeung et al. 1996). TBO is a cationic dye that binds to negatively charged groups in cells and gives two main spectra of reaction, i.e. pinkish purple for carboxylated polysaccharides such as pectin and greenish blue for aromatic substances such as lignin (Pradhan Mitra and Loqué 2014). Under fluorescence microscopy, if TBO staining causes the fluorescence quenching of the cell wall by UV-excitation, we would suggest the accumulation of lignin rather than cuticular material in the cell wall. In this study, the compressed seed coat stained greenish blue with TBO, indicating the lignified cell wall of the seed coat (Figs. 6h, 7h). In orchids, the lignified seed coat could protect the minute globular embryo at the time of seed dispersal.

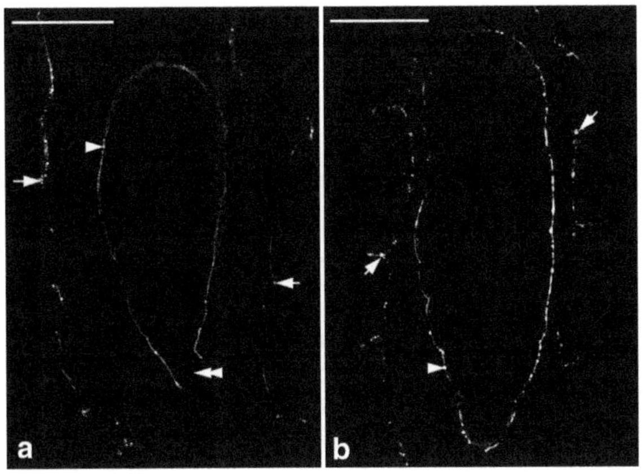

Fig. 8 Nile red staining fluorescence micrograph of developing embryos of *G. nantoensis*. **a** Nile red staining fluorescence micrograph of a late globular embryo at the same stage as that seen in Fig. 7g. The innermost walls of the seed coat (*arrows*) and the surface wall (*arrowhead*) of the embryo proper react positively to the stain. The fluorescence is absent in the suspensor wall (*double arrowhead*). *Scale bar* 50 μm. **b** Nile red staining fluorescence micrograph of a mature seed at the same stage as that seen in Fig. 7h. The seed coat (*arrow*) and the surface wall (*arrowhead*) of the embryo proper react positively to the stain. *Scale bar* 50 μm

Conclusions

The present study illustrates the key anatomical events in the seed development of *G. elata* and *G. nantoensis* and identifies a number of features that are not yet described clearly in the previous studies, such as the initial differentiation of ovule at the time of anthesis, the histochemical characters of the cell wall of suspensor and seed coat, and the changes of storage products within the embryo proper. This study provides basic knowledge for further molecular studies on embryo development and symbiotic germination of *Gastrodia* species.

Authors' contributions

LYY carried out the experimental work and drafted the manuscript; CXM, GSX and LYI provided funding, planned the study and revised the manuscript. All authors read and approved the final manuscript.

Author details

[1] Institute of Medicinal Plant Development, Chinese Academy of Medical Sciences & Peking Union Medical College, Beijing 100193, People's Republic of China. [2] Biology Department, National Museum of Natural Science, No 1, Kuan-Chien Rd, Taichung 40453, Taiwan. [3] Department of Life Sciences, National Chung Hsing University, Taichung 40227, Taiwan.

Acknowledgements

This work was supported by grants from the National Natural Science Foundation of China to Shun-Xing Guo; from National Museum of Natural Science, Taiwan, to Yung-I Lee.

Competing interests

The authors declare that they have no competing interests.

References

Abe K (1972) Contributions to the embryology of the family Orchidaceae VII. A comparative study of the orchid embryo sac. Sci Rep Tohoku Univ Ser IV (Biol) 36:179–201

Abe K (1976) A reinvestigation of the development of the embryo sac in *Gastrodia elata* Blume (Orchidaceae). Ann Bot 40:99–102

Afzelius K (1954) Embryo sac development in *Epipogium aphyllum*. Svensk Botanisk Tidskrift 48:513–520

Arditti J (1992) Fundamentals of orchid biology. Wiley, New York

Arekal GD, Karanth KA (1981) The embryology of *Epipogium roseum* (Orchidaceae). Pl Syst Evol 138:1–7

Duncan RE, Curtis JT (1942a) Intermittent growth of fruits of *Phalaenopsis*. A correlation of growth phases of an orchid fruit with their internal development. Bull Torrey Bot Club 69:167–183

Duncan RE, Curtis JT (1942b) Intermittent growth of fruits of *Cypripedium* and *Paphiopedilum*. A correlation of growth phases of an orchid fruit with their internal development. Bull Torrey Bot Club 69:353–359

Harling G (1950) Embryological studies in Compositae I. Anthemideae-Anthemidine. Acta Hort Berg 15:135–168

Holloway PJ (1982) Structure and histochemistry of plant cuticular membranes: an overview. In: Cutler DF, Alvin KL, Price CE (eds) The plant cuticle. Academic Press, London, pp 1–32

Hsu TC (2008) Taxonomy of *Gastrodia* (Orchidaceae) in Taiwan. Master thesis. Institute of Ecology and Evolutionary Biology, College of Life Science, National Taiwan University, Taiwan

Hsu TC, Kuo CM (2010) Supplements to the orchid flora of Taiwan (IV): four additions to the genus *Gastrodia*. Taiwania 55:243–248

Kawashima T, Goldberg RB (2010) The suspensor: not just suspending the embryo. Trend Plant Sci 15:23–30

Kusano S (1915) Experimental studies on the embryonal development in an angiosperm. J Col Agric, Tokyo Imp Univ 6:7–1209

Law SK, Yeung EC (1993) Embryology of *Cypripedium passerinum* (Orchidaceae): ovule development. Lindleyana 8:139–147

Leake JR (1994) The biology of myco-heterotrophic ('saprophytic') plants. New Phytol 127:171–216

Lee YI, Yeung CE (2010) The osmotic property and fluorescent tracer movement of developing orchid embryos of *Phaius tankervilliae* (Aiton) Bl. Sex Plant Reprod 23:337–341

Lee YI, Yeung EC (2012) Embryology of the lady's slipper orchid, *Paphiopedilum delenatii*: ovule development. Bot Stud 53:97–104

Lee YI, Lee N, Yeung CE, Chung MC (2005) Embryo development of *Cypripedium formosanum* in relation to seed germination in vitro. J Am Soc Hortic Sci 130:747–753

Lee YI, Yeung EC, Lee N, Chung MC (2006) Embryo development in the lady's slipper orchid, *Paphiopedilum delenatii*, with emphasis on the ultrastructure of the suspensor. Ann Bot 98:1311–1319

Lee YI, Yeung EC, Lee N, Chung MC (2008) Embryology of *Phalaenopsis amabilis* var. *formosa*: embryo development. Bot Stud 49:139–146

Lee YI, Yang CK, Gebauer G (2015) The importance of associations with saprotrophic non-Rhizoctonia fungi among fully mycoheterotrophic orchids is currently under-estimated: novel evidence from sub-tropical Asia. Ann Bot 116:423–435

Liang HX (1981) The micro- and mega-sporogenesis and the formation of male and female gametophytes in *Gastrodia elata* Blume. Acta Botanica Sinica 23:186–191

Liang HX (1984) Embryological studies of *Gastrodia elata* Blume. Acta Botanica Sinica 26:466–472

Lin TP (1977) Native orchids of Taiwan, vol 3. Southern Materials Ctr., Taipei

Mayer JLS, Carmello-Guerreiro SM, Appezzato-da-Glória B (2011) Anatomical development of the pericarp and seed of *Oncidium flexuosum* Sims (Orchidaceae). Flora 206:601–609

Merckx VSFT (2013) Mycoheterotrophy: an introduction. In: Merckx VSFT (ed) Mycoheterotrophy: the biology of plants living on fungi. Springer, Berlin, pp 1–17

Pradhan Mitra P, Loqué D (2014) Histochemical staining of *Arabidopsis thaliana* secondary cell wall elements. J Vis Exp 87:e51381

Rasmussen HN (1995) Terrestrial orchids-from seed to mycotrophic plant. Cambridge University Press, Cambridge

Sato Y (1974) Embryological studies in the Japanese *Epipactis* (Orchidaceae). Sci Rep Tohoku Univ Ser IV (Biol) 37:33–45

Swamy BGL (1942) Female gametophyte and embryogeny in Cymbidium bicolor Lindl. Proc Ind Acad Sci Sec B 15:194–201
Swamy BGL (1943) Embryology of Orchidaceae. Curr Sci 12:13–17
Swamy BGL (1949) Embryological studies in the Orchidaceae II. Embryology. Am Midl Nat 41:202–232
Tohda H (1967) An embryological study of *Hetaeria skikokiana*, a saprophytic orchid in Japan. Sei Rep Tohoku Univ Ser IV (Biol) 33:83–95
Vij SP, Sharma M (1986) Embryo sac development in Orchidaceae. In: Vij SP (ed) Biology, conservation and culture of orchids. Affiliated East-West Press, New Delhi, pp 31–48
Wirth M, Withner CL (1959) Embryology and development in the Orchidaceae. In: Withner CL (ed) The Orchids. a scientific survey. The Ronald Press Company, New York, pp 155–188
Xu JT, Guo SX (2000) Retrospect on the research of the cultivation of *Gastrodia elata* Bl, a rare traditional Chinese medicine. Chin Med J 113:686–692
Yam TW, Yeung EC, Ye XL, Zee SY, Arditti J (2002) Orchid embryos. In: Kull T, Arditti J (eds) Orchid Biology: Reviews and Persectives VIII. Kluwer Academic Publisher, Dordrecht, pp 287–385
Yeung EC (1984) Histological and histochemical staining procedures. In: Vasil IK (ed) Cell culture and somatic cell genetics of plants. Academic Press, Orlando, pp 689–697
Yeung EC, Chan CKW (2015) The glycol methacrylate embedding resins-Technovit 7100 and 8100. Plant microtechniques and protocols. Springer International Publishing, Berlin, pp 67–82
Yeung EC, Law SK (1989) Embryology of *Epidendrum ibaguense* I. Ovule development. Can J Bot 67:2219–2226
Yeung EC, Law SK (1992) Embryology of *Calypso bulbosa*II. Embryo development. Can J Bot 70:461–468
Yeung EC, Law SK (1997) Ovule and megagametophyte development in orchids. In: Arditti J, Pridgeon AM (eds) Orchid Biology: Reviews and Perspectives VII. Kluwer Academic Pub, Dordrecht, pp 31–73
Yeung EC, Meink DW (1993) Embryogenesis in angiosperms: development of the suspensor. Plant Cell 5:1371–1381
Yeung EC, Zee SY, Ye XL (1996) Embryology of *Cymbidium sinense*: embryo development. Ann Bot 78:105–110
Zhang XS, O'Neill SD (1993) Ovary and gametophyte development are coordinately regulated by auxin and ethylene following pollination. Plant Cell 5:403–418
Zhang Y, Lee YI, Deng L, Zhao S (2013) Asymbiotic germination of immature seeds and the seedling development of *Cypripedium macranthos* Sw., an endangered lady's slipper orchid. Sci Hortic 164:130–136

19

Efficient and heritable transformation of *Phalaenopsis* orchids

Hong-Xian Hsing[1,2†], Yi-Jyun Lin[1,2†], Chii-Gong Tong[1,2], Min-Jeng Li[1,2], Yun-Jin Chen[1,2] and Swee-Suak Ko[1,2*]

Abstract

Background: *Phalaenopsis* orchid (*Phal.* orchid) is visually attractive and it is important economic floriculture species. *Phal.* orchids have many unique biological features. However, investigation of these features and validation on their biological functions are limited due to the lack of an efficient transformation method.

Results: We developed a heritable and efficient *Agrobacterium*- mediated transformation using protocorms derived from tetraploid or diploid *Phal.* orchids. A T-DNA vector construct containing eGFP driven by ubiquitin promoter was subjected to transformation. An approximate 1.2–5.2 % transformation rate was achieved. Genomic PCR confirmed that hygromycin selection marker, *HptII* gene and target gene *eGFP* were integrated into the orchid genome. Southern blotting indicated a low T-DNA insertion number in the orchid genome of the transformants. Western blot confirmed the expression of eGFP protein in the transgenic orchids. Furthermore, the GFP signal was detected in the transgenic orchids under microscopy. After backcrossing the pollinia of the transgenic plants to four different *Phal.* orchid varieties, the BC1 progenies showed hygromycin resistance and all surviving BC1 seedlings were *HptII* positive in PCR and expressed GFP protein as shown by western blot.

Conclusions: This study demonstrated a stable transformation system was generated for *Phal.* orchids. This useful transformation protocol enables functional genomics studies and molecular breeding.

Keywords: *Phalaenopsis aphrodite*, *Agrobacterium tumefaciens*, Transformation, Protocorm, eGFP, Hygromycin selection

Background

Phalaenopsis orchids, commonly known as moth orchid, is one of the most popular and profitable ornamental plants in the global floral market. *Phal.* orchids have very unique biological features such as crassulacean acid metabolism (CAM) photosynthesis that uptake CO_2 at night (Guo and Lee 2006), epiphytic habitat with high water and nutrient usage efficiency, unique flower pattern formation (Su et al. 2013b), symbiosis with mycorrhizae (Yumiko et al. 2007), immature embryos (seeds without endosperm) in mature capsules (Yu and Goh 2001). To elucidate these interesting and novel biological features, validation of gene function using stable transformation approach must be helpful. Moreover, the recent technological advances in next generation sequencing (NGS) combining bioinformatics have generated highly abundant sequence information. Several orchid databases based on the transcriptomic information are constructed and available in the web, such as Orchidstra (Su et al. 2013a) and OrchidBase (Fu et al. 2011). In the era of big data and functional genomics, efficient transformation of *Phal.* orchid can contribute to the new discoveries with strong impact.

Over the years consumers have sought ever more attractive and versatile morphological traits in flowers, such as colors, shapes, fragrance. Although conventional breeding has been fruitful in last decades, to meet the demands of the market for novel floral colors or fragrances, control of flowering time, resistant to diseases and environmental stresses and so on, transgenic technology is needed. Introgression of foreign genes from

*Correspondence: sweesuak@gate.sinica.edu
†Hong-Xian Hsing and Yi-Jyun Lin are co-first authors
[1] Academia Sinica Biotechnology Center in Southern Taiwan, Tainan 741, Taiwan
Full list of author information is available at the end of the article

distant species may create more dramatic outcomes in transgenic lines than conventional breeding.

According to previous studies, the most reliable genetic transformation platforms successfully applied to *Phal.* orchids are the *Agrobacterium*-mediated approach and particle bombardment (Chai et al. 2002; Mii and Chin 2010; Teixeira da Silva 2011). Target explants for transformation using protocorms (Chin et al. 2007; Mishiba et al. 2005; Semiarti et al. 2007, 2010), protocorm-like bodies (PLB) (Anzai et al. 1996; Anzai and Tanaka 2001; Chai et al. 2002; Chan and Lin 2005; Chen et al. 2011; Hsieh et al. 1997; Julkifle et al. 2010; Liao et al. 2004; Subramaniam et al. 2009; Subramaniam and Xavier 2010), callus (Belarmino and Mii 2000; Sjahril and Mii 2006); and pollen tube pathways (Hsieh and Huang 1995; Tsay et al. 2012) have been reported. So far, the two most-often used host orchid tissues, protocorms and PLBs have been genetically transformed equally well. PLBs are presumed to be genetically uniform and can be induced efficiently from various somatic tissues including flower stalks, young leaves and stem segments (Chen et al. 2010; Yee et al. 2008). However, protocorm transformations have been proposed to be simpler than subculture of PLBs (Mishiba et al. 2005; Semiarti et al. 2007).

In the present study, we developed an alternative transformation procedure using protocorms generated by germination of seeds derived from diploid *Phal. aphrodite* and tetraploid *Phal.* cultivars. A T-DNA vector construct containing eGFP driven by ubiquitin promoter was used for transformation. The transformed protocorms were selected by hygromycin and subsequently regenerated successfully. We backcrossed the transgenic line and the BC1 progenies showed hygromycin resistant and the transgene is heritable. Finally, molecular analyses using PCR and western blot showed that all surviving backcross F1 explants were positive transformants.

Methods

Plant materials and growth condition

Three different *Phalaenopsis* orchid varieties were used for *Agrobacterium*- mediated transformation in this study. A popular big white flower tetraploid orchid of Dpts. Join Angel 'TH274-1' was bought from Join Orchids. It was self-pollinated and single seed descendant line (denoted L3) was used in our transformation. A commercial tetraploid orchid of *Phal. aphrodite* cv. M1663 (denoted Ama 4n) was bought from Chain Port Orchid Co. Ltd. A native Ama diploid of *Phal. aphrodite* subsp. *formosana* (denoted Ama 2n) was bought from the market of Shanhua. Plant-type and flower morphology of the three *Phalaenopsis* orchids is shown in the Additional file 1: Figure S1. *Phal.* orchid L3 with large white flowers more than 10 cm in diameter; whilst Ama (4n) and Ama (2n) have medium size white flowers around 7 cm in diameter. When blooming, flowers were self-pollinated to produce capsules. It took about 4–5 months for capsules to mature. Capsules were harvested when they turned slightly yellowish green at the tip of the capsule (Fig. 1a). Mature capsules could be stored at room temperature or 4 °C for 2 weeks and seeds still maintained good viability. Orchid seeds were sown on a 1/2 MS (Duchefa Bi ochemia) agar plate containing 0.1 % activated charcoal and placed in a tissue culture room at constant temperature of 24 °C under a 12 h photoperiod.

Plasmid vectors and bacterial strain

A green fluorescence marker gene eGFP driven by ubiquitin promoter was constructed into pZP200 backbone (http://www.biotech.unl.edu/pzp200) containing the selection marker *HptII* gene that encodes *hygromycin phosphotransferase,* was driven by cauliflower mosaic virus (CaMV) 35S promoter (Fig. 2a). The vector, pCAMBIA1305.1, which contained a β-glucuronidase (*GUS*) was driven by CaMV 35S promoter (http://www.cambia.org/daisy/cambia/585.html). The plasmid was separately subjected to transformation and subsequent hygromycin selections. *Agrobacterium tumefaciens* strain EHA105 was used for transfection in this study.

Inoculation, co-cultivation, and selection

Agrobacterium strain EHA105 containing eGFP target gene was spread on lysogeny broth (LB) plate containing 50 ppm kanamycin and 100 ppm rifampicin, and cultured overnight. Single colonies were selected and then cultured in 5 mL LB solution containing 50 ppm kanamycin and 100 ppm rifampicin, shaking at 250 rpm in a 28 °C incubator for 2 days. Agrobacteria were transferred to 20 mL LB containing 50 ppm kanamycin and 100 ppm rifampicin, and incubated at 28 °C to reach an optical density (OD) of a suspension of cells at 600 nm of around 1.0. Inoculum was centrifuged at 3500 rpm for 15 min, then the pellet was re-suspended with 1/2 MS containing 5 % sucrose, adjusted to pH 5.7, then 100 μM acetosyringone (AS) was added to the inoculum.

Healthy *Phal.* protocorm of size ~1.2 mm were selected, placed on a petri dish, and immersed in the *Agrobacterium* inoculum for 25 min. *Agrobacterium* inoculum was removed and the infected protocorms were spread on New Dogashima medium (NDM) (Belarmino and Mii 2000) containing 10 g/l maltose, 0.1 ppm naphthaleneacetic acid, 0.4 ppm benzyladenine (BA), 100 μM acetosyringone (AS), and 0.29 % phytagel. Infected protocorms were co-cultivated in the dark for 2–3 days until growth of *A. tumefaciens* was observed. Protocorms were washed with distilled water containing 40 ppm meropenem (China Chemical

Fig. 1 Procedures for *Agrobaterium*-mediated transformation of protocorms in *Phalaenopsis* orchid. **a** Mature orchid capsule at 4 months after pollination. **b** Stage of mature seeds ready for sowing. **c** Growth of protocorms sown on 1/2 MS agar plates for 1 month. **d** Healthy protocorm ready for infection. **e** *Agrobacteria* infection. **f** The 1st round selection for hygromycin resistance on T2 media containing 25 ppm hygromycin and 40 ppm meropenem (T2 + MH). **g** The 2nd round selection for hygromycin resistance on T2 + MH. **h** Growth of explants on 1/2 MS plate. **i** Growth of transgenic lines in jar. **j** Transplanted transgenic orchid to 1.7″ pot containing sphagnum moss

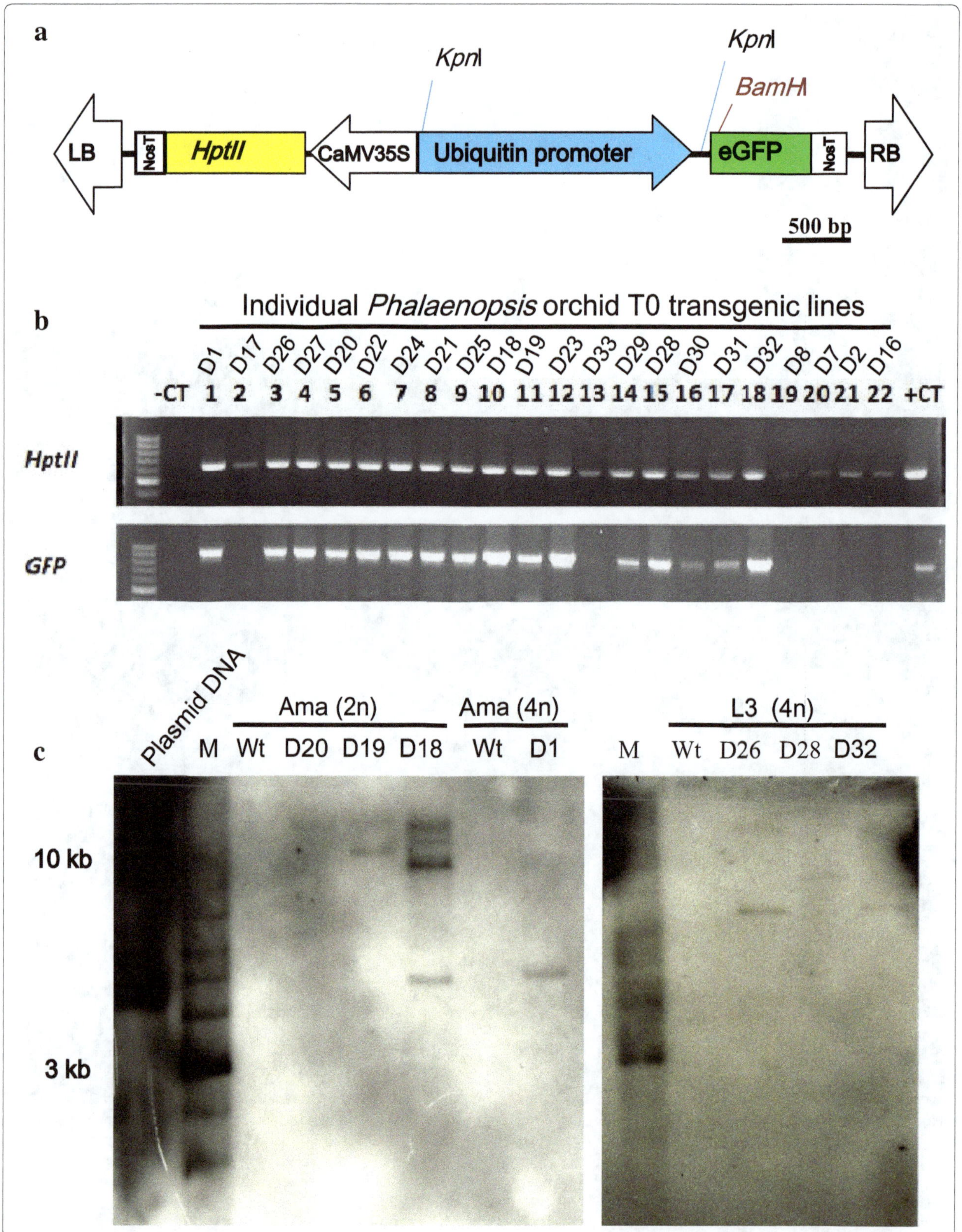
a
KpnI
KpnI
BamHI
LB
NosT
HptII
CaMV35S
Ubiquitin promoter
eGFP
NosT
RB
500 bp
b
Individual Phalaenopsis orchid T0 transgenic lines
D1 D17 D26 D27 D20 D22 D24 D21 D25 D18 D19 D23 D33 D29 D28 D30 D31 D32 D8 D7 D2 D16
-CT 1 2 3 4 5 6 7 8 9 10 11 12 13 14 15 16 17 18 19 20 21 22 +CT
HptII
GFP
c
Plasmid DNA
M
Ama (2n)
Wt D20 D19 D18
Ama (4n)
Wt D1
M
L3 (4n)
Wt D26 D28 D32
10 kb
3 kb

(See figure on previous page.)
Fig. 2 Confirmation of *Phalaenopsis* transgenic lines. **a** Construction map of Ubi:eGFP used in this study. Schematic structure of the T-DNA region in the vector of pZp200. CaMV 35S, promoter of CaMV 35S; *HptII*, coding region of *hygromycin phosphotransferase* gene; eGFP, coding region of *eGFP*; NosT, terminator of *nopaline synthase* gene; *LB* left border; *RB* right border. **b** Genomic PCR analysis of *eGFP* and *HptII* genes. −CT, negative control without addition of the genomic DNA template; +CT, with added plasmid DNA. **c** Southern blot analysis of *HptII* showing integration of T-DNA in *Phalaenopsis* orchid genome. Transgenic lines and their respective Wt plants were analyzed side by side

& Pharmaceutical) 3–4 times, and afterwards the protocorms were spread on T2 plates (Chen et al. 2009) containing 25 ppm hygromycin and 40 ppm meropenem (T2 + MH). Antibiotic selection of putative transgenic explants was conducted on T2 + MH three times, at 2-week intervals.

Transformation rate

To test the amenability of this protocol, a batch of experiments was carried out in triplicate using Ubi:eGFP construct. The numbers of the initially germinated protocorms of TH274 × TH601 hybrid and surviving protocorms that were resistant to hygromycin from the first to the third round of selection were recorded. Parental lines of TH274 and TH601 were purchased from Join Orchids. The percentage transformation rate was calculated based on the explants surviving after the third round of selection.

DNA isolation

Genomic DNA was isolated from a small piece of leaf tissue about 2 × 1 cm taken from putative transgenic lines using the cetyltrimethylammonium bromide (CTAB) method (Lee et al. 2015). The genomic DNA was used for PCR as well as for Southern blot analysis.

PCR determined transgenic lines

Polymerase chain reaction (PCR) was used to determine transgenic lines using *HptII* primers 5′-GATGTA GGAGGGCGTGGATA-3′ and 5′-CGTCTGCTGC TC CATACAAG-3′ and the target gene was confirmed by using primers eGFP 5′-ATGGTGAGCAAGGGCG AGGA-3′ and Nos-R 5′-ATCGCAAGACCGGCAAC AGGA-3′. The amplicon sizes are predicted to be 621 and 930 bp, respectively. The PCR amplification was carried out using following conditions: 95 °C, 3 min (1 cycle), 95 °C, 1 min, 58 °C, 1 min, 72 °C, 1 min (34 cycles), 72 °C, 10 min. Annealing temperature for *HptII* and *eGFP* was 58 and 60 °C, respectively.

For early detection of gene insertion, a small portion of leaf tissue (1 × 1 mm) was excised from small seedlings after 2 months recovery from the 3rd round hygromycin selection. DNA was directly extracted using Phire Plant Direct PCR Kit (Thermo Scientific) and PCR was performed using the *HptII* primers.

Southern blot analysis

Genomic DNA was extracted from leaf tissues (1 g fresh weight) using the CTAB method. After quantification by gel electrophoresis, 10 μg of each DNA was digested with *BamHI* and fractionated on a 0.8 % agarose gel. Southern hybridization and detection were carried out using a digoxigenin-labeled *HptII* probe following the manufacturer's instructions (Roche, http://www.rocheappliedscience.com).

Western blot analysis

Total protein of leaf samples was extracted with Culture Cell Lysis Reagent (CCLR) buffer (100 mM K_2HPO_4, 100 mM KH_2PO_4, pH 7.8, containing 1 % Triton X-100, 10 % glycerol, 1 mM EDTA, and 7 mM 2-mercaptoethanol). The protein amount was measured by using the Bio-Rad Protein Assay Kit with bovine serum albumin as a standard. For western blot analysis, 30 μg total protein from each sample was loaded and separated by SDS-PAGE with a 10 % acrylamide gel and transferred onto polyvinylidene fluoride (PVDF) membrane for antibody probing. The following primary antibodies were used: anti-eGFP, rabbit polyclonal antibody (Yao-Hong Biotechnology, Cat#YH-80005) at 1:10,000 dilution and anti-Actin mouse monoclonal antibody (Sigma, A0480) at 1:2500 dilution. Horseradish peroxidase (HRP)-conjugated secondary antibodies were obtained from GeneTex. Secondary anti-rabbit antibody was diluted at 1:10,000; anti-mouse antibody was diluted at 1:12,000.

GFP fluorescence microscopy

Leaf and root tissues of the D20 transgenic line and the non-transgenic wild-type were used for GFP florescence observation. Root tissue excised about 8 mm from the root tips and leaves was sampled from the mid vein of young developing leaves. Tissues were embedded in 4 % agarose II low gelling temperature (AMERESCo, Lot# 1624co42). The agarose sample block was trimmed to a desired size and adhered to a specimen stage. Vertical and transverse sections of the roots were made, and leaves were transverse sectioned. Tissues were sectioned to ~120 μm thickness using a blade installed in the Vibratome 1000 plus sectioning system (Fedelco, SL). A GFP signal was detected and an image was taken using a Zeiss LSM710 confocal microscope equipped with a

T-PMT under an FITC filter at excitation of 488 nm and emission wavelength of 500–560 nm.

GUS assay

T0 explants of 35S:GUS and non-transgenic wild type were stained with X-Gluc and incubated at 37 °C overnight according to a previously described protocol (Jefferson et al. 1987). Chlorophyll was removed by immersing in 95 % ethanol for 24 h. Pictures of whole explants were taken using a Nikon SMZ1500 dissecting microscope at 7.5× magnification. For detailed observation of the GUS expression pattern in transgenic cells, explants were sectioned using a vibratome according to the protocol mentioned in the previous section.

Inheritance of transgene

Transgenic orchid was crossed to non-transgenic wild types to generate BC1 segregating populations to determine if the transgene is heritable. One T0 plant (line D1) bloomed giving two flowers. A total of four D1 pollinia were backcrossed to four independent female parental non-transgenic *Phal.* orchids in a distinct genetic background, i.e., Ama (4n, the original Wt of D1), Ama (2n), C2-4 hybrid (TH274 × *Phalaenopsis equestris*), and V65 × V15 (hybrid orchid), respectively. Parental lines of *Phalaenopsis equestris*, V65 and V15 were purchased from Jumbo Orchids, Kaohsiung Guanyinshan and Chain Port Orchid Co. Ltd., respectively. Approximately four months after pollination, orchid capsules were harvested and the BC1 seeds were sown on a 1/2 MS plate containing 25 ppm hygromycin. Two months after sowing, the numbers of explants that were resistant or susceptible to hygromycin were calculated. The surviving green seedlings were considered to be transgenic lines that were resistant to hygromycin, whilst those wilting or undeveloped protocorms were hygromycin susceptible wild-type like progenies. In order to confirm introgression of the transgene to the next generation, 22 BC1 s of Ama (4n)/D1, Ama (4n) Wt as a negative control and D1 (the original T0 transgenic plant) as a positive control were applied to PCR detection using *HptII* primers and western blot using eGFP antibody.

Results

Transformation processes and obtained transgenic seedlings

Mature orchid capsules were harvested (Fig. 1a) and surface sterilized with 70 % EtOH. The capsule was cut longitudinally with a sterilized scalpel and tiny seeds were spread evenly on 1/2 Murashige and Skoog (MS) agar plates. Seed germination and growth of protocorms took place on agar plates at constant temperature of 24 °C (Fig. 1b). Five weeks after sowing, protocorms had grown substantially and had turned light green in color (Fig. 1c). Healthy globular shape protocorms of approximately 1.2 mm in length (Fig. 1d) were selected and infected with *Agrobacterium* inoculum (Fig. 1e). Selection of putative transgenic lines was performed by three successive hygromycin selections at 2-week intervals per selection (Fig. 1f–h). The greenish explants after the 3rd round of antibiotic selection (Fig. 1h) were recovered in a 1/2 MS agar plate for 2 months and then transferred to a jar for approximately 3 months before transplanting the transgenic orchid to a 1.7" pot containing sphagnum moss. Finally, a total of 74 independent transgenic orchid seedlings overexpressing Ubi:eGFP were obtained and grown in a GMO greenhouse and all had apparently normal morphology. The phenotypes of the 20 transgenic lines are shown in the Additional file 1: Figure S2.

Timeline for transformation

It took about 32 weeks to obtain transformation results, starting from sowing seeds to PCR analysis on transformed seedlings right before transplanting to 1.7″ pots. However, the whole process could be reduced to 15 weeks with modifications. Firstly, always maintaining viable protocorms can save about 5 weeks, the amount of time needed for seeds to develop. Secondly, if seedlings are transferred to 1/2 MS not containing hygromycin; after the 3rd hygromycin selection, seedlings can grow rapidly. After 2 months of growth, a small piece of leaf tissue (1 × 1 mm) was excised for early PCR analysis. These modifications save time and increase efficiency of transformation.

Frequency of transformation

Transformation rate is critical for the success of a transformation system. In order to know the transformation efficiency of this protocol, three batches of experiments were tested using Ubi:eGFP vector. As can be seen in Additional file 1: Table S1, transformation rate varied in different batches ranging from 1.2 to 5.2 %. Normally, healthy *Phal.* orchid capsule contain tens of thousands tiny seeds that appear as dust. A single capsule of *Phal. aphrodite* can be sown on 8–10 agar plates of 9 cm diameter. Each plate can germinate more than 1000 protocorms for *Agrobacteria* infection. Thus, the selection is rather easy and less labor intensive. Therefore, increase initial *Agrobacteria* infected protocorms plate numbers can apparently obtain more transgenic *Phal.* orchid lines.

Molecular confirmation of transgenic lines

A total of 74 individual transgenic lines of *Phal.* orchids were obtained and planted in the greenhouse. Among them, 22 transgenic orchid pot seedlings (Additional file 1: Figure S2) were randomly selected, genomic DNA was isolated, and genomic PCR was performed. There was no *hygromycin phosphotransferase* gene (*HptII*) and enhanced green fluorescence protein (*eGFP*) amplification with the DNA from the Wt plant but obvious PCR banding was obtained in the positive control of the vector DNA (Fig. 2a). All of the 22 tested T0 transgenic lines showed *HptII* banding after the high stringency hygromycin selection (Fig. 2b). However, five T0 plants did not show *eGFP* bands (Fig. 2b). This might due to a problem with DNA insertion of eGFP in those lines. Overall, the T-DNA integration was accomplished normally in most of the transgenic lines.

To determine the T-DNA insertion number, Southern blot analysis was performed by randomly selecting some transgenic lines developed from the genetic background of Ama (2n) i.e., D18, D19 and D20; from Ama (4n) such as D1, and from L3 (4n) background such as D26, D28, and D32. Only the transgenic lines showed a hybridization signal, with no signal being seen in the Wt plants of Ama (2n), Ama (4n), and L3 (4n). Southern blot data indicated that most of the transgenic lines contained one or two copies of T-DNA insertion (Fig. 2c).

Total protein was isolated from leaves of Ubi:eGFP transgenic orchid lines and their respective three wild types, L3 (4n), Ama (4n), and Ama (2n). Thirty micrograms of total protein was loaded and western blot analysis was performed using anti-eGFP antibody. Results indicated that none of the three wild types expressed the GFP signal but all the transgenic lines expressed eGFP protein in the T0 transgenic orchids (Fig. 3a). The lower panel showed anti-Actin antibody to ensure equal protein loading (Fig. 3a).

Green florescent signal in orchid transgenic lines

Green florescent protein (GFP) is a marker for monitoring gene expression in transgenic plants (Hraska et al. 2006). Under confocal microscopy, as expected, there was an absence of GFP detection in the root and leaf of wild type but the GFP fluorescence signal was clearly observed in the D20 transgenic line, with more expression in the outer layers of the vertical root section (Fig. 3b). A strong GFP signal was seen in the cortex tissues in the transverse root and a very strong GFP signal was accumulated in the pericycle cells of the transgenic line as shown in Fig. 3c. The pericycle that was located between the endodermis and phloem is the tissue that enables plants to grow roots and facilitates plant development. Cells in the pericycle actively undergo cell division and proliferation (Dubrovsky et al. 2000). We observed more GFP signals around the pericycle under fluorescence microscope. Here we found the root pericycle close to the root tip has less endogenous GFP and it is good tissue to detect the expression of the GFP transgene (Fig. 3c).

Uniform GUS expression in transgenic tissues

Non-chimeric transgenic plants are essential for stable transformation. In order to check that the transformation protocol established in this study did not produce chimerism, T0 transgenic explants of 35S:GUS were stained with X-Gluc solution. All tissues showed strong GUS positive signals, including the leaf, root, and hypocotyl (Fig. 4a). Vibratome tissue section showed uniform GUS positive signals in all tissues and cells (Fig. 4b). These results confirmed that the transformation system established in this study did not produce chimeric transformants.

Stable inheritance of transgene

It is very important to develop transgenic orchids with stable inheritance and consistent expression of target genes in the next generation. Therefore, pollinia of the D1 transgenic line were backcrossed to four distinct genetic backgrounds, Ama (4n, the original Wt of D1), Ama (2n), and the hybrids of C2-4 and V65 × V15, respectively, and produced BC1 seeds. Some BC1 progenies showed resistance in hygromycin selection (Fig. 5a). The hygromycin resistant explants varied from 43 to 56 % in four BC1 progenies (Fig. 5b). Southern blotting data indicated that D1 is a single T-DNA insertion transgenic line (Fig. 2c). BC1 progenies were heritable (Fig. 5b). Genomic PCR analysis showed that 22 hygromycin resistant BC1 of Ama (4n)/D1 seedlings all contained *HptII* amplicons, indicating that the introgression of transgenes is stable in the next generation of *Phal.* orchid (Fig. 5c). Moreover, western blot also indicated that all hygromycin survival progenies positively expressed eGFP protein (Fig. 5d). We concluded that the transgenic lines generated using the *Agrobacterium*-mediated young protocorm method presented in this study is stable and efficient. To be noted, the expression and introgression of transgenes to the next generation was demonstrated in this study.

Discussion

Phalaenopsis orchids have been a mainstay of the global ornamental flower trade in recent years. Genetic engineering has emerged as a method with high potential to modify traits more precisely and enable functional genomics studies more efficiently. Although *Agrobacterium*-mediated transformation has been successfully applied to many agricultural or horticultural crops, it still remains difficult to apply to some plants that are not easily clonally propagated. Here, we reported an efficient method of *Agrobacterium*- mediated transformation using protocorms of *Phal. aphrodite*. Its advantages over previous methods can be summarized as follows:

We achieved a stable, efficient, yet less labor-intensive transformation method to genetically transform

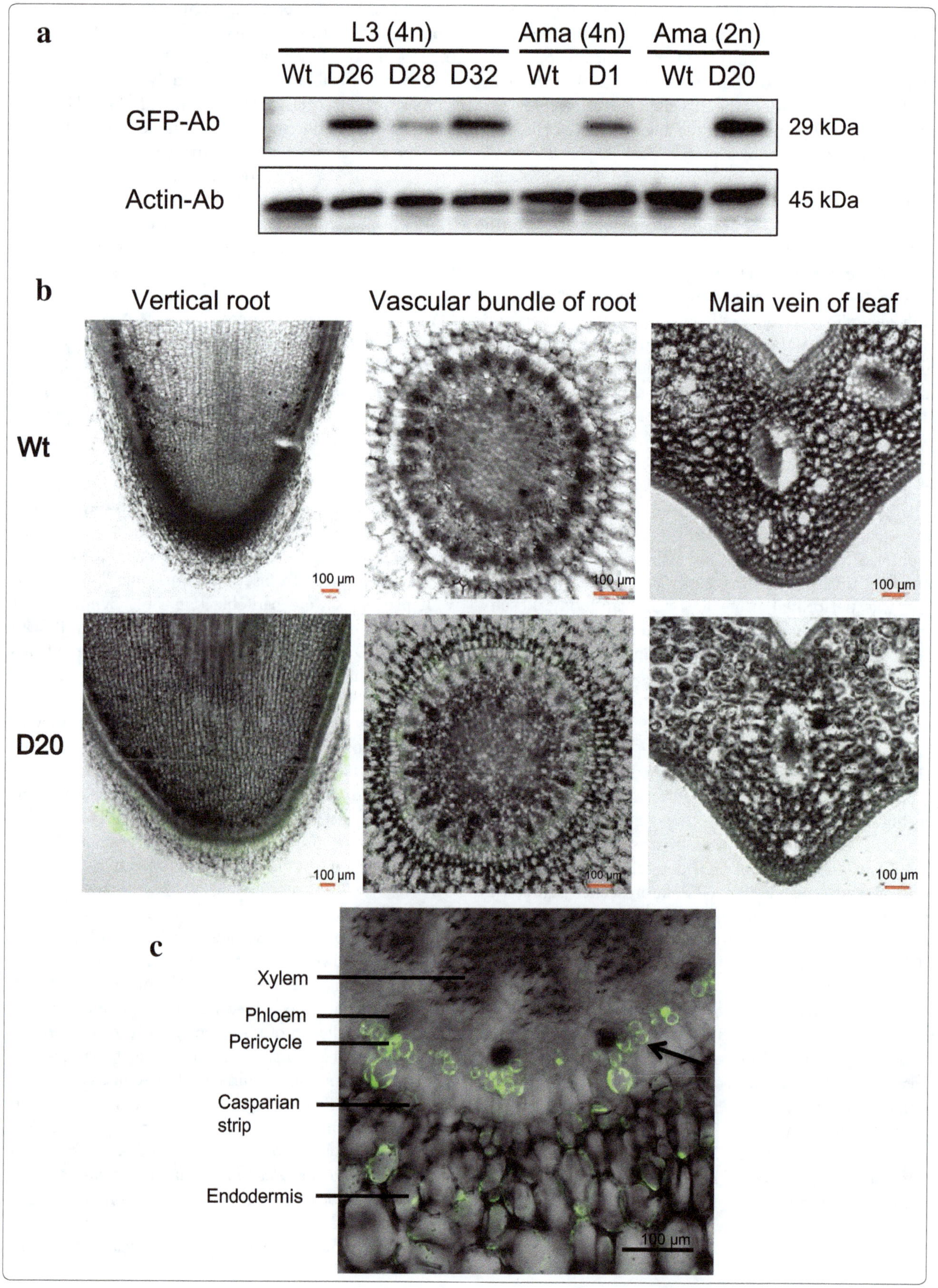
a
L3 (4n)
Ama (4n)
Ama (2n)
Wt D26 D28 D32 Wt D1 Wt D20
GFP-Ab
29 kDa
Actin-Ab
45 kDa
b
Vertical root
Vascular bundle of root
Main vein of leaf
Wt
D20
100 μm
c
Xylem
Phloem
Pericycle
Casparian strip
Endodermis
100 μm

(See figure on previous page.)
Fig. 3 Western blot and detection of the GFP fluorescence signal in transgenic orchids. **a** Western blot analysis of the expression of GFP protein in transgenic orchids. Total protein was isolated from the leaf of orchid and 30 µg total protein was loaded into 10 % acrylamide gel, electrophoresed, and transferred onto polyvinylidene fluoride (PVDF) membrane for antibody probing. The membrane was probed using an anti-GFP antibody and anti-actin antibody served as the loading control. *Wt* wild-type; *4n* tetraploid; *2n* diploid. **b** GFP fluorescence signal was detected in D20 transgenic *Phalaenopsis aphrodite*. **c** A magnified view of the GFP signal in partial tissue of **b**. Root tissues of the D20 transgenic line and Wt were vertical and transversely sectioned, and leaves were transverse-sectioned using a vibratome and an image was obtained using a confocal microscope

monocotyledonous *Phal.* orchids. Stable integration, expression, and inheritance of the transgenes were confirmed using molecular and genetic studies. To date, the transformation rates reported for *Phal.* orchids have been rather low and have never reached 2 % (Mishiba et al. 2005; Semiarti et al. 2007). However, this protocol could generate up to 5.2 % transformation efficiency (Additional file 1: Table S1).

Fig. 4 Histochemical Gus staining in transgenic *Phalaenopsis* orchid lines. Transgenic orchid explant harboring 35S:GUS and a Wt non-transgenic control were stained with X-Gluc and incubated at 37 °C overnight. **a** Image obtained using a Nikon SMZ1500 dissecting microscope (×7.5 magnification). **b** Explant tissue sections of approximately 120 µm in thickness were photographed on a Zeiss Axio Scope A1 microscope equipped with an Axio-Cam HRc camera (Zeiss, Germany) (×25 magnification)

Choose of appropriate size of target explants for transformation is very important for stable transformation. The orchid protocorms which are similar to immature embryos with extensive cell proliferation are good target for transformation. Chimerism could occur if large orchid embryos or PLBs are used as target tissues (Kuehnle and Sugii 1992). In the present study, we found small-size protocorms of 1.2 mm are suitable for transformation and we did not observe chimerism in transgenic explants after GUS staining (Fig. 4). Moreover, the inheritance of transgenes is stable in all the BC1 transformation lines tested in our genetic study as shown in Fig. 5. These data demonstrates that transformation protocol developed in this study is stable and heritable.

To speed up transformation process, we adopted three successive 25 ppm hygromycin selections at 2-week interval in this protocol. The total of 6 weeks in hygromycin selection media is much shorter than other transformation methods that usually take up to several months or longer. Selection under high concentration of hygromycin can reduce false positive transformants. In addition, using the Direct PCR Kit enables us to detect gene insertion in early stage to validate the positive transformants. These modifications significantly save time and increase efficiency of transformation. It is known that slow in growth is one of the bottlenecks for *Phal.* orchid production as well as for researches. Therefore, it is suggested once PCR confirms the transformants, immediately recover the transgenic seedlings in 1/2 MS agar without subjecting the pressure from antibiotics could speed up their growth dramatically.

It is known that GFP fluorescence has some weak points, such as the diminishing green fluorescence in older tissues, variation in fluorescence levels among different tissues, and interference with autofluorescence in plants (Hraska et al. 2006; Molinier et al. 2000). In the present study, we successfully detected the GFP fluorescent signal of transgenic *Phal.* orchid using a florescence microscope and enabled the determination of the presence of the transgene (Fig. 3b, c). Moreover, the transgene can be monitored non-destructively using a dissecting fluorescence microscope. The root tip can be partially cut to identify the GFP signal and will not affect the survival

a

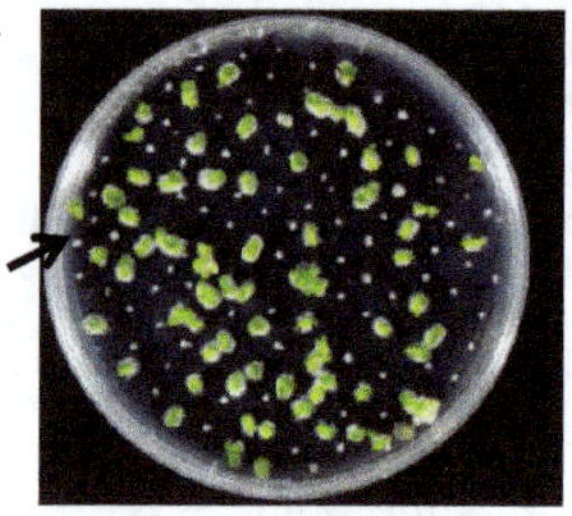

b

BC1	Number of seeds	Hygromycin resistant	Transgenic line (%)
Ama (2n)/ D1	110	52	47%
Ama (4n)/ D1	159	68	43%
C2-4/ D1	137	74	54%
V65xV15/ D1	194	109	56%

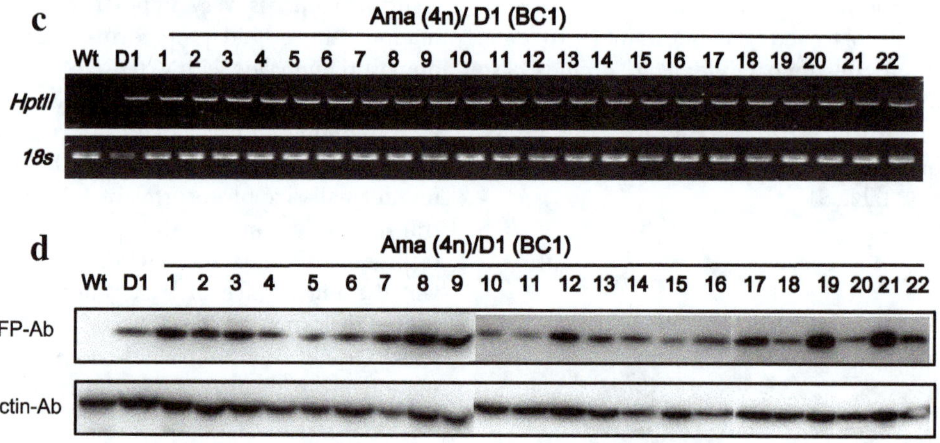

Fig. 5 Stable inheritance of transgene in BC1 orchid progeny. **a** Pollinia of D1 transgenic line was crossed-pollinated to four distinct genetic backgrounds of Wt orchids and generated BC1 progenies. Ama (4n)/D1 BC1 seeds were sown on 1/2 MS agar plates containing 25 ppm hygromycin. Some BC1 progenies were resistant to hygromycin and some were susceptible showing white and wilt symptoms. *Red arrows* indicate hygromycin susceptible explants. **b** Record of survival rate in BC1 transgenic lines. **c** Genomic PCR showed *HptII* gene integration in the hygromycin-resistant BC1 progenies. **d** Western blot showed GFP protein expression in the hygromycin-resistant BC1 progenies. Ama (4n) was used as the non-transgenic wild-type negative control. D1, the original T0 transgenic parent was used as a positive control

of the seedling. Interestingly, we found that pericycle cells close to root tip express high levels of GFP in this study (Fig. 3c). Higher cytoplasmic density in young tissues causing higher levels of expression of GFP than in older tissues has been previously described (Molinier et al. 2000). Therefore, it is suggested that the pericycle close to the root tip of orchid that contains dense proliferating cells could be a good tissue type to detect introgression of transgenes. We suggest that eGFP is a suitable reporter system that provides reliable results of introgression of transgenes and it is a good marker for the genetic transformation of *Phal.* orchids.

The *Agrobacterium*-mediated transformation system in this study may provide a practical tool for studying functional genomics, which can further assist the gene discovery and molecular breeding processes. GMO of ornamental crops that are not edible is more acceptable to consumers than GMO of food. Genome editing technology such as CRISPR (Liu et al. 2016; Zaidi et al. 2016; Zhu et al. 2016) provides the prospect of modification of orchids using highly efficient transformation techniques in *Phal.* orchids.

Conclusions

This study established an *Agrobacterium*-mediated transformation system using orchid protocorm as a target explant. This protocol works well in tetraploid as well as in diploid *Phal. aphrodite* varieties. Moreover, this method is simple, stable, efficient, heritable, time saving, and less labor-intensive. In the era of big data and functional genomics, this transformation method can significantly contribute to the gene discoveries, genetic engineering and general R&D of *Phal.* orchid.

Additional file

Additional file 1: Table S1. Transformation efficiency in different batches of experiments. TH274 × TH601 hybrid seeds were sown in 1/2 MS containing 25 ppm hygromycin and the surviving protocorms that were recorded from the first to the third round of selections. **Figure S1.** *Phalaenopsis* orchid cultivars used for transformation in this study. a Plant type of *Phalaenopsis* orchid cultivars at flowering stage. b Flower morphology of *Phalaenopsis* orchid cultivars. 4n, tetraploid; 2n, diploid. Bars = 5 cm (a), 2 cm (b). **Figure S2.** Phenotype of T0 transgenic orchid seedlings overexpressing Ubi:GFP. D1 to D33 represented the individual T0 transgenic lines overexpressing Ubi:GFP. The respectively wild type is written in the right hand side.

Authors' contributions

SSK designed the experiments, HXH, YJL, MJL, SSK, YJL conducted the experiments, SSK and CGT wrote the manuscript. All authors read and approved the final manuscript.

Author details

[1] Academia Sinica Biotechnology Center in Southern Taiwan, Tainan 741, Taiwan. [2] Agricultural Biotechnology Research Center, Academia Sinica, Taipei 115, Taiwan.

Acknowledgements

We are grateful to the support of the Academia Sinica Innovative Translational Agricultural Research Grant to Swee-Suak Ko project (iTAR#2015CP02-3). We express our sincere appreciation for technical support from Ms. Chi Fang, Ms. Whe-re Shen, and Ms. Yi-Chen Lien. We thank Ms. Miranda Loney for English editing.

Competing interests

The authors declare that they have no competing interests.

References

Anzai H, Tanaka T (2001) Transgenic Phalaenopsis (a moth orchid). In: Bajaj YPS (ed) Biotechnology in agriculture and forestry transgenic crops III 48. Springer, Berlin, pp 249–264. doi:10.1007/978-3-662-10603-7_18

Anzai H, Ishii Y, Shichinohe M, Katsumata K, Nojiri C, Morikawa H, Tanaka M (1996) Transformation of Phalaenopsis by particle bombardment. Plant Tissue Cult Lett 13:265–272. doi:10.5511/plantbiotechnology1984.13.265

Belarmino MM, Mii M (2000) Agrobacterium-mediated genetic transformation of a Phalaenopsis orchid. Plant Cell Rep 19:435–442. doi:10.1007/s002990050752

Chai ML, Xu CJ, Senthil KK, Kim JY, Kim DH (2002) Stable transformation of protocorm-like bodies in Phalaenopsis orchid mediated by *Agrobacterium tumefaciens*. Sci Hort 96:213–224. doi:10.16/S0304-4238(02)00084-5

Chan YL, Lin KH, Liao LJ, Chen WH, Chan MT (2005) Gene stacking in Phalaenopsis orchid enhances dual tolerance to pathogen attack. Transgenic Res 14:279–288. doi:10.1007/s11248-005-0106-5

Chen WH, Kao YL, Tang CY (2009) Method for producing polyploid plants of orchids. US Patent 8,383,881 B2, 26 Feb 2013

Chen YC, Yang YC, Chen YH, Chao YP, Chang LZ, Chen YW (2010) Modification of Phalaenopsis metabolism by genetic engineering. Acta Hort 878:473–480. doi:10.17660/ActaHortic.2010.878.61

Chen WH, Hsu CY, Cheng HY, Chang H, Chen HH, Ger MJ (2011) Downregulation of putative UDP-glucose: flavonoid 3-*O*-glucosyltransferase gene alters flower coloring in Phalaenopsis. Plant Cell Rep 30:1007–1017. doi:10.1007/s00299-011-1006-1

Chin DP, Mishiba K, Mii M (2007) Agrobacterium-mediated transformation of protocorm-like bodies in Cymbidium. Plant Cell Rep 26:735–743. doi:10.1007/s00299-006-0284-5

Dubrovsky JG, Doerner PW, Colon-Carmona A, Rost TL (2000) Pericycle cell proliferation and lateral root initiation in *Arabidopsis*. Plant Physiol 124:1648–1657. doi:10.1104/pp.124.4.1648

Fu CH et al (2011) OrchidBase: a collection of sequences of the transcriptome derived from orchids. Plant Cell Physiol 52:238–243. doi:10.1093/pcp/pcq201

Guo WJ, Lee N (2006) Effect of leaf and plant age, and day/night temperature on net CO2 uptake in *Phalaenopsis amabilis* var. formosa. J Amer Soc Hort Sci 131:320–326

Hraska M, Rakousky S, Curn V (2006) Green fluorescent protein as a vital marker for non-destructive detection of transformation events in transgenic plants. Plant Cell Tissue Organ Cult 86:303–318. doi:10.1007/s11240-006-9131-1

Hsieh RM, Huang PL (1995) Studies on genetic transformation of Phalaenopsis via pollen tube pathway. J Chin Soc Hort Sci 41:309–324

Hsieh RM et al (1997) *Agrobacterium tumefaciens*-mediated transformation of Phalaenopsis orchid. Rept Taiwan Sugar Inst 155:41–54

Jefferson RA, Kavanagh TA, Bevan MW (1987) GUS fusions: beta-glucuronidase as a sensitive and versatile gene fusion marker in higher plants. EMBO J 6:3901–3907

Julkifle AL, Rathinam X, Sinniah UR, Subramaniam S (2010) Optimisation of transient green fluorescent protein (GFP) gene expression in *Phalaenopsis violacea* orchid mediated by *Agrobacterium tumefaciens*-mediated transformation system. Aust J Basic Appl Sci 4:3424–3432

Kuehnle AR, Sugii N (1992) Transformation of Dendrobium orchid using particle bombardment of protocorms. Plant Cell Rep 11:484–488. doi:10.1007/bf00232696

Lee SH, Li CW, Liau CH, Chang PY, Lioa LJ, Lin CS, Chan MT (2015) Establishment of an Agrobacterium-mediated genetic transformation procedure for the experimental model orchid *Erycina pusilla*. Plant Cell Tiss Organ Cult 120:211–220. doi:10.1007/s11240-014-0596-z

Liao LJ, Pan IC, Chan YL, Hsu YH, Chen WH, Chan MT (2004) Transgene silencing in Phalaenopsis expressing the coat protein of Cymbidium mosaic virus is a manifestation of RNA-mediated resistance. Mol Breed 13:229–242. doi:10.1023/B:MOLB.0000022527.68551.30

Liu D, Hu R, Palla KJ, Tuskan GA, Yang X (2016) Advances and perspectives on the use of CRISPR/Cas9 systems in plant genomics research. Curr Opin Plant Biol 30:70–77. doi:10.1016/j.pbi.2016.01.007

Mii M, Chin DP (2010) Genetic transformation of orchids. Acta Hort 878:461–466. doi:10.17660/ActaHortic.2010.878

Mishiba K, Chin DP, Mii M (2005) Agrobacterium-mediated transformation of Phalaenopsis by targeting protocorms at an early stage after germination. Plant Cell Rep 24:297–303. doi:10.1007/s00299-005-0938-8

Molinier J, Himber C, Hahne G (2000) Use of green fluorescent protein for detection of transformed shoots and homozygous offspring. Plant Cell Rep 19:219–223. doi:10.1007/s002990050002

Semiarti E et al (2007) Agrobacterium-mediated transformation of the wild orchid species *Phalaenopsis amabilis*. Plant Biotech 24:265–272. doi:10.5511/plantbiotechnology.24.265

Semiarti E et al (2010) High-frequency genetic transformation of *Phalaenopsis amabilis* orchid using tomato extract-enriched medium for the pre-culture of protocorms. J Hort Sci Biotechnol 85:205–210. doi:10.1080/14620316.2010.11512655

Sjahril R, Mii M (2006) High-efficiency Agrobacterium-mediated trasnformation of Phalaenopsis using meropenem, a novel antibiotic to eliminate Agrobacterium. J Hort Sci Biotech 81:458–464. doi:10.1080/14620316.2006.11512088

Su CL, Chao YT, Yen SH, Chen CY, Chen WC, Chang YC, Shih MC (2013a) Orchidstra: an integrated orchid functional genomics database. Plant Cell Physiol 54:e11. doi:10.1093/pcp/pct004

Su CL, Chen WC, Lee AY, Chen CY, Chang YC, Chao YT, Shih MC (2013b) A modified ABCDE model of flowering in orchids based on gene expression profiling studies of the moth orchid Phalaenopsis aphrodite. PLoS ONE 8:e80462. doi:10.1371/journal.pone.0080462

Subramaniam S, Xavier R (2010) Emerging factors that influence efficiency of T-DNA gene transfer into *Phalaenopsis violacea* orchid via *Agrobacterium tumefaciens*—mediated transformation system Int. J Biol 2:64–73

Subramaniam S, Samian R, Midrarullah Xavier R (2009) Preliminary factors influencing transient expression of Gusa in Dendrobium Savin White protocorm-like bodies (PLBs) using Agrobacterium-mediated transformation system. World Appl Sci J 7:1295–1307

Teixeira da Silva JA, Tanaka M (2011) Optimization of particle bombardment conditions for hybrid Cymbidium: Part II. Transgenic Plant J 5:78–82

Tsay HS, Ho HM, Gupta SK, Wang CS, Chen PT, Chen ECF (2012) Development of pollen mediated activation tagging system for Phalaenopsis and Doritaenopsis Electron. J Biotechnol. doi:10.2225/vol15-issue4-fulltext-1

Yee NC, Abdullah JA, Mahmood M, Basiron N (2008) Co-transfer of gfp, CHS and hptII genes into Oncidium Sharry Baby PLB using the biolistic gun. Afr J Biotechnol 7:2605–2617. doi:10.5897/AJB08.458

Yu H, Goh CJ (2001) Molecular genetics of reproductive biology in orchids. Plant Physiol 127:1390–1393. doi:10.1104/pp.010676

Yumiko B, Brundrettb M, Batty A, Dixon K, Kochd J, Sivasithamparam K (2007) Diversity of mycorrhizal fungi of terrestrial orchids: compatibility webs, brief encounters, lasting relationships and alien invasions. Mycol Res 111:51–61. doi:10.1016/j.mycres.2006.11.006

Zaidi SS, Mansoor S, Ali Z, Tashkandi M, Mahfouz MM (2016) Engineering plants for Geminivirus resistance with CRISPR/Cas9 system. Trends Plant Sci 21:279–281. doi:10.1016/j.tplants.2016.01.023

Zhu J, Song N, Sun S, Yang W, Zhao H, Song W, Lai J (2016) Efficiency and inheritance of targeted mutagenesis in maize using CRISPR-Cas9. J Genet Genomics 43:25–36. doi:10.1016/j.jgg.2015.10.006

20

Developmental program impacts phenological plasticity of spring wheat under drought

Marwa N. M. E. Sanad[1,2*], Kimberley Garland Campbell[1,3†] and Kulvinder S. Gill[1†]

Abstract

Background: Developing drought-tolerant crops critically depends on the efficient response of a genotype to the limited water availability, a trait known as phenological plasticity. Our understanding of the phenological plasticity remains limited, in particular, about its relationships with plant developmental program. Here, we examined the plastic response of spring wheat at tillering, booting, heading, and anthesis stages to constant or periodic drought stress. The response was assessed by morphological and physiological parameters including symptoms.

Results: The dynamics of morphological symptoms were indicators of the plasticity identification of drought. We found that spring wheat exhibits higher phenological plasticity during tillering stage followed by the heading stage, while booting and anthesis stages are the most sensitive. Also, the adaptive response is thought to be influenced with the plant height genes. Furthermore, periodic stress caused more pronounced inhibition of yield than the constant stress, with limited resistance resolution under long period.

Conclusions: Our study shows the importance of considering the phenological plasticity in designing screens for drought tolerance in spring wheat and proposes tillering as the most informative stage for capturing genotypes with tolerance to limit water availability.

Keywords: Phenological plasticity, Developmental plasticity, Tillering stage, Drought, Spring wheat

Background

Low precipitation is a serious obstacle to sustainable farming. The negative impact of drought on the quality and quantity of major crop such as wheat *(Triticum aestivum* L.) yields threatens global food security. Several strategies are used by plant species to achieve drought tolerance including escape, avoidance, and resilience also known as phenotypic plasticity; (Farooq et al. 2009). However, the phenotypic plasticity can be limited by several environmental factors (Nicotra and Davidson 2010). Some of these factors are: the developmental plasticity, phenological plasticity (a predisposition for phenotypic plasticity), or length of the stress period.

Initially, phenotypic responses can result in an acclimation to drought. Continuing water deficit can then lead to more stable tolerance through adaptation mechanisms (adaptive plasticity) (Nicotra and Davidson 2010). The frequency and duration of the drought episodes play a major role in defining the degree of tolerance. The accumulated history of adaptive plasticity events generates a stresses imprint that can be revoked during the subsequent stress episodes. This phenomenon was defined as "plant memory" (Bruce et al. 2007; Aubin-Horth and Renn 2009; Walter et al. 2011) replacing previous term "immunological memory" used to explain damage-induced signaling (Baldwin and Schmelz 1996).

Plant evolution and plant ecology are determined to be influenced by the developmental plasticity, which contribute to improving yield stability in agriculture (de Jong and

*Correspondence: marwa.sanad@wsu.edu; drmarwanauib24@gmail.com
†Kimberley Garland Campbell, Kulvinder S. Gill contributed equally to this work
[1] Crop and Soil Sciences Department, Washington State University, Pullman, WA, USA
Full list of author information is available at the end of the article

Leyser, 2012). Because the efficiency of the phenotypic plasticity is determined by the developmental stage, stress duration and stress severity in the genotype-dependent manner (Sivamani et al. 2000; Chaves et al. 2003; Rizza et al. 2004; Milad et al. 2011). Thus, apart from defining the stress protocol, screening for drought tolerance requires a careful consideration of the growth stage at which the tolerance will be scored. Three consequences of phenotypic plasticity as a drought response were observed in plants: 1) phenotypic changes after single stress exposure (Noormets et al. 2008), 2) retrieving the positive phenotype after recurrent stress exposure (Bruce et al. 2007), and 3) overcoming rewatering stress (Xu et al. 2010). Successive drought and rewatering cycles can shift the time frame of plant development or phenological plasticity (Vitasse et al. 2010), and change the final biomass or leaf area (Xu et al. 2010). These factors highlight the importance of assessing the phenological plasticity for capturing the drought tolerant genotypes.

Development of "plant memory" for response to drought is determined by the time interval between consecutive stress periods (Bruce et al. 2007). It has been proposed that plants need sufficiently long exposure to drought stress before the appropriate cellular and molecular signals can be generated and then converted into the transcription of stress genes (Trewavas 2005; Vinocur and Altman 2005; Goswami et al. 2010). As an integral component of the developmental plasticity "plant memory" would also impact the genotype-by-environment (GxE) interaction because developmental stage predetermines which genes can be affected in response to stress (De Leonardis et al. 2007). Developmental plasticity and GxE interactions are irreversible. GxE interaction can cause permanent developmental change (West-Eberhard 2003). Thus, optimizing developmental plasticity would benefit the breeding programs to control the GxE interaction.

In this work, we hypothesize that some developmental stages would exhibit higher phenotypic plasticity in wheat. Also, it is highly preferred to detect a phenotypic plasticity during a long developmental stage that needs high water requirements. Because it would be sufficient for building an adaptive mechanism and optimizing drought-tolerant genotypes capture process.

Therefore, we aimed to determine the impact of the main developmental stages on the phenotypic plasticity in response to drought stress; which would minimize the GxE interaction, and optimize the developmental stage to be better suited for capturing drought tolerant genotypes. An ideal stage should satisfy the following criteria: (1) resistance symptoms to drought stress, (2) capable of assessing the tolerance and susceptible responses, (3) high yield fitness, (4) normal germination rate of seeds collected from the drought-stressed plants, (5) normal seedling growth, and (6) withstand the prolonged drought for adaptive response.

Methods

Planting conditions

Eight genotypes of spring wheat with different genetic background were evaluated (Table 1). The whole experiment included 480 pots arranged as three replications of eight genotypes in a total of 20 stress regimes (five growth stages by four stress durations) arranged as a completely random design. The experiment was conducted in a walk-in growth chamber (Conviron; Controlled Environments Ltd., Winnipeg, Manitoba, Canada). Plants were grown on a diurnal cycle of 16/8 h at 418 µmol PAR light. Growth chamber temperature was maintained at 22–23 °C/16 °C, and the CO_2 was approximately 404 ppm of 17% humidity. Seeds were embedded in moist containers, which were filled and compacted with Sunshine Mix LC1 potting soil (Sun Gro Horticulture, Bellevue, WA). An application of 14-14-14 Osmacote (Scotts Miracle-Gro Company, Marysville, OH) slow release fertilizer was incorporated at planting, stress induction, and stress recovery.

Stress regime description and duration

Constant stress was imposed at Zadoks growth stages: 21 for tillering, 45 for booting, 50 for beginning of head emergence (heading) and 60 for beginning of flowering (anthesis) (Zadoks et al. 1974), or not imposed in the case of the control treatment for each of the five growth stage treatments.

All plants were watered uniformly reaching the saturation prior to planting, until the stress imposition was reached for a given pot.

The stress was maintained for duration of 0 (control or watered group), 7, 14, and 21 days. For the 7-, 14-, and 21-day treatments, soil saturated was moisture uniformly

Table 1 The genotype name, scientific name, ID number, origin, and type of wheat

Genotype	Scientific name	ID No	Origin	Type
Perigee	*T. aestivum*	NASA	NASA-USA	S
Vandal	*T. aestivum*	PI546056	Idaho AES; USDA-ARS	HRS
PWB343	*T. aestivum*	BW26864	India	S
Klein Dragon	*T. aestivum*	CM64693	Argentina	S
PotamS-70	*T. aestivum*	BW623	CIMMYT	S
Indian	*T. aestivum*	Cltr4489	USA-Utah	SWS
Onas	*T. aestivum*	Cltr6221	South Australia	SWS
Edmore	*T. turgidum var. durum*	Cltr17748	USA-North Dakota	DS

S spring, *HRS* hard red spring, *SWS* soft white spring, and *DS* durum spring

prior to planting until stress started, then water was withheld for the length of the stress duration, and then plants were re-watered until soil saturation as needed. The 0-day stress treatments were watered throughout the experiment. Stress treatments were also imposed as periodic drought and re-watering cycles. In these periodic stress treatments water was withheld when plants reach a given growth stage for 0, 7, 14, or 21 days, then plants were watered until they reached the next growth stage when stress was initiated again for 0, 7, 14 or 21 days (Fig. 1). The 21-day periodic stress treatments experienced almost constant drought stress from tillering on.

Measurements

Phenotypic measurements

The yield component traits plant height (PH), tiller number (TN), total spike number (TSN), grain number (GN), and grain weight (GW) were measured at the end of the experiment for each genotype/stage/stress duration/replicate. All the morphological traits were measured according to Pask et al. (2012).

Chlorophyll fluorescence

The chlorophyll fluorescence represents the maximum efficiency of PhotosystemII (PSII). The effective photochemical quantum yield of PSII (YII) (Genty et al. 1989; Maxwell et al. 1994; Maxwell and Johnson 2000) was measured in light using, the portable photosynthesis yield analyzer (MINI-PAM-II, Heinz Walz GmbH, Germany). The efficiency of photosystem II was calculated as $YII = (F_m - F_o)/F_m$ based on the steady-state photosynthesis under lighting conditions (Genty et al. 1989). where: F_m: The ground fluorescence of light adapted samples. F_o: The maximal fluorescence of light adapted samples. YII: The yield of photochemical energy conversion.

The yield of photochemical energy measurement monitors any changes in photosynthetic activity during recovery. The YII was measured at the end of the stress treatment that was imposed at tillering stage for the 21-day duration and it was re-measured ten days after re-watering plants. The corresponding control plants were measured on the same day duration even though they hadn't experienced stress. Three reads were taken from each fully expanded leaf from each replicate and genotype at the tip, mid and bottom of each wheat leaf, and averaged. The measurements were taken in the light between 12 a.m. and 14 p.m.

Germination efficiency

After harvest of grain from each plant, ten seeds were chosen randomly from each treated plant (excluding the treatments which failed to produce seeds) to test the seed dormancy. All the grains were synchronized by imbibing with water at 4 °C for 24 h on sterilized CBD3.5 steel blue germination blotters (Ancho paper, St. Paul MN) in Petri dishes covered with aluminum foil and then placed in an incubator at 22 °C. On the 5th day, the germination

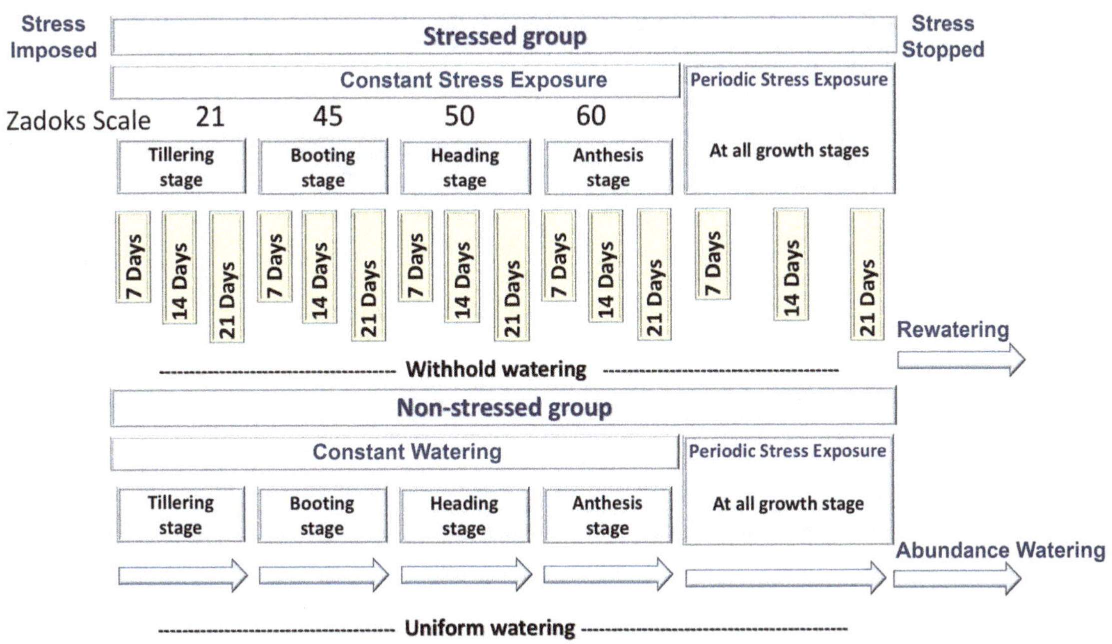

Fig. 1 The workflow of the experimental design. It describes the design of the stressed (non-watered) and non-stressed (watered) groups for both of single stress and periodic exposure

efficiency was estimated according to Guo et al. (2012) for each plate as follows: $Ge\% = n/N \times 100$, where n is a number of seeds germinated in 5 days; N is the total number of seeds.

Seedling growth

Another set of ten seeds was used for each treated plant (excluding the treatments that failed to produce seeds) to measure the coleoptile length and root lengths of the harvested seeds. The seeds were germinated as above. On the 10th day, both the coleoptile and root lengths were measured according to Pask et al. (2012).

KASP assay for *Rht* genes

We also studied the relationship between the interaction of the developmental stages and drought stress duration with genes for reduced plant height. The genomic DNA was extracted from leaves from the eight genotypes using the BioSprint DNA Plant Kit (Qiagen, CA, USA) according to the manufacturer's instructions. The alleles of the *Rht-B1* and *Rht-D1* genes were assayed using the Kompetitive allele-specific PCR (KASP) protocol. The KASP markers were developed in 2013 part of the MASWheat funding project based on the Rht markers that published previously (Ellis et al. 2002) (http://maswheat.ucdavis.edu/protocols/Dwarf/). The amplification was carried out using the thermal cycler T-100Tm (BioRad Laboratories, Inc.). The assay was genotyped through the endpoint method using the Light Cycler ® 480 instrument II (Roche Diagnostics Ltd. Forrenstrasse, Switzerland). According to the LGC Genomics, Ltd. (Middlesex, UK) guide manual, the reaction components, preparation, and PCR program were performed.

Statistical analysis

The statistical analyses were done using the SAS-Enterprise and SAS-JMP software. The charts were plotted using GraphPad PRISM version 7.00. The Mean and standard error were calculated, and treatments were compared by an analysis of variance utilizing the SAS GLM procedure with all effects fixed using the model: $Y_{ijk} = \mu + \alpha_i + \beta_j + \varepsilon_{ijk}$ where Y_{ijk} is the grain number (yield) of ijkth observation, μ is the overall mean plus the effects of the (α_i) ith genotype, the $\beta_{i(j)}$ is the effect of jth stress duration treatment, nested within the developmental stage, and, ε_{ijk}, is the random deviation associated with each observation. The replicate variance was omitted from the model after finding the non-significant variance between the replicates. The LS-means and their 95% confidence intervals for each treatment were calculated from each analysis and were compared using Tukey's and Dunnett's comparison tests. The relationships among yield traits were measured using the Pearson correlation coefficient with the pairwise estimation method.

Results

Dynamics of plant morphology in response to drought stress

During the constant stress exposure, leaf rolling was the earliest morphological characteristic observed at the end of the 7th day of drought. The next morphological symptom was wilting of leaves that became apparent at 14–21-day of drought. Typically, leaf yellowing was observed by the 14th day of drought followed wilting. Leaf drying was the last leaf symptom before death. Dryness started with brownish spots at the leaf tip appearing by the end of 14th day, which gradually expanded towards the leaf base by the day 21.

In the periodic stress exposure, the rolling and wilting recovered after re-watering of 7-day periodic treatment. The chance of recovery was less for the 14-day periodic stress when leaves were yellowing and drying. Surprisingly, plants could survive after 21-days of periodic stress treatment regardless of the developmental stage applied even after all the leaf symptoms (Fig. 2a–d). We observed generation of new leaves in plants after re-watering. In the constant stress treatment, most plants with yellowing and dryness symptoms failed to recover. The genotypes of Klein Dragon, Onas, and Indian were capable of recovering by generating new leaves. The susceptible genotypes such as PotamS-70 could not recover. Although Vandal and Perigee showed the most tolerant symptoms at 7-day treatment but gradually both showed less performance with prolonged drought.

Plants at all developmental stages exhibited dynamics of leaf symptoms described above regarding the stress duration (Fig. 2a–d). Additionally, unique plant and seed morphology were observed at certain developmental stages. When plants exposed to stress at the tillering stage were able to recover faster after re-watering. Leaf dryness at the tip without expanding, and generating new leaves under prolonged drought were observed only at tillering stage. Application of drought during the booting stage resulted in partial or complete inhibition of boots opening for all the genotypes, when awns with only half of the spikelet were seen (Fig. 2e). At the heading stage, we observed decline in the number of spikes, especially that are followed by the main spike. The most severe impact of the drought was at the booting and anthesis stages, causing empty spikes or shrunken seeds (Fig. 2f), especially with the prolonged duration that has obstructed the seed development.

Fig. 2 The figure illustrates the development of drought morphological symptoms. It shows the main and unique symptoms were shown during drought exposure. **a** Leaf rolling, **b** Leaf wilting, **c** Leaf yellowing, **d** Leaf dryness, **e** Spike damage (booting stage) and **f** shrunken seeds (anthesis stage)

Evaluating the phenological plasticity

Yield components and plant height

In this study, the grain number was chosen to represent the yield fitness. There was a high significant correlation ($r^2 = 0.9$) between the grain number and grain weight. Other yield components were the number of tillers and the number of spikes, in addition to the plant height. The primary effect of drought was associated with the grain number and the plant height parameters. Through all treatments, a significant reduction (49.19%) in the grain yield occurred between the stressed plants compared with the non-stressed plants.

An insignificant loss in grain number was found from the plants that were exposed to the drought stress during tillering stage compared with the non-stressed plants. Although there was a similar impact of the drought stress imposed at the heading, anthesis and tillering stages, the grain number from those plants which were drought stress at heading and anthesis stages were different from their control groups (Fig. 3A and Table 2). A significant negative impact was observed at the booting stage in response to drought. At the booting stage, the plant height and total spike number were significantly reduced under drought stress compared with non-stressed groups (Table 3). The total tiller number did not show a significant reduction (Table 3).

In consideration of the stress duration and in comparison to the non-stressed groups (0-day), all the stress durations (7-, 14-, and 21-day) reduced the yield fitness of the plants significantly. Overall, in the treatments, the largest reduction in yield fitness occurred between 0 and 7 days of drought (Fig. 3B and Table 2). When the stress was extended to 21 days, the highest reduction in yield occurred when drought was imposed at booting stage. Exclusive to tillering stage, plants that were exposed to stress for 21 days produced higher grain number than the group of 14-days of stress (Table 2).

The genetic effect on the drought tolerance is summarized in Fig. 3C. The grain number was reduced substantially in PWB343 and Edmore and to a lesser extent in Onas, Indian, Klein Dragon, and Potam-S70. Grain number was reduced only slightly in Vandal and Perigee. Based

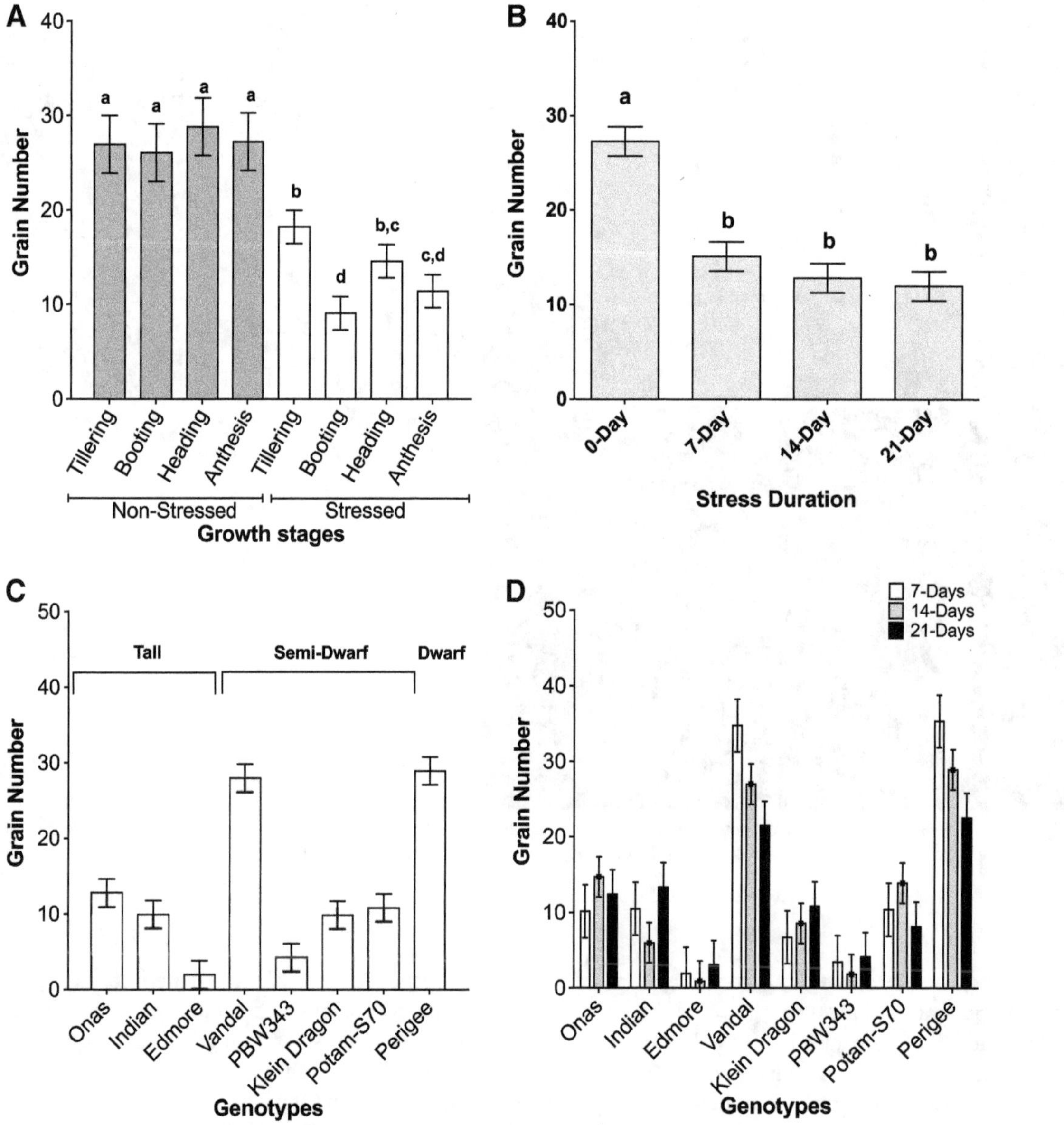

Fig. 3 The phenotypic plasticity was measured by the grain number in all charts. The Charts shows that environmental variance of GxE under drought stress was promoted by the adaptive developmental plasticity more than the stress duration only. **A** the developmental plasticity, in respect to the genetic and stress duration variances. An insignificant difference was found between the non-stressed plants and the plants that were exposed to drought stress during the tillering stage. **B** A significant reduction in grain number in response to any of drought stress durations of 7, 14 and 21 days. For **A** and **B**, the displayed letters; *a*, *b* and *c* indicate the significance ($P < 0.05$; one-way ANOVA and Tukey's multiple comparisons tests). **C** The distribution of genotype variance in response to drought stress treatments, classified by alleles at the Rht loci. **D** Genotypes demonstrated dynamic responses to prolonged drought. Onas, Indian, Edmore, Klein Dragon and PWB343 had increased grain number for 21 days of drought while others had decreased grain number

on the results of genotyping, the diagnostic alleles of *Rht* genes, Onas, Edmore and Indian are tall plants (Rht-B1a/Rht-D1a), Vandal, PBW-343, Klein Dragon and PotamS-70 are semi-dwarf (Rht-B1a/Rht-D1b), and Perigee represents a severe dwarf plant (Rht-B1b/Rht-D1b). Exposure to drought for 7 or 14 days had a stronger impact on the yield in the taller genotypes than on the yield of dwarf and semi-dwarf genotypes. However, two tall genotypes Onas and Indian could recover more efficiently after exposure to drought for 21 days than the semi-dwarf genotypes. The tetraploid genotype (Edmore) was more sensitive than Onas and Indian.

Table 2 Grain number losses when drought stress was imposed at various developmental stages either as constant or periodic stress, including all the genotypes under 7, 14, 21-days of drought

Stress exposure	Developmental stage	0-day	7-day	14-day	21-day
Constant stress	Tillering stage	26.96 ± 4.02	20.92 ± 4.70	14.33 ± 3.76	19.37 ± 2.82
Constant stress	Booting stage	26.08 ± 3.50	14.67 ± 3.95*	10.17 ± 2.26*	2.42 ± 1.42**
Constant stress	Heading stage	28.83 ± 3.58	12.75 ± 1.79*	16.21 ± 2.72*	14.83 ± 2.52*
Constant stress	Anthesis stage	27.25 ± 3.01	12.21 ± 2.20*	10.62 ± 1.88*	11.42 ± 2.17*
Periodic stress	All the stages	24.29 ± 2.25	6.29 ± 1.31*	3.46 ± 1.90*	9.25 ± 1.90*

* Significant at the P < 0.05 probability by the LSD test compared to the non-stressed group (0-day), **Significant reduction at P < 0.05 compared to the non-stressed groups and other treatments

Table 3 Reductions in plant yield components from the non-stress (control) due to drought imposed as constant stress or periodic stress, including the measurements of all the genotypes

Stress duration	Constant stress exposure				Periodic stress exposure			
	0	7	14	21	0	7	14	21
Traits								
Plant height	8.93	0.029	−3.58*	−5.38*	10.8	−4.16	−0.30	−0.36
Tiller number	0.122	0.19	−0.11	−0.21	0.26	−0.16	−0.24	0.14
Total spike number	0.39	−0.027	−0.11	−0.25*	0.54	−0.29*	−0.29*	0.04
Grain number	10.46	−0.168	−3.98*	−4.8*	13.47	−4.53*	−7.36*	−1.57
Grain weight	0.20	−0.03	−0.08*	−0.09*	0.23	−0.032	−0.12*	−0.07

* Significant at the P < 0.05 probability by the LSD test

In Fig. 3D, the interaction between the genetic variance and the stress duration was examined to identify any adaptive pattern if shown under the prolonged stress. Some genotypes had better grain number from plants with 21 days of stress treatment than with shorter periods (Fig. 3D). Furthermore, periodic stress reduced the yield fitness dramatically (Fig. 4). Grain number of the stressed treatments was significantly reduced compared to the non-stressed treatments (Fig. 4; Table 2). The constant stress exposure resulted in the lowest grain yield (Table 2). In our work, the genotypes showed a varying range of responses to periodic stress (Fig. 4). The grain yield of Vandal and Perigee was reduced less than the other genotypes, with respect to the difference comparing with the non-stressed groups. Onas, Indian and Klein dragon were more unwavering and adaptive. In contrast to constant stress exposure, prolonged periodic stress-induced the adaptive response only in the tall plants (Table 4).

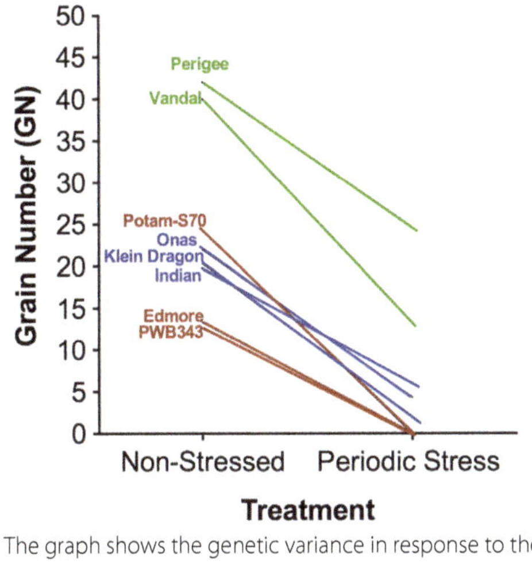

Fig. 4 The graph shows the genetic variance in response to the periodic stress. The periodic drought stress has a negative impact on the grain number, and was varied across the different genotyes

In general, the plant height was reduced significantly from the non-stressed control when drought imposed at all developmental stages (Table 3). Our results indicate that plant height was most affected by stress at the booting stage (Table 4). Also, drought inhibited plant height when the stress was applied during the tillering stage. The impact was higher than the heading or anthesis stages showing the greater impact on the tall and semi-dwarf genotypes (Table 4).

In our study, plant height was reduced significantly during tillering, booting and anthesis stages by the tall plants associating with an adaptive response to drought. The height of the semi-dwarf plants reduced significantly

Table 4 Means and standard errors for plant height show the impact of the growth stages (tillering, booting, flowering and anthesis), the stress durations (0, 7, 14, 21-days), and the periodic stress in response to drought for specific alleles of the reduced height genes

Description	Tall, Rht-a/a	Semi-dwarf, Rht-a/b	Dwarf, Rht-b/b
Growth stage effect			
Tillering-nonstress (control)	70.67 ± 3.05	50.21 ± 2.64	16.33 ± 5.28
Tillering imposed stress	50.63 ± 2.5*	41.01 ± 2.17*	17.28 ± 4.35
Booting-nonstress (control)	70.44 ± 2.03	48.67 ± 1.76	17.67 ± 3.5
Booting imposed stress	32.63 ± 3.0**	30.82 ± 2.6**	13.83 ± 5.2
Heading-nonstress (control)	67.0 ± 2.06	46.83 ± 1.78	16.17 ± 3.57
Heading imposed stress	62.78 ± 1.82	44.42 ± 1.58	12.28 ± 3.16*
Anthesis-nonstress (control)	67.11 ± 1.71	46.33 ± 1.48	14.0 ± 2.9
Anthesis imposed stress	52.15 ± 1.90*	40.25 ± 1.64*	13.05 ± 3.29
Stress duration effect			
0-day	68.80 ± 1.11	48.01 ± 0.96	16.04 ± 1.92
7-day	53.28 ± 2.33**	42.36 ± 2.02*	14.0 ± 4.03
14-day	47.55 ± 2.10**	39.22 ± 1.82**	14.87 ± 3.64
21-day	47.80 ± 2.63**	35.79 ± 2.28**	13.46 ± 4.56
Periodic stress effect			
Periodic-nonstress control (0 day)	63.77 ± 1.65	44.33 ± 1.43	14.17 ± 2.86
Periodic-7 day	38.89 ± 3.55**	33.08 ± 3.08*	14.17 ± 6.16
Periodic-14 day	44 ± 3.98*	31.92 ± 3.4*	12.67 ± 6.89
Periodic-21 day	47.44 ± 4.4*	28.92 ± 3.81*	9.5 ± 7.6*

* and ** Significant at $P < 0.1$ and 0.05 level of probability, respectively by the LSD test

but with limited loss in yield at low drought stress conditions. The only dwarf plant (Perigee) that did not show a significant reduction in plant height with booting stage exception. Interestingly, the heading stage has insignificant impact on plant height by the tall and semi-dwarf plants with the only severe dwarf plant exception. All the tall and semi-dwarf plants showed significant reduction of plant height under periodic stress while the reduction was insignificant of the dwarf plant (Perigee) except under 21 days of periodic stress.

Germination efficiency and seedling development

Exposure to drought stress at the anthesis stages for 7, 14, or 21 days of drought resulted in shrunken seeds, with germination efficiency 65 55, and 45%, respectively (Fig. 5). Reduced grain number following drought application at the booting stage was accompanied by an average germination of 53.3%. Though, seeds from plants stressed at tillering and heading stages had a germination percentage of 77, 55% with 7-days; 60, 70% with 14-days and 70, 71% with the 21-days stress, respectively. The length of coleoptiles and roots of 10 days old seedlings were measured to assess the effect of drought on seedling growth.

The coleoptile and root lengths were reduced significantly in the seeds harvested from the stressed treatments (Fig. 5). Across all the developmental stages, the root length correlated significantly with the coleoptile length ($r^2 = 0.83$), and the germination percent correlated with the coleoptile and root length ($r^2 = 0.84, 0.82$), respectively. Due to the high phenotypic plasticity of tillering and heading stages, the growth slopes of coleoptile length was reduced insignificantly even after drought stress for 21 days. A significant reduction was observed in the slope of the root length decline only at the heading stage, while at the tillering stage was not. The results revealed an association of fitness of grain yield and germination efficiency with seedling growth.

Analysis of chlorophyll fluorescence and pigments

The maximum yield of photosystem II was used to assess the photosynthetic activity. Only plants under the adaptive response of the least sensitive stage (tillering stage) were measured at two time points, at the end of 21 days of drought stress (sPSII) and 10 days after re-watering (rPSII). We found a significant reduction of both parameters (Fig. 6). Interestingly, the re-watering period of 10 days had no effect on the rPSII suggesting that optimization for the photosynthesis water deficiency remained stable. The chlorophyll fluorescence of the non-stressed group of the 21 days treatment decreased dramatically from the tillering stage to the re-watering stage. The watering of the non-stressed plants was a scheduled amount until the stressed plants were re-watered. At that

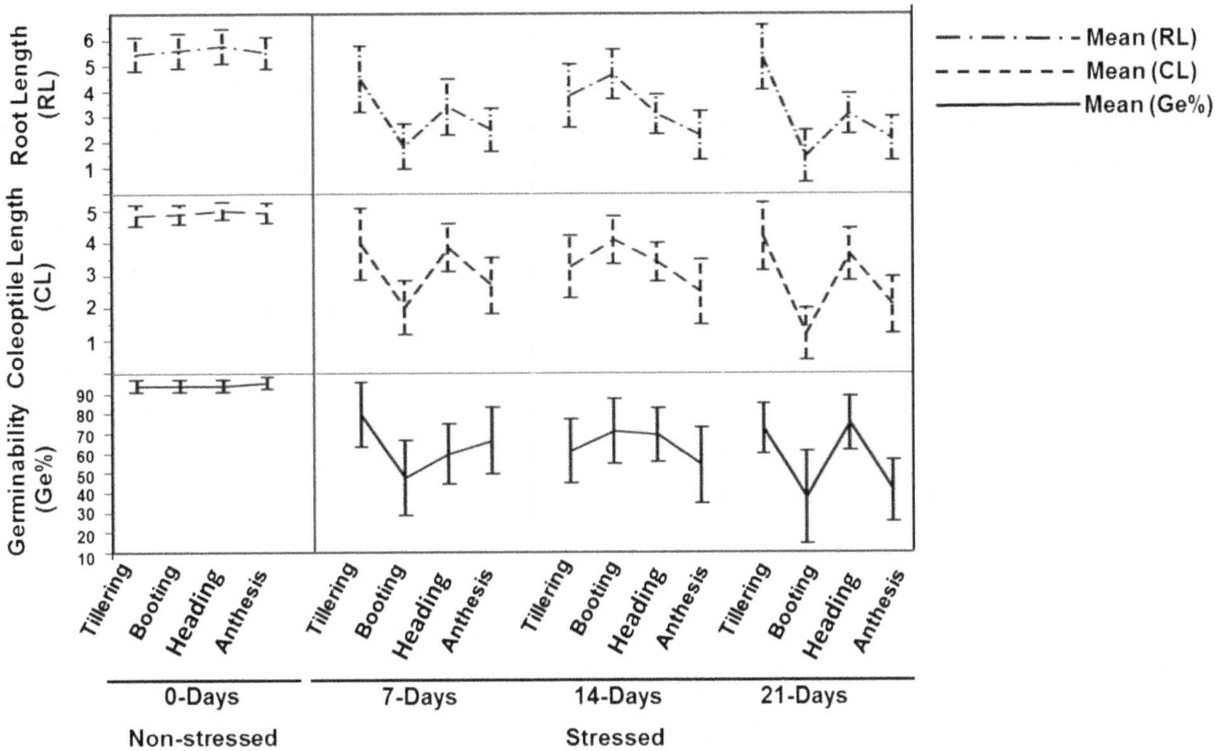

Fig. 5 The in vitro germination percentage of the produced seeds. Means and standard errors of in vitro germination percentage for the produced seeds of the used genotypes of spring wheat show the drought response, summarized by the impact of the growth stages for each stress duration

time, both non-stressed and stressed treatments were watered to full watering capacity (Fig. 6).

Discussion

Tillering stage is a plastic stage for drought screening

The hypothesis was built considering that high plasticity is preferable to be during a long stage that needs high water requirements. In a case of plasticity existence, minimizing this amount of water at long duration would be sufficient for building adaptive mechanism. Touching that topic, some studies stated that wheat would uptake higher portion of the total life water requirement during the vegetative stage, and wheat adapted to less amount of rainfall over years, due to the climate changes (Schillinger et al. 2008). So we assumed that vegetative stage might be carried an adaptive mechanism of water uptake in wheat plants. Because stress duration is an important factor to evaluate the stress severity, we imposed the stress for different drought stress durations (7, 14, and 21 days) at each developmental stages.

As we described in the results, we obtained some dynamic changes in leaf, spike, and seed symptoms, which associated with either a certain developmental stage, stress duration, or genetic response. Apparently,

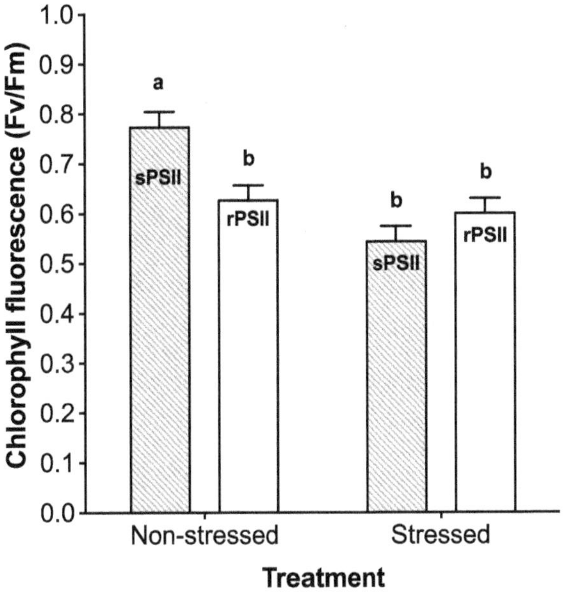

Fig. 6 The impact of drought stress on the chlorophyll fluorescence. The photosynthetic activity was measured at the end of the drought stress exposure after 21 days at tillering stage, the PSII compared with non-stressed plants during the stress (sPSII) and after re-watering (rPSII)

some leaf symptoms were well documented with drought; but, defining the morphological dynamics during the developmental stages or drought regimes are poorly described, although it is important traits and easy to record. These changes in plant morphology under different environments reflect the plastic response (De Jong and Leyser 2012), and are thought to be important to indicate specific physiological changes. Accounting the symptoms with the phenotypic measurements, the plant response pipeline could be drawn. Leaf wilting and rolling are indicated the start of a lack of available moisture in the soil and a consequent shortage of water transport to stem and leaves, then follow by correlation with reduced photosynthesis and pigment content. Also, leaf rolling occurs typically in most grasses including rice, maize, and sorghum during perception of drought induced by increased production of salicylic acid signal (Kadioglu and Terzi 2007). Those symptoms would not be used to indicate the plastic response to drought, because they are common symptoms through the developmental stages, especially if plants would recover after re-watering. As a result of a reduction of the chlorophyll content, leaf yellowing was observed, which can be used as a measurement of the damaging effect of drought (Moore and Lovell 1970). While leaf dryness reflects the cell programmed death. When the plants experienced dryness through the whole leaf under drought stress, this should be a non-recoverable symptom unless there is an adaptive mechanism induced to generate new leaves. Tolerant stressed plants at tillering stage experienced leaf dryness through leaf tip, but through the whole leaf only in the sensitive genotypes. Exclusively, at the tillering stage, the generation of new leaves after a prolonged drought indicated the adaptive response to drought. Booting stage was the least plastic developmental stage, crediting unique negative symptom (inhibition of boots opening) resulted in short spikes, few seeds, empty spikes and low seed vigor. At heading stage, limited regeneration would happen especially after the prolonged drought. The plant would focus to survive the spikes that already were formed. Also, drought stress during the anthesis stage drove the stressed plant to maturation faster, shortening the milk and dough stage resulting in shrunken seeds without generating new leaves. Heading and anthesis stages promoted less correlation between the grain number and grain weight, which might be due to the significant impact of drought on the fertile tillers at the spike formation during the grain-filling. Regarding the productive stages, the perturbation in seed development under drought stress may produce shrunken seeds, which contain low protein and starch content, reducing the germination efficiency and causing abnormal seedling development (Wood et al. 1977; Guo et al. 2012). Fábián et al. (2011) highlighted a greater significant reduction in the size of the mature embryos in the sensitive genotype. In addition, the low germinability was correlated with the size, development and genetic variability of plant population (Li et al. 2014). However, taking credits of high germinability and seedling development would value the phenological plasticity of developmental stages. Also, the association found between germinability of the produced seeds and seedling development effected plant morphology and yield uniquely at different developmental stages indicated the dynamics of developmental plasticity.

Interestingly, although under the prolonged drought, the tillering group recovering to adaptive symptoms were seen. The maximum quantum efficiency (PSII) was reduced by drought. The significant loss of light energy was constantly either during the stress (sPSII) or after re-watering (rPSII). It seems that lowering PSII yield could be energy leakage due to the damaging impact of drought on the chlorophyll (Huang et al. 2009). In the meantime, the constant loss suggesting the possibility of two reasons. First, that optimization for photosynthesis water deficiency remained stable under the adaptive mechanism. Second, abundant watering after constant stress for a long time might be considered as a stress. Fortuitously, we observed a significant reduction of the photochemical efficiency in the non-stressed plants after changing the watering pattern from scheduled watering to water abundant. However, re-watering seems to be a mechanism or perturb mechanism. When plants experience complete recovery following re-watering, the recovery may depend on the pre-drought intensity or duration (Xu and Zhou 2007; Xu et al. 2009, 2010). Unfortunately, scant pieces of literature discussed that phenomenon, especially in cereals. These changes in the phenotypic response caused by the repeated and by prolonged drought may indicate induction of acclimation mechanism, such as desiccation tolerance mechanism (Berjak 2006; Tobias 2012). Periodic exposure to drought stress may alter the developmental program of spring wheat by changing the duration of the individual stages and scaling down the fitness of grain yield.

Exclusively, during tillering stage, plants are capable of recovering after re-watering producing normal seeds and seedling growth. Plants are capable of adapting with less water requirements under long development stage at the prolonged drought compare to the short stress exposure. Under our experimental conditions, we concluded total grain number reduction of 32.46% in tillering stage, 65.18% in booting stage, 49.39% in heading stage, and 58.09% in anthesis stage under the constant stress exposure. While a reduction of 73.92% under the periodic stress exposure. Veesar et al. (2007) reported in *Triticum aestivum* that in order to increase the numbers of spikelet per spike, stress should be avoided at tillering stage. Although they stated under field conditions yield declines of 20.74, 46.85 and 101.23% when drought stress was subjected at tillering, booting and at grain formation respectively. Moreover, Alghabari et al. (2014) stated high sensitivity to drought at booting and anthesis stages were mainly due to grains per spikelet, and mean grain weight, respectively. The tillering stage is taken to give a stable and good yield followed by the heading stage while booting and anthesis had the lowest yield.

Assessment the resistance to periodic stress is a vital aspect

Moreover, the integrated effect of the periodic exposure to drought through all the stages caused a harsh combination of general and state-specific symptoms. Resistant plants capacity to hold out the physiological and biochemical changes under the periodical drought stress may demonstrate the adaptive response explained by the phenological plasticity. In previous reports, a tolerant genotype produced 18–44% more grain than a susceptible genotype under severe cyclic water limiting conditions due to osmotic adjustment (Izanloo et al. 2008). Similarly, in our research and through the entire experiment, the most tolerant genotype produced 10.6–30.15% grains more than the most susceptible genotype under constant stress exposure and periodic stress exposure, respectively. Which indicates that considering resistance response under periodic drought stress helps to build a robust drought tolerance assessment and capture superior drought genotypes.

Plant height genes may contribute to the adaptive response

In general, it is documented that changes in plant height is one of the mechanisms for adaptive modulation of plant growth and is mediated by the nuclear transcriptional regulators "DELLA proteins"; which plays the central role in gibberellin (GA) signaling under drought stress. Also, DELLA proteins accumulation restrain the plant growth and increase the capability of survival under drought (Achard and Genschik 2009). After observing a general significant reduction of plant height during tillering, booting and anthesis stages of the tall and semi-dwarf genotypes with a dwarf exception. Meanwhile, tall plants (Onas, Indian and Edmore) have better adaptability only under prolonged drought. We concurred with the previous reports that the impact of height alleles varied with genetic backgrounds and environments, where the tall and semi-dwarf plants exhibited similarly and adaptable to stress under prolonged drought. While the shortest tall wheat genotypes may be useful in marginal drought stress environments (Mathews et al. 2006). Although, Alghabari et al. (2015) suggested that severe dwarf Rht alleles are better able to enhance tolerance to high temperature and drought stress in wheat. Therefore, that observation is worth further studies, especially when the association of plant height's impact on the developmental plasticity in wheat is not discuss enough in the literature.

Conclusions

Our results concluded that, (1) the developmental stages accommodate phenotypic plasticity, while the timing factor appeared to be resolving the adaptive plasticity interacting with the developmental stage and genotypes. However, the environmental variance of GxE under drought stress in wheat was promoted by the developmental plasticity. (2) Intense drought stress shifted the phenology of the developmental stages wheat. In depth, tillering stage scored the highest phenological plasticity. Heading stage seemed to be similar to tillering but the low germination and seedling development of produced seeds hindered that plastic response. While, booting and anthesis stages showed a maladaptive response to drought stress. In this study, we recommend considering the resistance response to the periodic drought stress as an important aspect in screening methods. Also, investigating in the impact of plant height alleles on the adaptive response is recommended. We propose tillering stage is to be a promising stage for better suited for capturing drought-tolerant genotypes through a screening method. Resistance to periodic drought stress is an important aspect in screening methods.

Authors' contributions

A significant contribution was produced in the statistical analysis, and editing the manuscript from KC. Also, she facilitated the analysis of germination rate of the produced seeds and the seedling measurements, as well as the genotyping the plant height. KG is provided the seeds and supported the greenhouse facilities under his supervision at WSU, Pullman, WA, USA. All authors read and approved the final manuscript.

Author details

[1] Crop and Soil Sciences Department, Washington State University, Pullman, WA, USA. [2] Genetics and Cytology Department, National Research Center, Giza, Egypt. [3] Wheat Genetics, Quality, Physiology, Diseases Research, USDA-ARS, Pullman, WA, USA.

Acknowledgements

We thank Dr. Andrei Smertenko for his valuable comments on the final version of the manuscript. Also, our appreciation to Dr. Camille Steber, USDA, Pullman, WA, USA, for supporting the use of the Mini-PAM.

Funding

Majority of this work was done through a postdoctoral fellowship for Dr. Marwa Sanad from the Egyptian government in Dr. Kulvinder Gill's lab at Washington State University, Pullman, WA, USA.

References

Achard P, Genschik P (2009) Releasing the brakes of plant growth: how GAs shutdown DELLA proteins. J Exp Bot 60:1085–1092. doi:10.1093/jxb/ern301

Alghabari F, Lukac M, Jones HE, Gooding MJ (2014) Effect of Rht Alleles on the Tolerance of wheat grain set to high temperature and drought stress during booting and anthesis. J Agron Crop Sci 200:36–45. doi:10.1111/jac.12038

Alghabari F, Ihsan MZ, Hussain S, Aishia G, Daur I (2015) Effect of Rht alleles on wheat grain yield and quality under high temperature and drought stress during booting and anthesis. Environ Sci Pollut Res Int 22:15506–15515. doi:10.1007/s11356-015-4724-z

Aubin-Horth N, Renn SCP (2009) Genomic reaction norms: using integrative biology to understand molecular mechanisms of phenotypic plasticity. Mol Ecol 18:3763–3780

Baldwin IT, Schmelz EA (1996) Immunological "memory" in the induced accumulation of nicotine in wild tobacco. Ecology 77:236–246. doi:10.2307/2265673

Berjak P (2006) Unifying perspectives of some mechanisms basic to desiccation tolerance across life forms. Seed Sci Res 16:1–15. doi:10.1079/SSR2005236

Bruce TJA, Matthes MC, Napier JA, Pickett JA (2007) Stressful "memories" of plants: evidence and possible mechanisms. Plant Sci 173:603–608. doi:10.1016/j.plantsci.2007.09.002

Chaves MM, Maroco JP, Pereira JS (2003) Understanding plant responses to drought from genes to the whole plant. Funct Plant Biol 30:239–264. doi:10.1071/FP02076

De Jong M, Leyser O (2012) Developmental plasticity in plants. Cold Spring Harb Symp Quant Biol 77:63–73. doi:10.1101/sqb.2012.77.014720

De Leonardis AM, Marone D, Mazzucotelli E, Neffar F, Rizza F, Fonzo ND, Cattivelli L, Mastrangelo AM (2007) Durum wheat genes up-regulated in the early phases of cold stress are modulated by drought in a developmental and genotype dependent manner. Plant Sci 172:1005–1016. doi:10.1016/j.plantsci.2007.02.002

Ellis MH, Spielmeyer W, Gale KR, Rebetzke G, Richards R (2002) "Perfect" markers for the Rht-B1b and Rht-D1b dwarfing genes in wheat. Theor Appl Genet 105:1038–1042. doi:10.1007/s00122-002-1048-4

Fábián A, Jäger K, Rakszegi M, Barnabás B (2011) Embryo and endosperm development in wheat (*Triticum aestivum* L.) kernels subjected to drought stress. Plant Cell Rep 30:551–563. doi:10.1007/s00299-010-0966-x

Farooq M, Wahid A, Kobayashi N, Fujita S, Basra SMA (2009) Plant drought stress: effects, mechanisms and management. Agron Sustain Dev 29:185–212. doi:10.1051/agro:2008021

Genty B, Briantais J-M, Baker NR (1989) The relationship between the quantum yield of photosynthetic electron transport and quenching of chlorophyll fluorescence. Biochim Biophys Acta Gen Subj 990:87–92. doi:10.1016/S0304-4165(89)80016-9

Goswami A, Banerjee R, Raha S (2010) Mechanisms of plant adaptation/memory in rice seedlings under arsenic and heat stress: expression of heat-shock protein gene *HSP70*. AoB Plants. doi:10.1093/aobpla/plq023

Guo R, Hao WP, Gong DZ (2012) Effects of water stress on germination and growth of linseed seedlings (*Linum usitatissimum* L.), photosynthetic efficiency and accumulation of metabolites. J Agric Sci 4:p253. doi:10.5539/jas.v4n10p253

Huang XY, Chao DY, Gao JP, Zhu MZ, Shi M, Lin HX (2009) A previously unknown zinc finger protein, DST, regulates drought and salt tolerance in rice via stomatal aperture control. Genes Dev 23:1805–1817. doi:10.1101/gad.1812409

Izanloo A, Condon AG, Langridge P, Tester M, Schnurbusch T (2008) Different mechanisms of adaptation to cyclic water stress in two South Australian bread wheat cultivars. J Exp Bot 59:3327–3346. doi:10.1093/jxb/ern199

Kadioglu A, Terzi R (2007) A dehydration avoidance mechanism: leaf rolling. Bot Rev 73:290–302. doi:10.1663/0006-8101(2007)73[290:ADAMLR]2.0.CO;2

Li X, Mu C, Lin J (2014) The germination and seedlings growth response of wheat and corn to drought and low temperature in spring of Northeast China. J Anim Plant Sci 21:3212–3222

Mathews KL, Chapman SC, Trethowan R, Singh RP, Crossa J, Pfeiffer W, Ginkel MV, DeLacy I (2006) Global adaptation of spring bread and durum wheat lines near-isogenic for major reduced height genes. Crop Sci 46:603. doi:10.2135/cropsci2005.05-0056

Maxwell K, Johnson GN (2000) Chlorophyll fluorescence—a practical guide. J Exp Bot 51:659–668. doi:10.1093/jexbot/51.345.659

Maxwell DP, Falk S, Trick CG, Huner NPA (1994) Growth at low temperature mimics high-light acclimation in *Chlorella vulgaris*. Plant Physiol 105:535–543. doi:10.1104/pp.105.2.535

Milad SI, Wahba LE, Barakat MN (2011) Identification of RAPD and ISSR markers associated with flag leaf senescence under water-stressed conditions in wheat (*Triticum aestivum* L.). *Aust.* J Crop Sci 5:337–343

Moore K, Lovell P (1970) Chlorophyll content and the pattern of yellowing in senescent leaves. Ann Bot 34:1097–1100

Nicotra AB, Davidson A (2010) Adaptive phenotypic plasticity and plant water use. Funct Plant Biol 37:117–127. doi:10.1071/FP09139

Noormets A, McNulty SG, DeForest JL, Sun G, Li Q, Chen J (2008) Drought during canopy development has lasting effect on annual carbon balance in a deciduous temperate forest. New Phytol 179:818–828. doi:10.1111/j.1469-8137.2008.02501.x

Pask AJD, Pietragalla J, Mullan DM, Reynolds MP (2012) Physiological breeding II: a field guide to wheat phenotyping. International maize and wheat improvement center, CIMMYT, Maxico. 1–132

Rizza F, Badeck FW, Cattivelli L, Lidestri O, Fonzo ND, Stanca AM (2004) Use of a water stress index to identify barley genotypes adapted to rainfed and irrigated conditions. Crop Sci 44:2127–2137. doi:10.2135/cropsci2004.2127

Schillinger WF, Schofstoll SE, Alldredge JR (2008) Available water and wheat grain yield relations in a Mediterranean climate. F Crop Res 109:45–49. doi:10.1016/j.fcr.2008.06.008

Sivamani E, Bahieldin A, Wraith JM, Al-Niemi T, Dyer WE, Ho TD, Qu R (2000) Improved biomass productivity and water use efficiency under water deficit conditions in transgenic wheat constitutively expressing the barley *HVA1* gene. Plant Sci 155:1–9. doi:10.1016/S0168-9452(99)00247-2

Tobias MN (2012) Drought and desiccation-tolerance and sensitivity in plants. In: Mworia JK (ed) Botany. InTech, Rijeka. doi:10.5772/33584

Trewavas A (2005) Plant intelligence. Naturwissenschaften 92:401–413. doi:10.1007/s00114-005-0014-9

Veesar NF, Channa AN, Rind MJ, Larik AS (2007) Influence of water stress imposed at different stages on growth and yield attributes in bread wheat genotypes (*Triticum aestivum* L.). Wheat Inf Serv 104:15–19 http://www.shigen.nig.ac.jp/ewis/article/html/18/article.html

Vinocur B, Altman A (2005) Recent advances in engineering plant tolerance to abiotic stress: achievements and limitations. Curr Opin Biotechnol 16:123–132. doi:10.1016/j.copbio.2005.02.001

Vitasse Y, Bresson CC, Kremer A, Michalet R, Delzon S (2010) Quantifying phenological plasticity to temperature in two temperate tree species. Funct Ecol 24:1211–1218. doi:10.1111/j.1365-2435.2010.01748.x

Walter J, Nagy L, Hein R, Rascher U, Beierkuhnlein C, Willner E, Jentsch A (2011) Do plants remember drought? Hints towards a drought-memory in grasses. Environ Exp Bot 71:34–40. doi:10.1016/j.envexpbot.2010.10.020

West-Eberhard MJ (2003) Developmental plasticity and evolution. Oxford University Press, New York, pp 101–102

Wood DW, Longden PC, Scott RK (1977) Seed size variation; its extent, source and significance in field crops. Seed Sci Technol 5:337–352

Xu ZZ, Zhou GS (2007) Photosynthetic recovery of a perennial grass *Leymus chinensis* after different periods of soil drought. Plant Prod Sci 10:277–285. doi:10.1626/pps.10.277

Xu Z, Zhou G, Shimizu H (2009) Are plant growth and photosynthesis limited by pre-drought following rewatering in grass? J Exp Bot 60:3737–3749. doi:10.1093/jxb/erp216

Xu Z, Zhou G, Shimizu H (2010) Plant responses to drought and rewatering. Plant Signal Behav 5:649–654. doi:10.4161/psb.5.6.11398

Zadoks JC, Chang TT, Konzak CF (1974) A decimal code for the growth stages of cereals. Weed Res 14:415–421. doi:10.1111/j.1365-3180.1974.tb01084.x

Permissions

All chapters in this book were first published in BS, by Springer; hereby published with permission under the Creative Commons Attribution License or equivalent. Every chapter published in this book has been scrutinized by our experts. Their significance has been extensively debated. The topics covered herein carry significant findings which will fuel the growth of the discipline. They may even be implemented as practical applications or may be referred to as a beginning point for another development.

The contributors of this book come from diverse backgrounds, making this book a truly international effort. This book will bring forth new frontiers with its revolutionizing research information and detailed analysis of the nascent developments around the world.

We would like to thank all the contributing authors for lending their expertise to make the book truly unique. They have played a crucial role in the development of this book. Without their invaluable contributions this book wouldn't have been possible. They have made vital efforts to compile up to date information on the varied aspects of this subject to make this book a valuable addition to the collection of many professionals and students.

This book was conceptualized with the vision of imparting up-to-date information and advanced data in this field. To ensure the same, a matchless editorial board was set up. Every individual on the board went through rigorous rounds of assessment to prove their worth. After which they invested a large part of their time researching and compiling the most relevant data for our readers.

The editorial board has been involved in producing this book since its inception. They have spent rigorous hours researching and exploring the diverse topics which have resulted in the successful publishing of this book. They have passed on their knowledge of decades through this book. To expedite this challenging task, the publisher supported the team at every step. A small team of assistant editors was also appointed to further simplify the editing procedure and attain best results for the readers.

Apart from the editorial board, the designing team has also invested a significant amount of their time in understanding the subject and creating the most relevant covers. They scrutinized every image to scout for the most suitable representation of the subject and create an appropriate cover for the book.

The publishing team has been an ardent support to the editorial, designing and production team. Their endless efforts to recruit the best for this project, has resulted in the accomplishment of this book. They are a veteran in the field of academics and their pool of knowledge is as vast as their experience in printing. Their expertise and guidance has proved useful at every step. Their uncompromising quality standards have made this book an exceptional effort. Their encouragement from time to time has been an inspiration for everyone.

The publisher and the editorial board hope that this book will prove to be a valuable piece of knowledge for researchers, students, practitioners and scholars across the globe.

List of Contributors

Chih-Li Wang and Yi-Hong Cheng
Department of Plant Pathology, National Chung Hsing University, Taichung 40227, Taiwan

Ji-hong Zhang, He-long Sun, Shao-yang Chen and Li Zeng
School of Chemical Engineering, Xiangtan University, Xiangtan 411105, People's Republic of China

Tao-tao Wang
College of Horticulture Forestry & Sciences, Huazhong Agricultural University, Wuhan 430070, People's Republic of China

Yi-Yin Chang
Department of Life Sciences, National Taiwan University, Taipei, Taiwan

Yu-Ming Ju
Institute of Plant and Microbial Biology, Academia Sinica, Taipei, Taiwan

Bi-Xian Wu, Xu-Jian Xu and Chun-Ping Yang
Department of Forestry, Sichuan Agricultural University, Chengdu 611130, China

Guo-Xing Cao
Department of Forestry, Sichuan Agricultural University, Chengdu 611130, China
Xishuangbanna Tropical Botanical Garden, Chinese Academy of Sciences, Mengla 666303, China

Xie Wang
Soil and Fertilizer Research Institute, Sichuan Academy of Agricultural Sciences, Chengdu 610066, Sichuan, China

Muhammad Khalid, Danial Hassani and Danfeng Huang
School of Agriculture and Biology, Shanghai Jiao Tong University, Shanghai 200240, People's Republic of China

Muhammad Bilal
State Key Laboratory of Microbial Metabolism, School of Life Sciences and Biotechnology, Shanghai Jiao Tong University, Shanghai 200240, China

Fayaz Asad
Key Laboratory of Alpine Ecology and Biodiversity, Key Laboratory of Tibetan Environment, Changes and Land Surface Processes, Institute of Tibetan Plateau Research, Chinese Academy of Sciences, Beijing 100101, China

Huei-Mei Hsieh, Mei-Chu Chung, Pao-Yang Chen, Fei-Man Hsu, Wen-Wei Liao, Ai-Ning Sung, Chun-Ru Lin, Chung-Ju Rachel Wang, Yu-Hsin Kao, Mei-Jane Fang and Yu-Ming Ju
Institute of Plant and Microbial Biology, Academia Sinica, Taipei 11529, Taiwan

Chi-Yung Lai
Graduate Institute of Biotechnology, National Changhua University of Education, Changhua 50074, Taiwan

Chieh-Chen Huang
Department of Life Sciences, National Chung Hsing University, Taichung 40227, Taiwan

Jyh-Ching Chou
Department of Natural Resources and Environmental Studies, National Dong Hwa University, Hualien 97401, Taiwan

Wen-Neng Chou
National Museum of Natural Science, Taichung 40453, Taiwan

Bill Chia-Han Chang
Yourgene Bioscience, New Taipei 23863, Taiwan

Yingang Li, Dongyue Jiang and Xinhong Liu
Zhejiang Academy of Forestry, Hangzhou 310023, China

Wuqin Xu
Key Laboratory of Conservation Biology for Endangered Wildlife of Ministry of Education, College of Life Sciences, Zhejiang University, Hangzhou 310058, China

Wentao Zou
Institute of Tropical Forestry, Chinese Academy of Forestry, Guangzhou 510520, China

Mukesh Meena, Manish Kumar Dubey, Mohd. Aamir and R. S. Upadhyay
Laboratory of Mycopathology and Microbial Technology, Centre of Advanced Study in Botany, Banaras Hindu University, Varanasi, Uttar Pradesh 221 005, India

Andleeb Zehra
Laboratory of Mycopathology and Microbial Technology, Centre of Advanced Study in Botany, Banaras Hindu University, Varanasi, Uttar Pradesh 221 005, India
Department of Botany, Banaras Hindu University, Varanasi 221 005, India

Ya-Zhu Ko and Hsing-Hua Ho
Department of Biological Sciences, National Sun Yat-sen University, Kaohsiung 804, Taiwan

Huei-Chuan Shih
Department of Nursing, Meiho University, Pingtung 912, Taiwan

Chi-Chu Tsai
Kaohsiung District Agricultural Research and Extension Station, Pingtung 900, Taiwan
National Pingtung University of Science and Technology, Pingtung 912, Taiwan

Pei-Chun Liao
Department of Life Science, National Taiwan Normal University, Taipei, Taiwan

Yu-Chung Chiang
Department of Biological Sciences, National Sun Yat-sen University, Kaohsiung 804, Taiwan
Department of Biomedical Science and Environment Biology, Kaohsiung Medical University, Kaohsiung, Taiwan

Lawrence W. Zettler and Andrew L. Stice
Department of Biology, Illinois College, 1101 West College Ave., Jacksonville, IL 62650, USA

Landy Rajaovelona
Kew Madagascar Conservation Centre, Lot #J131B, Ambodivaonjo-Ivandry, 101, Antananarivo, Madagascar

Kazutomo Yokoya, Jonathan P. Kendon and Viswambharan Sarasan
Royal Botanic Gardens Kew, Richmond, Surrey TW9 3AB, UK

Amanda E. Wood
Department of Biology, Illinois College, 1101 West College Ave., Jacksonville, IL 62650, USA
Environmental Studies Program, University of Pennsylvania, Philadelphia, PA 19104, USA

Jeng-Shyan Deng and Ming-Ming Lee
Department of Food Nutrition and Health Biotechnology, Asia University, Taichung, Taiwan

Jui-Shu Chang
School of Chinese Medicine, Graduate Institute of Integrated Medicine College of Chinese Medicine, China Medical University, Taichung, Taiwan

Jung-Chun Liao
School of Pharmacy, College of Pharmacy, China Medical University, Taichung, Taiwan

Wei Chao and Guan-Jhong Huang
School of Chinese Pharmaceutical Sciences and Chinese Medicine Resources, College of Chinese Medicine, China Medical University, Taichung, Taiwan

Chien-Hua Cheng
Department of Visual Communication Design, Asia University, Taichung, Taiwan

Guo-Zhang M. Song
Department of Soil and Water Conservation, National Chung Hsing University, Taichung 40227, Taiwan

Kuo-Jung Chao
International Master Program of Agriculture, National Chung Hsing University, Taichung 40227, Taiwan

David Doley and David Yates
Centre for Mined Land Rehabilitation, The University of Queensland, Brisbane, QLD 4072, Australia

Hữu Đăng Trần, Quốc Đạt Nguyễn and Hiếu Cường Nguyễn
Southern Institute of Ecology, Vietnam Academy of Science and Technology, 1 Mạc Đĩnh Chi Street, District 1, Ho Chi Minh City, Vietnam

Parusuraman Athen and K. M. Wong
Singapore Botanic Gardens, National Parks Board, 1 Cluny Road, Singapore 259569, Singapore

Hồng Trường Lưu
Southern Institute of Ecology, Vietnam Academy of Science and Technology, 1 Mạc Đĩnh Chi Street, District 1, Ho Chi Minh City, Vietnam
Graduate University of Science and Technology, Vietnam Academy of Science and Technology, 18 Hoàng Quốc Việt, Cầu Giấy, Hanoi, Vietnam

Dariusz L. Szlachetko, Magdalena Dudek and Piotr Rutkowski
Department of Plant Taxonomy and Nature Conservation, The University of Gdańsk, ul. Wita Stwosza 59, 80-308 Gdańsk, Poland

Marta Kolanowska
Department of Plant Taxonomy and Nature Conservation, The University of Gdańsk, ul. Wita Stwosza 59, 80-308 Gdańsk, Poland
Department of Biodiversity Research, Global Change Research Institute AS CR, Bělidla 4a., 60300 Brno, Czech Republic

Aleksandra Naczk and Marcin Górniak
Department of Molecular Evolution, The University of Gdańsk, Wita Stwosza 59, 80-308 Gdańsk, Poland

Guy Chiron
Herbiers, Université de Lyon I, 69622 Villeurbanne Cedex, France

Jyh-Nong Tsai and Pao-Jen Ann
Division of Plant Pathology, Taiwan Agricultural Research Institute, Wufeng, Taichung, Taiwan

Ruey-Fen Liou
Department of Plant Pathology and Microbiology, National Taiwan University, Taipei, Taiwan

Wen-Hsui Hsieh and Wen-Hsiung Ko
Department of Plant Pathology, National Chung Hsing University, Taichung, Taiwan

Li-Ru Chen
Department of Horticulture and Biotechnology, Chinese Culture University, Taipei 11114, Taiwan

Chia-Yun Ko
Institute of Plant and Microbial Biology, Academia Sinica, Taipei 11529, Taiwan

William R. Folk
Department of Biochemistry, University of Missouri, Columbia, MO 65211, USA

Tsai-Yun Lin
Department of Life Science & Institute of Bioinformatics and Structural Biology, National Tsing Hua University, Hsinchu 30013, Taiwan

Mao-Yin Sheng
Karst Research Institute, Guizhou Normal University, Baoshanbei Road 180, Guiyang 550001, P. R. China
School of Life Science, Nanjing University, Hankou Road 32, Nanjing 210093, P. R. China
National Engineering Research Center for Karst Rocky Desertification Control, Guizhou Normal University, Baoshanbei Road 180, Guiyang 550001, P. R. China

Lin-Jiao Wang
Karst Research Institute, Guizhou Normal University, Baoshanbei Road 180, Guiyang 550001, P. R. China
National Engineering Research Center for Karst Rocky Desertification Control, Guizhou Normal University, Baoshanbei Road 180, Guiyang 550001, P. R. China

Jia-Ying Du
National Engineering Research Center for Karst Rocky Desertification Control, Guizhou Normal University, Baoshanbei Road 180, Guiyang 550001, P. R. China
State Key Laboratory Incubation Base for Karst Mountain Ecology Environment of Guizhou Province, Baoshaobei Road 180, Guiyang 550001, P. R. China

Pei-Cai Wen
Karst Research Institute, Guizhou Normal University, Baoshanbei Road 180, Guiyang 550001, P. R. China
State Key Laboratory Incubation Base for Karst Mountain Ecology Environment of Guizhou Province, Baoshaobei Road 180, Guiyang 550001, P. R. China

Yuan-Yuan Li, Xiao-Mei Chen and Shun-Xing Guo
Institute of Medicinal Plant Development, Chinese Academy of Medical Sciences & Peking Union Medical College, Beijing 100193, People's Republic of China

Hong-Xian Hsing, Yi-Jyun Lin, Chii-Gong Tong, Min-Jeng Li, Yun-Jin Chen and Swee-Suak Ko
Academia Sinica Biotechnology Center in Southern Taiwan, Tainan 741, Taiwan
Agricultural Biotechnology Research Center, Academia Sinica, Taipei 115, Taiwan

Kulvinder S. Gill
Crop and Soil Sciences Department, Washington State University, Pullman, WA, USA

Marwa N. M. E. Sanad
Crop and Soil Sciences Department, Washington State University, Pullman, WA, USA
Genetics and Cytology Department, National Research Center, Giza, Egypt

Kimberley Garland Campbell
Crop and Soil Sciences Department, Washington State University, Pullman, WA, USA
Wheat Genetics, Quality, Physiology, Diseases Research, USDA-ARS, Pullman, WA, USA

Index

www.ingramcontent.com/pod-product-compliance
Lightning Source LLC
LaVergne TN
LVHW080921210926
848485LV00004B/702
* 9 7 8 1 6 3 9 8 9 4 6 5 9 *